Classical and Multilinear Harmonic Analysis

This two-volume text in harmonic analysis introduces a wealth of analytical results and techniques. It is largely self-contained and is intended for graduates and researchers in pure and applied analysis. Numerous exercises and problems make the text suitable for self-study and the classroom alike.

This first volume starts with classical one-dimensional topics: Fourier series; harmonic functions; Hilbert transforms. Then the higher-dimensional Calderón–Zygmund and Littlewood–Paley theories are developed. Probabilistic methods and their applications are discussed, as are applications of harmonic analysis to partial differential equations. The volume concludes with an introduction to the Weyl calculus.

The second volume goes beyond the classical to the highly contemporary and focuses on multilinear aspects of harmonic analysis: the bilinear Hilbert transform; Coifman–Meyer theory; Carleson's resolution of the Lusin conjecture; Calderón's commutators and the Cauchy integral on Lipschitz curves. The material in this volume has not been collected previously in book form.

Camil Muscalu is Associate Professor in the Department of Mathematics at Cornell University.

Wilhelm Schlag is Professor in the Department of Mathematics at the University of Chicago.

CAMBRIDGE STUDIES IN ADVANCED MATHEMATICS

All the titles listed below can be obtained from good booksellers or from Cambridge University Press. For a complete series listing visit: www.cambridge.org/mathematics.

Already published

93 D. Applebaum *Lévy processes and stochastic calculus (1st Edition)*
94 B. Conrad *Modular forms and the Ramanujan conjecture*
95 M. Schechter *An introduction to nonlinear analysis*
96 R. Carter *Lie algebras of finite and affine type*
97 H. L. Montgomery & R. C. Vaughan *Multiplicative number theory, I*
98 I. Chavel *Riemannian geometry (2nd Edition)*
99 D. Goldfeld *Automorphic forms and L-functions for the group GL(n,R)*
100 M. B. Marcus & J. Rosen *Markov processes, Gaussian processes, and local times*
101 P. Gille & T. Szamuely *Central simple algebras and Galois cohomology*
102 J. Bertoin *Random fragmentation and coagulation processes*
103 E. Frenkel *Langlands correspondence for loop groups*
104 A. Ambrosetti & A. Malchiodi *Nonlinear analysis and semilinear elliptic problems*
105 T. Tao & V. H. Vu *Additive combinatorics*
106 E. B. Davies *Linear operators and their spectra*
107 K. Kodaira *Complex analysis*
108 T. Ceccherini-Silberstein, F. Scarabotti & F. Tolli *Harmonic analysis on finite groups*
109 H. Geiges *An introduction to contact topology*
110 J. Faraut *Analysis on Lie groups: an introduction*
111 E. Park *Complex topological K-theory*
112 D. W. Stroock *Partial differential equations for probabilists*
113 A. Kirillov, Jr *An introduction to Lie groups and Lie algebras*
114 F. Gesztesy *et al. Soliton equations and their algebro-geometric solutions, II*
115 E. de Faria & W. de Melo *Mathematical tools for one-dimensional dynamics*
116 D. Applebaum *Lévy processes and stochastic calculus (2nd Edition)*
117 T. Szamuely *Galois groups and fundamental groups*
118 G. W. Anderson, A. Guionnet & O. Zeitouni *An introduction to random matrices*
119 C. Perez-Garcia & W. H. Schikhof *Locally convex spaces over non-Archimedean valued fields*
120 P. K. Friz & N. B. Victoir *Multidimensional stochastic processes as rough paths*
121 T. Ceccherini-Silberstein, F. Scarabotti & F. Tolli *Representation theory of the symmetric groups*
122 S. Kalikow & R. McCutcheon *An outline of ergodic theory*
123 G. F. Lawler & V. Limic *Random walk: a modern introduction*
124 K. Lux & H. Pahlings *Representations of groups*
125 K. S. Kedlaya *p-adic differential equations*
126 R. Beals & R. Wong *Special functions*
127 E. de Faria & W. de Melo *Mathematical aspects of quantum field theory*
128 A. Terras *Zeta functions of graphs*
129 D. Goldfeld & J. Hundley *Automorphic representations and L-functions for the general linear group, I*
130 D. Goldfeld & J. Hundley *Automorphic representations and L-functions for the general linear group, II*
131 D. A. Craven *The theory of fusion systems*
132 J. Väänänen *Models and games*
133 G. Malle & D. Testerman *Linear algebraic groups and finite groups of Lie type*
134 P. Li *Geometric analysis*
135 F. Maggi *Sets of finite perimeter and geometric variational problems*
136 M. Brodmann & R. Y. Sharp *Local cohomology (2nd Edition)*
137 C. Muscalu & W. Schlag *Classical and multilinear harmonic analysis, I*
138 C. Muscalu & W. Schlag *Classical and multilinear harmonic analysis, II*
139 B. Helffer *Spectral theory and its applications*

Classical and Multilinear Harmonic Analysis

Volume I

CAMIL MUSCALU
Cornell University

WILHELM SCHLAG
University of Chicago

CAMBRIDGE
UNIVERSITY PRESS

University Printing House, Cambridge CB2 8BS, United Kingdom

One Liberty Plaza, 20th Floor, New York, NY 10006, USA

477 Williamstown Road, Port Melbourne, VIC 3207, Australia

4843/24, 2nd Floor, Ansari Road, Daryaganj, Delhi - 110002, India

79 Anson Road, #06-04/06, Singapore 079906

Cambridge University Press is part of the University of Cambridge.

It furthers the University's mission by disseminating knowledge in the pursuit of education, learning and research at the highest international levels of excellence.

www.cambridge.org
Information on this title: www.cambridge.org/9780521882453

First published 2013
Reprinted 2016

A catalogue record for this publication is available from the British Library

Library of Congress Cataloging in Publication data
Muscalu, C. (Camil), author.
Classical and multilinear harmonic analysis / C. Muscalu and W. Schlag.
volumes cm. – (Cambridge studies in advanced mathematics ; 137–)
Includes bibliographical references.
ISBN 978-0-521-88245-3 (v. 1 : hardback)
1. Harmonic analysis. I. Schlag, Wilhelm, 1969– author. II. Title.
QA403.M87 2013
515′.2422 – dc23 2012024828

ISBN 978-0-521-88245-3 Hardback
ISBN 978-1-107-47159-7 Paperback

Contents

Preface

Harmonic analysis is an old subject. It originated with the ideas of Fourier in the early nineteenth century (which were preceded by work of Euler, Bernoulli, and others). These ideas were revolutionary at the time and could not be understood by means of the mathematics available to Fourier and his contemporaries. However, it was clear even then that the idea of representing any function as a superposition of elementary harmonics (sine and cosine) from an arithmetic sequence of frequencies had important applications to the partial differential equations of physics that were being investigated at the time, such as the heat and wave equations. In fact, it was precisely the desire to solve these equations that led to this bold idea in the first place.

Research into the precise mathematical meaning of such Fourier series consumed the efforts of many mathematicians during the entire nineteenth century as well as much of the twentieth century. Many ideas that took their beginnings and motivations from Fourier series research became disciplines in their own right. Set theory (Cantor) and measure theory (Lebesgue) are clear examples, but others, such as functional analysis (Hilbert and Banach spaces), the spectral theory of operators, and the theory of compact and locally compact groups and their representations, all exhibit clear and immediate connections with Fourier series and integrals. Furthermore, soon after Fourier proposed representing every function on a compact interval as a trigonometric series, his idea was generalized by Sturm and Liouville to expansions with respect to the eigenfunctions of very general second-order differential operators subject to natural boundary conditions – a groundbreaking result in its own right.

Not surprisingly harmonic analysis is therefore a vast discipline of mathematics, which continues to be a vibrant research area to this day. In addition, over the past 60 years Euclidean harmonic analysis, as represented by the schools associated with A. Calderón and A. Zygmund at the University of Chicago as well as these associated with C. Fefferman and E. Stein at Princeton University, has been inextricably linked with partial differential equations (PDEs). While applications to the theory of elliptic PDEs and pseudodifferential operators were a driving force in the development of the Calderón–Zygmund school from the very beginning, the past 25 years have also seen an influx of harmonic analysis techniques to the theory of nonlinear dispersive equations such as the Schrödinger and wave equations. These developments continue to this day.

The basic "divide and conquer" idea of harmonic analysis can be stated as follows: that we should study those classes of functions that arise in interesting contexts (for example, as solutions of differential equations; as measured data; as audio or video signals on DVDs, CDs, or possibly transmitted across glass fiber cables or great distances such as that between Mars and Earth; as samples of random processes) by breaking them into basic constituent parts and that (a) these basic parts are both as simple as possible and amenable to study and (b) ideally, reflect some structure inherent to the problem at hand.

In a classical manifestation these basic parts are given by the standing waves used in Fourier series, but over the past 30 years wavelets (as well as curvelets and ringlets) have revolutionized applied harmonic analysis, especially as used in image processing.

This fundamental idea is ubiquitous in science and engineering. Examples where it arises and is put to use include the following: all types of medical imaging (such as magnetic resonance imaging (MRI), computed tomography, ultrasound, echocardiography, and positron emission tomography (PET)); signal processing, especially through the methods used in compressed sensing; and inverse problems such as those that arise in remote sensing, medical imaging, geophysics, and oil exploration. In addition, advances in electrical engineering – and with it essentially the whole of modern industry as we know it – have only been possible through the systematic use and implementation of mathematics, often in the form of Fourier analysis and its ramifications.

To go even further, Nature appears to carry in herself the blueprint of the *basic harmonics*. Indeed, electrons and other elementary particles are understood as *spherical waves*, and the discrete energy levels so characteristic of quantum mechanics are dictated by the necessity of fitting such a standing wave onto a two-dimensional spherical surface. String theory takes this concept to an entirely different level of abstraction by reducing *everything* to the vibration of tiny strings.

Harmonic analysis has the advantage, over other subjects in mathematics, that it has never been completely isolated or divorced from applications; rather, a significant part of it has been steadily guided and inspired by them. For example, the study of the Cauchy kernel on Lipschitz curves might have arisen as a seemingly academic exercise and a rather mindless generalization but for the fact that there are so many important problems in materials science where Nature produces just such non-smooth boundaries.

Conceptual developments in harmonic analysis are at the center of many important scientific and technological advances. It is rather remarkable that wavelets are actually used in the JPEG 2000 standard. The technical details will be of interest mainly to specialists, but the conceptual framework has a much greater reach and perhaps significance. Theorems have hypotheses that may not be exactly satisfied in many real physical situations; in fact, engineering is not an exact science but rather one of good enough approximation. So, while theorems may not apply in a strict sense the thinking that went into them can still be extremely useful. It is precisely this thinking that our book wishes to present.

Classic monographs and textbooks in harmonic analysis include those by Stein [108], [110], and Stein and Weiss [112]; amongst the older literature there is the timeless encyclopedic work on Fourier series by Zygmund [131] and the more accessible introduction to Fourier analysis by Katznelson [65]. Various excellent, more specialized texts are also available, such as Folland [41], which focuses on phase-space analysis and the Weyl calculus, as well as Sogge [105], which covers oscillatory integrals. Wolff's lecture notes [128] can serve as an introduction to the ideas associated with the Kakeya problem. This is a more geometric, as well as combinatorial, aspect of harmonic analysis that is still rather poorly understood.

Our intention was not to compete with any of these well-known texts. Rather, our book is designed as a teaching tool, both in a traditional classroom setting as well as in the setting of independent study by an advanced undergraduate or beginning graduate student. In addition, the authors hope that it will also be useful for any mathematician or mathematically inclined scientist who wishes to acquire a working knowledge of select topics in (mostly Euclidean) Fourier analysis.

The two volumes of our book are different in both scope and character, although they should be perceived as forming a natural unit. In this first volume we introduce the reader to a broad array of results and techniques, starting from the beginnings of the field in Fourier series and then developing the theory along what the authors hope are natural avenues motivated by certain basic questions. The selection of the material in this volume is of course partly a reflection of the authors' tastes, but it also follows a specific purpose: to introduce the reader to sufficiently many topics in classical Fourier analysis, and in enough detail, to allow them to continue a guided study of more advanced material possibly leading to original research in analysis, pure or applied.

All the material in this volume should be considered basic. It can be found in many texts but to the best of our knowledge not in a single place. The authors feel that this volume presents the course that they should have taken in graduate school, on the basis of hindsight. To what extent the entirety of this volume constitutes a reasonable course is up to the individual teacher to decide. It is more likely that selections will need to be made, and it is of course also to be expected that lecturers will wish to supplement the material with certain topics of their own choosing that are not covered here. However, the authors feel, particularly since this has been tested on individual students, that Vol. I could be covered by a beginning graduate student over the course of a year in independent, but guided, study. This would then culminate in some form of qualifying or "topic" exam, after which the student would be expected to begin independent research.

Particular emphasis has been placed on the inclusion of exercises and problems. The former are dispersed throughout the main body of the text and are for the most part an integral part of the theory. As a rule they are less difficult than the end-of-chapter problems. The latter serve to develop the theory further and to give the reader the opportunity to try his or her hand at the occasional hard problem. An old and commonplace principle, to which the authors adhere, is that any piece of mathematics can only be learned by active work, and this is reflected in our book. In addition, the authors have striven to emphasize intuition and ideas over both generality and technique, without, however, sacrificing rigor, elegance, or for that matter, relevance.

While Vol. I presents developments in harmonic analysis up to the mid to late 1980s, Vol. II picks up from there and focuses on more recent aspects. In comparison with the first volume the second is of necessity much more selective, and many topics of current interest could not be included. Examples of the many omitted areas that come to mind are the oscillatory integrals related to the Kakeya, restriction, and Bochner–Riesz conjectures, multilinear Strichartz estimates and geometric measure theory and its relations to combinatorics and number theory.

The selection of topics in Vol. II can roughly be described as phase-space oriented; it comprises material that either grew out of or is closely related to the David–Journé $T(1)$ theorem, the Cauchy integral on Lipschitz curves, and the Calderón commutators on the one hand and the solution to Lusin's problem by Carleson on the other hand. The latter work, also through Fefferman's reworking of Carleson's proof, greatly influenced the resolution of the bilinear Hilbert-transform boundedness problem by Lacey and Thiele in the mid 1990s. Generally speaking, Carleson's work has had a profound influence on the more combinatorially oriented analysis of phase space that lies at the heart of Vol. II.

To be more specific, Vol. II concentrates on paraproducts (which make an appearance in Vol. I), the bilinear Hilbert transform, Carleson's theorem, and Calderón commutators and the Cauchy integral on Lipschitz curves. In fact, the analysis of paraproducts can be seen as the most foundational material for Vol. II; paraproducts are developed on polydisks and as flag paraproducts in their own right. In this sense, Vol. II is more research oriented in character, as some results are quite recent and the organization and development of the material are in part original and specific to Vol. II.

In terms of presentation, in the second volume the authors have generally chosen not the shortest proofs but those that are most robust as well as (to their taste) most illuminating. For example, there are simpler ways of approaching the Coifman–Meyer theorem on paraproducts but these do not carry over to other contexts such as flag paraproducts. Much emphasis has been placed on motivation, and so the authors have included applications to PDEs, for example in the form of Strichartz estimates and their use in nonlinear equations in Vol. I and in the form of fractional Leibnitz rules with application to the KdV equation in Vol. II. The chapter entitled "Iterated Fourier series and physical reality" in Vol. II is entirely devoted to motivation and to an explanation of the larger framework in which much of that volume sits.

Throughout both volumes the authors have striven to emphasize intuition and ideas, and often figures have been used as an important part of an explanation or proof. This is particularly the case in Vol. II, which is more demanding in terms of technique and with often longer and more complicated proofs than those in Vol. I.

Volume II overlaps considerably with the recent research literature, especially with papers by Lacey and Thiele on the one hand and by the first author, Tao, and Thiele on the other hand. We shall not give an account of these papers here, since a discussion can be found in the end-of-chapter notes (we have avoided placing citations and references in the main body of the text, so as not to distract the reader).

We shall now comment briefly on the relation of Vol. II to classic textbooks in the area. The influential work by Coifman and Meyer [24] overlaps the first and fourth chapters of Vol. II. However, not only is the technical approach developed in [24] completely different from that in this book but it is also designed for a different purpose: it is less a textbook than a rapid review of many deep topics such as Hardy spaces on Lipschitz domains, Murai's proof of the Cauchy integral boundedness, and commutators and the Cauchy integral. Finally, [24] develops wavelets, which are an essential tool in many real-world applications but make no appearance in this book.

Another well-known text that overlaps Vol. II contains Christ's CBMS lectures [21]. These are centered around the $T(1)$ and $T(b)$ theorems and their applications. While some of this material does make an appearance in our book, Vol. II does not use $T(1)$ or $T(b)$.

We conclude this preface with a discussion of prerequisites. Essential is a grounding in basic analysis, beginning with multivariable calculus (including writing a hypersurface as a graph and the notion of Gaussian curvature on a hypersurface) going on to measure theory and integration and the functional analysis relevant to basic Hilbert spaces (orthogonal projections, bases, completeness) as well as to Banach spaces (weak and strong convergence, bases, Hahn–Banach, uniform boundedness), and finally the basics of complex analysis (including holomorphic functions and conformal maps). In Vol. I probability theory also makes an appearance, and it might be helpful if the reader has had some exposure to the notions of independence, expected value, variance, and distribution functions. The second volume requires very little more in terms of preparation other than a fairly mature understanding of the above topics. The authors do not recommend, however, that one should attempt Vol. II before Vol. I.

As is customary in analysis, interpolation is frequently used. To be more specific, we rely on both the Riesz–Thorin and Marcinkiewicz interpolation theorems. We state these facts at the end of the first chapter but omit the proofs, as they can readily be found in the texts of Stein and Weiss [112] and Katznelson [65]. Finally, in Chapter 11 of the present volume, on the restriction theorem, we also make use of Stein's generalization of the Riesz–Thorin theorem to analytic families of operators. Volume II uses multilinear interpolation, and the facts needed are collected in an appendix. A standard reference for interpolation theory, especially as it relates to Besov, Sobolev, and Lorentz spaces, is the text by Bergh and Löfström [7].

As we have already pointed out, many important topics are omitted from our two volumes. Apart from classical topics such as inner and outer functions, which we could not include in Vol. I owing to space considerations, an aspect of modern harmonic analysis that is not covered here is the vast area dealing with oscillatory integrals. This field touches upon several other areas including geometric measure theory, combinatorial geometry, and number theory and is relevant to nonlinear dispersive PDEs in several ways, such as through bilinear restriction estimates (see for example Wolff's paper [127] as well as his lecture notes [128]). This material would naturally comprise a third volume, which would need to present the research that has been done since the work of Sogge [105] and Stein [109].

How to use this volume. Ideally, the reader should work through the chapters in order. A reader or class familiar with the Fourier transform in $\mathbb{R}^d$ could start with Chapter 7, then move on to Chapter 8 and subsequently choose any of the remaining four chapters in this volume according to taste and time constraints. Chapters 7 through 11 constitute the backbone of real-variable harmonic analysis. Of those, Chapter 10 can be regarded as an optional extra; however, the authors feel that it is of importance and that students should be exposed to this material. As an application of the Logvinenko–Sereda theorem of Section 10.3, we prove the local solvability of constant coefficient PDEs (the Malgrange–Ehrenpreis theorem) in Section 10.4.

Another such outlier is Chapter 12, in which we introduce the reader to an area that is itself the subject of many books; see for example Taylor's book on pseudodifferential operators [120] as well as Hörmander's treatise [58]. The present authors decided to include a very brief account of this story, since it is an essential part of harmonic analysis and also since it originates in Calderón's work as part of his investigations of singular integrals and the Cauchy problem for elliptic operators. In principle, Chapter 12 can be read separately by a mature reader who is familiar with Cotlar's lemma from Chapter 9. In the writing of this chapter a difficult decision had to be made, namely which of the two main incarnations of pseudodifferential operators to use, that of Kohn–Nirenberg or that of Weyl. While the former is somewhat simpler technically, and therefore often used for elliptic PDEs, the older Weyl quantization is very natural owing to its symmetry and is the one that is normally used in the so-called semiclassical calculus. We therefore chose to follow the latter route; Kohn–Nirenberg pseudo-differential operators make only a very brief appearance in this text.

In Chapter 5 we introduce the reader to probability theory, which is also often omitted in a more traditional harmonic analysis presentation. However, the authors felt that the ideas developed in that chapter (which are very elementary for the most part) are an essential part of modern analysis and of mathematics in general. They appear in many different settings and should be in the toolbox of any working analyst, pure or applied. Chapter 6 contains several examples of how probabilistic thinking and results appear in harmonic analysis. Section 6.3 on Sidon sets can be omitted on first reading, as it is somewhat specialized (it contains, in particular, Rider's theorem, which we prove there).

The first four chapters of this volume are intended for a reader who has had no or very little prior exposure to Fourier series and integrals, harmonic functions, and their conjugates. A basic introductory advanced-undergraduate or beginning-graduate course would cover the first three chapters, omitting

Section 2.5, and would then move on to the first section of the fourth chapter (here, the material on locally compact Abelian groups could be omitted entirely, since it is used only in a non-Euclidean setting in the proof of Rider's theorem in Section 6.3, and the stationary phase method is used only in the final two chapters of this volume). After that, an instructor could then select any topics from Chapters 5–8 as desired, the most traditional choice being the Calderón–Zygmund, Mikhlin, and Littlewood–Paley theorems. The last of these theorems, at least as presented here, does require a minimal knowledge of probability, namely in the form of Khinchine's inequality.

Feedback. The authors welcome comments on this book and ask that they be sent to harmonic@math.uchicago.edu.

Acknowledgements

Wilhelm Schlag expresses his gratitude to Rowan Killip for detailed comments on his old harmonic analysis notes from 2000, from which the first volume of this book eventually emerged. Furthermore, he thanks Serguei Denissov, Charles Epstein, Burak Erdogan, Patrick Gérard, David Jerison, Carlos Kenig, Andrew Lawrie, Gerd Mockenhaupt, Paul Müller, Casey Rodriguez, Barry Simon, Chris Sogge, Wolfgang Staubach, Eli Stein, and Bobby Wilson for many helpful suggestions and comments on a preliminary version of Vol. I. Finally, he thanks the many students and listeners who attended his lectures and classes at Princeton University, the California Institute of Technology, the University of Chicago, and the Erwin Schrödinger Institute in Vienna over the past ten years. Their patience, interest, and helpful comments have led to numerous improvements and important corrections.

The second volume of the book is partly based on two graduate courses given by Camil Muscalu at Şcoala Normală Superioară, Bucureşti, in the summer of 2004 and at Cornell University in the fall of 2007. First and foremost he would like to thank Wilhelm Schlag for the idea of writing this book together. Then, he would like to thank all the participants of those classes for their passion for analysis and for their questions and remarks. In addition, he would like to thank his graduate students Cristina Benea, Joeun Jung, and Pok Wai Fong for their careful reading of the manuscript and for making various corrections and suggestions and Pierre Germain and Raphaël Côte for their meticulous comments. He would also like to thank his collaborators Terry Tao and Christoph Thiele. Many ideas that came out of this collaboration are scattered through the pages of the second volume of the book. Last but not least, he would like to express his gratitude to Nicolae Popa from the Institute of Mathematics of the Romanian Academy for introducing him to the world of harmonic analysis and for his unconditional support and friendship over the years.

Many thanks go to our long-suffering editors at Cambridge University Press, Roger Astley and David Tranah, who continued to believe in this project and support it even when it might have been more logical not to do so. Their cheerful patience and confidence is gratefully acknowledged. Barry Simon at the California Institute of Technology deserves much credit for first suggesting to David Tranah roughly ten years ago that Wilhelm Schlag's harmonic analysis notes should be turned into a book.

The authors were partly supported by the National Science Foundation during the preparation of this book.

1

Fourier series: convergence and summability

1.1. The basics: partial sums and the Dirichlet kernel

1.1.1. Definitions

We begin with a basic object in analysis, namely the Fourier series associated with a function or a measure on the circle. To be specific, let $\mathbb{T} = \mathbb{R}/\mathbb{Z}$ be the one-dimensional torus (in other words, the circle). We will consider various function spaces on the torus $\mathbb{T}$, namely the space of continuous functions $C(\mathbb{T})$, the space of Hölder continuous functions $C^\alpha(\mathbb{T})$ where $0 < \alpha \leq 1$, and the Lebesgue spaces $L^p(\mathbb{T})$ where $1 \leq p \leq \infty$. The space of complex Borel measures on $\mathbb{T}$ will be denoted by $\mathcal{M}(\mathbb{T})$. Any $\mu \in \mathcal{M}(\mathbb{T})$ has associated with it a *Fourier series*

$$\mu \sim \sum_{n=-\infty}^{\infty} \hat{\mu}(n) e(nx) \tag{1.1}$$

where $e(x) := e^{2\pi i x}$ and

$$\hat{\mu}(n) := \int_0^1 e(-nx)\,\mu(dx) = \int_{\mathbb{T}} e(-nx)\,\mu(dx).$$

The symbol $\sim$ in (1.1) is formal and simply means that the series on the right-hand side is associated with μ. If $\mu(dx) = f(x)\,dx$ where $f \in L^1(\mathbb{T})$, then we may write $\hat{f}(n)$ instead of $\hat{\mu}(n)$.

The central question which we wish to explore in this chapter is the following: when does μ equal the right-hand side in (1.1), that is, when does it represent f in a suitable sense? Note that if we start from a *trigonometric polynomial*

$$f(x) = \sum_{n=-\infty}^{\infty} a_n e(nx)$$

where all but finitely many a_n are zero, then we see that

$$\hat{f}(n) = a_n \quad \forall n \in \mathbb{Z}. \tag{1.2}$$

In other words, we have the pointwise equality

$$f(x) = \sum_{n=-\infty}^{\infty} \hat{f}(n) e_n(x),$$

with $e_n(x) := e(nx)$. Property (1.2) is equivalent to the basic *orthogonality relation*

$$\int_{\mathbb{T}} e_n(x) \overline{e_m(x)}\, dx = \delta_0(n-m), \tag{1.3}$$

where $\delta_0(j) = 1$ if $j = 0$ and $\delta_0(j) = 0$ otherwise.

It is therefore natural to explore the question of the *convergence* of the Fourier series for more general functions. Of course, precise meaning of the convergence of infinite series needs to be specified before this fundamental question can be answered. It is fair to say that much modern analysis (including functional analysis) arose out of the struggle with this question. For example, the notion of the Lebesgue integral was developed in order to overcome the deficiencies in the older Riemannian definition of the integral that had been revealed through the study of Fourier series. The reader will note the recurring theme of the convergence of Fourier series throughout both volumes of this book.

1.1.2. Dirichlet kernel

It is natural to start from the most basic notion of convergence, namely that of *pointwise convergence*, in the case where the measure μ is of the form $\mu(dx) = f(x)\, dx$ with $f(x)$ continuous or of even better regularity. The partial sums of $f \in L^1(\mathbb{T})$ are defined as

$$\begin{aligned} S_N f(x) &= \sum_{n=-N}^{N} \hat{f}(n) e(nx) = \sum_{n=-N}^{N} \int_{\mathbb{T}} e(-ny) f(y)\, dy\; e(nx) \\ &= \int_{\mathbb{T}} \sum_{n=-N}^{N} e(n(x-y)) f(y)\, dy = \int_{\mathbb{T}} D_N(x-y) f(y)\, dy, \end{aligned}$$

where $D_N(x) := \sum_{n=-N}^{N} e(nx)$ is the *Dirichlet kernel*. In other words, we have shown that the partial sum operator S_N is given by convolution with the *Dirichlet kernel* D_N:

$$S_N f(x) = (D_N * f)(x). \tag{1.4}$$

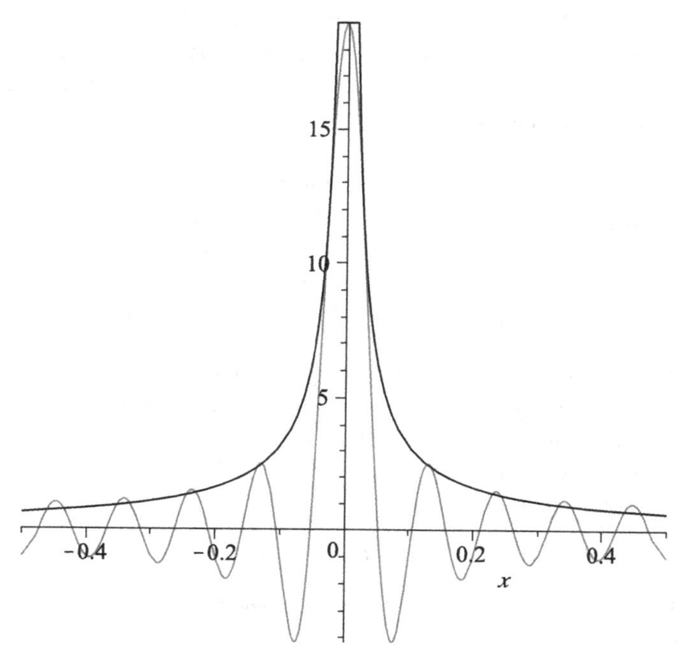

Figure 1.1. The Dirichlet kernel D_N and the upper envelope $\min(2N+1, |\pi x|^{-1})$ for $N = 9$. See Exercise 1.1.

In order to understand basic properties of this convolution, we first sum the geometric series defining $D_N(x)$ to obtain an explicit expression for the Dirichlet kernel.

Exercise 1.1 Verify that, for each integer $N \geq 0$,

$$D_N(x) = \frac{\sin((2N+1)\pi x)}{\sin(\pi x)} \tag{1.5}$$

and draw the graph of D_N for several different values of N, say $N = 2$ and $N = 5$; cf. Figure 1.1. Prove the bound

$$|D_N(x)| \leq C \min\left(N, \frac{1}{|x|}\right) \tag{1.6}$$

for all $N \geq 1$ and some absolute constant C. Finally, prove the bound

$$C^{-1} \log N \leq \|D_N\|_{L^1(\mathbb{T})} \leq C \log N \tag{1.7}$$

for all $N \geq 2$, where C is another absolute constant.

The growth of the bound in (1.7), as well as the oscillatory nature of D_N as given by (1.5), indicates that to understand the pointwise or almost everywhere convergence properties of $S_N f$ may be a very delicate matter. This will become clearer as we develop the theory.

1.1.3. Convolution

In order to study (1.4) we need to establish some basic properties of the convolution of two functions f, g on $\mathbb{T}$. If f and g are continuous, say, then define

$$(f * g)(x) := \int_{\mathbb{T}} f(x-y)g(y)\,dy = \int_{\mathbb{T}} g(x-y)f(y)\,dy. \tag{1.8}$$

It is helpful to think of $f * g$ as an average of translates of f by the measure $g(y)\,dy$ (or the same statement but with f and g interchanged). In particular, convolution commutes with the translation operator τ_z, which is defined for any $z \in \mathbb{T}$ by its action on functions, i.e., $(\tau_z f)(x) = f(x-z)$. Indeed, one may immediately verify that

$$\tau_z(f * g) = (\tau_z f) * g = f * (\tau_z g). \tag{1.9}$$

In passing, we mention the important relation between the Fourier transform and translations:

$$\widehat{(\tau_z \mu)}(n) = e(-zn)\hat{\mu}(n) \quad \forall n \in \mathbb{Z}.$$

In what follows, we will abbreviate *almost everywhere* or *almost every* by *a.e.*

Lemma 1.1 *The operation of convolution as defined in (1.8) satisfies the following properties.*

(i) *Let $f, g \in L^1(\mathbb{T})$. Then, for a.e. $x \in \mathbb{T}$, one has that $f(x-y)g(y)$ is L^1 in y. Thus, the integral in* (1.8) *is well defined for a.e. $x \in \mathbb{T}$ (but not necessarily for every x), and the bound $\|f * g\|_1 \le \|f\|_1 \|g\|_1$ holds.*

(ii) *More generally, $\|f * g\|_r \le \|f\|_p \|g\|_q$ for all $1 \le r, p, q \le \infty$,*

$$1 + \frac{1}{r} = \frac{1}{p} + \frac{1}{q}, \; f \in L^p, g \in L^q.$$

This is called Young's inequality.

(iii) *If $f \in C(\mathbb{T})$, $\mu \in \mathcal{M}(\mathbb{T})$ then $f * \mu$ is well defined. For $1 \le p \le \infty$,*

$$\|f * \mu\|_p \le \|f\|_p \|\mu\|;$$

this allows one to extend $f * \mu$ to arbitrary $f \in L^p$.

(iv) *If $f \in L^p(\mathbb{T})$ and $g \in L^{p'}(\mathbb{T})$, where $1 \le p \le \infty$, and*

$$\frac{1}{p} + \frac{1}{p'} = 1$$

To bound the second term in (1.12) one needs to invoke the oscillation of $D_N(y)$. In fact, we have

$$
\begin{aligned}
B &:= \int_{1/2>|y|>\delta} (f(x-y)-f(x))D_N(y)\,dy \\
&= \int_{1/2>|y|>\delta} \frac{f(x-y)-f(x)}{\sin(\pi y)} \sin((2N+1)\pi y)\,dy \\
&= -\int_{1/2>|y|>\delta} h_x(y) \sin\left((2N+1)\pi\left(y+\frac{1}{2N+1}\right)\right) dy
\end{aligned}
$$

where

$$h_x(y) := \frac{f(x-y)-f(x)}{\sin(\pi y)}.$$

Therefore, with all integrals understood to be in the interval $(-\frac{1}{2}, \frac{1}{2})$,

$$
\begin{aligned}
2B &= \int_{|y|>\delta} h_x(y) \sin((2N+1)\pi y)\,dy \\
&\quad - \int_{|y-1/(2N+1)|>\delta} h_x\left(y-\frac{1}{2N+1}\right) \sin((2N+1)\pi y)\,dy \\
&= \int_{|y|>\delta} \left(h_x(y) - h_x\left(y-\frac{1}{2N+1}\right)\right) \sin((2N+1)\pi y)\,dy \\
&\quad - \int_{[-\delta,-\delta+1/(2N+1)]} h_x\left(y-\frac{1}{2N+1}\right) \sin((2N+1)\pi y)\,dy \\
&\quad + \int_{[\delta,\delta+1/(2N+1)]} h_x\left(y-\frac{1}{2N+1}\right) \sin((2N+1)\pi y)\,dy.
\end{aligned}
$$

These integrals are estimated by putting absolute values inside. To do so we use the bounds

$$
\begin{aligned}
|h_x(y)| &< C\frac{\|f\|_\infty}{\delta}, \\
|h_x(y) - h_x(y+\tau)| &< C\left(\frac{|\tau|^\alpha [f]_\alpha}{\delta} + \frac{\|f\|_\infty}{\delta^2}|\tau|\right),
\end{aligned}
$$

if $|y| > \delta > 2\tau$.

In view of the preceding discussion, one obtains

$$|B| \le C\left(\frac{N^{-\alpha}[f]_\alpha}{\delta} + \frac{N^{-1}\|f\|_\infty}{\delta^2}\right), \tag{1.14}$$

provided that $\delta > 1/N$. Choosing $\delta = N^{-\alpha/3}$ one concludes from (1.12), (1.13), and (1.14) that

$$|(S_N f)(x) - f(x)| \le C\left(N^{-\alpha^2/3} + N^{-2\alpha/3} + N^{-1+2\alpha/3}\right) \tag{1.15}$$

for any function with $\|f\|_\infty + [f]_\alpha \le 1$, which proves the theorem. □

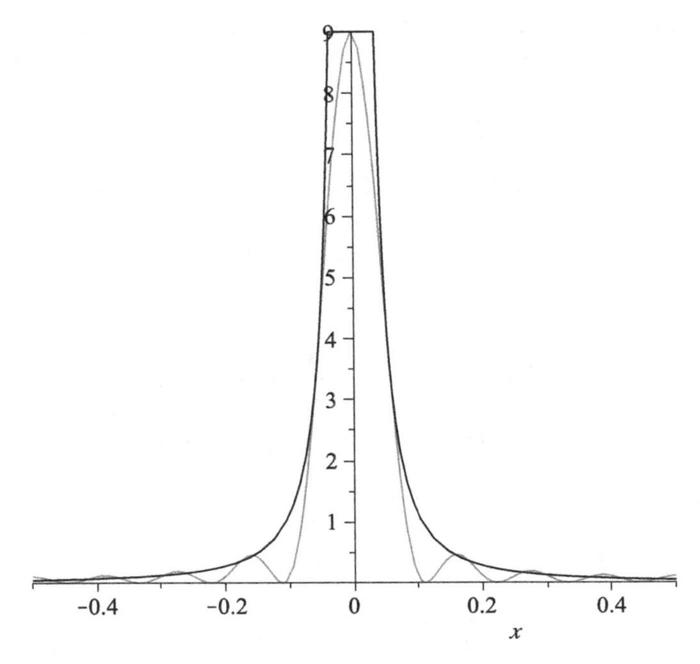

Figure 1.2. The Fejér kernel K_N and the upper envelope $\min(N, (N(\pi x)^2)^{-1})$ for $N = 9$.

The reader is invited to optimize the rate of decay derived in (1.15). In other words, the challenge is to obtain the largest $\beta > 0$ in terms of α such that the bound in (1.15) becomes $CN^{-\beta}$ for any f with $\|f\|_\infty + [f]_\alpha \leq 1$.

1.2. Approximate identities, Fejér kernel

1.2.1. Cesáro means of partial sums

The difficulties with the Dirichlet kernel (see Figure 1.1), such as its slow $1/|x|$ decay, can be regarded as a result of the "discontinuity" in $\widehat{D_N} = \chi_{[-N,N]}$: this indicator function on the lattice $\mathbb{Z}$ jumps at $\pm N$. Therefore we may hope to obtain a kernel that is easier to analyze – in a sense that will be made precise below by means of the notion of *approximate identity* – by substituting for D_N a suitable average whose Fourier transform does not exhibit such jumps.

An elementary way of carrying this out is given by the Cesàro mean, i.e.,

$$\sigma_N f := \frac{1}{N} \sum_{n=0}^{N-1} S_n f.$$

Setting

$$K_N := \frac{1}{N}\sum_{n=0}^{N-1} D_n,$$

where K_N is called the *Fejér kernel*, one therefore has $\sigma_N f = K_N * f$.

Exercise 1.3 Let K_N be a Fejér kernel with N a positive integer.
(a) Verify that $\widehat{K}_N$ looks like a triangle (see Figure 1.3), i.e., for all $n \in \mathbb{Z}$,

$$\widehat{K}_N(n) = \left(1 - \frac{|n|}{N}\right)^+. \tag{1.16}$$

(b) Show that

$$K_N(x) = \frac{1}{N}\left(\frac{\sin(N\pi x)}{\sin(\pi x)}\right)^2. \tag{1.17}$$

(c) Conclude that

$$0 \le K_N(x) \le C\, N^{-1} \min(N^2, x^{-2}). \tag{1.18}$$

We remark that the square and thus the positivity in (1.17) are not entirely surprising, since the triangle in (1.16) can be written as the convolution of two rectangles (the convolution is now at the level of the Fourier coefficients on the lattice $\mathbb{Z}$). Therefore we should expect K_N to have the form of the square of a version of D_M, where M is about half the size of N, suitably normalized.

The properties established in Exercise 1.3 ensure that the K_N form what is called an *approximate identity*.

Definition 1.3 The family $\{\Phi_N\}_{N=1}^\infty \subset L^\infty(\mathbb{T})$ forms an approximate identity provided that
(A1) $\int_0^1 \Phi_N(x)\, dx = 1$ for all N,
(A2) $\sup_N \int_0^1 |\Phi_N(x)|\, dx < \infty$,
(A3) for all $\delta > 0$ one has $\int_{|x|>\delta} |\Phi_N(x)|\, dx \to 0$ as $N \to \infty$.

The term "approximate identity" derives from the fact that $\Phi_N * f \to f$ as $N \to \infty$ in any reasonable sense; see Proposition 1.5. In other words, $\Phi_N \rightharpoonup \delta_0$ in the weak-$*$ sense of measures. Clearly, the so-called *box kernels*

$$\Phi_N(x) = N\chi_{[|x|N<1/2]}, \quad N \ge 1,$$

satisfy (A1)–(A3) and as a family consistitute the most basic example of an approximate identity. Note that the set $\{D_N\}_{N\ge 1}$ does *not* form an approximate identity. Finally, we remark that Definition 1.3 applies not just to the torus $\mathbb{T}$ but equally well to the line $\mathbb{R}$, the tori $\mathbb{T}^d$, or the Euclidean spaces $\mathbb{R}^d$.

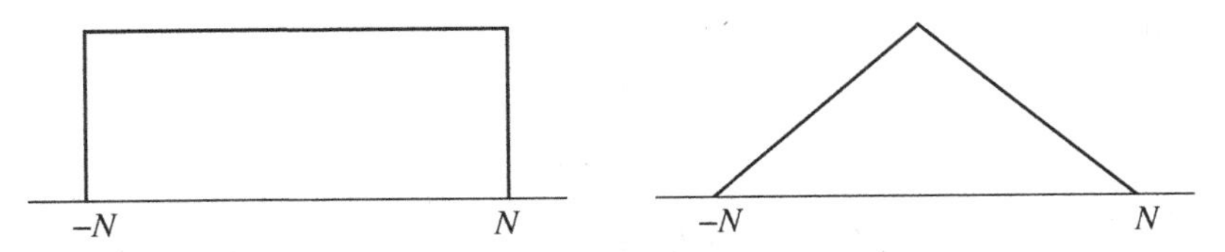

Figure 1.3. The Dirichlet and Fejér kernels as Fourier multipliers.

Next, we verify that the K_N belong to this class.

Lemma 1.4 *The Fejér kernels $\{K_N\}_{N=1}^{\infty}$ form an approximate identity.*

Proof We clearly have $\int_0^1 K_N(x)\,dx = 1$ and $K_N(x) \geq 0$, so that (A1) and (A2) hold. Condition (A3) follows from the bound (1.18). □

1.2.2. Convergence properties of approximate identities

Now we establish the basic convergence property of families that form approximate identities.

Proposition 1.5 *For any approximate identity $\{\Phi_N\}_{N=1}^{\infty}$ one has the following.*

(i) *If $f \in C(\mathbb{T})$ then $\|\Phi_N * f - f\|_\infty \to 0$ as $N \to \infty$.*
(ii) *If $f \in L^p(\mathbb{T})$, where $1 \leq p < \infty$, then $\|\Phi_N * f - f\|_p \to 0$ as $N \to \infty$.*
(iii) *For any measure $\mu \in \mathcal{M}(\mathbb{T})$, one has*

$$\Phi_N * \mu \rightharpoonup \mu, \qquad N \to \infty,$$

in the weak-$*$ sense.

Proof We begin with the uniform convergence. Since $\mathbb{T}$ is compact, f is uniformly continuous. Given $\varepsilon > 0$, let $\delta > 0$ be such that

$$\sup_x \sup_{|y|<\delta} |f(x-y) - f(x)| < \varepsilon.$$

Then, by Definition 1.3,

$$\begin{aligned}
|(\Phi_N * f)(x) - f(x)| &= \left| \int_{\mathbb{T}} (f(x-y) - f(x))\Phi_N(y)\,dy \right| \\
&\leq \sup_{x\in\mathbb{T}} \sup_{|y|<\delta} |f(x-y) - f(x)| \int_{\mathbb{T}} |\Phi_N(t)|\,dt + \int_{|y|\geq\delta} |\Phi_N(y)| 2\|f\|_\infty\,dy \\
&< C\varepsilon,
\end{aligned}$$

provided that N is large.

Fix any $f \in L^p(\mathbb{T})$ and let $g \in C(\mathbb{T})$ be such that $\|f - g\|_p < \varepsilon$. Then

$$\|\Phi_N * f - f\|_p \leq \|\Phi_N * (f - g)\|_p + \|f - g\|_p + \|\Phi_N * g - g\|_p$$
$$\leq \big(\sup_N \|\Phi_N\|_1 + 1\big)\|f - g\|_p + \|\Phi_N * g - g\|_\infty,$$

where we have used Young's inequality, see Lemma 1.1(ii), to obtain the first term on the right-hand side. Using (A2) the assumption on g and Young's inequality finish the proof.

The final statement is immediate from (i) and duality. □

This simple convergence result applied to the Fejér kernels implies some basic analytical properties. A *trigonometric polynomial* is a series $\sum_n a_n e(nx)$ in which only finitely many a_n are nonzero. Clearly, this class forms a vector space over the complex numbers. In fact, the trigonometric polynomials form an algebra under both multiplication and convolution.

Corollary 1.6 *The exponential family $\{e(nx)\}_{n\in\mathbb{Z}}$ satisfies the following properties.*

(i) *The trigonometric polynomials are dense in $C(\mathbb{T})$ in the uniform topology and in $L^p(\mathbb{T})$ for any $1 \leq p < \infty$.*

(ii) *For any $f \in L^2(\mathbb{T})$,*

$$\|f\|_2^2 = \sum_{n\in\mathbb{Z}} |\hat{f}(n)|^2.$$

(iii) *The exponentials $\{e(nx)\}_{n\in\mathbb{Z}}$ form an orthonormal basis in $L^2(\mathbb{T})$.*

(iv) *For all $f, g \in L^2(\mathbb{T})$ one has* Parseval's identity,

$$\int_{\mathbb{T}} f(x)\bar{g}(x)\, dx = \sum_{n\in\mathbb{Z}} \hat{f}(n)\overline{\hat{g}(n)}.$$

Proof By Lemma 1.4, the $\{K_N\}_{N=1}^\infty$ form an approximate identity and Proposition 1.5 applies. Since $\sigma_N f = K_N * f$ is a trigonometric polynomial, property (i) follows.

Properties (ii)–(iv) are equivalent by basic Hilbert space theory. We will start from the orthogonality relations (1.3). One thus has Bessel's inequality

$$\sum |\hat{f}(n)|^2 \leq \|f\|_2^2$$

and equality is equivalent to the linear span of $\{e_n\}$ being dense in $L^2(\mathbb{T})$. That, however, is guaranteed by property (i). □

We remark that Corollary 1.6 is only possible with the Lebesgue integral since, unlike the Riemann integral, it guarantees that $L^2(\mathbb{T})$ is a complete space and thus a Hilbert space. We record two further basic facts, which are immediate consequences of the approximate-identity property of Fejér kernels.

Corollary 1.7 *One has the following uniqueness property. If* $f \in L^1(\mathbb{T})$ *and* $\hat{f}(n) = 0$ *for all* $n \in \mathbb{Z}$ *then* $f = 0$. *More generally, if* $\mu \in \mathcal{M}(\mathbb{T})$ *satisfies* $\hat{\mu}(n) = 0$ *for all* $n \in \mathbb{Z}$ *then* $\mu = 0$.

Proof By assumption $\sigma_N f = 0$ for all N; we now apply the convergence property of Proposition 1.5. □

The following is the *Riemann–Lebesgue* lemma.

Corollary 1.8 *If* $f \in L^1(\mathbb{T})$ *then* $\hat{f}(n) \to 0$ *as* $n \to \infty$.

Proof Given $\varepsilon > 0$, let N be such that $\|\sigma_N f - f\|_1 < \varepsilon$. Then $|\hat{f}(n)| = |\widehat{\sigma_N f}(n) - \hat{f}(n)| \leq \|\sigma_N f - f\|_1 < \varepsilon$ for all $|n| > N$. □

Using the Fejér kernel we can now give a simpler proof of Theorem 1.2.

Second proof of Theorem 1.2 For simplicity, we take $\alpha < 1$. The case $\alpha = 1$ is left to the reader. We use the quantitative estimates

$$\|\sigma_N f - f\|_\infty \leq \int_0^1 \min\left(N, (Ny^2)^{-1}\right)|y|^\alpha \, dy \, [f]_\alpha \leq CN^{-\alpha} \, [f]_\alpha.$$

Since $S_N \sigma_N f = \sigma_N f$ we have

$$S_N f - f = S_N(f - \sigma_N f) + \sigma_N f - f,$$

whence

$$\|S_N f - f\|_\infty \leq (\|D_N\|_1 + 1)\|f - \sigma_N f\|_\infty \leq CN^{-\alpha} \log N \, [f]_\alpha,$$

and we are done. □

1.3. The L^p convergence of partial sums

We now turn our attention to whether the partial sums $S_N f$ converge in the sense of $L^p(\mathbb{T})$ or in the sense of $C(\mathbb{T})$. Observe that it makes no sense to ask about the uniform convergence of $S_N f$ for general $f \in L^\infty(\mathbb{T})$, because the uniform limits of continuous functions are continuous.

Proposition 1.9 *The following statements are equivalent for any* $1 \leq p \leq \infty$:

(i) *for every* $f \in L^p(\mathbb{T})$ *(or* $f \in C(\mathbb{T})$ *if* $p = \infty$*) one has*

$$\|S_N f - f\|_p \to 0 \text{ as } N \to \infty;$$

(ii) $\sup_N \|S_N\|_{p \to p} < \infty$.

Proof The implication (ii) $\Longrightarrow$ (i) follows from the fact that trigonometric polynomials are dense in the respective norms; see Corollary 1.6. The implication (i) $\Longrightarrow$ (ii) can be deduced immediately from the uniform boundedness principle of functional analysis. Alternatively, one can prove this directly by the method of the "gliding hump". Suppose that $\sup_N \|S_N\|_{p\to p} = \infty$. For every positive integer ℓ one can therefore find a large integer N_ℓ such that

$$\|S_{N_\ell} f_\ell\|_p > 2^\ell,$$

where f_ℓ is a trigonometric polynomial with $\|f_\ell\|_p = 1$. Now let

$$f(x) = \sum_{\ell=1}^{\infty} \frac{1}{\ell^2} e(M_\ell x) f_\ell(x),$$

with integers $\{M_\ell\}$ to be specified. Notice that

$$\|f\|_p \le \sum_{\ell=1}^{\infty} \frac{1}{\ell^2} \|f_\ell\|_p < \infty.$$

Now choose a sequence $\{M_\ell\}$ tending to infinity so rapidly that the Fourier support of

$$e(M_\ell x) f_\ell(x)$$

lies to the right of the Fourier support of

$$g_\ell(x) := \sum_{j=1}^{\ell-1} \frac{1}{j^2} e(M_j x) f_j(x)$$

for every $j \ge 2$ (here "Fourier support" means those integers for which the corresponding Fourier coefficients are nonzero). We also demand that $M_\ell - N_\ell \to \infty$ as $\ell \to \infty$ and that $S_{M_\ell - N_\ell - 1} f = g_\ell$ and $S_{M_\ell + N_\ell} f = g_{\ell+1}$. Then

$$\|(S_{M_\ell+N_\ell} - S_{M_\ell-N_\ell-1}) f\|_p = \frac{1}{\ell^2} \|S_{N_\ell} f_\ell\|_p > \frac{2^\ell}{\ell^2},$$

which tends to ∞ as $\ell \to \infty$. However, since $N_\ell + M_\ell \to \infty$ and $M_\ell - N_\ell - 1 \to \infty$ the left-hand side tends to 0 as $\ell \to \infty$. This contradiction finishes the proof. □

Exercise 1.4 Let $\{c_n\}_{n\in\mathbb{Z}}$ be an arbitrary sequence of complex numbers and associate with it formally a Fourier series

$$f(x) \sim \sum_n c_n e(nx).$$

Show that there exists $\mu \in \mathcal{M}(\mathbb{T})$ with the property that $\hat{\mu}(n) = c_n$ for all $n \in \mathbb{Z}$ if and only if $\{\sigma_n f\}_{n\ge1}$ is bounded in $\mathcal{M}(\mathbb{T})$. Discuss also the case of $L^p(\mathbb{T})$ with $1 \le p < \infty$ and $C(\mathbb{T})$.

1.3.1. Failure of uniform convergence

We can now settle the question of the uniform convergence of S_N on functions in $C(\mathbb{T})$.

Corollary 1.10 *Fourier series do not converge on $C(\mathbb{T})$ and $L^1(\mathbb{T})$, i.e., there exists $f \in C(\mathbb{T})$ such that $\|S_N f - f\|_\infty \nrightarrow 0$ and $g \in L^1(\mathbb{T})$ such that $\|S_N g - g\|_1 \nrightarrow 0$ as $n \to \infty$.*

Proof By Proposition 1.9 it suffices to verify the limits

$$\sup_N \|S_N\|_{\infty\to\infty} = \infty, \tag{1.19}$$

$$\sup_N \|S_N\|_{1\to 1} = \infty.$$

Both properties follow from the fact that

$$\|D_N\|_1 \to \infty$$

as $N \to \infty$; see (1.7). To deduce (1.19) from this, notice that

$$\begin{aligned} \|S_N\|_{\infty\to\infty} &= \sup_{\|f\|_\infty = 1} \|D_N * f\|_\infty \\ &\geq \sup_{\|f\|_\infty = 1} |(D_N * f)(0)| = \|D_N\|_1. \end{aligned}$$

We remark that the inequality sign here can be replaced with an equality, in view of the translation invariance (1.9). Furthermore, with Fejér kernels $\{K_M\}_{M=1}^\infty$,

$$\|S_N\|_{1\to 1} \geq \|D_N * K_M\|_1 \to \|D_N\|_1$$

as $M \to \infty$. □

Exercise 1.5 The previous proof was indirect. Construct $f \in C(\mathbb{T})$ and $g \in L^1(\mathbb{T})$ such that $S_N f$ does not converge to f uniformly as $N \to \infty$ and such that $S_N g$ does not converge to g in $L^1(\mathbb{T})$. *Hint:* The method of the gliding hump, from the proof of Proposition 1.9, may be used.

We shall see below that, for $1 < p < \infty$,

$$\sup_N \|S_N\|_{p\to p} < \infty.$$

Thus, by Corollary 1.10, for any $f \in L^p(\mathbb{T})$ with $1 < p < \infty$,

$$\|S_N f - f\|_p \to 0 \text{ as } N \to \infty. \tag{1.20}$$

The case $p = 2$ is clear, see Corollary 1.6, but the result for $p \neq 2$ is much deeper. We will need to develop the theory of the conjugate function to obtain it; see Chapter 3. Note that, unlike in Theorem 1.2, here there is no explicit rate of convergence in terms of some expression involving N; clearly this cannot be expected, given only the size of $\|f\|_p$, since $g(x) := e(-mx)f(x)$ has the same L^p norm as f but $\hat{g}(n) = \hat{f}(n+m)$ for all $n \in \mathbb{Z}$. This suggests that we may hope to obtain such a rate if we remove this freedom of translation in the Fourier coefficients. One way of accomplishing this is by imposing a little regularity, as expressed for example in terms of the standard *Sobolev spaces*. Since we do not yet have at our disposal the L^p convergence in (1.20) for general p, we need to restrict ourselves to $p = 2$; this case is particularly simple.

Exercise 1.6 For any $s \in \mathbb{R}$ define the Hilbert space $H^s(\mathbb{T})$ by means of the norm

$$\|f\|_{H^s}^2 := |\hat{f}(0)|^2 + \sum_{n\in\mathbb{Z}} |n|^{2s} |\hat{f}(n)|^2. \tag{1.21}$$

Obtain the following quantitative improvements in certain qualitative convergence properties.

(a) Show that for any $0 \le s \le 1$ one has $\|f(\cdot + \theta) - f\|_2 \le 2\pi \|f\|_{H^s} |\theta|^s$.
(b) Derive a rate of convergence for $\|S_N f - f\|_2$ in terms of N alone, assuming that $\|f\|_{H^s} \le 1$ where $s > 0$ is fixed.

1.4. Regularity and Fourier series

1.4.1. Bernstein's inequality

We now investigate further the connection between the regularity of a function and its associated Fourier series. The following estimate, which builds upon the fact that $(d/dx)e(nx) = 2\pi i n e(nx)$, is known as *Bernstein's inequality*.

Proposition 1.11 *Let f be a trigonometric polynomial with $\hat{f}(k) = 0$ for all $|k| > n$. Then*

$$\|f'\|_p \le Cn\|f\|_p$$

for any $1 \le p \le \infty$. The constant C is absolute.

Proof Let

$$V_n(x) := (1 + e(nx) + e(-nx))K_n(x) \tag{1.22}$$

be de la Vallée Poussin's kernel. We leave it to the reader to check that

$$\widehat{V_n}(j) = 1 \quad \text{if } |j| \le n$$

as well as

$$\|V_n'\|_1 \leq Cn.$$

Then $f = V_n * f$ and thus $f' = V_n' * f$ so that, by Young's inequality,

$$\|f'\|_p \leq \|V_n'\|_1 \|f\|_p \leq Cn\|f\|_p$$

as claimed. □

1.4.2. Convex Fourier coefficients

It is interesting to note that there exists a converse to Bernstein's inequality due to Bohr. We ask the reader to verify a special case of this converse in Exercise 1.7 below; see Problem 1.8 at the end of the chapter for a more difficult case. For the proofs of these converses it is useful to invoke the following general fact.

Lemma 1.12 *Let $\{a_n\}_{n\in\mathbb{Z}}$ be an even sequence of nonnegative numbers that tend to zero, which is convex in the following sense:*

$$a_{n+1} + a_{n-1} - 2a_n \geq 0 \quad \forall n > 0.$$

Then there exists $f \in L^1(\mathbb{T})$ with $f \geq 0$ and $\hat{f}(n) = a_n$.

Proof To understand the construction, we start from the simple observation (based on integration by parts) that

$$\int_x^\infty \varphi''(y)(y - x)\, dy = \varphi(x)$$

for any C^2 function on the line such that $\varphi'(x)x \to 0$ and $\varphi(x) \to 0$ as $x \to \infty$. In the discrete setting, this becomes

$$\sum_{n>m}(a_{n+1} + a_{n-1} - 2a_n)(n - m) = a_m \tag{1.23}$$

for every sequence such that $a_n \to 0$ and $n(a_n - a_{n-1}) \to 0$ as $n \to \infty$. In particular, for such sequences (1.23) can be rewritten as (with Fejér kernel K_n)

$$\sum_{n>m} n(a_{n+1} + a_{n-1} - 2a_n)\widehat{K_n}(m) = a_m,$$

which means that the required function f is given by

$$\sum_{n=1}^{\infty} n(a_{n+1} + a_{n-1} - 2a_n)K_n =: f.$$

Note that, by the convexity assumption, $a_n - a_{n+1}$ is a decreasing sequence, whence $n(a_n - a_{n+1}) \to 0$ and

$$\sum_{n=1}^{N} n(a_{n+1} + a_{n-1} - 2a_n) = a_0 - a_N - N(a_N - a_{N+1}) \to a_0$$

as $N \to \infty$. □

Exercise 1.7 Suppose that $f \in L^1(\mathbb{T})$ satisfies $\hat{f}(j) = 0$ for all j with $|j| < n$. Then show that

$$\|f''\|_p \geq Cn^2 \|f\|_p$$

holds for all $1 \leq p \leq \infty$, where C is independent of $n \in \mathbb{Z}^+$ and of the choices of f and p.

1.4.3. Smoothness and Fourier coefficients

Now we will turn to the question how the smoothness of a function is reflected in the decay of its Fourier coefficients. We begin with the easy observation, based on integration by parts, that for any $f \in C^1(\mathbb{T})$ one has

$$\widehat{f'}(n) = 2\pi i n \hat{f}(n) \quad \forall n \in \mathbb{Z}. \tag{1.24}$$

This shows that we have not only $\hat{f}(n) = O(n^{-1})$ but also $C^1(\mathbb{T}) \hookrightarrow H^1(\mathbb{T})$, where H^1 is the Sobolev space defined in (1.21). If $f \in C^k(\mathbb{T})$ with $k \geq 2$ then we may iterate this relation to obtain a decay of the form $O(n^{-2})$. The following exercise establishes the connection between rapid decay of Fourier coefficients and the infinite regularity of the corresponding function.

Exercise 1.8 Let $f \in L^1(\mathbb{T})$.

(a) Show that $f \in C^\infty(\mathbb{T})$ if and only if $\hat{f}$ decays rapidly, i.e., for every $M \geq 1$ one has $\hat{f}(n) = O(|n|^{-M})$ as $|n| \to \infty$.

(b) Show that $f(x) = F(e(x))$, where F is analytic on some neighborhood of $\{|z| = 1\}$, if and only if $\hat{f}(n)$ decays exponentially, i.e., $\hat{f}(n) = O(e^{-\varepsilon|n|})$ as $|n| \to \infty$ for some $\varepsilon > 0$.

1.4.4. Smoothness and decay

In the previous subsection we used integration by parts to find the decay of $\hat{f}(n)$ provided that $f \in C^1(\mathbb{T})$. If f lies in a Hölder continuous class $C^\alpha(\mathbb{T})$ with $0 \leq \alpha \leq 1$ then the decay may be obtained as follows. Starting from

$$\hat{f}(n) = -\int_{\mathbb{T}} e\left(-n\left(y + \frac{1}{2n}\right)\right) f(y)\, dy = -\int_{\mathbb{T}} e(-ny) f\left(y - \frac{1}{2n}\right) dy$$

we obtain

$$\hat{f}(n) = \frac{1}{2}\int_{\mathbb{T}} e(-ny)\left(f(y) - f\left(y - \frac{1}{2n}\right)\right) dy,$$

which implies that if $f \in C^\alpha(\mathbb{T})$ then

$$\hat{f}(n) = O(n^{-\alpha}) \quad \text{as } |n| \to \infty. \tag{1.25}$$

Note that (1.24) is strictly stronger than this bound, since it allows for L^2 summation and thus an embedding into $H^1(\mathbb{T})$, whereas (1.25) does not. However, since we have already observed that $C^1(\mathbb{T}) \hookrightarrow H^1(\mathbb{T})$ and since clearly $C^0(\mathbb{T}) \hookrightarrow H^0(\mathbb{T}) = L^2(\mathbb{T})$, one could invoke some interpolation machinery at this point to conclude that $C^\alpha(\mathbb{T}) \hookrightarrow H^\alpha(\mathbb{T})$ for all $0 \le \alpha \le 1$; nevertheless we shall not make use of this fact.

Next, we ask how much regularity on f it takes to achieve

$$\sum_{n=-\infty}^{\infty} |\hat{f}(n)| < \infty. \tag{1.26}$$

In other words, we ask for a sufficient condition for f to lie in the Wiener algebra $\mathbb{A}(\mathbb{T})$ of Exercise 1.2. There are a number of ways to interpret the requirement of "regularity". It is easy to settle this embedding question at the level of the spaces $H^s(\mathbb{T})$. In fact, applying Cauchy–Schwarz yields

$$\sum_{n \neq 0} |\hat{f}(n)| \le \left(\sum_{n\neq 0} |\hat{f}(n)|^2 |n|^{1+\varepsilon}\right)^{1/2} \left(\sum_{n \neq 0} |n|^{-1-\varepsilon}\right)^{1/2},$$

so that

$$\sum_{n=-\infty}^{\infty} |\hat{f}(n)|^2 |n|^{1+\varepsilon} < \infty \tag{1.27}$$

is a sufficient condition for (1.26) to hold. In other words, we have shown that $H^s(\mathbb{T}) \hookrightarrow \mathbb{A}(\mathbb{T})$ for any $s > \frac{1}{2}$. We leave it to the reader to check that this fails for $s = \frac{1}{2}$.

1.4.5. Sobolev spaces and embeddings

Next, we would like to address the more challenging question as to the minimal value of $\alpha > 0$ such that $C^\alpha(\mathbb{T}) \hookrightarrow \mathbb{A}(\mathbb{T})$. In fact, we first ask, for which values of α does $C^\alpha(\mathbb{T}) \hookrightarrow H^s(\mathbb{T})$ for some $s > \frac{1}{2}$? We have already observed that $\alpha > \frac{1}{2}$ suffices for this, but this observation required some interpolation theory. Instead, we prefer to give a direct proof in Theorem 1.13 below. It introduces an important idea that we shall see repeatedly throughout this book, namely the grouping of the Fourier coefficients $\hat{f}(n)$ into blocks having approximately the same n; more precisely, we introduce the partial sums

$$(P_j f)(x) := \sum_{2^{j-1} \leq |n| < 2^j} \hat{f}(n) e(nx) \tag{1.28}$$

for all $j \geq 1$.

The following result is known as *Bernstein's theorem.*

Theorem 1.13 *For any $1 \geq \alpha > 0$ one has $C^\alpha(\mathbb{T}) \hookrightarrow H^\beta(\mathbb{T})$ for arbitrary $0 < \beta < \alpha$. In particular, for any $f \in C^\alpha(\mathbb{T})$ with $\alpha > \frac{1}{2}$ one has*

$$\sum_{n \in \mathbb{Z}} |\hat{f}(n)| < \infty$$

and thus $C^\alpha(\mathbb{T}) \hookrightarrow \mathbb{A}(\mathbb{T})$ for any $\alpha > \frac{1}{2}$.

Proof As in the second proof of Theorem 1.2, for simplicity, we take $\alpha < 1$; the case $\alpha = 1$ is left to the reader. Let $[f]_\alpha \leq 1$. We claim that, for every $j \geq 0$,

$$\sum_{2^j \leq |n| < 2^{j+1}} |\hat{f}(n)|^2 \leq C 2^{-2j\alpha}. \tag{1.29}$$

The reader should note the similarity between the bound (1.29) and the estimate (1.25), although (1.29) is strictly stronger. What allows this improvement is the use of orthogonality or, in other words, Parseval's identity (see Corollary 1.6). If (1.29) is true then, for any $\beta < \alpha$,

$$\sum_{n \neq 0} |\hat{f}(n)|^2 |n|^{2\beta} \leq C \sum_{j=0}^{\infty} 2^{2j\beta} \sum_{2^j \leq |n| < 2^{j+1}} |\hat{f}(n)|^2 \leq C \sum_{j=0}^{\infty} 2^{-2j(\alpha - \beta)} < \infty.$$

To prove (1.29) we choose a kernel φ_j such that

$$\widehat{\varphi_j}(n) = 1 \quad \text{if } 2^j \leq |n| \leq 2^{j+1} \tag{1.30}$$

and

$$\widehat{\varphi_j}(n) = 0 \quad \text{if } |n| \ll 2^j \text{ or } |n| \gg 2^{j+1}. \tag{1.31}$$

Equation (1.30) implies that

$$\sum_{2^j \leq |n| < 2^{j+1}} |\hat{f}(n)|^2 \leq \|\varphi_j * f\|_2^2, \tag{1.32}$$

so that we still need to bound, at least implicitly, the right-hand side for which (1.31) will be decisive. The explicit properties of φ_j that allow us to obtain this bound are *cancellation*, meaning that $\widehat{\varphi_j}(0) = 0$, and bounds on the kernel $\varphi_j(x)$ that resemble those of the Fejér kernel K_N with $N = 2^j$.

There are various ways to construct φ_j. We use de la Vallée Poussin's kernel from (1.22) for this purpose. Set

$$\varphi_j(x) := V_{2^{j-1}}(x)(e(3 \times 2^{j-1}x) + e(-3 \times 2^{j-1}x)) \tag{1.33}$$

We leave it to the reader to check that

$$\widehat{\varphi}_j(n) = 1 \quad \text{for } 2^j \leq |n| \leq 2^{j+1}$$

and that $\widehat{\varphi}_j(0) = 0$, which is the same as

$$\int_{\mathbb{T}} \varphi_j(x)\, dx = 0.$$

Moreover, since the φ_j are constructed from Fejér kernels one has the bounds

$$|\varphi_j(x)| \leq C\, 2^{-j} \min(2^{2j}\,,\ |x|^{-2}).$$

Therefore

$$\begin{aligned}
|(\varphi_j * f)(x)| &= \left| \int_{\mathbb{T}} \varphi_j(y)(f(x-y) - f(x))\, dy \right| \\
&\leq \int_{\mathbb{T}} |\varphi_j(y)||f(x-y) - f(x)|\, dy \\
&\leq C \int_0^1 |\varphi_j(y)||y|^\alpha\, dy \\
&\leq C2^{-j} \int_{|y|>2^{-j}} |y|^{\alpha-2}\, dy + C2^j \int_{|y|\leq 2^{-j}} |y|^\alpha\, dy \\
&\leq C2^{-\alpha j},
\end{aligned}$$

as claimed. In summary we see that, for any $0 \leq \beta < \alpha \leq 1$,

$$\|f\|_{H^\beta} \leq C(\alpha, \beta)\|f\|_{C^\alpha}$$

and the theorem follows. □

1.4.6. Optimality of the embedding

Next, we show one cannot take $\beta > \alpha$ in Theorem 1.13.

Exercise 1.9 Considering a *lacunary series* of the form, with $\alpha < 1$,

$$f(x) := \sum_{k\geq 1} 2^{-\alpha k} e(2^k x),$$

show that $C^\alpha(\mathbb{T})$ does not embed into H^β for any $\beta > \alpha$. *Hint:* To bound $|f(x+y) - f(x)|$ distinguish the cases $2^k|y| < 1$ and $2^k|y| \geq 1$.

The second statement of Theorem 1.13, concerning $\mathbb{A}(\mathbb{T})$, is more difficult.

Proposition 1.14 *There is no embedding from $C^{1/2}(\mathbb{T})$ into $\mathbb{A}(\mathbb{T})$. In fact, there exists a function $f \in C^{1/2}(\mathbb{T}) \setminus \mathbb{A}(\mathbb{T})$.*

Proof We claim that there exists a sequence of trigonometric polynomials $P_n(x) = \sum_{\ell=0}^{2^n-1} a_{n,\ell}\, e(\ell x)$ such that, with $N = 2^n$,

$$\begin{aligned} \|P_n\|_\infty &\simeq \sqrt{N}, \\ \|\widehat{P_n}\|_{\ell^1} &\simeq N \end{aligned} \tag{1.34}$$

for each $n \geq 1$ (the $P_n(x)$ are the Rudin–Shapiro polynomials). Here $a \simeq b$ means $C^{-1}a \leq b \leq Ca$, where C is an absolute constant, and $\widehat{P_n}$ refers to the sequence of Fourier coefficients.

Assuming such a sequence for now, we set

$$\begin{aligned} T_n(x) &:= 2^{-n} e(2^n x) P_n(x), \\ f &:= \sum_{n=1}^{\infty} T_n. \end{aligned}$$

Note that the above series converges uniformly to $f \in C(\mathbb{T})$ since $\|T_n\|_\infty \leq C2^{-n/2}$. Moreover, the Fourier supports of the T_n are pairwise disjoint for distinct n. Thus, $\|f\|_{\mathbb{A}(\mathbb{T})} = \infty$ and $f \notin \mathbb{A}(\mathbb{T})$. Finally, let $h \simeq 2^{-m}$ for some positive integer m. Then

$$\begin{aligned} |f(x+h) - f(x)| &\leq \sum_{n=1}^{m} |T_n(x+h) - T_n(x)| + \sum_{n>m} \|T_n\|_\infty \\ &\leq \sum_{n=1}^{m} C|h|\|T_n'\|_\infty + \sum_{n>m} C\, 2^{-n/2} \\ &\leq C\left(\sum_{n=1}^{m} 2^{n/2}\, |h| + 2^{-m/2}\right) \leq C|h|^{1/2} \end{aligned}$$

and thus $f \in C^{1/2}(\mathbb{T})$ as desired. To pass to the last line we used Bernstein's inequality, Proposition 1.11.

It thus remains to establish the existence of the Rudin–Shapiro polynomials (1.34). Define them inductively by $P_0(x) = Q_0(x) = 1$ and

$$\begin{aligned} P_{n+1}(x) &= P_n(x) + e(2^n x) Q_n(x), \\ Q_{n+1}(x) &= P_n(x) - e(2^n x) Q_n(x), \end{aligned}$$

for each $n \geq 0$. Since

$$|P_{n+1}(x)|^2 + |Q_{n+1}(x)|^2 = 2(|P_n(x)|^2 + |Q_n(x)|^2) = 2^{n+1}$$

we see that $\|P_{n+1}\|_\infty$ is of the desired magnitude. Furthermore, all coefficients of both P_n and Q_n are either $+1$ or -1, and the only exponentials with nonzero coefficients are $e(\ell x)$ with $1 \le \ell \le 2^n - 1$. Hence $\{P_n\}_{n=0}^\infty$ is the desired family of polynomials. □

Exercise 1.10 Show that any trigonometric polynomial P with $\|\hat{P}\|_{\ell^1} = N$ and the property that the cardinality of its Fourier support is at most N satisfies $\sqrt{N} \le \|P\|_2 \le \|P\|_\infty \le N$. Hence the polynomials constructed in the previous proof are "extremal" for the lower bound on $\|P\|_\infty$, whereas the Dirichlet kernel (or Fejér kernel) is extremal for the upper bound.

1.5. Higher dimensions

We conclude this chapter with a brief discussion of the Fourier series associated with functions on $\mathbb{T}^d = \mathbb{R}^d/\mathbb{Z}^d$ with $d \ge 2$. The exponential basis in this case is $\{e(\nu \cdot x)\}_{\nu \in \mathbb{Z}^d}$, where the dot indicates a scalar product in $\mathbb{R}^d$, and this is an orthonormal family in the usual $L^2(\mathbb{T}^d)$ sense. Thus, with every measure $\mu \in \mathcal{M}(\mathbb{T}^d)$ we associate a Fourier series

$$\sum_{\nu \in \mathbb{Z}^d} \hat{\mu}(\nu)\, e(\nu \cdot x), \qquad \hat{\mu}(\nu) := \int_{\mathbb{T}^d} e(-\nu \cdot x)\, \mu(dx).$$

As before, a special role is played by the trigonometric polynomials

$$\sum_{\nu \in \mathbb{Z}^d} a_\nu e(\nu \cdot x),$$

where all but finitely many a_ν vanish. In contrast with the one-dimensional torus, in higher dimensions we face a nontrivial ambiguity in the definition of partial sums. In fact we can pose the convergence problem relative to any exhaustion of $\mathbb{Z}^d$ by finite sets A_k that are increasing and whose union is the whole of $\mathbb{Z}^d$. A possible choice here is the squares $[-k, k]^d$, but one could also take more general rectangles, or balls relative to the Euclidean metric, or other shapes. Even though this distinction may seem innocuous, it has given rise to very important developments in harmonic analysis such as Fefferman's ball multiplier theorem and the Bochner–Riesz conjecture, which is still unresolved in dimensions 3 and higher. For now, we content ourselves with some basic results.

Proposition 1.15 *The space of trigonometric polynomials is dense in $C(\mathbb{T}^d)$, and one has Parseval's identity for $L^2(\mathbb{T}^d)$, i.e.,*

$$\|f\|_2^2 = \sum_{\nu \in \mathbb{Z}^d} |\hat{f}(\nu)|^2 \quad \forall f \in L^2(\mathbb{T}^d).$$

If $f \in C^\infty(\mathbb{T}^d)$ then the Fourier series associated with f converges uniformly to f irrespective of the way in which the partial sums are formed.

Proof We base the proof on the fact that the products of Fejér kernels with respect to the individual coordinate axes form an approximate identity. In other words, the family

$$K_{N,d}(x) := \prod_{j=1}^{d} K_N(x_j) \quad \forall N \geq 1$$

forms an approximate identity on $\mathbb{T}^d$. Here $x = (x_1, \ldots, x_d)$. This follows immediately from the one-dimensional analysis above. Hence, for any $f \in C(\mathbb{T}^d)$ one has

$$\|K_{N,d} * f - f\|_\infty \to 0 \quad \text{as } N \to \infty.$$

By inspection $K_{N,d} * f$ is a trigonometric polynomial, which implies the claimed denseness and thus also the Plancherel theorem.

Suppose that $f \in C^\infty(\mathbb{T}^d)$. Repeated integration by parts yields the estimate $\hat{f}(\nu) = O(|\nu|^{-m})$ as $|\nu| \to \infty$ for arbitrary but fixed $m \geq 1$. The convergence statement now follows from $K_{N,d} * f \to f$ in $C(\mathbb{T}^d)$ as $N \to \infty$ and the rapid decay of the coefficients. □

A useful corollary to the previous result is that tensor functions are dense in $C(\mathbb{T}^d)$. A *tensor function* on $\mathbb{T}^d$ is a linear combination of functions of the form $\prod_{j=1}^d f_j(x_j)$ where $f_j \in C(\mathbb{T})$. In particular, trigonometric polynomials are tensor functions, whence the claim.

1.6. Interpolation of operators

Throughout this book we shall make use of the following two fundamental interpolation theorems. We will merely state the results and for more details refer the reader to standard references; see the notes below. The first result is due to Riesz and Thorin. Here L^p spaces are scalar-valued (real or complex), and throughout the book all measure spaces are σ-finite.

Theorem 1.16 *Let (X, μ) be a measure space. Suppose that we have $1 \leq p_1, p_2 \leq \infty$ and assume that $Y \subset L^{p_1}(X, \mu) \cap L^{p_2}(X, \mu)$ is a subspace that is dense in both $L^{p_1}(X, \mu)$ and $L^{p_2}(X, \mu)$. Let T be a linear operator defined on Y that takes its values in the measurable functions on some other space $(\tilde{X}, \tilde{\mu})$ in such a way that for all $f \in Y$ one has*

$$\|Tf\|_{L^{q_j}(\tilde{X},\tilde{\mu})} \leq A_j \|f\|_{L^{p_j}(X,\mu)}, \quad j = 1, 2,$$

where $1 \le q_1, q_2 \le \infty$. *Then*

$$\|Tf\|_{L^q(\tilde{X},\tilde{\mu})} \le A_1^\theta A_2^{1-\theta} \|f\|_{L^p(X,\mu)} \quad \forall f \in Y,$$

where

$$\frac{1}{p} = \frac{\theta}{p_1} + \frac{1-\theta}{p_2}, \quad \frac{1}{q} = \frac{\theta}{q_1} + \frac{1-\theta}{q_2} \tag{1.35}$$

for all $0 \le \theta \le 1$.

Like Hölder's inequality this interpolation result is based on convexity, to be precise *log-convexity*. The standard proof relies on the three-lines theorem from complex analysis, which states that the maximum of the absolute value along vertical lines of a function holomorphic in a vertical strip is log-convex.

In addition to operators that are bounded on Lebesgue spaces, we shall also need the following weak-type boundedness property. Let T be a map from $L^p(X, \mu)$ to the measurable functions on $(\tilde{X}, \tilde{\mu})$. For $1 \le q < \infty$ we say that T is weak-type (p, q) if and only if

$$\tilde{\mu}\big(\big\{x \in \tilde{X} \mid |(Tf)(x)| > \lambda\big\}\big) \le A\lambda^{-q} \|f\|_{L^p(X,\mu)}^q \quad \forall \lambda > 0$$

for all $f \in L^p(X, \mu)$. Further, we define weak-type (p, ∞) to be the same as strong-type (p, ∞), which simply means "bounded in the usual Lebesgue sense".

In the following *Marcinkiewicz interpolation theorem* we allow the operators T to be quasilinear, which means that for some constant $\kappa > 0$ one has

$$|T(f_1 + f_2)| \le \kappa(|f_1| + |f_2|)$$

for all step functions f_1, f_2.

Theorem 1.17 *Suppose that* $1 \le p_1 < p_2 \le \infty$ *and* $q_j \ge p_j$ *with* $q_1 \ne q_2$. *Let* (p, q) *be as in* (1.35) *with* $0 < \theta < 1$. *If* T *is a quasilinear operator that is weak-type* (p_j, q_j) *for* $j = 1, 2$ *then* T *is strong-type* (p, q).

This is proved by breaking functions up according to the sizes of their level sets. It applies in the wider context of Lorentz spaces. To be more specific, one can weaken the hypotheses of Theorem 1.17 further by requiring only weak-type bounds when T is applied to indicator functions of sets. This is referred to as *restricted weak-type* (p, q) and the resulting interpolation theorem can be very helpful in applications.

Notes

An encyclopedic treatment of Fourier series is found in the classic treatise by Zygmund [133]. A less formidable but still comprehensive account of many fundamental results is given in Katznelson [65]. Both these references contain the interpolation results in Section 1.6; see in particular [65, Chapter IV]. A comprehensive account of interpolation theory is given in the monograph by Bergh and Löfström [7], which goes far beyond what we need here. Stein and Weiss [112, Chapter V] presents restricted weak-type interpolation as well as Lorentz spaces. For the construction in Proposition 1.14, see [65, p. 36, Exercise 6.6]. For gap series, as well as other constructions involving Fourier series with an arithmetic flavor, see the classic paper of Rudin [94], which introduced the Λ_p set problem later solved by Bourgain [10].

Problems

Problem 1.1 Suppose that $f \in L^1(\mathbb{T})$ and that $\{S_n f\}_{n=1}^{\infty}$ (the sequence of partial sums of the Fourier series for f) converges in $L^p(\mathbb{T})$ to g for some $p \in [1, \infty]$ and some $g \in L^p$. Prove that $f = g$. If $p = \infty$ conclude that f is continuous.

Problem 1.2 Let $T(x) = \sum_{n=0}^{N}(a_n \cos(2\pi nx) + b_n \sin(2\pi nx))$ be an arbitrary trigonometric polynomial with real coefficients $a_0, \ldots, a_N, b_0, \ldots, b_N$. Show that there is a polynomial $P(z) = \sum_{\ell=0}^{2N} c_\ell z^\ell \in \mathbb{C}[z]$ such that

$$T(x) = e^{-2\pi i N x} P(e^{2\pi i x})$$

and such that $P(z) = z^{2N}\, \overline{P(\bar{z}^{-1})}$. How are the zeros of P distributed in the complex plane?

Problem 1.3 Suppose that $T(x) = \sum_{n=0}^{N}(a_n \cos(2\pi nx) + b_n \sin(2\pi nx))$ is such that $T \geq 0$ everywhere on $\mathbb{T}$ and $a_n, b_n \in \mathbb{R}$ for all $n = 0, 1, \ldots, N$. Show that there are $c_0, \ldots, c_N \in \mathbb{C}$ such that

$$T(x) = \left| \sum_{n=0}^{N} c_n e^{2\pi i n x} \right|^2. \tag{1.36}$$

Find the c_n for the Fejér kernel.

Problem 1.4 Suppose that $T(x) = a_0 + \sum_{h=1}^{H} a_h \cos(2\pi hx)$ satisfies $T(x) \geq 0$ for all $x \in \mathbb{T}$ and $T(0) = 1$. Show that, for any complex numbers $y_1, \ldots, y_N$,

$$\left| \sum_{n=1}^{N} y_n \right|^2 \leq (N + H) \left(a_0 \sum_{n=1}^{N} |y_n|^2 + \sum_{h=1}^{H} |a_h| \left| \sum_{n=1}^{N-h} y_{n+h} \bar{y}_n \right| \right)$$

Hint: Write (1.36) with $\sum_n c_n = 1$. Then apply the Cauchy–Schwarz inequality to $\sum_n y_n = \sum_{m,n} y_n c_m$. The above result is called van der Corput's inequality; see Montgomery [84], p. 18. It plays a role in the theory of the uniform distribution of sequences; see the Chapter 6 problems.

Problem 1.5 Suppose that $\sum_{n=1}^{\infty} n\, |a_n|^2 < \infty$ and $\sum_1^{\infty} a_n$ is Cesàro summable. Show that $\sum_{n=1}^{\infty} a_n$ converges. Use this to prove that any $f \in C(\mathbb{T}) \cap H^{1/2}(\mathbb{T})$ satisfies $S_n f \to f$ uniformly. Note that $H^{1/2}(\mathbb{T})$ does not embed into $\mathbb{A}(\mathbb{T})$, so this convergence does not follow trivially.

Problem 1.6 Show that there exists an absolute constant C such that

$$C^{-1}\sum_{n\neq 0}|n|\,|\hat{f}(n)|^2 \leq \int_{\mathbb{T}^2}\frac{|f(x)-f(y)|^2}{\sin^2(\pi(x-y))}\,dxdy \leq C\sum_{n\neq 0}|n|\,|\hat{f}(n)|^2$$

for any $f \in H^{1/2}(\mathbb{T})$. Does this generalize to $H^s(\mathbb{T})$ and, if so, for which values of s?

Problem 1.7 Use the previous two problems to prove the following theorem of Pal and Bohr: for any real function $f \in C(\mathbb{T})$ there exists a homeomorphism $\varphi : \mathbb{T} \to \mathbb{T}$ such that

$$S_n(f \circ \varphi) \longrightarrow f \circ \varphi$$

uniformly. *Hint:* Without loss of generality let $f > 0$. Consider the domain defined in terms of polar coordinates by means of $r(\theta) = f(\theta/2\pi)$. Then apply the Riemann mapping theorem to the unit disk.

Problem 1.8 Suppose that $f \in L^1(\mathbb{T})$ satisfies $\hat{f}(j) = 0$ for all j with $|j| < n$. Show that

$$\|f'\|_p \geq Cn\|f\|_p$$

for all $1 \leq p \leq \infty$, where C is independent of $n \in \mathbb{Z}^+$ and of the choices of f and p.

Problem 1.9 Show that

$$\|f * g\|_{L^2(\mathbb{T})}^2 \leq \|f * f\|_{L^2(\mathbb{T})}\,\|g * g\|_{L^2(\mathbb{T})}$$

for all $f, g \in L^2(\mathbb{T})$.

Problem 1.10 Show that for every $p > 0$ there exists an approximate identity $K_{N,p}$ on $\mathbb{T}$ with the following properties:

- $\widehat{K_{N,p}}(\nu) = 1$ for all $|\nu| \leq N$,
- $\widehat{K_{N,p}}(\nu) = 1$ for all $|\nu| > CN$,
- $|K_{N,p}(\theta)| \leq CN^{1-p}\min(N^p, |\theta|^{-p})$,

where $C = C(p)$ is some constant and $N \geq 1$ is arbitrary.

Problem 1.11 Given N disjoint arcs $\{I_\alpha\}_{\alpha=1}^N \subset \mathbb{T}$, set $f = \sum_{\alpha=1}^N \chi_{I_\alpha}$. Show that

$$\sum_{|\nu|>k}|\hat{f}(\nu)|^2 \leq \frac{CN}{k}.$$

Hint: The bound N^2/k is much easier and should be obtained first. Going from N^2 to N then requires one to exploit orthogonality in a suitable fashion.

Problem 1.12 Given any function $\psi : \mathbb{Z}^+ \to \mathbb{R}^+$ such that $\psi(n) \to 0$ as $n \to \infty$, show that one can find a measurable set $E \subset \mathbb{T}$ for which

$$\limsup_{n\to\infty}\frac{|\widehat{\chi_E}(n)|}{\psi(n)} = \infty.$$

Problem 1.13 In this problem the reader is asked to analyze some well-known partial differential equations in terms of Fourier series.

(a) Solve the heat equation $u_t - u_{\theta\theta} = 0$, $u(0) = u_0$ (the data at time $t = 0$) on $\mathbb{T}$ using a Fourier series. Show that if $u_0 \in L^2(\mathbb{T})$ then $u(t, \theta)$ is analytic in θ for every $t > 0$ and solves the heat equation. Prove that $\|u(t) - u_0\|_2 \to 0$ as $t \to 0$. Write $u(t) = G_t * u_0$ and show that G_t is an approximate identity for $t > 0$. Conclude that $u(t) \to 0$ as $t \to 0$ in the L^p or $C(\mathbb{T})$ sense. Repeat for higher-dimensional tori.

(b) Solve the Schrödinger equation $iu_t - u_{\theta\theta} = 0$, $u(0) = u_0$ on $\mathbb{T}$ with $u_0 \in L^2(\mathbb{T})$, using a Fourier series. In what sense does this Fourier series "solve" the equation? Show that $\|u(t)\|_2 = \|u_0\|_2$ for all t. Discuss the limit $u(t)$ as $t \to 0$. Repeat for higher-dimensional tori.

(c) Solve the wave equation $u_{tt} - u_{\theta\theta} = 0$ on $\mathbb{T}$ by Fourier series. Discuss the Cauchy problem as in (a) and (b). Show that if $u_t(0) = 0$ then, with $u(0) = f$,

$$u(t, \theta) = \tfrac{1}{2}(f(\theta + t) + f(\theta - t)).$$

2

Harmonic functions; Poisson kernel

2.1. Harmonic functions

In this chapter we shall investigate harmonic functions on the disk. This class of functions is not only of fundamental importance to analysis in general but also essential to the resolution of the L^p-convergence problem of Fourier series, i.e., to the question whether $\|S_N f - f\|_p \to 0$ as $N \to \infty$ when $1 < p < \infty$ (see Chapter 3).

We now briefly review some basic facts about harmonic functions on general domains $\Omega \subset \mathbb{R}^2$ (in fact, we could also consider $\mathbb{R}^d$ here for most properties, with the exception of any reference to holomorphic functions). As usual, a domain is open and connected. We say that $u \in C^2(\Omega)$ is *harmonic* provided that $\Delta u = 0$ on Ω, where Δ is the Laplacian, i.e., we require that $u_{xx} + u_{yy} = 0$ on Ω. Examples of such functions are easily constructed; they include all linear functions and $u(x, y) = x^2 - y^2$. More generally, for any holomorphic F on Ω, both real and imaginary parts of F are harmonic, as follows from the *Cauchy–Riemann equations*

$$F = u + iv, \quad u_x - v_y = 0, \quad u_y + v_x = 0.$$

We recall the important converse from complex function theory: if u is harmonic on a *simply connected* domain Ω then there exists a holomorphic function F on Ω with $\operatorname{Re} F = u$. To see this, define

$$v(z) = \int_\gamma -u_y \, dx + u_x \, dy,$$

where γ is any path connecting a fixed point $z_0 \in \Omega$ with the variable point $z \in \Omega$. By the harmonicity of u and the simple connectivity of Ω we see that the path integral does not depend on the specific choice of γ. By inspection, $F := u + iv$ is C^2 and the Cauchy–Riemann equations hold, whence F is holomorphic.

This implies that *harmonic functions u are analytic* in the sense that their infinite Taylor series converge locally and are equal to u. In particular, harmonic functions are C^∞.

In a discrete setting, harmonic functions u on the lattice $\mathbb{Z}^2$ are given by

$$u(n,m) = \tfrac{1}{4}\big(u(n+1,m) + u(n-1,m) + u(n,m+1) + u(n,m-1)\big) \quad \forall (n,m) \in \mathbb{Z}^2.$$

Note that this means that the value of u at every point (n,m) is the average of the values of u at the four nearest neighbors of (n,m). This motivates the *mean-value property* enjoyed by harmonic functions in the continuum.

2.1.1. Mean-value property and maximum principle

The mean value property is given as follows.

Lemma 2.1 *Let u be harmonic on some domain $\Omega \subset \mathbb{R}^2$. Then, for every $z \in \Omega$ and for all $0 < r < \operatorname{dist}(z, \partial\Omega)$, one has*

$$u(z) = \int_{\mathbb{T}} u(z + re(\theta))\, d\theta = \frac{1}{\pi r^2} \int_{z+r\mathbb{D}} u(x,y)\, dx dy. \tag{2.1}$$

Proof By the divergence theorem, with dm the Lebesgue measure in the plane,

$$\frac{d}{dr} \int_{\mathbb{T}} u(z + re(\theta))\, d\theta = \frac{1}{2\pi r} \int_{z+r\partial\mathbb{D}} \partial_n u(w)\, \sigma(dw) = \frac{1}{2\pi r} \int_{z+r\mathbb{D}} \Delta u\, dm = 0.$$

Thus, the circular means (with σ the arc length)

$$M(r) := \int_{\mathbb{T}} u(z + re(\theta))\, d\theta = \frac{1}{2\pi r} \int_{|w-z|=r} u(w)\, \sigma(dw)$$

are constant and, since $M(r) \to u(z)$ as $r \to 0$, we see that $u(z) = M(r)$ as claimed. Integrating $su(z) = sM(s)$ over $0 < s < r$ implies the mean-value property over solid disks. □

The special role of circles here is no coincidence. In fact, not only is the Laplacian Δ invariant under rotations (which means that for every rotation ρ in the plane and every C^2 function u one has $\Delta(u \circ \rho) = (\Delta u) \circ \rho$) but Δ uniquely has this property. If $L = a\partial_{xx} + b\partial_{xy} + c\partial_{yy}$ enjoys this commutation property with constant a, b, c then $a = c$ and $b = 0$.

An immediate consequence of the mean-value property is the *maximum principle*.

Corollary 2.2 *Let u be harmonic in $\Omega \subset \mathbb{R}^2$. If u attains an extremum in Ω then u must be constant.*

Proof Suppose that $u(z) \le u(z_0)$ for all $z \in \Omega$, where $z_0 \in \Omega$ is fixed. Define

$$\mathcal{S} := \{z \in \Omega \mid u(z) = u(z_0)\}.$$

Then $\mathcal{S} \neq \emptyset$, and $\mathcal{S}$ is closed. Moreover, by the mean-value property, $\mathcal{S}$ is also an open set, whence $\mathcal{S} = \Omega$ as claimed. □

Exercise 2.1 Show that it suffices to assume that u attains a local extremum in the previous result.

Another common formulation of the maximum principle *on bounded domains* relies on boundary values.

Corollary 2.3 *Let Ω be bounded and suppose that $u \in C(\overline{\Omega})$ is harmonic on Ω. Then*

$$\min_{\overline{\Omega}} u \le u(z) \le \max_{\overline{\Omega}} u \quad \forall z \in \Omega,$$

and equality can occur only if u is a constant.

Proof Since $\overline{\Omega}$ is compact, u attains both its maximum and its minimum on that set. We may assume that u is not constant, but then Corollary 2.2 implies that the extrema are not attained in Ω, whence the claim. □

2.2. The Poisson kernel

There is a close connection between Fourier series and analytic or harmonic functions on the disk $\mathbb{D} := \{z \in \mathbb{C} \mid |z| \le 1\}$. In fact, at least formally, Fourier series can be viewed as the "boundary values" of a Laurent series

$$\sum_{n=-\infty}^{\infty} a_n z^n; \tag{2.2}$$

this can be seen by setting $z = x + iy = e(\theta)$. Alternatively, suppose that we are given a continuous function f on $\mathbb{T}$ and wish to find the harmonic extension u of f into $\mathbb{D}$, i.e., a solution to

$$\begin{aligned} \triangle u &= 0 \quad \text{in } \mathbb{D}, \\ u &= f \quad \text{on } \partial\mathbb{D} = \mathbb{T}. \end{aligned} \tag{2.3}$$

The term "solution" here refers to a pointwise solution, i.e., a function $f \in C^2(\mathbb{D}) \cap C(\bar{\mathbb{D}})$. However, as we shall see, it is also important to investigate other notions of solutions of (2.3), with less regular functions f.

2.2.1. Derivation of the Poisson kernel

Note that we cannot use negative powers of z in (2.2) in an ansatz for u. However, we can use complex conjugates instead. Indeed, since $\Delta z^n = 0$ and $\Delta \bar{z}^n = 0$ for every integer $n \geq 0$, we are led to define

$$u(z) = \sum_{n=0}^{\infty} \hat{f}(n) z^n + \sum_{n=-\infty}^{-1} \hat{f}(n) \bar{z}^{|n|}, \tag{2.4}$$

which, at least formally, satisfies $u(e(\theta)) = \sum_{n=-\infty}^{\infty} \hat{f}(n) e(n\theta) = f(\theta)$. Inserting $z = re(\theta)$ and

$$\hat{f}(n) = \int_{\mathbb{T}} e(-n\varphi) f(\varphi)\, d\varphi$$

into (2.4) yields

$$u(re(\theta)) = \int_{\mathbb{T}} \sum_{n\in\mathbb{Z}} r^{|n|} e(n(\theta - \varphi)) f(\varphi)\, d\varphi.$$

This resembles the derivation of the Dirichlet kernel in Chapter 1, and we now ask the reader to find a closed form for the sum.

Exercise 2.2 Check that, for $0 \leq r < 1$,

$$P_r(\theta) := \sum_{n\in\mathbb{Z}} r^{|n|} e(n\theta) = \frac{1 - r^2}{1 - 2r\cos(2\pi\theta) + r^2}.$$

This is the *Poisson kernel.*

2.2.2. The Poisson kernel as an approximate identity

Based on our formal calculation above, we therefore expect to obtain the harmonic extension of a sufficiently well-behaved function f on $\mathbb{T}$ by means of the convolution

$$u(re(\theta)) = \int_{\mathbb{T}} P_r(\theta - \varphi) f(\varphi)\, d\varphi = (P_r * f)(\theta)$$

for $0 \leq r < 1$.

Note that $P_r(\theta)$, for $0 \leq r < 1$, is a harmonic function of the variables $x + iy = re(\theta)$. Moreover, for any finite measure $\mu \in \mathcal{M}(\mathbb{T})$ the expression $(P_r * \mu)(\theta)$ is not only well defined but in fact defines a harmonic function on $\mathbb{D}$.

The remainder of this chapter will therefore be devoted to analyzing the boundary behavior of $P_r * \mu$. Clearly Proposition 1.5 will play an important role in this investigation, but the fact that we are dealing with harmonic functions will enter in a crucial way (such as through the maximum principle).

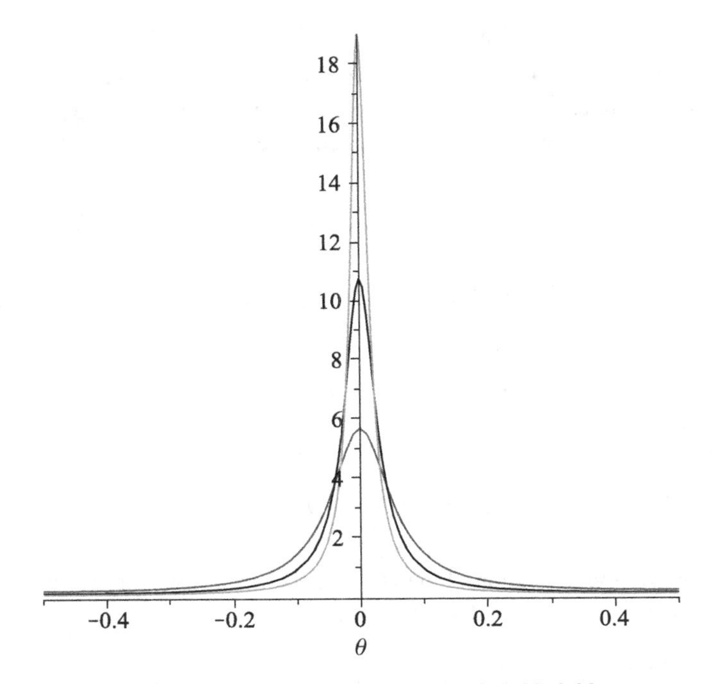

Figure 2.1. Graphs of P_r for $r = 0.70, 0.83, 0.90$.

In the following we use the notion of *approximate identity* in a more general form than in Chapter 1. However, the reader will have no difficulty in transferring this notion, including Proposition 1.5, to the present context.

Exercise 2.3 Check that $\{P_r\}_{0<r<1}$ is an approximate identity. The role of $N \in \mathbb{Z}^+$ in Definition 1.3 is played here by $0 < r < 1$ and $N \to \infty$ is replaced by $r \to 1$.

An important role is that of the kernel $Q_r(\theta)$, which is the *harmonic conjugate* of $P_r(\theta)$. Recall that this means that $P_r(\theta) + i\,Q_r(\theta)$ is analytic in $z = re(\theta)$ with $Q_0 = 0$. In this case it is easy to find $Q_r(\theta)$, since

$$P_r(\theta) = \operatorname{Re}\left(\frac{1+z}{1-z}\right)$$

and therefore

$$Q_r(\theta) = \operatorname{Im}\left(\frac{1+z}{1-z}\right) = \frac{2r\sin(2\pi\theta)}{1-2r\cos(2\pi\theta)+r^2}\,.$$

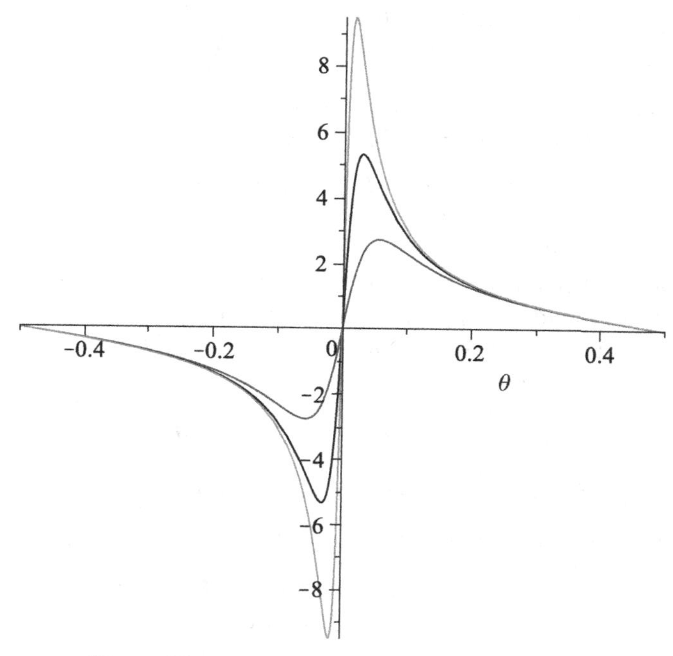

Figure 2.2. Graphs of Q_r for $r = 0.70, 0.83, 0.90$.

Exercise 2.4

(a) Show that $\{Q_r\}_{0<r<1}$ is *not* an approximate identity.

(b) Check that $Q_1(\theta) = \cot(\pi\theta)$. Draw the graph of $Q_1(\theta)$. What is the asymptotic behavior of $Q_1(\theta)$ for θ values close to zero?

We will study conjugate harmonic functions later. First, we clarify in what sense the harmonic extension $P_r * f$ of f attains f as $r \to 1-$.

2.2.3. The boundary-value problem

Definition 2.4 For any $1 \leq p < \infty$ define

$$h^p(\mathbb{D}) := \left\{ u : \mathbb{D} \to \mathbb{C} \text{ harmonic} \;\middle|\; \sup_{0<r<1} \int_{\mathbb{T}} |u(re(\theta))|^p \, d\theta < \infty \right\}$$

with an obvious modification for $p = \infty$. These are the "little" Hardy spaces with norm

$$|||u|||_p := \sup_{0<r<1} \|u(re(\cdot))\|_{L^p(\mathbb{T})}.$$

It is important to observe that $P_r(\theta) \in h^1(\mathbb{D})$. This function has "boundary values" δ_0 (the Dirac mass at $\theta = 0$) since $P_r = P_r * \delta_0$.

Theorem 2.5 *There is a one-to-one correspondence between $h^1(\mathbb{D})$ and $\mathcal{M}(\mathbb{T})$, given by $\mu \in \mathcal{M}(\mathbb{T}) \mapsto F_r(\theta) := (P_r * \mu)(\theta)$. Furthermore,*

$$\|\mu\| = \sup_{0<r<1} \|F_r\|_1 = \lim_{r\to 1} \|F_r\|_1 \tag{2.5}$$

and one has the following.

(i) *The measure μ is absolutely continuous with respect to Lebesgue measure ($\mu \ll d\theta$) if and only if $\{F_r\}$ converges in $L^1(\mathbb{T})$. If this is the case then $d\mu = f\,d\theta$, where f is the L^1 limit of F_r.*

(ii) *The following are equivalent for $1 < p \le \infty$:*

$$d\mu = f\,d\theta \text{ with } f \in L^p(\mathbb{T}) \iff \{F_r\}_{0<r<1} \text{ is } L^p\text{-bounded;}$$

$$d\mu = f\,d\theta \text{ with } f \in L^p(\mathbb{T}) \iff \{F_r\} \text{ converges in } L^p \text{ if } 1 < p < \infty$$

and in the weak-$*$ sense in L^∞ if $p = \infty$ as $r \to 1$.

(iii) *That the above function f is continuous $\iff$ F extends to a continuous function on $\overline{\mathbb{D}}$ $\iff$ F_r converges uniformly as $r \to 1-$.*

This theorem identifies $h^1(\mathbb{D})$ with $\mathcal{M}(\mathbb{T})$ and $h^p(\mathbb{D})$ with $L^p(\mathbb{T})$ for $1 < p \le \infty$. Moreover, $h^\infty(\mathbb{D})$ contains the subclass of harmonic functions that can be extended continuously onto $\overline{\mathbb{D}}$; this subclass is the same as $C(\mathbb{T})$. Before proving the theorem we present two simple lemmas. In what follows we use the notation $F_r(\theta) := F(re(\theta))$.

Lemma 2.6

(i) *If $F \in C(\overline{\mathbb{D}})$ and $\Delta F = 0$ in $\mathbb{D}$ then $F_r = P_r * F_1$ for any $0 \le r < 1$.*

(ii) *If $\Delta F = 0$ in $\mathbb{D}$ then $F_{rs} = P_r * F_s$ for any $0 \le r, s < 1$.*

(iii) *As a function of $r \in (0,1)$, the norms $\|F_r\|_p$ are nondecreasing for any $1 \le p \le \infty$.*

Proof (i) Let $u(re(\theta)) := (P_r * F_1)(\theta)$ for any $0 \le r < 1$, θ. Then $\Delta u = 0$ in $\mathbb{D}$. By Proposition 1.5 and Exercise 2.3, $\|u_r - F_1\|_\infty \to 0$ as $r \to 1$. Hence u extends to a continuous function on $\overline{\mathbb{D}}$ with the same boundary values as F. By the maximum principle, $u = F$ as claimed.

(ii) Rescaling the disc $s\mathbb{D}$ to $\mathbb{D}$ reduces (ii) to (i).

(iii) By (ii) and Young's inequality we have

$$\|F_{rs}\|_p \le \|P_r\|_1 \|F_s\|_p = \|F_s\|_p,$$

as claimed. □

Lemma 2.7 *Let $F \in h^1(\mathbb{D})$. Then there exists a unique measure $\mu \in \mathcal{M}(\mathbb{T})$ such that $F_r = P_r * \mu$.*

Proof Since the unit ball of $\mathcal{M}(\mathbb{T})$ is weak-$*$ compact there exists a subsequence $r_j \to 1$ with $F_{r_j} \to \mu$ in the weak-$*$ sense to some $\mu \in \mathcal{M}(\mathbb{T})$. Then, for any $0 < r < 1$,

$$P_r * \mu = \lim_{j\to\infty} (F_{r_j} * P_r) = \lim_{j\to\infty} F_{rr_j} = F_r$$

by Lemma 2.6(ii). Let $f \in C(\mathbb{T})$. Then $\langle F_r, f\rangle = \langle P_r * \mu, f\rangle = \langle \mu, P_r * f\rangle \to \langle \mu, f\rangle$ as $r \to 1$, where we have again used Proposition 1.5. This shows that in the weak-$*$ sense

$$\mu = \lim_{r\to 1} F_r, \tag{2.6}$$

which implies the uniqueness of μ. □

Proof of Theorem 2.5 If $\mu \in \mathcal{M}(\mathbb{T})$ then $P_r * \mu \in h^1(\mathbb{D})$. Conversely, given $F \in h^1(\mathbb{D})$, then by Lemma 2.7 there is a unique μ such that $F_r = P_r * \mu$. This gives the one-to-one correspondence. Moreover, (2.6) and Lemma 2.6(iii) show that

$$\|\mu\| \le \limsup_{r\to 1} \|F_r\|_1 = \sup_{0<r<1} \|F_r\|_1 = \lim_{r\to 1} \|F_r\|_1.$$

Since clearly we also have

$$\sup_{0<r<1} \|F_r\|_1 \le \sup_{0<r<1} \|P_r\|_1 \|\mu\| = \|\mu\|,$$

(2.5) follows. If $f \in L^1(\mathbb{T})$ and $\mu(d\theta) = f(\theta)\, d\theta$ then Proposition 1.5 shows that $F_r \to f$ in $L^1(\mathbb{T})$. Conversely, if $F_r \to f$ in the sense of $L^1(\mathbb{T})$ then, because of (2.6), it is necessarily true that $\mu(d\theta) = f(\theta)\, d\theta$, which proves (i). By analogous arguments one can derive (ii) and (iii). We leave the details to the reader (simply invoke Proposition 1.5 again). □

As a consequence of the solvability of the boundary value problem for continuous data, we can now characterize harmonic functions via the mean value property.

Corollary 2.8 *Let $\Omega \subset \mathbb{R}^2$ be a domain and suppose that $u \in C(\Omega)$ satisfies the mean value property, i.e., (2.1). Then $u \in C^\infty(\Omega)$ and u is harmonic in Ω.*

Proof Let $D \subset \Omega$ be any disk compactly contained in Ω. By Theorem 2.5, there exists $v \in C(\bar{D})$ and harmonic on D with $u = v$ on ∂D. Clearly, $u - v$ satisfies the mean-value property and therefore also the maximum principle on D. Hence, $u - v$ cannot attain any extrema in D. Since it vanishes on ∂D we must have $u - v = 0$ on D, whence the result. □

2.3. The Hardy–Littlewood maximal function

Next, we turn to the question of the almost everywhere convergence of $P_r * f$ to f as $r \to 1$. By Theorem 2.5 we have pointwise convergence if f is continuous. If $f \in L^1(\mathbb{T})$ then we write $f = \lim_n g_n$ as a limit in L^1 with $g_n \in C(\mathbb{T})$, but then we face two limits, namely one as $n \to \infty$ and the other as $r \to 1-$, and we wish to interchange them. It is a common feature that such an interchange requires some form of *uniform control.* To be specific, we will use the *Hardy–Littlewood maximal function* Mf, associated with any $f \in L^1(\mathbb{T})$, to dominate the convolution with the Poisson kernel $P_r * f$ *uniformly in* $0 < r < 1$. This turns out to be an instance of the general fact that *radially bounded approximate identities* are dominated by the Hardy–Littlewood maximal function; see condition (A4) below.

The Hardy–Littlewood maximal function is defined by

$$(Mf)(\theta) = \sup_{\theta \in I \subset \mathbb{T}} \frac{1}{|I|} \int_I |f(\varphi)|\, d\varphi,$$

where $I \subset \mathbb{T}$ is an open interval and $|I|$ denotes the length of I. The basic fact about Mf that we shall need in this context is given by the following result. We say that g belongs to weak-L^1 if there exists a finite constant C such that

$$\left|\left\{\theta \in \mathbb{T} \,\middle|\, |g(\theta)| > \lambda\right\}\right| \le C\lambda^{-1}$$

for all $\lambda > 0$. Henceforth, the notation $|\cdot|$ applied to sets will denote the Lebesgue measure. Note that any $g \in L^1$ belongs to weak-L^1. This is an instance of *Markov's inequality*,

$$\left|\left\{\theta \in \mathbb{T} \,\middle|\, |f(\theta)| > \lambda\right\}\right| \le \lambda^{-p} \|f\|_p^p$$

for any $1 \le p < \infty$ and any $\lambda > 0$.

2.3.1. The boundedness properties of the maximal function

We refer to the map $f \to Mf$ as the Hardy–Littlewood maximal operator M. It is sublinear, which means that $M(f + g) \le Mf + Mg$. One also has $M(cf) = |c|M(f)$ for any scalar c.

Proposition 2.9 *The Hardy–Littlewood maximal operator M satisfies the following bounds:*

(i) *M is bounded from L^1 to weak-L^1, i.e.,*

$$|\{\theta \in \mathbb{T} \mid Mf(\theta) > \lambda\}| \leq \frac{3}{\lambda}\|f\|_1$$

for all $\lambda > 0$.

(ii) *For any $1 < p \leq \infty$, M is bounded on L^p.*

Proof (i) Fix some $\lambda > 0$ and any compact

$$K \subset \{\theta \in \mathbb{T} \mid Mf(\theta) > \lambda\}. \tag{2.7}$$

There exists a finite cover $\{I_j\}_{j=1}^N$ of $\mathbb{T}$ by open arcs I_j such that

$$\int_{I_j} |f(\varphi)|\, d\varphi > \lambda |I_j| \tag{2.8}$$

for each j. We now apply Wiener's covering lemma to pass to a more convenient sub-cover. Select an arc of maximal length from $\{I_j\}$; call it J_1. Observe that any I_j such that $I_j \cap J_1 \neq \emptyset$ satisfies $I_j \subset 3J_1$ where $3J_1$ is the arc with the same center as J_1 and three times the length (if $3J_1$ has length larger than 1 then set $3J_1 = \mathbb{T}$). Now remove all arcs from $\{I_j\}_{j=1}^N$ that intersect J_1. Let J_2 be a remaining are having maximal length. Continuing in this fashion we obtain arcs $\{J_\ell\}_{\ell=1}^L$ that are pairwise disjoint and such that

$$\bigcup_{j=1}^N I_j \subset \bigcup_{\ell=1}^L 3J_\ell.$$

Therefore, in view of (2.7) and (2.8),

$$\begin{aligned} |K| &\leq \left|\bigcup_{\ell=1}^L 3J_\ell\right| \leq 3\sum_{\ell=1}^L |J_\ell| \\ &\leq \frac{3}{\lambda}\sum_{\ell=1}^L \int_{J_\ell} |f(\varphi)|\, d\varphi \leq \frac{3}{\lambda}\|f\|_1 \end{aligned}$$

as claimed.

To prove part (ii), one interpolates the bound from (i) with the trivial L^∞ bound

$$\|Mf\|_\infty \leq \|f\|_\infty,$$

by means of Marcinkiewicz's interpolation theorem. □

We remark that the same argument based on Wiener's covering lemma yields the corresponding statement for $M\mu$ when $\mu \in \mathcal{M}(\mathbb{T})$. In other words, if we define

$$(M\mu)(\theta) = \sup_{\theta \in I \subset \mathbb{T}} \frac{|\mu|(I)}{|I|}$$

then, for all $\lambda > 0$,

$$|\{\theta \in \mathbb{T} \mid (M\mu)(\theta) > \lambda\}| \leq \frac{3}{\lambda}\|\mu\|, \tag{2.9}$$

where $\|\mu\|$ is the total variation of μ.

Exercise 2.5 Find, within a multiplicative constant, $M\delta_0$ where δ_0 is the Dirac measure at 0. Use this to prove that the weak-L^1 bound on $M\mu$ cannot be improved in general, even when μ is absolutely continuous.

For a considerable strengthening of the result given in the above exercise, the reader should consult Problem 3.5 at the end of the next chapter.

2.4. Almost everywhere convergence

We now introduce the class of approximate identities dominated by the maximal function.

Definition 2.10 Let $\{\Phi_n\}_{n=1}^{\infty}$ be an approximate identity as in Definition 1.3. We say that it is *radially bounded* if there exist functions $\{\Psi_n\}_{n=1}^{\infty}$ on $\mathbb{T}$ such that the following additional property holds:
(A4) $|\Phi_n| \leq \Psi_n$ for Ψ_n even and decreasing, i.e., $\Psi_n(\theta) \leq \Psi_n(\varphi)$ for $0 \leq \varphi \leq \theta \leq \frac{1}{2}$, for all $n \geq 1$. Finally, we require that $\sup_n \|\Psi_n\|_1 < \infty$.

All the basic examples of approximate identities that we have encountered so far (Fejér, Poisson, and box kernels) are of this type. The following lemma establishes the aforementioned uniform control of the convolution with radially bounded approximate identities in terms of the Hardy–Littlewood maximal function.

Lemma 2.11 *If* $\{\Phi_n\}_{n=1}^{\infty}$ *satisfies* (A4) *then for any* $f \in L^1(\mathbb{T})$ *one has*

$$\sup_n |(\Phi_n * f)(\theta)| \leq \sup_n \|\Psi_n\|_1 \, Mf(\theta)$$

for all $\theta \in \mathbb{T}$.

Proof It clearly suffices to show the following. Let

$$K : \left[-\tfrac{1}{2}, \tfrac{1}{2}\right] \to \mathbb{R}^+ \cup \{0\}$$

be even and decreasing; then, for any $f \in L^1(\mathbb{T})$, we have

$$|(K * f)(\theta)| \leq \|K\|_1 Mf(\theta). \tag{2.10}$$

Indeed, assume that (2.10) holds. Then

$$\sup_n |(\Phi_n * f)(\theta)| \leq \sup_n (\Psi_n * |f|)(\theta) \leq \sup_n \|\Psi_n\|_1 \, Mf(\theta),$$

and the lemma follows. The idea behind (2.10) is to show that K can be written as an average of box kernels, i.e., that for some positive measure μ we have

$$K(\theta) = \int_0^{1/2} \chi_{[-\varphi,\varphi]}(\theta)\, \mu(d\varphi). \tag{2.11}$$

We leave it to the reader to check that $d\mu = -dK + K\left(\frac{1}{2}\right)\delta_{1/2}$ is a suitable choice. Here dK refers to the measure with decreasing distribution function K, as is customary in Stieltjes integrals. Notice that (2.11) implies that

$$\int_0^1 K(\theta)\, d\theta = \int_0^{1/2} 2\varphi\, d\mu(\varphi).$$

Moreover, by (2.11),

$$\begin{aligned}
|(K * f)(\theta)| &= \left| \int_0^{1/2} \left(\frac{1}{2\varphi}\chi_{[-\varphi,\varphi]} * f \right)(\theta)\, 2\varphi\, \mu(d\varphi) \right| \\
&\le Mf(\theta) \int_0^{1/2} 2\varphi\, \mu(d\varphi) \\
&= Mf(\theta)\, \|K\|_1,
\end{aligned}$$

which is (2.10). □

The following theorem establishes the main almost everywhere (a.e.) convergence result of this chapter. The reader should note how the maximal function enters precisely in order to interchange the aforementioned double limit.

Theorem 2.12 *If $\{\Phi_n\}_{n=1}^{\infty}$ satisfies (A1)–(A4) then, for any $f \in L^1(\mathbb{T})$, one has $\Phi_n * f \to f$ a.e. as $n \to \infty$.*

Proof Pick $\varepsilon > 0$ and let $g \in C(\mathbb{T})$ with $\|f - g\|_1 < \varepsilon$. By Proposition 1.5 one has, with $h = f - g$,

$$\begin{aligned}
&\left| \left\{ \theta \in \mathbb{T} \;\middle|\; \limsup_{n\to\infty} |(\Phi_n * f)(\theta) - f(\theta)| > \sqrt{\varepsilon} \right\} \right| \\
&\le \left| \left\{ \theta \in \mathbb{T} \;\middle|\; \limsup_{n\to\infty} |(\Phi_n * h)(\theta)| > \tfrac{1}{2}\sqrt{\varepsilon} \right\} \right| + \left| \left\{ \theta \in \mathbb{T} \;\middle|\; |h(\theta)| > \tfrac{1}{2}\sqrt{\varepsilon} \right\} \right| \\
&\le \left| \left\{ \theta \in \mathbb{T} \;\middle|\; \sup_n |(\Phi_n * h)(\theta)| > \tfrac{1}{2}\sqrt{\varepsilon} \right\} \right| + \left| \left\{ \theta \in \mathbb{T} \;\middle|\; |h(\theta)| > \tfrac{1}{2}\sqrt{\varepsilon} \right\} \right| \\
&\le \left| \left\{ \theta \in \mathbb{T} \mid CMh(\theta) > \tfrac{1}{2}\sqrt{\varepsilon} \right\} \right| + \left| \left\{ \theta \in \mathbb{T} \;\middle|\; |h(\theta)| > \tfrac{1}{2}\sqrt{\varepsilon} \right\} \right| \\
&\le C\sqrt{\varepsilon}.
\end{aligned}$$

To pass to the final inequality we have used Proposition 2.9 as well as Markov's inequality (recall that $\|h\|_1 < \varepsilon$); see the start of Section 2.3. □

As a corollary we obtain not only the classic Lebesgue differentiation theorem (by considering the box kernel) but also the a.e. convergence of the Cesáro means $\sigma_N f$ (via the Fejér kernel) as well as of the Poisson integrals $P_r * f$ to f for any $f \in L^1(\mathbb{T})$. A theorem of Kolmogorov states that this *fails* for the partial sums $S_N f$ on $L^1(\mathbb{T})$. We will present this example in Chapter 6.

2.4.1. The case of measures

It is natural to ask whether there is an analogue of Theorem 2.12 for measures $\mu \in \mathcal{M}(\mathbb{T})$. We begin with the following general fact from measure theory.

Lemma 2.13 *If $\mu \in \mathcal{M}(\mathbb{T})$ is a positive measure that is singular with respect to Lebesgue measure m (symbolically, $\mu \perp m$) then for a.e. $\theta \in \mathbb{T}$ with respect to Lebesgue measure we have*

$$\frac{\mu([\theta - \varepsilon, \theta + \varepsilon])}{2\varepsilon} \to 0 \quad \text{as } \varepsilon \to 0.$$

Proof For every $\lambda \geq 0$ let

$$E(\lambda) := \left\{ \theta \in \mathbb{T} \;\middle|\; \limsup_{\varepsilon \to 0} \frac{\mu([\theta - \varepsilon, \theta + \varepsilon])}{2\varepsilon} > \lambda \right\}.$$

By assumption there exists a Borel set $A \subset \mathbb{T}$ with $|A| = 0$ and such that $\mu(E) = \mu(E \cap A)$ for every Borel set $E \subset \mathbb{T}$. Suppose first that A is compact. Then it follows that $E(0) \subset A$, whence $|E(0)| = 0$ as desired.

In general A does not need to be compact, but, for every $\delta > 0$, there exists a compact K_δ such that $\mu(A \setminus K_\delta) < \delta$. Denote by μ_δ the measure μ localized to $A \setminus K_\delta$. Then, by the preceding, we have for every $\lambda > 0$

$$|E(\lambda)| \leq |\{\theta \in \mathbb{T} \mid M(\mu_\delta)(\theta) > \lambda\}|.$$

However, by the weak-L^1 estimate for the Hardy–Littlewood maximal function, see (2.9), one has that the measure of the set on the right-hand side satisfies

$$|\{\theta \in \mathbb{T} | M(\mu_\delta)(\theta) > \lambda\}| < \frac{3}{\lambda} \|\mu_\delta\| < \frac{3}{\lambda} \delta.$$

Since $\delta > 0$ is arbitrary, we are done. □

Exercise 2.6 Let $\{\Phi_n\}_{n=1}^{\infty}$ satisfy (A1)–(A3) of Definition 1.3 and also (A4) from Definition 2.10, and assume that the $\{\Psi_n\}_{n=1}^{\infty}$ from Definition 2.10 also satisfy

$$\sup_{\delta < |\theta| < 1/2} |\Psi_n(\theta)| \to 0 \quad \text{as } n \to \infty$$

for all $\delta > 0$. Under these assumptions show that, for any $\mu \in \mathcal{M}(\mathbb{T})$,

$$\Phi_n * \mu \to f \text{ a.e.} \quad \text{as } n \to \infty,$$

where $\mu(d\theta) = f(\theta)\,d\theta + \nu_s(d\theta)$ is the Lebesgue decomposition, i.e., $f \in L^1(\mathbb{T})$ and $\nu_s \perp m$.

2.5. Weighted estimates for maximal functions

2.5.1. Changing measures

As the Hardy–Littlewood maximal function Mf is clearly an object of fundamental importance, we shall explore it further. For one thing, there is no reason to limit ourselves to the circle or even the line; it is natural to work on $\mathbb{R}^d$ with $d \geq 1$. Moreover, we will allow for more general measures than the Lebesgue measure. Thus, let μ be a positive Borel measure on $\mathbb{R}^d$ and define

$$(M_\mu f)(x) := \sup_{B \ni x} \frac{1}{\mu(B)} \int_B |f(y)|\, \mu(dy), \tag{2.12}$$

where the supremum ranges over all balls containing x. This definition is meaningful for any $f \in L^1_{\text{loc}}(\mu)$ (which means that $\chi_B f \in L^1(\mu)$ for any ball $B \subset \mathbb{R}^d$). Naturally, we would like to preserve the property that $M_\mu f$ lies in weak-L^1 for any $f \in L^1(\mu)$. Inspection of the proof of this property for the circle case shows that, owing to the Wiener covering lemma, we need to impose the *doubling property*: there exists a constant A such that

$$\mu(B^*) \leq A\mu(B) \quad \forall B \subset \mathbb{R}^d, \tag{2.13}$$

where B is a ball and B^* is its double (with the same center but, twice the radius). In view of this doubling property we see that it makes no difference if we use cubes instead of balls in (2.12) or if we require x to be the center of B; the resulting maximal functions will be comparable by multiplicative constants.[1] First, we record the weak-L^1 bound and the resulting strong L^p bound.

Proposition 2.14 *Let μ be a doubling measure. Then*

$$\begin{aligned} \mu(\{x \in \mathbb{R}^d \mid (M_\mu f)(x) > \lambda\}) &\leq C(A)\lambda^{-1}\|f\|_{L^1(\mu)} \quad \forall \lambda > 0, \\ \|M_\mu f\|_{L^p(\mu)} &\leq C(p, A)\|f\|_{L^p(\mu)} \quad \forall 1 < p \leq \infty, \end{aligned} \tag{2.14}$$

for any functions f for which the respective right-hand sides are finite.

[1] If the two maximal functions are denoted $(Mf)_1$ and $(Mf)_2$ then "comparable by multiplicative constants" means that $c(Mf)_1 < (Mf)_2 < C(Mf)_1$, where c, C given by $c < 1 < C$ are constants.

Proof This is essentially identical with the proof of Proposition 2.9. The Wiener covering lemma applies equally well in $\mathbb{R}^d$, and we have $\mu(3B) \le A^2\mu(B)$ by (2.13). This gives the weak-L^1 bound, and the L^p bound follows by interpolation as before. □

Note that this establishes the Lebesgue differentiation theorem for general doubling measures.

2.5.2. Weighted estimates for the maximal function

Next we wish to explore a somewhat different question, namely whether the standard maximal operator remains bounded on weighted spaces. More precisely, we would like to characterize all measurable functions $w \ge 0$ in $\mathbb{R}^d$ with the property that, for fixed $1 < p < \infty$, one has the bound

$$\int_{\mathbb{R}^d} (Mf)^p(x)w(x)\,dx \le C(p)\int_{\mathbb{R}^d} |f(x)|^p w(x)\,dx, \tag{2.15}$$

with constant $C(p)$ and all $f \in L^p(w)$, or even those functions having the weak-L^p version (now also allowing $p = 1$),

$$w(\{Mf > \lambda\}) \le C(p)\lambda^{-p}\int |f(x)|^p\, w(x)\,dx, \tag{2.16}$$

where $w(E) := \int_E w(x)\,dx$. Here M is as in (2.12) but with μ the Lebesgue measure.

Assume that (2.16) holds with $1 \le p < \infty$ fixed. Let B be any ball and $f \ge 0$ be such that $f(B) := \int_B f(y)\,dy > 0$. Let $\lambda \in (0, f(B)/|B|)$. Then

$$B \subset \{x \mid M(f\chi_B)(x) > \lambda\}$$

and thus, from the weak-L^p bound, we have

$$w(B) \le C(p)\lambda^{-p}\int_B |f(y)|^p w(y)\,dy.$$

Maximizing over λ implies that

$$w(B)\left(\frac{f(B)}{|B|}\right)^p \le C(p)\int_B |f(y)|^p w(y)\,dy. \tag{2.17}$$

Setting $f := \chi_E$ for some measurable $E \subset B$ implies that

$$w(B)\left(\frac{|E|}{|B|}\right)^p \le C(p)\,w(E). \tag{2.18}$$

Exercise 2.7 Deduce the following dichotomies from (2.18):

(a) either $w > 0$ a.e. or $w = 0$ a.e;

(b) either $w \in L^1_{\text{loc}}(\mathbb{R}^d)$ or $w = \infty$ a.e.

Also verify that any w satisfying (2.18) defines a doubling measure.

We shall now deduce conditions on w from (2.17); they are called A_p conditions on w. As we shall see, they are also *sufficient* for (2.16) (and, in fact, the strong bounds when $1 < p < \infty$) to hold. In what follows, $⨍_B := |B|^{-1} \int_B$, and we shall often omit the infinitesimal dx from integrals relative to Lebesgue measure.

Proposition 2.15 *If the estimate* (2.16) *holds with* $p = 1$ *then*

$$Mw \le C(1)w \quad a.e., \tag{2.19}$$

where $C(1)$ *is the constant from* (2.16) *with* $p = 1$. *If (2.16) holds for* $1 < p < \infty$ *then*

$$⨍_B w \left(⨍_B w^{1-p'} \right)^{p-1} \le C(p) \quad \forall B \subset \mathbb{R}^d, \tag{2.20}$$

where B is a ball.

Proof Because $w\,dx$ is a doubling measure, the Lebesgue differentiation theorem holds, as we verified above. Hence, taking x to be a Lebesgue point of w, we infer from (2.18) with $p = 1$ that $w(B)/|B| \le C(1)w(x)$ for every ball $B \ni x$. This is equivalent with $Mw(x) \le Cw(x)$.

If $1 < p < \infty$ then we deduce from (2.17), by setting $f = w^{1-p'}\chi_B$, that

$$w(B) \left(⨍_B w^{1-p'} \right)^p \le C(p) \int_B w^{1-p'},$$

which implies (2.20). Strictly speaking, we need to replace w here with $\min(w, n)$ and then let $n \to \infty$. □

Exercise 2.8 Show that, for $p > 1$, any $w > 0$ that satisfies (2.20) also satisfies (2.17) and therefore (2.18). *Hint:* Use Hölder's inequality.

The conditions (2.19) and (2.20) are called A_1 and A_p *conditions*, respectively. Any w satisfying these conditions is referred to as an A_p weight (with $1 \le p < \infty$); we write $w \in A_p$. These conditions are not only necessary for (2.16), as we have just shown, but are also *sufficient*. We shall now prove the following theorem.

Theorem 2.16 *If $w \in A_p$ then* (2.16) *holds.*

Exercise 2.9 Verify the following properties of A_p classes:

(a) $A_p \subset A_q$ if $1 \le p < q$;

(b) $w \in A_p$ if and only if $w^{1-p'} \in A_{p'}$;

(c) if $w_0, w_1 \in A_1$ then $w_0 w_1^{1-p} \in A_p$.

Hint: These properties follow immediately from the definitions of the quantities and Hölder's inequality (the latter is needed for the first property only if $p > 1$).

For $p = 2$ the A_p-condition takes the following form:

$$\fint_B w \fint_B w^{-1} \le C \qquad \forall\, B \subset \mathbb{R}^d. \tag{2.21}$$

2.5.3. The Calderón–Zygmund decomposition

To prove Theorem 2.16, we shall use a fundamental device, the *Calderón–Zygmund decomposition* in $L^1(\mathbb{R}^d)$. The construction is an example of a *stopping-time argument.*

Lemma 2.17 *Let $f \in L^1(\mathbb{R}^d)$ and $\lambda > 0$. Then one can write $f = g + b$ with $|g| \le \lambda$ and $b = \sum_Q \chi_Q f$, where the sum runs over a collection $\mathcal{B} = \{Q\}$ of disjoint cubes such that for each Q one has*

$$\lambda < \frac{1}{|Q|}\int_Q |f| \le 2^d \lambda. \tag{2.22}$$

Furthermore,

$$\left| \bigcup_{Q \in \mathcal{B}} Q \right| < \frac{1}{\lambda}\|f\|_1. \tag{2.23}$$

Proof For each $\ell \in \mathbb{Z}$ we define a collection $\mathcal{D}_\ell$ of dyadic cubes by

$$\mathcal{D}_\ell = \left\{ \Pi_{i=1}^d [2^\ell m_i, 2^\ell(m_i + 1)) \mid m_1, \ldots, m_d \in \mathbb{Z} \right\}. \tag{2.24}$$

Notice that if $Q \in \mathcal{D}_\ell$ and $Q' \in \mathcal{D}_{\ell'}$ then $Q \cap Q' = \emptyset$ or $Q \subset Q'$ or $Q' \subset Q$. In other words, either any two dyadic cubes are disjoint or one cube is contained inside the other. Now pick ℓ_0 large enough that

$$\frac{1}{|Q|}\int_Q |f|\, dx \le \lambda$$

for every $Q \in \mathcal{D}_{\ell_0}$. For each such cube consider its 2^d "children" of size $2^{\ell_0 - 1}$. Any such cube Q' will then have the property that

$$\text{either} \quad \frac{1}{|Q'|}\int_{Q'} |f(x)|\, dx \le \lambda \quad \text{or} \quad \frac{1}{|Q'|}\int_{Q'} |f(x)|\, dx > \lambda. \tag{2.25}$$

In the latter case we stop and include Q' in a family $\mathcal{B}$. Observe that in this case

$$\frac{1}{|Q'|}\int_{Q'} |f(x)|\, dx \le \frac{2^d}{|Q|}\int_Q |f(x)|\, dx \le 2^d \lambda,$$

where Q denotes the parent of Q'. Thus (2.22) holds.

If, however, the first inequality in (2.25) holds then subdivide Q' again into its children, with half the size of Q'. Continuing in this fashion produces a collection of disjoint (dyadic) cubes $\mathcal{B}$ satisfying (2.22). Consequently (2.23) also holds, since

$$\left|\bigcup_{\mathcal{B}} Q\right| \leq \sum_{\mathcal{B}} |Q| < \sum_{\mathcal{B}} \frac{1}{\lambda} \int_Q |f(x)|\, dx \leq \frac{1}{\lambda} \int_{\mathbb{R}^d} |f(x)|\, dx.$$

Now let $x_0 \in \mathbb{R}^d \setminus \bigcup_{\mathcal{B}} \overline{Q}$. Then x_0 is contained in a decreasing sequence $\{Q_j\}$ of dyadic cubes, each of which satisfies

$$\frac{1}{|Q_j|} \int_{Q_j} |f(x)|\, dx \leq \lambda.$$

By Lebesgue's theorem, $|f(x_0)| \leq \lambda$ for a.e. such x_0. Since, moreover, $\mathbb{R}^d \setminus \cup_{\mathcal{B}} Q$ and $\mathbb{R}^d \setminus \cup_{\mathcal{B}} \overline{Q}$ differ only by a set of measure zero, we can set

$$g := f - \sum_{Q \in \mathcal{B}} \chi_{\overline{Q}} f$$

so that $|g| \leq \lambda$ a.e. as desired. □

We refer to λ in the above decomposition as the *height*. In view of (2.22) there exists a strong connection between the Hardy–Littlewood maximal function and this decomposition; in fact,

$$\{Mf > c(d)\lambda\} \supset \bigcup_{Q \in \mathcal{B}} Q.$$

In the following exercise, we ask the reader to establish a reverse inclusion.

Exercise 2.10 Given $f \in L^1(\mathbb{R}^d)$, perform a Calderón–Zygmund decomposition at height λ that results in the collection $\mathcal{B}$ of cubes. Show that there exists a constant $C(d)$ such that

$$\{Mf > C(d)\lambda\} \subset \bigcup_{Q \in \mathcal{B}} Q^*, \tag{2.26}$$

where Q^* is the double of Q.

In the proof of Theorem 2.16, the following corollary to (2.26) will play a key role.

Proposition 2.18 *For any measurable $w \geq 0$ and $1 \leq p < \infty$ there exist constants C_p, depending also on the dimension, such that*

$$\begin{aligned} \int_{\{Mf>\lambda\}} w(x)\, dx &\leq C_1 \lambda^{-1} \int_{\mathbb{R}^d} |f(x)|\, (Mw)(x)\, dx, \\ \int_{\mathbb{R}^d} (Mf)^p(x) w(x)\, dx &\leq C_p \int_{\mathbb{R}^d} |f(x)|^p (Mw)(x)\, dx; \end{aligned} \tag{2.27}$$

the latter estimate requires that $1 < p < \infty$.

Proof We shall prove only the weak-L^1 part of (2.27), as the other part follows by Marcinkiewicz interpolation with an easy L^∞ estimate. We write

$$\begin{aligned}\int_{\{Mf>C(d)\lambda\}} w(x)\,dx &\le \sum_{Q\in\mathcal{B}} 2^d|Q| \fint_{Q^*} w(x)\,dx \\ &\le 2^d\lambda^{-1}\sum_{Q\in\mathcal{B}}\int_Q |f(y)|\left(\fint_{Q^*} w(x)\,dx\right)dy \\ &\le 2^d\lambda^{-1}\int_{\mathbb{R}^d}|f(y)|(Mw)(y)\,dy,\end{aligned}$$

as claimed. The first inequality follows from (2.26), the second from (2.22), and the third from the definition of the maximal function and the disjointness of the cubes in $\mathcal{B}$. □

Exercise 2.11 Given $f\in L^1(\mathbb{R}^d)$ and $\lambda>0$ show that there exists $E\subset\mathbb{R}^d$ with $|E|\le\lambda^{-1}$ and such that

$$\int_{\mathbb{R}^d\setminus E}|f(x)|^2\,dx\le\lambda\|f\|_1^2.$$

State the analogous result for the torus.

2.5.4. The weak bound for A_p weights

We can now prove Theorem 2.16.

Proof of Theorem 2.16 If $p=1$ then combining the first estimate in (2.27) with the condition (2.19) concludes the proof.

If $1<p<\infty$ then we know from Exercise 2.8 that property (2.17), and therefore also (2.18), holds. Performing a Calderón–Zygmund decomposition of f at height $c(d)\lambda$, where $c(d)$ is a small constant, one argues as in Exercise 2.10 that

$$\{Mf>\lambda\}\subset\bigcup_{Q\in\mathcal{B}}Q^*.$$

Therefore

$$\begin{aligned}w(\{Mf>\lambda\}) &\le \sum_{Q\in\mathcal{B}} w(Q^*)\le C\sum_{Q\in\mathcal{B}}w(Q) \\ &\le C\sum_{Q\in\mathcal{B}}\left(\frac{|Q|}{f(Q)}\right)^p\int_Q|f|^p\,w \\ &\le C\lambda^{-p}\int_{\mathbb{R}^d}|f|^p\,w.\end{aligned}$$

Here, we used (2.18) to obtain the second inequality, then (2.17), (2.22), and the disjointness of the cubes in $\mathcal{B}$. □

In fact, one has a *strong* L^p bound for $p > 1$. The standard proof of the above theorem is typically based on the *reverse Hölder* inequality. This remarkable property essentially means that any $w \in A_p$ with $p > 1$ also belongs to some A_q where $1 < q < p$. By interpolation, Theorem 2.16 then implies the above strong bound on $L^p(\mathbb{R}^d)$. We shall present the reverse Hölder inequality in Chapter 7 in the context of singular integral bounds relative to A_p weights. For the maximal function, however, we shall now give an argument based on a Calderón–Zygmund decomposition and essentially no more than the definition of A_p weights. Before presenting the details, we introduce another maximal function, namely the *dyadic maximal function*:

$$(M_{\text{dyad}} f)(x) := \sup_{Q \ni x} \fint_Q |f(y)|\, dy,$$

where the supremum ranges over all *dyadic cubes* defined in (2.24). By inspection of the proof of the Calderón–Zygmund decomposition at height λ, we see that

$$\{M_{\text{dyad}} f > \lambda\} = \bigcup_{Q \in \mathcal{B}} Q, \tag{2.28}$$

where $\mathcal{B}$ is the family of "bad" cubes.

Note that the dyadic maximal function is not comparable in the pointwise sense with the usual maximal function, since the dyadic maximal function of a function that vanishes on $x_1 > 0$ also vanishes on that half-space. In other words, while Mf dominates $M_{\text{dyad}} f$ the converse does not hold. However, one has the following property of the level sets, which is sufficient for many purposes.

Exercise 2.12 Show that there exists a constant $C(d)$ such that

$$|\{Mf > C(d)\lambda\}| \le C(d)|\{M_{\text{dyad}} f > \lambda\}|$$

for all $\lambda > 0$. In fact, for any doubling measure μ (such as $w\,dx$ where $w \in A_p$), one has

$$\mu(\{Mf > C(d)\lambda\}) \le C(d)\mu(\{M_{\text{dyad}} f > \lambda\})$$

for all $\lambda > 0$.

2.5.5. The strong bound for A_p weights

Now we can formulate and prove the strong L^p bound for A_p weights.

Theorem 2.19 *Any weight $w \in A_p$ with $1 < p < \infty$ satisfies the strong L^p-boundedness property* (2.15).

Proof By Exercise 2.12 it suffices to prove

$$\int_{\mathbb{R}^d} (M_{\mathrm{dyad}} f)^p(x)\, w(x)\, dx \le C \int_{\mathbb{R}^d} |f(x)|^p w(x)\, dx, \tag{2.29}$$

where the constant $C = C(p)$ is bounded for $p_0 \le p \le p_1$ with $1 < p_0 < p_1 < \infty$ arbitrary but fixed. To this end we perform a Calderón–Zygmund decomposition of f at height C_0^k for each integer $k \in \mathbb{Z}$, where $C_0 := C_0(d)$ is a suitable constant. Denote the totality of the "bad" cubes generated in this process for all k by $\mathcal{B}$. With each $Q \in \mathcal{B}$ associate $E(Q)$, defined by

$$E(Q) := Q - \bigcup_{\substack{Q' \in \mathcal{B} \\ Q' \subsetneq Q}} Q'.$$

Then we have

$$\int_{\mathbb{R}^d} (M_{\mathrm{dyad}} f)^p(x)\, w(x)\, dx \le C \sum_{Q \in \mathcal{B}} \left(⨍_Q |f| \right)^p w(E(Q)). \tag{2.30}$$

By construction, if C_0 is large enough then $|E(Q)| > \frac{1}{2}|Q|$ for each $Q \in \mathcal{B}$. Hence, if we set $\sigma := w^{1-p'}$ then, by Exercises 2.9 and 2.8 as well as (2.18), we may conclude that $\sigma(E(Q)) > c_0 \sigma(Q)$ for all cubes, with some constant c_0. Hence we can bound the right-hand side of (2.30) by

$$\begin{aligned}
&C \sum_{Q \in \mathcal{B}} \left(⨍_Q |f| \right)^p w(E(Q)) \\
&\le \sum_{Q \in \mathcal{B}} \left(\frac{1}{\sigma(Q)} ⨍_Q |f| \sigma^{-1} \sigma \right)^p \sigma(E(Q)) \left[\frac{w(E(Q))}{\sigma(E(Q))} \left(\frac{\sigma(Q)}{|Q|} \right)^p \right] \\
&\le C \sum_{Q \in \mathcal{B}} \left(\frac{1}{\sigma(Q)} ⨍_Q |f| \sigma^{-1} \sigma \right)^p \sigma(E(Q)) \left[\frac{w(Q)}{|Q|} \left(\frac{\sigma(Q)}{|Q|} \right)^{p-1} \right] \\
&\le C \int_{\mathbb{R}^d} \left(M_\sigma (f \sigma^{-1}) \right)^p \sigma.
\end{aligned}$$

To pass to the third line we removed the expression in brackets since, by the definition of the A_p-class, it is bounded uniformly in Q. However, we can now invoke Proposition 2.14 to bound the last line by

$$\int_{\mathbb{R}^d} |f|^p \sigma^{1-p} = \int_{\mathbb{R}^d} |f|^p w,$$

as desired. □

Notes

Standard references on harmonic functions on the disk and on the Poisson kernel are the classic book by Hoffman [55], as well as those by Garnett [46] and Koosis [71]. For a more systematic development of A_p weights, see for example Stein's book [111]; we shall return to them later in Chapter 7 in the context of singular integrals. The proof of Theorem 2.19 is due to Christ and Fefferman [22]. Harmonic functions play a very important role in higher dimensions also, and notions such as the mean value property, the maximum principle, and the Poisson kernel carry over to $\mathbb{R}^d$ for $d > 2$. The book by Han and Lin [52] begins with a discussion of harmonic functions in general dimensions and then continues with a rapid development of scalar second-order elliptic PDEs.

Problems

Problem 2.1 Let (X, μ) be a general measure space. We say a bounded sequence $\{f_n\}_{n=1}^{\infty}$ in $L^1(\mu)$ is *uniformly integrable* if for every $\varepsilon > 0$ there exists $\delta > 0$ such that for any measurable E one has

$$\mu(E) < \delta \quad \Longrightarrow \quad \sup_n \int_E |f_n|\, d\mu < \varepsilon$$

and there exists $E_0 \subset X$ with $\sup_{n\geq 1} \int_{X\setminus E_0} |f_n|\, d\mu < \varepsilon$. The same applies to any subset of L^1, not just to sequences. For simplicity, suppose now that μ is a finite measure.

(a) Let $\phi : [0, \infty) \to [0, \infty)$ be a continuous increasing function with $\lim_{t\to\infty} \phi(t)/t = +\infty$. Prove that

$$\sup_n \int \phi(|f_n(x)|)\, \mu(dx) < \infty$$

implies that $\{f_n\}$ is uniformly integrable. Conversely, show that this inequality is also necessary for uniform integrability and in particular that bounded sequences in $L^p(\mu)$ with $p > 1$ are uniformly integrable.

(b) Show that $\{f_n\}_{n=1}^{\infty}$ is uniformly integrable if and only if

$$\lim_{A\to\infty} \sup_{n\geq 1} \int_{[|f_n|>A]} |f_n(x)|\, \mu(dx) = 0.$$

(c) Show that for an arbitrary sequence $\{f_n\}_{n=1}^{\infty}$ in $L^1(\mu)$ the following are equivalent: (i) $f_n \to f$ in $L^1(\mu)$ as $n \to \infty$; (ii) $f_n \to f$ in measure, with $\{f_n\}_{n\geq 1}$ uniformly integrable.

Problem 2.2 Suppose that $\{f_n\}_{n=1}^{\infty}$ is a sequence in $L^1(\mathbb{T})$. Show that there is a subsequence $\{f_{n_j}\}_{j=1}^{\infty}$ and a measure μ with $f_{n_j} \xrightarrow{\sigma^*} \mu$ provided that $\sup_n \|f_n\|_1 < \infty$. Here σ^* is the weak-* convergence of measures. Show that in general $\mu \notin L^1(\mathbb{T})$. However, if we assume in addition that $\{f_n\}_1^{\infty}$ is uniformly integrable then $\mu(d\theta) = f(\theta)\, d\theta$ for some $f \in L^1(\mathbb{T})$. Can we conclude anything about strong convergence (i.e., in the L^1 norm) of $\{f_n\}$? Consider the analogous question on $L^p(\mathbb{T})$, $p > 1$.

Problem 2.3 Let μ be a positive finite Borel measure on $\mathbb{R}^d$ or $\mathbb{T}^d$. Set

$$M\mu(x) := \sup_{r>0} \frac{\mu(B(x,r))}{m(B(x,r))},$$

where m is the Lebesgue measure. This problem examines the behavior of $M\mu$ when μ is a singular measure; see Problem 3.5 for more on this case.

(a) Show that $\mu \perp m$ implies $\mu(\{x \mid M\mu(x) < \infty\}) = 0$.

(b) Show that if $\mu \perp m$ then

$$\limsup_{r\to 0} \frac{\mu(B(x,r))}{m(B(x,r))} = \infty$$

for μ-*a.e.* x. Also show that this limit vanishes m-almost-everywhere.

Problem 2.4 For any $f \in L^1(\mathbb{R}^d)$ and $1 \le k \le d$, let

$$M_k f(x) = \sup_{r>0} r^{-k} \int_{B(x,r)} |f(y)|\, dy.$$

Show that, for every $\lambda > 0$,

$$m_L(\{x \in L \mid M_k f(x) > \lambda\}) \le \frac{C}{\lambda} \|f\|_{L^1(\mathbb{R}^d)},$$

where L is an arbitrary affine k-dimensional subspace and m_L stands for Lebesgue measure (i.e., k-dimensional measure) on this subspace; C is an absolute constant.

Problem 2.5 Prove the Besicovitch covering lemma on $\mathbb{T}$. Suppose that $\{I_j\}$ are finitely many arcs with $|I_j| < 1$. Then there is a sub-collection $\{I_{j_k}\}$ such that the following properties hold:

(a) $\cup_k I_{j_k} = \cup_j I_j$;

(b) No point belongs to more than C of the I_{j_k}, where C is an *absolute* constant. What is the optimal value of C?

What can you say about the situation for higher dimensions (see for example Füredi and Loeb [44] and the references cited therein)?

Problem 2.6 Let $F \ge 0$ be a harmonic function on $\mathbb{D}$. Show that there exists a positive measure μ on $\mathbb{T}$ with $F(re(\theta)) = (P_r * \mu)(\theta)$ for $0 \le r < 1$ and with $\|\mu\| = F(0)$.

Problem 2.7 A sequence of complex numbers $\{a_n\}_{n\in\mathbb{Z}}$ is called positive definite if

(a) $\overline{a_n} = a_{-n}$ for all $n \in \mathbb{Z}$,

(b) $\sum_{n,k} a_{n-k}\, z_n \overline{z_k} \ge 0$ for all complex sequences $\{z_n\}_{n\in\mathbb{Z}}$.

Show that any positive definite sequence satisfies $|a_n| \le a_0$ for all $n \in \mathbb{Z}$. Show that if μ is a positive measure then $a_n = \int_{\mathbb{T}} e(-n\theta)\mu(d\theta)$ is such a sequence. Now prove that every positive definite sequence is of this form. *Hint:* Apply the previous problem to the harmonic function $\sum_n a_n r^{|n|} e(n\theta)$. Check the positivity of this sum by representing it as in property (a) above.

Problem 2.8 Verify the mean value property and the maximum principle for harmonic functions on domains in $\mathbb{R}^d$ for any $d \ge 2$.

Problem 2.9 Let $w \in A_p$ for any $1 \le p < \infty$. Show that for $f \in L^1_{\mathrm{loc}}$ with $f \ge 0$ one has

$$\fint_Q f \le C\left(\frac{1}{w(Q)}\int_Q |f(x)|^p\, w(x)\, dx\right)^{1/p}$$

for any cube Q.

Problem 2.10 Show that $w(x) = |x|^a$ belongs to $A_p(\mathbb{R}^d)$ with $p > 1$ if and only if $-d < a < d(p-1)$. Also show that $w(x)\,dx$ is a doubling measure if and only if $a > -d$.

Problem 2.11 Verify Green's identity for a bounded domain $\Omega \subset \mathbb{R}^d$ with C^∞ boundary and functions $u, v \in C^\infty(\bar{\Omega})$:

$$\int_\Omega (u\Delta v - v\Delta u)\, dx = \int_{\partial\Omega}\left(u\frac{\partial v}{\partial n} - v\frac{\partial u}{\partial n}\right) d\sigma, \tag{2.31}$$

where σ is the surface measure on $\partial\Omega$, and $\partial/\partial n$ refers to the normal derivative with respect to the outward-pointing normal vector.

Use this identity to verify that a solution to $\Delta u = f$ with $f \in \mathcal{S}(\mathbb{R}^d)$ is given by

$$u(z) = \frac{1}{2\pi}\int_{\mathbb{R}^2} \log|z-\zeta|\, f(\zeta)\, m(d\zeta)$$

in $\mathbb{R}^2$ with m the Lebesgue measure in the plane and

$$u(x) = c(d)\int_{\mathbb{R}^d} |x-y|^{2-d} f(y)\, dy$$

in $\mathbb{R}^d$, for $d \ge 3$ with a suitable constant $c(d)$. Discuss the uniqueness of these solutions. *Hint:* Apply (2.31) to a Ω equal to a big ball minus a small ball around the point at which u is being evaluated. Then pass to suitable limits.

Problem 2.12 Let $f \in L^1(\mathbb{T})$ satisfy $\hat{f}(n) = 0$ if $|n| > N$, where N is some positive integer. Show that there exists $E \subset \mathbb{T}$ with $|E| < \lambda^{-1}$ and with

$$\int_{\mathbb{T}\setminus E}\int_{\mathbb{T}} k_N(\theta)|f(\varphi-\theta)|^2\, d\theta\, d\varphi \le C\lambda \|f\|_1^2,$$

where $k_N(\theta) := N\chi_{[2|\theta|N\le 1]}$ is the box kernel and C is some absolute constant. Compare this with Exercise 2.11.

3

Conjugate harmonic functions; Hilbert transform

3.1. Hardy spaces of analytic functions

In the previous chapter, we dealt with harmonic functions on the disk satisfying various boundedness properties. We now turn to functions $F = u + iv \in h^1(\mathbb{D})$ that are *analytic* in $\mathbb{D}$. The usage of h^1 here is legitimate, since analytic functions form a subset of the class of functions that are complex-valued and harmonic. Thus analytic functions in $h^1(\mathbb{D})$ form the class $\mathbb{H}^1(\mathbb{D})$, the "big" Hardy space. We showed in the previous chapter that, for functions in this class, $F_r = P_r * \mu$ for some $\mu \in \mathcal{M}(\mathbb{T})$. It is important to note that, by analyticity $\hat{\mu}(n) = 0$ if $n < 0$. A theorem by F. and M. Riesz asserts that such measures μ are absolutely continuous. From the example

$$F(z) := \frac{1+z}{1-z} = P_r(\theta) + i\,Q_r(\theta), \qquad z = re(\theta),$$

one sees an important difference between the analytic and the harmonic cases. Indeed, while $P_r \in h^1(\mathbb{D})$, clearly $F \notin h^1(\mathbb{D})$. The boundary measure associated with P_r is δ_0, whereas F is not associated with any boundary measure in the sense of the previous chapter.

3.1.1. Subharmonic functions

An important technical device that will allow us to obtain more information in the analytic case is provided by *subharmonic functions*. Loosely speaking, they allow us to exploit algebraic properties of analytic functions that harmonic functions do not have (such as the fact that F^2 is analytic if F is analytic, a property that fails for harmonic functions).

Definition 3.1 Let $\Omega \subset \mathbb{R}^2$ be an open and connected region and let $f : \Omega \to \mathbb{R} \cup \{-\infty\}$, where we extend the topology to $\mathbb{R} \cup \{-\infty\}$ in an obvious way. We say that f is *subharmonic* if:

(i) f is continuous;
(ii) for all $z \in \Omega$ there exists $r_z > 0$ such that

$$f(z) \leq \int_0^1 f(z + re(\theta))\, d\theta$$

for all $0 < r < r_z$; we refer to this as the (local) sub-mean-value property.

Usually one requires only that f be upper semicontinuous but the stronger condition (i) is sufficient for our purposes. It is helpful to keep in mind that in one dimension harmonic implies linear and subharmonic implies convex. Subharmonic functions derive their name from the fact that they lie below harmonic functions, in the same way that convex functions lie below linear functions. We will make this precise by means of the important *maximum principle*, which subharmonic functions obey. We begin by deriving some basic properties of this class.

Lemma 3.2 *Subharmonic functions satisfy the following properties.*
(i) *If f and g are subharmonic then $f \vee g := \max(f, g)$ is subharmonic.*
(ii) *If $f \in C^2(\Omega)$ then f is subharmonic implies that $\Delta f \geq 0$ in Ω and vice versa.*
(iii) *That F is analytic implies that $\log |F|$ and $|F|^\alpha$ with $\alpha > 0$ are subharmonic.*
(iv) *If f is subharmonic and φ is increasing and convex then $\varphi \circ f$ is subharmonic (we set $\varphi(-\infty) := \lim_{x\to-\infty} \varphi(x)$).*

Proof (i) is immediate. For (ii) use *Jensen's formula*,

$$\int_{\mathbb{T}} f(z + re(\theta))\, d\theta - f(z) = \iint_{D(z,r)} \log \frac{r}{|w - z|} \Delta f(w)\, m(dw), \qquad (3.1)$$

where m stands for two-dimensional Lebesgue measure and

$$D(z, r) = \big\{ w \in \mathbb{C} \,\big|\, |w - z| < r \big\}.$$

The reader is asked to verify this formula in the following exercise. If $\Delta f \geq 0$, then the sub-mean-value property holds. If $\Delta f(z_0) < 0$ then let $r_0 > 0$ be sufficiently small that $\Delta f < 0$ on $D(z_0, r_0)$ Since $\log r_0/|w - z_0| > 0$ on this disk, Jensen's formula implies that the sub-mean-value property fails. Next we verify (iv) by means of Jensen's inequality for convex functions:

$$\varphi(f(z)) \leq \varphi\left(\int_{\mathbb{T}} f(z + re(\theta))\, d\theta \right) \leq \int_{\mathbb{T}} \varphi(f(z + re(\theta))\, d\theta.$$

The first inequality uses the fact that φ is increasing, whereas the second uses the convexity of φ. If F is analytic then $\log |F|$ is continuous, with values in

$\mathbb{R} \cup \{-\infty\}$. If $F(z_0) \neq 0$ then $\log|F(z)|$ is harmonic on some disk $D(z_0, r_0)$. Thus, one has the stronger, mean-value, property on this disk. If $F(z_0) = 0$ then $\log|F(z_0)| = -\infty$, and there is nothing to prove. To see that $|F|^\alpha$ is subharmonic, apply (iv) to $\log|F(z)|$ with $\varphi(x) = \exp(\alpha x)$. □

Exercise 3.1 Prove Jensen's formula (3.1) for C^2 functions.

Now we can derive the aforementioned domination of subharmonic functions by harmonic functions.

Lemma 3.3 *Let Ω be a bounded region. Suppose that f is subharmonic on Ω, $f \in C(\overline{\Omega})$, and let u be harmonic on Ω, $u \in C(\overline{\Omega})$. If $f \leq u$ on $\partial\Omega$ then $f \leq u$ on Ω.*

Proof We may take $u = 0$, so that $f \leq 0$ on $\partial\Omega$. Let $M = \max_{\overline{\Omega}} f$ and assume that $M > 0$. Set

$$S = \{z \in \overline{\Omega} \mid f(z) = M\}.$$

Then $S \subset \Omega$ and S is closed in Ω. If $z \in S$ then, by the sub-mean-value property, there exists $r_z > 0$ such that $D(z, r_z) \subset S$. Hence S is also open. Since Ω is assumed to be connected, one obtains $S = \Omega$. This is a contradiction. □

3.1.2. Sub-mean-value property

The following lemma shows that the sub-mean-value property holds on any disk in Ω. The point here is that we are upgrading the local sub-mean-value property, to a true sub-mean-value property, using the largest possible disks.

Lemma 3.4 *Let f be subharmonic in Ω, $z_0 \in \Omega$, $\overline{D(z_0, r)} \subset \Omega$. Then*

$$f(z_0) \leq \int_{\mathbb{T}} f(z_0 + re(\theta))\, d\theta.$$

Proof Let $g_n = \max(f, -n)$, where $n \geq 1$. Without loss of generality, set $z_0 = 0$. Define $u_n(z)$ to be the harmonic extension of g_n restricted to $\partial D(z_0, r)$, where $r > 0$ is as in the statement of the lemma. By the previous lemma

$$f(0) \leq g_n(0) \leq u_n(0) = \int_{\mathbb{T}} u_n(re(\theta))\, d\theta,$$

the last equality expressing the mean-value property of harmonic functions. Since

$$\max_{|z| \leq r} u_n(z) \leq \max_{|z| \leq r} f(z),$$

the monotone convergence theorem for decreasing sequences yields

$$f(0) \leq \int_{\mathbb{T}} f(re(\theta))\, d\theta$$

as claimed. □

Corollary 3.5 *If g is subharmonic on $\mathbb{D}$ then, for all θ,*

$$g(rse(\theta)) \leq (P_r * g_s)(\theta)$$

for any $0 < r, s < 1$.

Proof If $g > -\infty$ everywhere on $\mathbb{D}$ then this follows from Lemma 3.3. If not then set $g_n = g \vee n := \max(g, n)$. Thus

$$g(rse(\theta)) \leq g_n(rse(\theta)) \leq (P_r * (g_n)_s)(\theta)$$

and consequently

$$g(rse(\theta)) \leq \limsup_{n \longrightarrow \infty} (P_r * (g_n)_s)(\theta) \leq (P_r * g_s)(\theta),$$

where the final inequality follows from Fatou's lemma (which can be applied in the "reverse form" here since the g_n have a uniform upper bound). □

3.1.3. Maximal function F^*

Note that if $g_s \notin L^1(\mathbb{T})$ then $g \equiv -\infty$ on $D(0, s)$ and so $g \equiv -\infty$ on $D(0, 1)$. We now introduce the radial maximal function associated with any function on the disk. It, and the more general *nontangential maximal function*, where the supremum is taken over a cone in $\mathbb{D}$, are of central importance in the analysis discussed in this chapter.

Definition 3.6 Let F be any complex-valued function on $\mathbb{D}$; then $F^* : \mathbb{T} \to \mathbb{R}$ is defined as follow:

$$F^*(\theta) = \sup_{0<r<1} |F(re(\theta))|. \tag{3.2}$$

We showed in the previous chapter (see Lemma 2.11) that any $u \in h^1(\mathbb{D})$ satisfies $u^* \leq CM\mu$, where μ is the boundary measure of u, i.e., $u_r = P_r * \mu$. In particular, one has

$$\big|\{\theta \in \mathbb{T} \mid u^*(\theta) > \lambda\}\big| \leq C\lambda^{-1} |||u|||_1.$$

We shall prove an analogous bound for subharmonic functions that are L^1-bounded.

Proposition 3.7 *Suppose that g is subharmonic on $\mathbb{D}$, $g \geq 0$, and g is L^1-bounded, i.e.,*

$$|||g|||_1 := \sup_{0<r<1} \int_{\mathbb{T}} g(re(\theta))\, d\theta < \infty.$$

Then

(i) $\left|\{\theta \in \mathbb{T} \mid g^*(\theta) > \lambda\}\right| \leq \frac{3}{\lambda}|||g|||_1$ for all $\lambda > 0$.

(ii) *If g is L^p-bounded, i.e.,*

$$|||g|||_p^p := \sup_{0<r<1} \int_{\mathbb{T}} g(re(\theta))^p \, d\theta < \infty$$

with $1 < p \leq \infty$, then

$$\|g^*\|_{L^p(\mathbb{T})} \leq C_p |||g|||_p.$$

Proof (i) Let $g_{r_n} \rightharpoonup \mu \in \mathcal{M}(\mathbb{T})$ in the weak-$*$ sense. Then $\|\mu\| \leq |||g|||_1$ and

$$g_s \longleftarrow g_{r_n s} \leq g_{r_n} * P_s \longrightarrow P_s * \mu$$

Thus, by Lemma 2.11,

$$g^* \leq \sup_{0<s<1} P_s * \mu \leq M\mu$$

and the desired bound now follows from Proposition 2.9.

(ii) If $|||g|||_p < \infty$ then $d\mu/d\theta \in L^p(\mathbb{T})$ with $\|d\mu/d\theta\|_p \leq |||g|||_p$ and thus

$$g^* \leq CM\left(\frac{d\mu}{d\theta}\right) \in L^p(\mathbb{T})$$

by Proposition 2.9, as claimed. □

3.2. Riesz theorems

In this section we present two theorems due to F. and M. Riesz. The first will be presented in three versions. It is important to note that this result fails without analyticity.

3.2.1. The three versions of the first F. and M. Riesz theorem

Theorem 3.8 (Version 1) *Suppose that $F \in h^1(\mathbb{D})$ is analytic. Then $F^* \in L^1(\mathbb{T})$.*

Proof We have that $|F|^{1/2}$ is subharmonic and L^2-bounded. By Proposition 3.7, therefore $|F|^{1/2*} \in L^2(\mathbb{T})$. But $|F|^{1/2*} = |F^*|^{1/2}$ and thus $F^* \in L^1(\mathbb{T})$. □

Exercise 3.2 Generalize Theorem 3.8 to a nontangential maximal function. This implies taking the maximum in (3.2) over cones symmetric relative to the radial segment with arbitrary but fixed opening angles $0 < \alpha < \pi$.

Let $F \in h^1(\mathbb{D})$. By Theorem 2.5, $F_r = P_r * \mu$ where $\mu \in \mathcal{M}(\mathbb{T})$ has Lebesgue decomposition $\mu(d\theta) = f(\theta)\, d\theta + \nu_s(d\theta)$, ν_s singular and $f \in L^1(\mathbb{T})$. By

Exercise 2.6(b) one has $P_r * \mu \to f$ a.e. as $r \to 1$. Thus, $\lim_{r\to 1} F(re(\theta)) = f(\theta)$ exists for a.e. $\theta \in \mathbb{T}$. This justifies the following statement of the theorem.

Theorem 3.8 (Version 2) *Assume that $F \in h^1(\mathbb{D})$ and that F is analytic. Let $f(\theta) = \lim_{r\to 1} F(re(\theta))$. Then $F_r = P_r * f$ for all $0 < r < 1$.*

Proof We have $F_r \to f$ a.e. and $|F_r| \le F^* \in L^1$ by the previous theorem. Therefore, $F_r \to f$ in $L^1(\mathbb{T})$ and Theorem 2.5 finishes the proof. □

Theorem 3.8 (Version 3) *Suppose that $\mu \in \mathcal{M}(\mathbb{T})$, $\hat{\mu}(n) = 0$ for all $n < 0$. Then μ is absolutely continuous with respect to Lebesgue measure.*

Proof Since $\hat{\mu}(n) = 0$ for $n \in \mathbb{Z}^-$ one has that $F_r = P_r * \mu$ is analytic on $\mathbb{D}$. By the second version of the theorem and the remark preceding it, one concludes that $\mu(d\theta) = f(\theta)\,d\theta$ with $f = \lim_{r\to 1-} F(re(\theta)) \in L^1(\mathbb{T})$, as claimed. □

The logic of this argument shows that if $\mu \perp m$ (where m is Lebesgue measure) then the harmonic extension u_μ of μ satisfies $u_\mu^* \notin L^1(\mathbb{T})$. It is possible to give a more quantitative version of this fact. Indeed, suppose that μ is a positive measure. Then, for some absolute constant C,

$$C^{-1} M\mu < u_\mu^* < CM\mu, \tag{3.3}$$

where the upper bound is given by Lemma 2.11 (applied to the Poisson kernel) and the lower bound follows from the assumption $\mu \ge 0$ and the fact that the Poisson kernel dominates the box kernel. Part (a) of Problem 3.5 at the end of the chapter therefore implies the quantitative non-L^1 statement

$$\left|\left\{\theta \in \mathbb{T} \,\middle|\, u_\mu^*(\theta) \ge \lambda\right\}\right| \ge \frac{C}{\lambda}\|\mu\|$$

for $\mu \perp m$, where μ is a positive measure.

The first F. and M. Riesz theorem raises the following question. Given $f \in L^1(\mathbb{T})$, how can one decide whether

$$P_r * f + iQ_r * f \in h^1(\mathbb{D})?$$

We know that necessarily $u_f^* = (P_r * f)^* \in L^1(\mathbb{T})$. A theorem by Burkholder, Gundy, and Silverstein states that this is also sufficient (they proved this for the *nontangential* maximal function). It is important to note the difference from (3.3), i.e., this is *not* the same as the Hardy–Littlewood maximal function satisfying $Mf \in L^1(\mathbb{T})$, owing to possible cancellation in f. In fact, it is known that

$$Mf \in L^1(\mathbb{T}) \iff |f|\log(2 + |f|) \in L^1;$$

see Problem 7.6.

3.2.2. Second F. and M. Riesz theorem

We conclude this section with another theorem due to the Riesz brothers, which can be seen as a generalization of the uniqueness theorem for analytic functions. In fact, it says that if an analytic function F on the disk does not have too wild a growth as one approaches the boundary (expressed through the L^1-boundedness condition) then the boundary values are well defined and cannot vanish on a set of positive measure (unless F vanishes identically).

Theorem 3.9 (Second F. & M. Riesz theorem) *Let F be analytic on $\mathbb{D}$ and L^1-bounded, i.e., $F \in h^1(\mathbb{D})$. Assume that $F \not\equiv 0$ and set $f = \lim_{r\to 1-} F_r$. Then $\log|f| \in L^1(\mathbb{T})$. In particular, f does not vanish on a set of positive measure.*

Proof The idea is that if $F(0) \neq 0$ then

$$\int_{\mathbb{T}} \log|f(\theta)|\, d\theta \geq \log|F(0)| > -\infty.$$

Since $\log_+|f| \leq |f| \in L^1(\mathbb{T})$ by Theorem 3.8, we should be done. Some care needs to be taken, though, as F attains the boundary values f only in the almost everywhere sense. This issue can easily be handled by means of Fatou's lemma. First, we have $F^* \in L^1(\mathbb{T})$, so $\log_+|F_r| \leq F^*$ implies that $\log_+|f| \in L^1(\mathbb{T})$ by Lebesgue dominated convergence. Second, by subharmonicity,

$$\int_{\mathbb{T}} \log|F_r(\theta)|\, d\theta \geq \log|F(0)|$$

so that

$$\begin{aligned}\int_{\mathbb{T}} \log|f(\theta)|\, d\theta &= \int_{\mathbb{T}} \lim_{r\to 1} \log|F_r(\theta)|\, d\theta \\ &\geq \limsup_{r\to 1} \int_{\mathbb{T}} \log|F_r(\theta)|\, d\theta \geq \log|F(0)|.\end{aligned}$$

If $F(0) \neq 0$ then we are done. If $F(0) = 0$ then choose another point $z_0 \in \mathbb{D}$ for which $F(z_0) \neq 0$. Now either one repeats the previous argument with the Poisson kernel instead of the sub-mean-value property or one composes F with an automorphism of the unit disk that moves 0 to z_0. Then the previous argument applies. □

Theorem 3.9 is a far-reaching generalization of the following fact from complex analysis, which is proved by Schwarz reflection and Riemann mapping: let F be analytic in Ω and continuous up to $\Gamma \subset \partial\Omega$, which is an open Lipschitz arc; if $F = 0$ on Γ then $F \equiv 0$ in Ω.

3.3. Definition and simple properties of the conjugate function

We shall now exhibit some basic properties of the kernel Q_r introduced in Chapter 2. We refer to it as the kernel *conjugate* to the Poisson kernel P_r. The study of the mapping properties of the convolution operator with kernel Q_r will allow us to answer the question of the L^p convergence of Fourier series with $1 < p < \infty$ that was raised in Chapter 1. We begin by recalling the definition of a *conjugate function.*

Definition 3.10 Let u be real-valued and harmonic in $\mathbb{D}$. Then we define $\tilde{u}$ to be that unique real-valued and harmonic function in $\mathbb{D}$ for which $u + i\tilde{u}$ is analytic and $\tilde{u}(0) = 0$. If u is complex-valued and harmonic then we set $\tilde{u} := (\operatorname{Re}\tilde{u}) + i(\operatorname{Im}\tilde{u})$.

The following lemma presents some simple but useful properties of the harmonic conjugate $\tilde{u}$.

Lemma 3.11 *Let $\tilde{u}$ be the harmonic conjugate as above.*

(i) *If u is constant then $\tilde{u} = 0$.*
(ii) *If u is analytic in $\mathbb{D}$ and $u(0) = 0$ then $\tilde{u} = -iu$. If u is co-analytic (meaning that $\bar{u}$ is analytic) then $\tilde{u} = iu$.*
(iii) *Any harmonic function u can be written uniquely as $u = c + f + \bar{g}$ with c a constant, f, g analytic, and $f(0) = g(0) = 0$.*

Proof (i) and (ii) follow immediately from the definition, whereas (iii) is given by

$$u = u(0) + \tfrac{1}{2}(u - u(0) + i\tilde{u}) + \tfrac{1}{2}(u - u(0) - i\tilde{u})$$

The uniqueness of c, f, g is also clear. □

There is a simple relation between the Fourier coefficients of u_r and that of its harmonic conjugate. This relation will be the key to expressing the partial-sum operator for Fourier series in terms of the harmonic conjugate. Henceforth, we shall occasionally use $\mathcal{F}$ to denote the Fourier transform, for notational clarity.

Lemma 3.12 *Suppose that u is harmonic on $\mathbb{D}$. Then, for all $n \in \mathbb{Z}$, $n \neq 0$,*

$$\mathcal{F}(\tilde{u}_r)(n) = -i\,\operatorname{sign}(n)\,\widehat{u_r}(n). \tag{3.4}$$

Proof By Lemma 3.11 it suffices to take u as constant, analytic, and co-analytic. We will present the case for which u is analytic and $u(0) = 0$. Then $\tilde{u} = -iu$, so that $\mathcal{F}(\tilde{u}_r)(n) = -i\widehat{u_r}(n)$ for all $u \in \mathbb{Z}$. But $\widehat{u_r}(n) = 0$ for $n \leq 0$ and thus (3.4) holds in this case. □

Together with the Plancherel theorem, the previous lemma immediately allows us to conclude the following relation between the L^2 norms of $\tilde{u}_r$ and u_r.

Proposition 3.13 *Let $u \in h^2(\mathbb{D})$ be real-valued. Then $\tilde{u} \in h^2(\mathbb{D})$. In fact,*

$$\|\tilde{u}_r\|_2^2 = \|u_r\|_2^2 - |u(0)|^2.$$

Proof As already mentioned, this follows from Lemma 3.12 and Parseval's identity. Alternatively, by Cauchy's theorem,

$$\int_{\mathbb{T}} (u_r + i\tilde{u}_r)^2(\theta)\, d\theta = (u + i\tilde{u})^2(0) = u^2(0).$$

Since the right-hand side is real-valued the left-hand side is also necessarily real, and thus

$$u^2(0) = \int_{\mathbb{T}} u_r^2(\theta)\, d\theta - \int_{\mathbb{T}} \tilde{u}_r^2(\theta)\, d\theta$$

as claimed. □

The previous L^2-boundedness result allows us to determine the boundary values of the conjugates of h^2 functions.

Corollary 3.14 *If $u \in h^2(\mathbb{D})$ then $\lim_{r\to 1} \tilde{u}(re(\theta))$ exists for a.e. $\theta \in \mathbb{T}$ as an $L^2(\mathbb{T})$ function, and the convergence is also in the sense of L^2.*

Proof By Proposition 3.13 we know that $\tilde{u} \in h^2(\mathbb{D})$ and so the results of Chapter 2 lead to the desired conclusion. □

3.4. The weak-L^1 bound on the maximal function

We now consider the case of $h^1(\mathbb{D})$. From the example of the Poisson kernel and its conjugate we see that there is no hope of obtaining a result as strong as in the case of h^2. However, we shall now show that the radial maximal function, as specified in Definition 3.6, of the conjugate to any $h^1(\mathbb{D})$ function is bounded in weak-$L^1(\mathbb{T})$.

We emphasize, however, that this has nothing to do with the weak-L^1 boundedness of the Hardy–Littlewood maximal function. In fact, it is not possible to control (bound) the convolutions with Q_r by means of that maximal function uniformly in $0 < r < 1$.

Nevertheless the following theorem, going back to Besicovitch and Kolmogorov, shows that the weak-L^1 property still holds. The proof presented here is based on the notion of *harmonic measure*.

Theorem 3.15 *Let $u \in h^1(\mathbb{D})$. Then*

$$\left|\{\theta \in \mathbb{T} \mid |\tilde{u}^*(\theta)| > \lambda\}\right| \leq \frac{C}{\lambda} |||u|||_1$$

with some absolute constant C.

Proof By Theorem 2.5, $u_r = P_r * \mu$. Splitting μ into real and imaginary parts, and then each piece into its positive and negative parts, we reduce ourselves to the case $u \geq 0$. Let

$$E_\lambda = \{\theta \in \mathbb{T} \mid \tilde{u}^*(\theta) > \lambda\}$$

and set $F = -\tilde{u} + iu$. Then F is analytic and $F(0) = iu(0)$. Define a function

$$\omega_\lambda(x, y) := \frac{1}{\pi} \int_{(-\infty,-\lambda)\cup(\lambda,\infty)} \frac{y}{(x-t)^2 + y^2}\, dt,$$

which is harmonic for $y > 0$ and nonnegative. The following two properties of ω_λ will be important:

(1) $\omega_\lambda(x, y) \geq \frac{1}{2}$ if $|x| > \lambda$;

(2) $\omega_\lambda(0, y) \leq \frac{2y}{\pi\lambda}$.

For the first property it suffices to note that

$$\int_0^\infty \frac{1}{\pi} \frac{y}{x^2 + y^2}\, dx = \tfrac{1}{2} \quad \forall y > 0.$$

For the second property, compute

$$\omega_\lambda(0, y) = \frac{1}{\pi} \int_{(-\infty,-\lambda)\cup(\lambda,\infty)} \frac{y}{t^2 + y^2}\, dt \leq \frac{2}{\pi} \int_{\lambda/y}^\infty \frac{dt}{1+t^2} \leq \frac{2y}{\pi\lambda},$$

as claimed.

Observe now that $\omega_\lambda \circ F$ is harmonic and that $\theta \in E_\lambda$ implies

$$(\omega_\lambda \circ F)(re(\theta)) \geq \tfrac{1}{2}$$

for some $0 < r < 1$. Thus

$$|E_\lambda| \leq \left|\{\theta \in \mathbb{T} \mid (\omega_\lambda \circ F)^*(\theta) \geq \tfrac{1}{2}\}\right| \leq 3 \times 2\, |||\omega_\lambda \circ F|||_1, \tag{3.5}$$

by Proposition 3.7. Since $\omega_\lambda \circ F \geq 0$, the mean-value property implies that

$$|||\omega_\lambda \circ F|||_1 = (\omega_\lambda \circ F)(0) = \omega_\lambda(iu(0)) \leq \frac{2}{\pi}\frac{u(0)}{\lambda} = \frac{2}{\pi}\frac{|||u|||_1}{\lambda}.$$

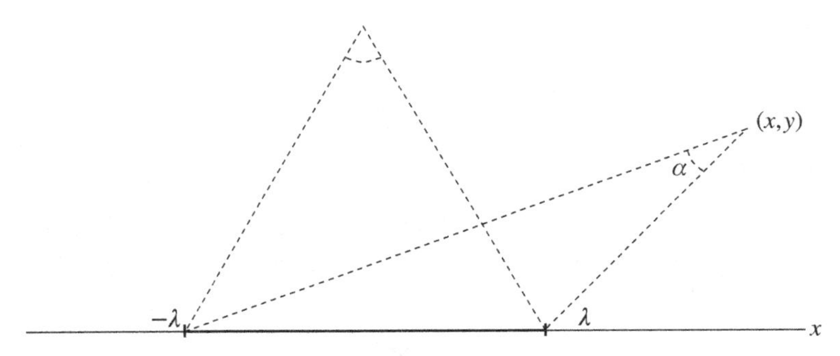

Figure 3.1. The angle α subtended by the line segment $[-\lambda, \lambda]$ is given by the harmonic measure $\omega_\lambda(x, y)$ in the proof of Theorem 3.15: $\omega_\lambda(x, y) = 1 - \alpha/\pi$.

Combining this with (3.5) yields

$$|E_\lambda| \leq \frac{12}{\pi} \frac{|||u|||_1}{\lambda},$$

and we are done. □

Exercise 3.3 Show that $\omega_\lambda(x, y) = 1 - \alpha/\pi$, where α is the angle subtended by the line segment $[-\lambda, \lambda]$ at the point $z = x + iy$. Figure 3.1 shows two possible positions of z. Use this to show that $\omega_\lambda(x, y) \geq \frac{1}{2}$ provided that (x, y) lies outside the semicircle with radius λ and center 0. Furthermore, show that ω_λ is the unique harmonic function in the upper half-plane that equals 1 on $\Gamma := (-\infty, -\lambda] \cup [\lambda, \infty)$ and 0 on $(-\lambda, \lambda) = \mathbb{R} \setminus \bar{\Gamma}$ and which is globally bounded. It is called the *harmonic measure* of $(-\infty, -\lambda] \cup [\lambda, \infty)$ and can be defined in the same fashion on general domains Ω and any open $\Gamma \subset \bar{\Omega}$ (in fact, more general Γ than that). It turns out that the harmonic measure of Γ relative to Ω equals the probability that Brownian motion starting at $z \in \Omega$ hits the boundary for the first time at Γ rather than at $\partial\Omega \setminus \Gamma$. Of particular importance is the conformal invariance of this notion, but this probabilistic connection lies beyond our scope in this book.

3.5. The Hilbert transform

In the following result we introduce the *Hilbert transform* and establish a weak-L^1 bound for it. Formally speaking, the Hilbert transform $H\mu$ of a measure $\mu \in \mathcal{M}(\mathbb{T})$ is defined by

$$\mu \mapsto u_\mu \mapsto \widetilde{u_\mu} \mapsto \lim_{r \to 1} (\widetilde{u_\mu})_r =: H\mu,$$

i.e., the Hilbert transform of a function on $\mathbb{T}$ equals the boundary values of the conjugate function of its harmonic extension. By Corollary 3.14 this is well defined if $\mu(d\theta) = f(\theta)\,d\theta$, $f \in L^2(\mathbb{T})$. We now consider the case $f \in L^1(\mathbb{T})$.

3.5.1. The weak-L^1 bound

Corollary 3.16 *Given $u \in h^1(\mathbb{D})$, the limit $\lim_{r\to 1} \tilde{u}(re(\theta))$ exists for a.e. θ. With $u = P_r * \mu$, $\mu \in M(\mathbb{T})$, this limit is denoted by $H\mu$. There exists the weak-L^1 bound*

$$\left|\{\theta \in \mathbb{T} \mid |H\mu(\theta)| > \lambda\}\right| \leq \frac{C}{\lambda}\|\mu\|.$$

Proof If $\mu(d\theta) = f(\theta)\,d\theta$ with $f \in L^2(\mathbb{T})$ then $\lim_{r\to 1} \widetilde{u_f}(re(\theta))$ exists for a.e. θ, by Corollary 3.14. If $f \in L^1(\mathbb{T})$ and $\varepsilon > 0$ then let $g \in L^2(\mathbb{T})$ such that $\|f - g\|_1 < \varepsilon$. Denote, for any $\delta > 0$,

$$E_\delta := \left\{\theta \in \mathbb{T} \,\middle|\, \limsup_{r,s\to 1} \left|\widetilde{u_f}(re(\theta)) - \widetilde{u_f}(se(\theta))\right| > \delta\right\}$$

and

$$F_\delta := \left\{\theta \in \mathbb{T} \,\middle|\, \limsup_{r,s\to 1} \left|\widetilde{u_h}(re(\theta)) - \widetilde{u_h}(se(\theta))\right| > \delta\right\},$$

where $h = f - g$. In view of the preceding theorem and the L^2 case,

$$\begin{aligned} |E_\delta| = |F_\delta| &\leq \left|\{\theta \in \mathbb{T} \mid (\widetilde{u_h})^*(\theta) > \tfrac{1}{2}\delta\}\right| \\ &\leq \frac{C}{\delta}|||u_h|||_1 \leq \frac{C}{\delta}\|f - g\|_1 \to 0 \end{aligned}$$

as $\varepsilon \to 0$. This finishes the case where μ is absolutely continuous relative to Lebesgue measure.

To treat singular measures, we first consider measures $\mu = \nu$ that satisfy $|\mathrm{supp}(\nu)| = 0$. Here

$$\mathrm{supp}(\nu) := \mathbb{T} \setminus \cup\{I \subset \mathbb{T} \mid \nu(I) = 0\},$$

I being an arc. Observe that for any $\theta \notin \mathrm{supp}(\nu)$ the limit $\lim_{r\to 1} \widetilde{u_r}(\theta)$ exists since the analytic function $u + i\tilde{u}$ can be continued across an interval J on $\mathbb{T}$ for which $\mu(J) = 0$ and which contains θ. Hence $\lim_{r\to 1} \widetilde{u_r}$ exists a.e. by the assumption $|\mathrm{supp}(\nu)| = 0$. If $\mu \in \mathcal{M}(\mathbb{T})$ is an arbitrary singular measure then use inner regularity to observe that for every $\varepsilon > 0$ there exists $\nu \in \mathcal{M}(\mathbb{T})$ with $\|\mu - \nu\| < \varepsilon$ and $|\mathrm{supp}(\nu)| = 0$. Indeed, set $\nu(A) := \mu(A \cap K)$ for all Borel sets A where K is compact and $|\mu|(\mathbb{T}\setminus K) < \varepsilon$. The theorem now follows on passing from the statement for ν to one for μ by means of the same argument as that used in the absolutely continuous case above. □

3.5.2. The L^p bound

It is now easy to obtain the L^p-boundedness of the Hilbert transform on $1 < p < \infty$. This result is due to Marcel Riesz, who gave a different proof; see the exercise following the theorem for his original argument.

Theorem 3.17 *If $1 < p < \infty$ then $\|Hf\|_p \le C_p \|f\|_p$. Consequently, if $u \in h^p(\mathbb{D})$ with $1 < p < \infty$ then $\tilde{u} \in h^p(\mathbb{D})$ and $|||\tilde{u}|||_p \le C_p |||u|||_p$.*

Proof By Proposition 3.13, $\|Hu\|_2 \le \|u\|_2$; the equality occurs if and only if

$$\int_{\mathbb{T}} u(\theta)\, d\theta = 0.$$

Interpolating this with the weak-L^1 bound from Corollary 3.16 by means of the Marcinkiewicz interpolation theorem finishes the case $1 < p \le 2$. If $2 < p < \infty$ then we use duality. More precisely, if $f, g \in L^2(\mathbb{T})$ then

$$\begin{aligned}\langle f, Hg\rangle &= \sum_{n\in\mathbb{Z}} \hat{f}(n)\overline{\widehat{Hg}(n)} = \sum_{n\in\mathbb{Z}} i\,\mathrm{sign}(n)\hat{f}(n)\overline{\hat{g}(n)} \\ &= \sum_{n\in\mathbb{Z}} -\widehat{Hf}(n)\overline{\hat{g}(n)} = -\langle Hf, g\rangle.\end{aligned}$$

This shows that $H^* = -H$. Hence, if $f \in L^p(\mathbb{T}) \subset L^2(\mathbb{T})$ and $g \in L^2(\mathbb{T}) \subset L^{p'}(\mathbb{T})$ then

$$|\langle Hf, g\rangle| = |\langle f, Hg\rangle| \le \|f\|_p \|Hg\|_{p'} \le C_{p'} \|f\|_p \|g\|_{p'}$$

and thus $\|Hf\|_p \le C_{p'} \|f\|_p$ as claimed. □

Exercise 3.4 Give a complex-variable proof of the $L^{2m}(\mathbb{T})$-boundedness of the Hilbert transform by applying the Cauchy integral formula to $(u + i\tilde{u})^{2m}$ when m is a positive integer; cf. the proof of Proposition 3.13. Obtain the Theorem 3.17 from this by interpolation and duality.

Consider the analytic mapping $F = u + iv$ that takes $\mathbb{D}$ onto the strip

$$\{z \mid |\mathrm{Re}\, z| < 1\}.$$

Then $u \in h^\infty(\mathbb{D})$ but clearly $v \notin h^\infty(\mathbb{D})$, so that Theorem 3.17 fails on $L^\infty(\mathbb{T})$. By duality, it also fails on $L^1(\mathbb{T})$. The correct substitute for L^1 in this context is the space of *real parts of functions in* $\mathbb{H}^1(\mathbb{D})$. This is a deep result that goes much further than the F. and M. Riesz theorem. The statement is that

$$\|Hf\|_1 \le C \|u_f^*\|_1, \tag{3.6}$$

where u_f^* is the nontangential maximal function of the harmonic extension u_f of f. Recall that by the Burkholder–Gundy–Silverstein theorem the right-hand side of (3.6) is finite if and only if f is the real part of an analytic L^1-bounded function. A more modern approach to (3.6) is given by the real-variable theory of Hardy spaces, which subsumes statements such as (3.6) by the boundedness of singular integral operators on such spaces.

3.5.3. Kernel representation of the Hilbert transform

Next, we turn to the problem of expressing Hf in terms of a kernel. By Exercise 2.4 it is clear that one would expect that

$$(H\mu)(\theta) = \int_{\mathbb{T}} \cot(\pi(\theta - \varphi))\, \mu(d\varphi) \tag{3.7}$$

for any $\mu \in \mathcal{M}(\mathbb{T})$. This, however, requires justification as the integral on the right-hand side is not necessarily convergent.

Proposition 3.18 *If $\mu \in \mathcal{M}(\mathbb{T})$ then*

$$\lim_{\epsilon \to 0} \int_{|\theta-\varphi|>\epsilon} \cot(\pi(\theta - \varphi))\, \mu(d\varphi) = (H\mu)(\theta) \tag{3.8}$$

for a.e. $\theta \in \mathbb{T}$. In other words, (3.7) *holds in the principal value sense.*

Exercise 3.5 Verify that the limit in (3.8) exists for all $d\mu = f\,d\theta + d\nu$ where $f \in C^1(\mathbb{T})$ and $|\mathrm{supp}(\nu)| = 0$. Furthermore, show that these measures are dense in $\mathcal{M}(\mathbb{T})$.

Proof of Proposition 3.18 The idea is to represent a general measure as a limit of measures of the kind given by Exercise 3.5. As we saw in the proof of the a.e. convergence result Theorem 2.12, the double limit appearing in such an argument requires a bound on an appropriate maximal function. In this case the natural bound is of the form

$$\left|\left\{\theta \in \mathbb{T} \,\middle|\, \sup_{0<\epsilon<1/2} \left|\int_{|\theta-\varphi|>\epsilon} \cot(\pi(\theta - \varphi))\, \mu(d\varphi)\right| > \lambda\right\}\right| \le \frac{C}{\lambda}\|\mu\| \tag{3.9}$$

for all $\lambda > 0$. We leave it to the reader to check that (3.9) implies the theorem. In order to prove (3.9) we invoke our strongest result on the conjugate function, namely Theorem 3.15. More precisely, we claim that

$$\sup_{0<r<1} \left|(Q_r * \mu)(\theta) - \int_{|\theta-\varphi|>1-r} \cot(\pi(\theta - \varphi))\, \mu(d\varphi)\right| \le CM\mu(\theta), \tag{3.10}$$

where $M\mu$ is the Hardy–Littlewood maximal function. Since

$$\sup_{0<r<1} |(Q_r * \mu)(\theta)| = \widetilde{u_\mu}^*(\theta),$$

(3.9) follows from (3.10) by means of Theorem 3.15 and Proposition 2.9. To verify (3.10), write the difference within the absolute-value signs as $(K_r * \mu)(\theta)$, where

$$K_r(\theta) = \begin{cases} Q_r(\theta) - \cot(\pi\theta) & \text{if } 1 - r < |\theta| < \frac{1}{2}, \\ Q_r(\theta) & \text{if } |\theta| \le 1 - r. \end{cases}$$

By means of calculus one can check that

$$|K_r(\theta)| \le \begin{cases} C\dfrac{(1-r)^2}{|\theta|^3} & \text{if } |\theta| > 1 - r, \\ C(1-r)^{-1} & \text{if } |\theta| \le 1 - r. \end{cases}$$

Denote the right-hand side by $\tilde{K}_r$. Up to a normalization constant, the $\{\tilde{K}_r\}_{0<r<1}$ form a radially bounded approximate identity, and (3.10) therefore follows from Lemma 2.11. □

Proposition 3.18 raises the question whether Theorem 3.17 can be proved on the basis of the representation (3.8) alone, i.e., without using complex variables or harmonic functions. Going even further, we may ask if the L^2-boundedness of the Hilbert transform can be proved without using the Fourier transform. We shall answer both these questions later in the context of Calderon–Zygmund theory, where they play an important role.

3.5.4. The Hilbert transform on $L^\infty(\mathbb{T})$

Exercise 3.6 Show by means of (3.8) that H is not bounded on $L^\infty(\mathbb{T})$. For example, consider $H\chi_{[0,1/2)}$. Also, deduce from the fact that $\cot(\pi\theta) \notin L^1(\mathbb{T})$ that H is not bounded on L^1.

The following proposition shows that the Hilbert transform Hf is exponentially integrable for bounded functions f. In Exercise 3.6, you should have found that $H\chi_{[0,1/2]}$ has logarithmic behavior at 0 and at $\frac{1}{2}$, which shows that the following result is optimal.

Proposition 3.19 *Let f be a real-valued function on $\mathbb{T}$ with $|f| \le 1$. Then, for any $0 \le \alpha < \frac{1}{2}\pi$,*

$$\int_{\mathbb{T}} e^{\alpha|Hf(\theta)|}\, d\theta \le \frac{2}{\cos\alpha}.$$

Proof Let $u = u_f$ be the harmonic extension of f to $\mathbb{D}$ and set $F = \tilde{u} - iu$. Then $|u| \le 1$ by the maximum principle, and hence $\cos(\alpha u) \ge \cos\alpha$. Therefore

$$\operatorname{Re}(e^{\alpha F}) = \operatorname{Re}(e^{\alpha\tilde{u}}e^{-i\alpha u}) = \cos(\alpha u)\, e^{\alpha\tilde{u}} \ge \cos\alpha\, e^{\alpha\tilde{u}}. \tag{3.11}$$

By the mean-value property,

$$\int_{\mathbb{T}} \text{Re}\, e^{\alpha F_r(\theta)}\, d\theta = \text{Re}\, e^{\alpha F(0)} = \text{Re}\, e^{-i\alpha(u(0))} = \cos(\alpha u(0)) \leq 1.$$

Combining this with (3.11) yields

$$\int_{\mathbb{T}} e^{\alpha \tilde{u}_r(\theta)}\, d\theta \leq \frac{1}{\cos\alpha}$$

and therefore, by Fatou's lemma,

$$\int_{\mathbb{T}} e^{\alpha(Hf)(\theta)}\, d\theta \leq \frac{1}{\cos\alpha}.$$

Since this inequality also holds for $-f$, the proposition follows. □

In later chapters we will develop the real-variable theory of singular integrals, which contains the results on the Hilbert transform obtained above. The basic theorem, due to Calderon and Zygmund, states that singular integrals are bounded on $L^p(\mathbb{R}^n)$ for $1 < p < \infty$, thus generalizing Theorem 3.17. In addition we shall investigate the boundedness of the Hilbert transform (and the more general singular integral operators) relative to the A_p classes from the previous chapter.

The analogue of Proposition 3.19 for singular integrals is given by the facts that they are bounded from L^∞ to BMO (bounded mean oscillation space) and that BMO functions are exponentially integrable (this bound is called the John–Nirenberg inequality). The dual question of what happens on L^1 leads into the real-variable theory of Hardy spaces. The analogue of (3.6) is then that singular integrals are bounded on $H^1(\mathbb{R}^n)$. The dual space of H^1 is BMO, a classic result of Charles Fefferman. We shall present this duality in a dyadic setting (given by the Haar system) in Chapter 8.

3.6. Convergence of Fourier series in L^p

We conclude the theory of the conjugate function by returning to the question of the $L^p(\mathbb{T})$ convergence of Fourier series. Recall from Chapter 1 that this convergence fails for $p = 1$ and $p = \infty$, but we will now deduce from Theorem 3.17 that it *does hold* for the full range $1 < p < \infty$.

Theorem 3.20 *Let S_N denote the partial-sum operator of Fourier series. Then for any $1 < p < \infty$ the S_N are uniformly bounded on $L^p(\mathbb{T})$, i.e.,*

$$\sup_N \|S_N\|_{p\to p} < \infty.$$

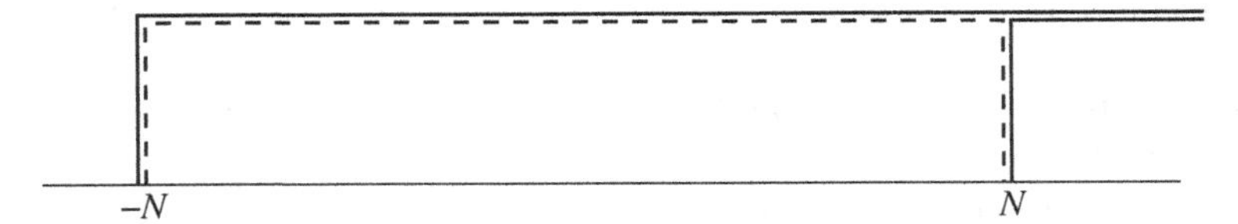

Figure 3.2. Reducing partial sums to the Hilbert transform.

By Proposition 1.9 this implies the convergence of $S_N f \to f$ in $L^p(\mathbb{T})$ with $1 < p < \infty$ for any $f \in L^p(\mathbb{T})$.

Proof The point is simply that S_N can be written in terms of the Hilbert transform. Indeed, recall that

$$\widehat{Hf}(n) = -i \,\mathrm{sign}(n) \hat{f}(n)$$

so that, with $e_n(x) = e(nx)$,

$$Tf := \tfrac{1}{2}(1 + iH)f = \textstyle\sum_n \chi_{(0,\infty)}(n) \hat{f}(n) e_n.$$

In other words, on the Fourier side T corresponds to multiplication by $\chi_{(0,\infty)}$ whereas S_N corresponds to multiplication by $\chi_{[-N,N]}$. It remains to write $\chi_{[-N,N]}$ as the difference of two shifted $\chi_{(0,\infty)}$, i.e.,

$$\chi_{[-N,N]} = \chi_{(-N-1,\infty)} - \chi_{(N,\infty)},$$

or, in terms of H and T,

$$S_N f = \overline{e_{N+1}}\, T(e_{N+1} f) - e_N\, T(\overline{e_N} f).$$

Hence, for $1 < p < \infty$,

$$\|S_N\|_{p \to p} \le 2\|T\|_{p \to p} \le 1 + \|H\|_{p \to p}$$

uniformly in N, as claimed. □

We invite the reader to investigate the analogous result for higher-dimensional Fourier series.

Exercise 3.7 Let $\underline{N} = (N_1, \ldots, N_d)$ denote a vector with positive integer entries. Prove that

$$\sup_{\underline{N}} \|S_{\underline{N}}\|_{p \to p} < \infty \quad \forall 1 < p < \infty,$$

where $S_{\underline{N}}$ is the partial sum operator for Fourier series on $\mathbb{T}^d$, defined as follows:

$$S_{\underline{N}} f(x) = \sum_{\nu \in \mathbb{Z}^d} \prod_{j=1}^{d} \chi_{[|\nu_j| < N_j]} \hat{f}(\nu) e(\nu \cdot x).$$

Conclude that $S_{\underline{N}} f$ converges in the L^p sense for any $f \in L^p(\mathbb{T}^d)$ and $1 < p < \infty$, provided that each component N_j tends to ∞.

We remark that the corresponding result, in which expanding rectangles are replaced with Euclidean balls in $\mathbb{Z}^d$, is *false*; in fact, Charlie Fefferman's ball multiplier theorem states that the analogue of the previous exercise only holds for $p = 2$ in that case. This is delicate, and relies on the existence of *Kakeya sets*; see Fefferman [37].

Notes

For the topics covered in this chapter see Hoffman's classic [55], Garnett [46], and especially Koosis [71]. The Burkholder–Gundy–Silverstein theorem, which is proved in [71], can be seen as a precursor to the Fefferman–Stein *grand maximal function*, which leads to the real-variable theory of Hardy spaces; see Stein [111]. Koosis [71] also presents some elements of the modern real-variable theory of H^1 spaces, such as the atomic decomposition of H^1. The full real-variable H^p theory in higher dimensions is developed in [111]. We shall present some elements of real-variable H^1 theory on the interval $[0, 1]$ in Section 8.4. In particular, we will establish the duality theorem and an atomic decomposition, defined in terms of the Haar basis, for the dyadic version of the Hardy space.

Many substantial pieces of classical theory, such as inner and outer functions, are omitted in this chapter. However, we shall not require those aspects of H^p theory in this book. They can be found in any of the aforementioned texts.

The conjugate function and the Hilbert transform provide the link between the complex-variable-based theory of Hardy spaces and the real-variable Calderón–Zygmund theory, to which we will turn in Chapter 7.

Problems

Problem 3.1 Establish the maximum principle for subharmonic functions:

(a) Show that if u is subharmonic on Ω and satisfies $u(z) \leq u(z_0)$ for all $z \in \Omega$ and some fixed $z_0 \in \Omega$ then u is a constant.

(b) Show that if Ω is bounded and u is continuous on $\overline{\Omega}$ then $u(z) \leq \max_{\partial\Omega} u$ for all $z \in \Omega$, equality being possibly only if u is a constant.

Problem 3.2 Let u be subharmonic on a domain Ω. Show that there exists a unique positive measure μ on Ω such that $\mu(K) < \infty$ for every $K \subset\subset \Omega$ (the double inclusion symbol means that K is a compact subset of Ω) and such that

$$u(z) = \iint \log|z-\zeta|\,d\mu(\zeta) + h(z),$$

where h is harmonic on Ω. This is called Riesz's representation of subharmonic functions. *Hint:* First assume that u is smooth, and use Green's formula. Then use the fact that $\mu(dx) = \Delta u(x)\,dx$, which is a positive measure. You may take Ω to be the unit disk for simplicity.

Problem 3.3 With u and μ as in the previous exercise, show that

$$\int_{\mathbb{T}} u(z + re(\theta))\,d\theta - u(z) = \int_0^r \frac{\mu(D(z,t))}{t}\,dt$$

for all $D(z,r) \subset \Omega$ (this generalizes Jensen's formula (3.1) to functions that are not twice continuously differentiable). Recover Jensen's formula (a result from complex analysis) by setting $u(z) := \log|f(z)|$, where f is analytic.

Problem 3.4 Let μ be a finite positive Borel measure on $\mathbb{C}$. Then for every $H \in (0,1)$ there exist disks $D_j := D(\zeta_j, r_j)$ with $\sum_j r_j \le 5H$ and such that for any $z \in \mathbb{C} \setminus \bigcup_j D_j$ one has

$$u(z) > n \log\left(\frac{H}{e}\right),$$

where $n = \mu(\mathbb{C})$. *Hint:* Let $n_z(t) := \mu(D(z,t))$. For a suitable constant $p > 0$ (depending on n and H) define a point z to be p-admissible if $n_z(t) \le pt$ for all $t > 0$. Define D_j by successively picking the largest possible disk associated with any p-nonadmissible points. Show that such maximal disks are attained. Then use

$$u(z) \ge \int_{|z-\zeta|\le 1} \log|z-\zeta|\,\mu(d\zeta) = \int_0^1 \log t\,dn_z(t).$$

Problem 3.5 This problem explores the behavior of the Hardy–Littlewood maximal function of measures that are singular relative to Lebesgue measure.

(a) Prove that if $\mu \in \mathcal{M}(\mathbb{T}) \setminus \{0\}$ satisfies $\mu \perp m$ then $M\mu \notin L^1$. In fact, show that

$$\left|\{\theta \in \mathbb{T} \mid M\mu(\theta) > \lambda\}\right| \ge \frac{c}{\lambda}\|\mu\|,$$

provided that $\lambda > \|\mu\|$ with an absolute constant $c > 0$.

(b) Prove that there is a numerical constant C such that if $\mu \in \mathcal{M}(\mathbb{T})$ is a *positive* measure and F is the associated harmonic function then $M\mu \le CF^*$. Conclude that if μ is singular then $F^* \notin L^1$.

Problem 3.6 Show that the class of analytic function $F(z)$ in the unit disk having positive real part is in one-to-one relation with the class of functions of the form

$$F(\zeta) = \int_{\mathbb{T}} \frac{1+\zeta e(-\theta)}{1-\zeta e(-\theta)}\,\mu(d\theta) + ic, \quad |\zeta| < 1,$$

where μ is a positive Borel measure on $\mathbb{T}$ and c is a real constant.

Problem 3.7 Find an expression for the harmonic measure of an arc on $\mathbb{T}$ relative to the disk $\mathbb{D}$. Relate it to the harmonic measure of an interval relative to the upper half-plane by means of a conformal mapping.

Problem 3.8 This problem provides a weak form of the a.e. convergence of Fourier series on $L^1(\mathbb{T})$. In the following chapter we will see via Kolmogorov's construction that the full form of the a.e. convergence fails for L^1 functions.

(a) Show that, for any $\lambda > 0$,

$$\sup_N \left|\left\{\theta \in \mathbb{T} \,\middle|\, |(S_N f)(\theta)| > \lambda\right\}\right| \leq \frac{C}{\lambda}\|f\|_1,$$

with C some absolute constant.

(b) Explain what such an inequality would mean if the position of the supremum were moved to give

$$\left|\left\{\theta \in \mathbb{T} \,\middle|\, \sup_N |(S_N f)(\theta)| > \lambda\right\}\right| \leq \frac{C}{\lambda}\|f\|_1$$

for all $\lambda > 0$? Could this be true? *Hint:* See Chapter 6.

(c) Using (a) show that for every $f \in L^1(\mathbb{T})$ there exists a subsequence $\{N_j\} \longrightarrow \infty$ depending on f such that

$$S_{N_j} f \longrightarrow f \quad a.e.$$

as $j \to \infty$.

Problem 3.9 Prove the following multiplier estimate. Let $\bar{m} := \{m_n\}_{n\in\mathbb{Z}}$ be a sequence in $\mathbb{C}$ satisfying

$$\sum_{n\in\mathbb{Z}} |m_n - m_{n-1}| \leq B, \qquad \lim_{n\to-\infty} m_n = 0.$$

Define a multiplier operator $T_{\bar{m}}$ by the rule

$$T_{\bar{m}} f(\theta) = \sum_n m_n \hat{f}(n)\, e(n\theta)$$

for trigonometric polynomials f. Prove that for any such f one has

$$\|T_{\bar{m}} f\|_p \leq C(p) B \|f\|_p$$

for any $1 < p < \infty$.

Problem 3.10 Let $\varphi \in C^\infty(\mathbb{T})$ and denote by A_φ the operator corresponding to multiplication by φ. With H the Hilbert transform on $\mathbb{T}$, show that

$$[A_\varphi, H] = A_\varphi \circ H - H \circ A_\varphi$$

is a *smoothing* operator, i.e., if $\mu \in \mathcal{M}(\mathbb{T})$ is an arbitrary measure then

$$[A_\varphi, H]\mu \in C^\infty(\mathbb{T}).$$

Problem 3.11 The complex-variable methods developed in Chapter 2 as well in the present chapter for the disk apply equally well to the upper half-plane. For example, the Poisson kernel is

$$P_t(x) = \frac{1}{\pi} \frac{t}{x^2 + t^2}, \quad x \in \mathbb{R},\ t > 0,$$

and its harmonic conjugate is

$$Q_t(x) = \frac{1}{\pi} \frac{x}{x^2 + t^2}.$$

Observe that

$$P_t(x) + i\,Q_t(x) = \frac{i}{\pi} \frac{1}{x + it} = \frac{i}{\pi z}$$

with $z = x + it$, which should remind the reader of Cauchy's formula. The Hilbert transform now reads

$$(Hf)(x) = \frac{1}{\pi} \int_{-\infty}^{\infty} \frac{f(y)}{x - y}\, dy$$

in the principal-value sense, the precise statement being just as in Proposition 3.18.

In this problem the reader is asked to develop the theory of the upper half-plane in the context of the previous chapters. The relation between the disk and upper half-plane given by conformal mapping should be considered as well.

Problem 3.12 By considering $F = (u + i\tilde{u}) \log(1 + u + i\tilde{u})$ show that if $u \geq 0$ and $\tilde{u} \in h^1(\mathbb{D})$ then $\int_{\mathbb{T}} u \log(1 + u)\, d\theta < \infty$.

4

The Fourier transform on $\mathbb{R}^d$ and on LCA groups

4.1. The Euclidean Fourier transform

4.1.1. Basic definitions: Fourier transform, Schwartz space

It is natural to extend the definition of the Fourier transform for the circle, given in Chapter 1, to Euclidean spaces. Thus, let $\mu \in \mathcal{M}(\mathbb{R}^d)$ be a complex-valued Borel measure and set

$$\hat{\mu}(\xi) = \int_{\mathbb{R}^d} e^{-2\pi i x\cdot\xi}\,\mu(dx) \quad \forall \xi \in \mathbb{R}^d.$$

The quantity $\hat{\mu}(\xi)$ is called the Fourier transform of μ. Clearly $\|\hat{\mu}\|_\infty \le \|\mu\|$, the total variation of μ. Moreover, it follows immediately from the dominated convergence theorem that $\hat{\mu}$ is continuous. If $\mu(dx) = f(x)\,dx$ then we may write $\hat{\mu}(\xi)$ as $\hat{f}(\xi)$. In particular, the Fourier transform is well defined on $L^1(\mathbb{R}^d)$ and defines a contraction into $C(\mathbb{R}^d)$, the space of continuous functions. Many elementary properties from Chapter 1 carry over to this setting, such as convolutions, Young's inequality, and the action of the Fourier transform on convolutions and translations. In order to be precise, we collect a few facts in the following lemma.

Lemma 4.1 *Let $\mu \in \mathcal{M}(\mathbb{R}^d)$ and let τ_y denote translation by $y \in \mathbb{R}^d$:*

$$(\tau_y\mu)(E) = \mu(E - y).$$

Then

$$\widehat{\tau_y\mu}(\xi) = e^{-2\pi i\xi\cdot y}\hat{\mu}(\xi) \quad \forall \xi \in \mathbb{R}^d.$$

Let $e_\eta(x) := e^{2\pi i x\cdot\eta}$. Then $\widehat{e_\eta\mu}(\xi) = \hat{\mu}(\xi - \eta)$. If $f, g \in L^1(\mathbb{R}^d)$ then the integral

$$\int_{\mathbb{R}^d} f(x-y)g(y)\,dy$$

is absolutely convergent for a.e. $x \in \mathbb{R}^d$ and is denoted $(f * g)(x)$. One has $f * g \in L^1(\mathbb{R}^d)$ and $\widehat{f * g} = \hat{f}\hat{g}$.

Finally, if $g \in L^p$ for $1 \le p \le \infty$ and $f \in L^1$ then $f * g \in L^p$ and $\|f * g\|_p \le \|f\|_1 \|g\|_p$.

We leave the proof, which is elementary, to the reader. We record one more very important fact, namely the *change of variables formula* for the Fourier transform. Let A be an invertible $d \times d$ matrix with real entries, and denote by $f \circ A$ the function $f(Ax)$. Then for any $f \in L^1(\mathbb{R}^d)$ one has

$$\widehat{f \circ A} = |\det A|^{-1} \hat{f} \circ A^{-t}, \tag{4.1}$$

where A^{-t} is the transpose of the inverse of A. A special case of this is the dilation identity. For any $\lambda > 0$ let $f_\lambda(x) = f(\lambda x)$. Then

$$\widehat{f_\lambda}(\xi) = \lambda^{-d} \hat{f}(\xi/\lambda). \tag{4.2}$$

In analogy with the Fourier transform on the circle, we expect that $L^2(\mathbb{R}^d)$ will play a special role. To be more precise, we expect the Plancherel theorem $\|\hat{f}\|_2 = \|f\|_2$ to hold. Note that as it stands this is only meaningful for $f \in L^1 \cap L^2(\mathbb{R}^d)$, since otherwise the defining integral is not absolutely convergent. In fact, it is convenient to work with a much smaller space, called *Schwartz* space. In what follows we use the standard multi-index notation: for any $\alpha = (\alpha_1, \ldots, \alpha_d) \in \mathbb{Z}_0^d$ (where $\mathbb{Z}_0$ is the set of nonnegative integers), we have

$$x^\alpha := \prod_{j=1}^d x_j^{\alpha_j}, \quad x = (x_1, \ldots, x_d),$$

$$\partial^\alpha := \prod_{j=1}^d \frac{\partial^{\alpha_j}}{\partial x_j^{\alpha_j}}.$$

Moreover, $|\alpha| = \sum_{j=1}^d \alpha_j$ denotes the *length* of the multi-index α.

Definition 4.2 The Schwartz space $\mathcal{S}(\mathbb{R}^d)$ is defined as the collection of all functions in $C^\infty(\mathbb{R}^d)$ that decay rapidly, together with all derivatives. In other words, $f \in \mathcal{S}(\mathbb{R}^d)$ if and only if $f \in C^\infty(\mathbb{R}^d)$ and

$$x^\alpha \partial^\beta f(x) \in L^\infty(\mathbb{R}^d) \quad \forall \alpha, \beta,$$

where α, β are arbitrary multi-indices. We introduce the following notion of convergence in $\mathcal{S}(\mathbb{R}^d)$: a sequence $f_n \in \mathcal{S}(\mathbb{R}^d)$ converges to $g \in \mathcal{S}(\mathbb{R}^d)$ if and only if

$$\|x^\alpha \partial^\beta (f_n - g)\|_\infty \to 0, \quad n \to \infty,$$

for all α, β.

For readers familiar with Fréchet spaces we remark that the *seminorms*

$$p_{\alpha\beta}(f) := \|x^\alpha \partial^\beta f\|_\infty, \qquad \alpha, \beta \in \mathbb{Z}_0^d,$$

turn $\mathcal{S}(\mathbb{R}^d)$ into such a space. Since Fréchet spaces are metrizable, it follows that the notion of convergence introduced in Definition 4.2 completely characterizes the topology in $\mathcal{S}(\mathbb{R}^d)$. However, we make no use of this fact and work merely with the sequential characterization.

Exercise 4.1 Let $f \in C^\infty(\mathbb{R}^d)$. Show that $f \in \mathcal{S}(\mathbb{R}^d)$ if and only if $x^\alpha D^\beta f \in L^\infty(\mathbb{R}^d)$ for all α, β.

One has that $\mathcal{S}(\mathbb{R}^d)$ is dense in $L^p(\mathbb{R}^d)$ for all $1 \le p < \infty$ since this is already the case for smooth functions with compact support. It is evident that the map $f \mapsto x^\alpha \partial^\beta f(x)$ is continuous on $\mathcal{S}(\mathbb{R}^d)$. Somewhat less obvious is that the Fourier transform has the same property.

Proposition 4.3 *The Fourier transform is a continuous operation from the Schwartz space into itself.*

Proof Fix any $f \in \mathcal{S}(\mathbb{R}^d)$. The proof hinges on the pointwise identities

$$\widehat{\partial^\alpha f}(\xi) = (2\pi i)^{|\alpha|}\, \xi^\alpha \hat{f}(\xi),$$
$$\partial^\beta \hat{f}(\xi) = (-2\pi i)^{|\beta|}\, \widehat{x^\beta f}(\xi),$$

valid for all multi-indices α, β. The first follows by integration by parts, whereas the second is obtained by differentiation under the integral sign. We leave the formal verification of these identities to the reader (by introducing difference quotients, justifying the limits using the dominated convergence theorem, etc.). This is easy, owing to the rapid decay and smoothness of f.

We first note from these identities that $\hat{f} \in C^\infty$ and, moreover, that

$$\xi^\alpha \partial_\xi^\beta \hat{f} \in L^\infty(\mathbb{R}^d) \quad \forall \alpha, \beta \in \mathbb{Z}_0^d,$$

which implies by definition that $\hat{f} \in \mathcal{S}(\mathbb{R}^d)$. For the continuity, let f_n be a sequence in $\mathcal{S}(\mathbb{R}^d)$ such that $f_n \to 0$ as $n \to \infty$. Then

$$\|\xi^\alpha \partial_\xi^\beta \hat{f}_n\|_\infty \le C_{\alpha,\beta}\, \|\partial_x^\alpha x^\beta f_n\|_1 \to 0$$

as $n \to \infty$. □

4.1.2. The inversion theorem

In view of the preceding fact, it is natural to ask whether the Fourier transform takes $\mathcal{S}(\mathbb{R}^d)$ *onto* itself. We will now prove that this does indeed hold; in fact it is a special case of the *Fourier inversion theorem*. The latter is the analogue of the Fourier summation problem for Fourier series. The Schwartz space is a convenient setting in which to formulate the inversion.

Proposition 4.4 *The Fourier transform takes the Schwartz space onto itself. In fact, for any $f \in \mathcal{S}(\mathbb{R}^d)$ one has*

$$f(x) = \int_{\mathbb{R}^d} e^{2\pi i x\cdot\xi} \hat{f}(\xi)\, d\xi \quad \forall x \in \mathbb{R}^d. \tag{4.3}$$

Proof We would like to proceed by inserting the expression

$$\hat{f}(\xi) = \int_{\mathbb{R}^d} e^{-2\pi i y\cdot\xi} f(y)\, dy$$

into (4.3) and showing that this recovers $f(x)$. The difficulty here is that interchanging the order of integration leads to the formal expression

$$\int_{\mathbb{R}^d} e^{2\pi i(x-y)\cdot\xi}\, d\xi$$

which needs to equal the measure $\delta_0(x-y)$ for the inversion formula (4.3) to hold. Although this is essentially the case, the problem here is of course that the previous integral is not well defined. We shall therefore need to be more careful.

The standard procedure is to introduce a Gaussian weight, since the Fourier transform has an explicit form on such weights. In fact, in Exercise 4.2 below we ask the reader to check the following identity:

$$\widehat{e^{-\pi|x|^2}}(\xi) = e^{-\pi|\xi|^2}. \tag{4.4}$$

Taking this for granted for now, we conclude from it and (4.2) that

$$\widehat{e^{-\pi\varepsilon^2|x|^2}}(\xi) = \varepsilon^{-d} e^{-\pi|\xi|^2/\varepsilon^2}$$

for any $\varepsilon > 0$.

Now fix $f \in \mathcal{S}(\mathbb{R}^d)$. We will commence the proof of (4.3) with the observation that

$$\int_{\mathbb{R}^d} e^{2\pi i x\cdot\xi} \hat{f}(\xi)\, d\xi = \lim_{\varepsilon\to 0} \int_{\mathbb{R}^d} e^{2\pi i x\cdot\xi} e^{-\pi\varepsilon^2|\xi|^2} \hat{f}(\xi)\, d\xi$$

for each $x \in \mathbb{R}^d$. This follows from the dominated convergence theorem since $\hat{f} \in L^1$. Next, one has

$$\begin{aligned}\int_{\mathbb{R}^d} & e^{2\pi i x\cdot\xi} e^{-\pi\varepsilon^2|\xi|^2}\, \hat{f}(\xi)\, d\xi \\ &= \int_{\mathbb{R}^d}\int_{\mathbb{R}^d} e^{2\pi i(x-y)\cdot\xi} e^{-\pi\varepsilon^2|\xi|^2}\, d\xi\, f(y)\, dy \\ &= \int_{\mathbb{R}^d} \varepsilon^{-d} e^{-\pi\varepsilon^{-2}|x-y|^2} f(y)\, dy.\end{aligned}$$

The final observation is that

$$\varepsilon^{-d} e^{-\pi\varepsilon^{-2}|x|^2}$$

is an approximate identity, in the sense of Chapter 1, with respect to the limit $\varepsilon \to 0$. Hence, since $f \in C(\mathbb{R}^d) \cap L^\infty(\mathbb{R}^d)$ one sees that

$$\lim_{\varepsilon\to 0} \int_{\mathbb{R}^d} \varepsilon^{-d} e^{-\pi\varepsilon^{-2}|x-y|^2} f(y)\, dy = f(x)$$

for all $x \in \mathbb{R}^d$, which concludes the proof. $\square$

Exercise 4.2 Prove the identity (4.4). *Hint:* Use contour integration in the complex plane.

Let us now return to the dilation identity (4.2) with f a smooth bump function. We note that this identity expresses an important normalization principle (at least heuristically), namely that the Fourier transform maps an L^1-normalized bump function onto an L^∞-normalized one, and vice versa.

Definition 4.5 The dual of $\mathcal{S}$, denoted by $\mathcal{S}'$, is the space of *tempered distributions*. In other words, any $u \in \mathcal{S}'$ is a continuous functional on $\mathcal{S}$, and we write $\langle u, \phi\rangle$ for u applied to $\phi \in \mathcal{S}$. The space $\mathcal{S}'$ is equipped with the weak-$*$ topology. Thus, $u_n \to u$ in $\mathcal{S}'$ if and only if $\langle u_n, \phi\rangle \to \langle u, \phi\rangle$ as $n \to \infty$ for every $\phi \in \mathcal{S}$. Naturally, $L^p(\mathbb{R}^d) \hookrightarrow \mathcal{S}'(\mathbb{R}^d)$ by the rule

$$\langle f, \phi\rangle = \int_{\mathbb{R}^d} f(x)\phi(x)\, dx, \qquad f \in L^p(\mathbb{R}^d),\ \phi \in \mathcal{S}(\mathbb{R}^d).$$

Using a little functional analysis it can be seen that a linear functional u on $\mathcal{S}$ is continuous if and only if there exists an N such that

$$|\langle u, \phi\rangle| \le C \sum_{|\alpha|,|\beta|\le N} \|x^\alpha \partial^\beta \phi\|_\infty \quad \forall \phi \in \mathcal{S}.$$

It follows that tempered distributions are invariant under differentiation and/or multiplication by any element of $C^\infty(\mathbb{R}^d)$ of at most polynomial growth. One has

$$\langle \partial^\alpha u, \phi \rangle = (-1)^{|\alpha|} \langle u, \partial^\alpha \phi \rangle \qquad \forall\, \phi \in \mathcal{S}(\mathbb{R}^d).$$

Any measure of finite total variation is an element of $\mathcal{S}'$, in particular the Dirac measure δ_0 is such an element.

The Fourier transform of $u \in \mathcal{S}'$ is defined by the relation $\langle \hat{u}, \phi \rangle := \langle u, \hat{\phi} \rangle$ for all $\phi \in \mathcal{S}(\mathbb{R}^d)$. Since the Fourier transform is an isomorphism on $\mathcal{S}(\mathbb{R}^d)$, this definition is meaningful and establishes the distributional Fourier transform as an isomorphism on $\mathcal{S}'$. For example, $\hat{1} = \delta_0$ since

$$\langle 1, \hat{\phi} \rangle = \int_{\mathbb{R}^d} \hat{\phi}(\xi)\, d\xi = \phi(0).$$

In particular, we can speak meaningfully of the Fourier transform of any function in $L^p(\mathbb{R}^d)$ with $1 \le p \le \infty$.

Exercise 4.3 Let $u \in \mathcal{S}'(\mathbb{R}^d)$. Suppose that $x^\alpha D^\beta u \in L^2(\mathbb{R}^d)$ for all α, β. Show that $u \in \mathcal{S}(\mathbb{R}^d)$. What can you conclude if, instead, $x^\alpha D^\beta u \in L^1(\mathbb{R}^d)$ for all α, β?

The next exercise presents a very useful fact, namely that an element of $\mathcal{S}'(\mathbb{R}^d)$ which is supported at a point, say $\{0\}$, is necessarily a linear combination of δ_0 and finitely many derivatives thereof. Here, "supported" refers to the property that $\langle u, \varphi \rangle = 0$ for all $\varphi \in \mathcal{S}(\mathbb{R}^d)$ that vanish near 0.

Exercise 4.4 Let $u, v \in \mathcal{S}'$ with $\langle u, \hat{\varphi} \rangle = \langle v, \hat{\varphi} \rangle$ for all $\varphi \in \mathcal{S}$ with $\operatorname{supp}(\hat{\varphi}) \subset \mathbb{R}^d \backslash \{0\}$. Then

$$u - v = P$$

for some polynomial P. *Hint:* Show that if a distribution u is supported at a point, say 0, then it is a finite linear combination of $\partial^\alpha \delta_0$. Use the fact that the continuity of a linear functional can be expressed by means of finitely many semi norms (in any Fréchet space). Then show that $\langle u, \varphi \rangle = 0$ if $\partial^\alpha \varphi(0) = 0$ for all $|\alpha| \le N$, where N is large enough. Finally, conclude by observing that a linear functional is a linear combination of a finite family of linear functionals if and only if it vanishes on the common kernel of that family.

The following exercise introduces an important example of a distributional Fourier transform. The functions $|x|^{-\alpha}$ are known as *Riesz potentials.*

Exercise 4.5 For any $0 < \alpha < d$ show that the distributional Fourier transform of $|x|^{-\alpha}$ in $\mathbb{R}^d$ equals $C(\alpha, d)|\xi|^{\alpha - d}$. *Hint:* Show that the Fourier transform equals a smooth function of $\xi \neq 0$. Use dilation and rotational symmetry to determine this smooth function. Nonzero multiplicative constants do not need to be determined. Finally, exclude any contributions from $\xi = 0$; see the previous exercise.

4.1.3. Poisson summation formula

An interesting example of a distributional identity is given by the Poisson summation formula.

Proposition 4.6 *For any $f \in \mathcal{S}(\mathbb{R}^d)$ one has*

$$\sum_{n\in\mathbb{Z}^d} f(n) = \sum_{\nu\in\mathbb{Z}^d} \hat{f}(\nu).$$

Equivalently, with convergence in the sense of $\mathcal{S}'$, one has

$$\sum_{\nu\in\mathbb{Z}^d} e^{2\pi i x\cdot\nu} = \sum_{n\in\mathbb{Z}^d} \delta_n(x),$$

where δ_n denotes the Dirac distribution at n.

Proof Given $f \in \mathcal{S}(\mathbb{R}^d)$, define $F(x) := \sum_{n\in\mathbb{Z}^d} f(x+n)$, which lies in $C^\infty(\mathbb{T}^d)$. The Fourier coefficients of F are, with $e(\theta) = e^{2\pi i\theta}$ as usual,

$$\begin{aligned}\hat{F}(\nu) &= \sum_{n\in\mathbb{Z}^d} \int_{[0,1]^d} f(x+n)e(-x\cdot\nu)\,dx \\ &= \sum_{n\in\mathbb{Z}^d} \int_{[0,1]^d+n} f(x)e(-x\cdot\nu)\,dx = \int_{\mathbb{R}^d} f(x)e(-x\cdot\nu)\,dx \\ &= \hat{f}(\nu).\end{aligned}$$

From Proposition 1.15 we therefore conclude that

$$F(x) = \sum_{\nu\in\mathbb{Z}^d} \hat{f}(\nu)e(x\cdot\nu),$$

which implies the first identity. The second follows simply by using definitions. □

4.1.4. Plancherel theorem

Henceforth, we shall use the notation $\check{f}$ for the integral in (4.3). It is now easy to establish the Plancherel theorem.

Corollary 4.7 *For any $f \in \mathcal{S}(\mathbb{R}^d)$ one has $\|\hat{f}\|_2 = \|f\|_2$. In particular, the Fourier transform extends unitarily to $L^2(\mathbb{R}^d)$.*

Proof We start from the following identity, which is of interest in its own right. For any $f, g \in \mathcal{S}(\mathbb{R}^d)$ one has

$$\int_{\mathbb{R}^d} f(\xi)\hat{g}(\xi)\,d\xi = \int_{\mathbb{R}^d} \hat{f}(x)g(x)\,dx. \tag{4.5}$$

The proof is an immediate consequence of the definition of the Fourier transform and Fubini's theorem. Therefore, one also has

$$\int_{\mathbb{R}^d} \hat{f}(\xi)\overline{\hat{g}(\xi)}\,d\xi = \int_{\mathbb{R}^d} f(x)\hat{\bar{\hat{g}}}(x)\,dx.$$

However,

$$\hat{\bar{\hat{g}}} = \bar{\check{\hat{g}}} = \bar{g}$$

by the previous proposition, whence we obtain

$$\int_{\mathbb{R}^d} \hat{f}(\xi)\overline{\hat{g}(\xi)}\,d\xi = \int_{\mathbb{R}^d} f(x)\bar{g}(x)\,dx,$$

which is equivalent to Plancherel's theorem and the unitarity of the Fourier transform. □

Exercise 4.6 Show that the Schwartz space is an algebra under both multiplication and convolution. In particular, prove that $\widehat{fg} = \hat{f} * \hat{g}$ for any $f, g \in \mathcal{S}(\mathbb{R}^d)$.

We now take another look at the proof of Proposition 4.4. Write

$$\Gamma_\varepsilon(x) := \varepsilon^{-d} e^{-\pi\varepsilon^{-2}|x|^2}.$$

Then we can make use of the identity

$$\int_{\mathbb{R}^d} e^{2\pi i x\cdot\xi}\, e^{-\pi\varepsilon^2|\xi|^2}\, \hat{f}(\xi)\,d\xi = (f * \Gamma_\varepsilon)(x), \tag{4.6}$$

valid pointwise for all Schwartz functions f and any $\varepsilon > 0$. Let $I_\varepsilon(f)$ be the operator defined by (4.6). Since Γ_ε is an approximate identity as $\varepsilon \to 0$, we can make the following useful observation, which parallels results we encountered for Fourier series on the circle.

Corollary 4.8 *For any $1 \le p < \infty$ and $f \in L^p(\mathbb{R}^d)$ one has $I_\varepsilon(f) \to f$ in $L^p(\mathbb{R}^d)$. If $f \in C(\mathbb{R}^d)$ and vanishes at infinity then $I_\varepsilon(f) \to f$ uniformly.*

In particular, one has *uniqueness*.

Corollary 4.9 *Suppose that $f \in L^1(\mathbb{R}^d)$ satisfies $\hat{f} = 0$ everywhere. Then $f = 0$.*

Exercise 4.7 Let μ be a compactly supported measure in $\mathbb{R}^d$. Show that $\hat{\mu}$ extends to an entire function in $\mathbb{C}^d$. Conclude that $\hat{\mu}$ cannot vanish on an open set in $\mathbb{R}^d$.

Finally, we note the following basic estimate, the *Hausdorff–Young* inequality.

Lemma 4.10 *Fix $1 \le p \le 2$. Then one has the property that $\hat{f} \in L^{p'}(\mathbb{R}^d)$ for any $f \in L^p(\mathbb{R}^d)$ and moreover, the bound $\|\hat{f}\|_{p'} \le \|f\|_p$.*

Proof Note that $f \in L^p(\mathbb{R}^d)$ can be written as $f = f_1 + f_2$, where

$$f_1 = f\chi_{[|f|<1]} \in L^2(\mathbb{R}^d), \quad f_2 = f\chi_{[|f|\ge 1]} \in L^1(\mathbb{R}^d).$$

Hence $\hat{f}_1 \in L^2$ is well defined by an L^2 extension of the Fourier transform, and $\hat{f}_2 \in L^\infty$ by integration. The estimate follows by interpolating between $p = 1$ and $p = 2$. □

4.1.5. Sobolev spaces

As in the case of the circle, one can easily express all degrees of smoothness via the Fourier transform. For example, one can do this at the level of the Sobolev spaces H^s and $\dot{H}^s$, which are defined in terms of the norms

$$\|f\|_{H^s} = \left\|\langle\xi\rangle^s \hat{f}\right\|_2, \quad \|f\|_{\dot{H}^s} = \left\||\xi|^s \hat{f}\right\|_2 \tag{4.7}$$

for any $s \in \mathbb{R}$ where $\langle\xi\rangle = \sqrt{1+|\xi|^2}$. These spaces are the completion of $\mathcal{S}(\mathbb{R}^d)$ under the norms in (4.7). For the homogeneous space $\dot{H}^s(\mathbb{R}^d)$ one needs the restriction $s > -d/2$ since otherwise $\mathcal{S}$ is not dense in this space.

Perhaps the most basic question relating to Sobolev spaces concerns their *embedding properties*. An example is given by the following result.

Lemma 4.11 *For any $f \in H^s(\mathbb{R}^d)$ one has*

$$\|f\|_p \le C(s)\|f\|_{H^s(\mathbb{R}^d)} \quad \forall 2 \le p \le \infty,$$

provided that $s > d/2$. In fact $H^s(\mathbb{R}^d) \hookrightarrow C(\mathbb{R}^d) \cap L^p(\mathbb{R}^d)$ for any such s and all $2 \le p \le \infty$.

Proof By Cauchy–Schwarz,

$$\|f\|_\infty \le \|\hat{f}\|_1 \le \|\langle\xi\rangle^{-s}\|_2 \, \|\langle\xi\rangle^s \hat{f}(\xi)\|_2$$

for any $s > d/2$. The continuity of f follows using approximation by Schwartz functions. Interpolation with L^2 implies the lemma. □

By duality one obtains, for any $s < -d/2$,

$$\|f\|_{H^s(\mathbb{R}^d)} \le C(s,d)\|f\|_{L^1(\mathbb{R}^d)}. \tag{4.8}$$

Exercise 4.8 Show that Lemma 4.11 fails at $s = d/2$. *Hint:* Argue by contradiction and use duality. Thus, show that (4.8) cannot hold at $s = -d/2$.

4.1.6. Trace lemma

A simple generalization of Lemma 4.11 is the following result, which is an example of what is called a *trace lemma.*

Lemma 4.12 *Let $V \subset \mathbb{R}^d$ be an affine subspace of dimension $d-k$, where $1 \le k \le d$. Then one has*

$$\|f\|_{L^2(V)} \le C(s)\,\|f\|_{H^s(\mathbb{R}^d)},$$

where $s > k/2$.

Proof This follows by choosing coordinates parallel to V as well as perpendicular to it. For the parallel coordinates one uses Plancherel on V, whereas for the others Cauchy–Schwarz leads to the desired conclusion as in the previous proof. □

An analogous statement also holds when V is not flat but a smooth manifold (or rather, a compact piece thereof).

It is also natural to ask whether the embeddings $\dot{H}^s(\mathbb{R}^d) \hookrightarrow L^p(\mathbb{R}^d)$ exist when $s \ge 0$. Of course, when $s = 0$ this is tautologically true for $p = 2$, but when $s > 0$ we expect an improvement over L^2, as expressed, for example, by L^p with $p > 2$. Note that the corresponding estimate

$$\|f\|_{L^p(\mathbb{R}^d)} \le C\|f\|_{\dot{H}^s(\mathbb{R}^d)} \tag{4.9}$$

is necessarily scaling invariant. Thus, if we replace f by $f(\lambda\cdot)$ then the right-hand side scales as $\lambda^{s-d/2}$ whereas the left-hand side scales as $\lambda^{-d/p}$. Hence we arrive at the relation

$$\frac{1}{2} - \frac{1}{p} = \frac{s}{d}. \tag{4.10}$$

This excludes the case $d/2$, by Exercise 4.8, and leaves open the possibility that for any $s \in (0, d/2)$ and p as in (4.10) the estimate (4.9) holds.

In order to approach (4.9) we first consider $f \in L^2(\mathbb{R}^d)$ with $\mathrm{supp}(\hat{f}) \subset \{R \le |\xi| \le 2R\}$ and some $R > 0$. Then, the following *Bernstein inequality* holds. The basic idea is the same as in Theorem 1.13: functions with bounded

Fourier support are arbitrarily smooth and this smoothness can be quantified in a number of ways, for example on the L^p scale. We remark that (4.11) below is scaling invariant.

Lemma 4.13 *Let $f \in \mathcal{S}(\mathbb{R}^d)$ satisfy* $\mathrm{supp}(\hat{f}) \subset \{|\xi| \le R\}$. *Then*

$$\|f\|_q \le C_d\, R^{d(1/p-1/q)}\, \|f\|_p, \tag{4.11}$$

for $1 \le p \le q \le \infty$.

Proof Since (4.11) is scaling invariant, it suffices to consider $R = 1$. Then one has $\hat{f} = \chi \hat{f}$, where χ is smooth and compactly supported and $\chi = 1$ on the unit ball. It follows that $f = \check{\chi} * f$ and, therefore, by Young's inequality,

$$\|f\|_q \le \|\check{\chi}\|_r \|f\|_p$$

with $1 + 1/q = 1/r + 1/p$, which is the same as $1/p' + 1/q = 1/r$. Note that $1 \le p \le q \le \infty$ guarantees that $1 \le r \le p'$. □

Exercise 4.9 Formulate and prove the general Young's inequality used in the previous proof.

In particular, if $\hat{f}$ is supported on $\{R \le |\xi| \le 2R\}$ then

$$\|f\|_p \le C R^{d(1/2-1/p)} \|f\|_2 = C \|f\|_{\dot{H}^s(\mathbb{R}^d)},$$

with s as in (4.10). This proves (4.9) for such functions, and the challenge is now to obtain it in the general case, which can indeed be done. One writes $f = \sum_{j\in\mathbb{Z}} P_j f$, where $P_j f$ has Fourier support on dyadic shells of size 2^j. While

$$\sum_{j\in\mathbb{Z}} \|P_j f\|_{\dot{H}^s}^2 \simeq \|f\|_{\dot{H}^s}^2$$

by definition, the problem is to show that

$$\|f\|_p^2 \le C \sum_j \|P_j f\|_p^2. \tag{4.12}$$

This is indeed true for finite $p \ge 2$ but some machinery is required to prove it. We shall develop the necessary tools in Chapter 8, which deals with Littlewood–Paley theory. There are alternative routes leading to (4.9), such as fractional integration.

4.2. Method of stationary or nonstationary phases

4.2.1. Fourier transform of surface-carried measures

We now take up a topic that has proven to be very important for various applications, especially in the theory of partial differential equations: the decay properties of measures with smooth compactly supported densities that live on smooth submanifolds of $\mathbb{R}^d$. Let us start with a concrete example, namely the Cauchy problem for the Schrödinger equation,

$$i\partial_t\psi + \Delta\psi = 0, \quad \psi(0) = \psi_0, \tag{4.13}$$

where ψ_0 is a fixed function, say a Schwartz function, and $\psi(0)$ refers to the function $\psi(x,t)$ evaluated at $t = 0$. Applying a Fourier transform converts (4.13) into

$$i\partial_t\hat{\psi}(\xi) - 4\pi^2|\xi|^2\hat{\psi}(\xi) = 0, \quad \hat{\psi}(0) = \widehat{\psi_0},$$

where $\hat{\psi}$ is the Fourier transform of ψ with respect to the space variable alone. The solution to this ordinary differential equation with respect to time is given by

$$\hat{\psi}(\xi,t) = e^{-4it\pi^2|\xi|^2}\widehat{\psi_0}(\xi),$$

whence the actual solution is, after Fourier inversion,

$$\psi(x,t) = \int_{\mathbb{R}^d} e^{2\pi i x\cdot\xi} e^{-4it\pi^2|\xi|^2}\,\widehat{\psi_0}(\xi)\,d\xi. \tag{4.14}$$

This formula is of course very relevant as it gives a solution to (4.13). For our purposes we would like to reinterpret the integral on the right-hand side in the following way. Consider the paraboloid

$$\mathcal{P} := \left\{(\xi, -2\pi|\xi|^2) \,\middle|\, \xi \in \mathbb{R}^d\right\} \subset \mathbb{R}^{d+1}$$

and place the measure μ onto $\mathcal{P}$, according to the formula (for given ψ_0 as above)

$$\mu(d\xi, d\tau) = \widehat{\psi_0}(\xi)\,d\xi,$$

using the relation

$$\int_{\mathbb{R}^{d+1}} F(\xi,\tau)\mu(d\xi,d\tau) = \int_{\mathbb{R}^d} F(\xi, -2\pi|\xi|^2)\,\widehat{\psi_0}(\xi)\,d\xi,$$

which is valid for any continuous and bounded F. Then we see that (4.14) can be interpreted as follows:

$$\psi(x,t) = \int_{\mathbb{R}^{d+1}} e^{2\pi i(x\cdot\xi + t\tau)}\mu(d\xi,d\tau).$$

In other words, the solution is given by the inverse Fourier transform of the measure μ that lives on the paraboloid $\mathcal{P}$. For a variety of reasons it turns out to be of fundamental importance to understand the decay properties of such Fourier transforms. Before investigating this question, we ask the reader to verify a few properties of (4.14).

Exercise 4.10 Let $\psi_0 \in \mathcal{S}(\mathbb{R}^d)$. Prove that (4.14) defines a function $\psi(x,t)$ that is smooth in t and belongs to the Schwartz class for all fixed times. Moreover, show that ψ solves (4.13) and satisfies

$$\|\psi(t)\|_{L^2} = \|\psi_0\|_2 \quad \forall t \in \mathbb{R}.$$

Finally, verify that for every $t > 0$

$$\psi(x,t) = c_d\, t^{-d/2} \int_{\mathbb{R}^d} e^{i|x-y|^2/4t}\, \psi_0(y)\, dy, \tag{4.15}$$

with a suitable constant c_d. Conclude that

$$\|\psi(t)\|_\infty \le c_d\, t^{-d/2} \|\psi_0\|_1$$

for all $t > 0$.

Now let $\mathcal{M} \subset \mathbb{R}^d$ be a smooth manifold of dimension $k < d$ with Riemannian measure μ. Fix some smooth, compactly supported, function ϕ on $\mathcal{M}$. The problem is to describe the asymptotic behavior of the Fourier transform $\widehat{\phi\,\mu}(\xi)$ for large ξ. For example, consider $\mathcal{M}$ to be a plane of dimension $d-1$. By translational and rotational symmetry we may take $\mathcal{M}$ to be $\mathbb{R}^{d-1}$, the subspace given by the first $d-1$ coordinates, and ϕ to be a Schwartz function of $x' := (x_1, \ldots, x_{d-1})$; μ is simply Lebesgue measure on $\mathbb{R}^{d-1}$. Therefore, with $\xi = (\xi', \xi_d)$,

$$\widehat{\phi\mu}(\xi) = \int_{\mathbb{R}^{d-1}} e^{-2\pi i x'\cdot\xi'}\, \phi(x')\, dx',$$

which is evidently a Schwartz function in ξ' but does not depend on ξ_d, whence also it does not decay in that direction. What we have shown is that the Fourier transform of a measure that lives on a plane does not decay in the direction normal to the plane. In contrast with this case, we shall demonstrate that if a hypersurface $\mathcal{M}$ has *nonvanishing Gaussian curvature*, this *guarantees decay* at a universal rate.

To facilitate computations, fix some $\mathcal{M}$ and a measure on it with smooth compactly supported density. Working in local coordinates and fixing the direction in which we wish to describe the decay reduces matters to *oscillatory integrals* of the form

$$I(\lambda) = \int_{\mathbb{R}^d} e^{i\lambda\phi(\xi)} a(\xi)\, d\xi, \tag{4.16}$$

with a compactly supported function $a \in C_0^\infty$ and smooth real-valued phase $\phi \in C^\infty$. The asymptotic behavior of the integral $I(\lambda)$ as $\lambda \to \infty$ is determined by the method of *stationary or nonstationary phases*. This method enables one to distinguish between the cases where ϕ does and does not have a critical point on supp(a).

4.2.2. Nonstationary phase

We begin with the easier case, that of a nonstationary phase. The reader should note that the standard Fourier transform is of this type.

Lemma 4.14 *If* $\nabla\phi \neq 0$ *on* supp(a) *then the integral* (4.16) *decays as*

$$\left| \int_{\mathbb{R}^d} e^{i\lambda\phi(\xi)} a(\xi)\, d\xi \right| \le C(N, a, \phi)\, \lambda^{-N}, \qquad \lambda \to \infty,$$

for arbitrary $N \ge 1$.

Proof Note that the exponential $e^{i\lambda\phi}$ is an eigenfunction of the operator

$$L := \frac{1}{i\lambda} \frac{\nabla\phi}{|\nabla\phi|^2} \nabla \tag{4.17}$$

Therefore we can apply L to the exponential within (4.16) as often as we wish. The adjoint of L is

$$L^* = \frac{i}{\lambda} \nabla \left(\frac{\nabla\phi}{|\nabla\phi|^2} \right).$$

Then, for any positive integer N, one has

$$\begin{aligned} \left| \int_{\mathbb{R}^d} e^{i\lambda\phi(\xi)} a(\xi)\, d\xi \right| &= \left| \int_{\mathbb{R}^d} L^N(e^{i\lambda\phi(\xi)}) a(\xi)\, d\xi \right| \\ &= \left| \int_{\mathbb{R}^d} e^{i\lambda\phi(\xi)} (L^*)^N (a(\xi))\, d\xi \right| \\ &\le \int_{\mathbb{R}^d} |(L^*)^N (a(\xi))|\, d\xi \\ &\le C(N, a, \phi)\lambda^{-N}, \end{aligned}$$

as claimed. □

4.2.3. Stationary phase

The situation changes when ϕ has a critical point inside supp(a). In that case one no longer has arbitrary decay, and it can be very difficult to determine the exact rate. However, if the critical point is *nondegenerate*, i.e., ϕ has a nondegenerate Hessian at that point, then there is a precise answer, as we will now show.

Lemma 4.15 *If* $\nabla\phi(\xi_0) = 0$ *for some* $\xi_0 \in \mathrm{supp}(a)$, $\nabla\phi \neq 0$ *away from* ξ_0, *and the Hessian of* ϕ *at the stationary point* ξ_0 *is nondegenerate, i.e.,* $\det D^2\phi(\xi_0) \neq 0$, *then for all* $\lambda \geq 1$

$$\left| \int_{\mathbb{R}^d} e^{i\lambda\phi(\xi)} a(\xi)\, d\xi \right| \leq C(d, a, \phi)\, \lambda^{-d/2}. \tag{4.18}$$

In fact,

$$\left| \partial_\lambda^k \left[e^{-i\lambda\phi(\xi_0)} \int_{\mathbb{R}^d} e^{i\lambda\phi(\xi)} a(\xi)\, d\xi \right] \right| \leq C(d, a, \phi, k)\, \lambda^{-d/2-k} \tag{4.19}$$

for any integer $k \geq 1$.

Proof Without loss of generality we can assume that $\xi_0 = 0$. By Taylor expansion,

$$\phi(\xi) = \phi(0) + \tfrac{1}{2}\langle D^2\phi(0)\xi, \xi\rangle + O(|\xi|^3).$$

Since $D^2\phi(0)$ is nondegenerate we have $|\nabla\phi(\xi)| \gtrsim |\xi|$ on $\mathrm{supp}(a)$. We split the integral into two parts:

$$\int_{\mathbb{R}^d} e^{i\lambda\phi(\xi)} a(\xi)\, d\xi = I + II$$

where the first part localizes the contribution near $\xi = 0$,

$$I := \int_{\mathbb{R}^d} e^{i\lambda\phi(\xi)} a(\xi) \chi_0(\lambda^{1/2}\xi)\, d\xi,$$

and the second part restores the original integral,

$$II = \int_{\mathbb{R}^d} e^{i\lambda\phi(\xi)} a(\xi)(1 - \chi_0(\lambda^{1/2}\xi))\, d\xi.$$

The first part has the desired decay:

$$|I| \leq C \int_{\mathbb{R}^d} |\chi_0(\lambda^{1/2}\xi)|\, d\xi \leq C\,\lambda^{-d/2}.$$

To bound the second part, we proceed as in the proof of the nonstationary case. Thus, let L be as in (4.17) and note that

$$\left| D^\alpha \left(\frac{\nabla\phi}{|\nabla\phi|^2} \right)(\xi) \right| \leq C_\alpha |\xi|^{-1-|\alpha|}$$

for any multi-index α. Thus, let $N > d$ be an integer. Then the Leibnitz rule yields

$$|II| \le C \int\limits_{[|\xi|>\lambda^{-1/2}]} |(L^*)^N (1 - \chi_0(\lambda^{1/2}\xi))a(\xi)|\, d\xi$$

$$\le C\lambda^{-N} \int\limits_{\lambda^{-1/2}}^{\infty} (\lambda^{N/2} r^{-N} + r^{-2N}) r^{d-1}\, dr \le C\,\lambda^{-d/2},$$

as claimed. The stronger bound (4.19) follows from the observation that

$$|\phi(\xi) - \phi(0)| \le C|\xi|^2$$

and the same argument as in the nonstationary case (with $N > d + 2k$) concludes the proof. □

It is possible to obtain asymptotic expansions of $I(\lambda)$ in inverse powers of large λ. In fact, the estimate provided by the stationary-phase lemma 4.15 is optimal.

Let us now return to the Fourier transform of $d\mu = \phi\, d\sigma$, where σ is the surface measure of a hypersurface $\mathcal{M}$ in $\mathbb{R}^d$ and ϕ is compactly supported and smooth. Denote the unit vector normal to $\mathcal{M}$ at $x \in \mathcal{M}$ by $N(x)$.

Corollary 4.16 *Let $\xi = \lambda\nu$ where $\nu \in S^{d-1}$ and $\lambda \ge 1$. If ν is not parallel to $N(x)$ for any $x \in$ supp(ϕ) then $\hat{\mu} = O(\lambda^{-k})$ as $\lambda \to \infty$ for any positive integer k. Assume that $\mathcal{M}$ has nonvanishing Gaussian curvature on* supp(ϕ). *Then*

$$|\hat{\mu}(\xi)| \le C(\mu, \mathcal{M})\, |\xi|^{-(d-1)/2}$$

for all $\lambda \ge 1$ and all ν. This is optimal if ν belongs to the normal bundle of $\mathcal{M}$ on supp(ϕ).

Proof One has (here and in Corollary 4.17 we drop the factor -2π from the exponents for simplicity)

$$\hat{\mu}(\xi) = \int_{\mathcal{M}} e^{i\lambda\langle\nu,x\rangle}\, \phi(x)\, \sigma(dx).$$

Note that $x \mapsto \langle\nu, x\rangle$ as a function on $\mathcal{M}$ has a critical point at $x_0 \in \mathcal{M}$ if and only if ν is perpendicular to $T_{x_0}\mathcal{M}$, the space tangent to $\mathcal{M}$ at x_0. Lemma 4.14 implies the first statement. For the second statement we note that the critical point at x_0 is nondegenerate if and only if the Gaussian curvature at x_0 does not vanish; thus the bound follows because the phase is stationary. □

Lemma 4.15 in fact allows for a more precise description of $\hat{\mu}$. We will illustrate this by means of $\widehat{\sigma_{S^{d-1}}}(x)$ for large x, where $\sigma_{S^{d-1}}$ is the surface measure of the unit sphere in $\mathbb{R}^d$. This representation is very useful in certain applications where the oscillatory nature of the Fourier transform of the surface measure is needed, and not just its decay.

Note that the function $\widehat{\sigma_{S^{d-1}}}(x)$ is smooth and bounded, so we are interested in its asymptotic behavior for large $|x|$, in terms of both the frequency of any oscillatory factor and the overall magnitude of the function. We need to understand the oscillations in order to bound the derivatives of $\widehat{\sigma_{S^{d-1}}}(x)$. We remark that the following result can also be obtained from the asymptotics of Bessel functions, but we will make no use of such special functions in this book.

Corollary 4.17 *One has the representation*

$$\widehat{\sigma_{S^{d-1}}}(x) = e^{i|x|}\omega_+(|x|) + e^{-i|x|}\omega_-(|x|), \qquad |x| \geq 1, \tag{4.20}$$

where $\omega_\pm$ are smooth and satisfy

$$|\partial_r^k \omega_\pm(r)| \leq C_k\, r^{-(d-1)/2-k} \quad \forall r \geq 1$$

and all $k \geq 0$.

Proof By definition,

$$\widehat{\sigma_{S^{d-1}}}(x) = \int_{S^{d-1}} e^{ix\cdot\xi}\, \sigma_{S^{d-1}}(d\xi),$$

which is rotationally invariant. Therefore we choose $x = (0, \ldots, 0, |x|)$ and denote the integral on the right-hand side by $I(|x|)$. Thus, with $\xi = (\xi_1, \ldots, \xi_d)$,

$$I(|x|) = \int_{S^{d-1}} e^{i|x|\xi_d}\, \sigma_{S^{d-1}}(d\xi).$$

Note that ξ_d, viewed as a function on S^{d-1}, has exactly two critical points, namely the north and south poles

$$(0, \ldots, 0, \pm 1),$$

which are, moreover, nondegenerate; this is illustrated in Figure 4.1, where ξ_d is the height function on the sphere. We therefore partition the sphere into three parts: a small neighborhood around each pole, and the remainder. The Fourier integral $I(|x|)$ then splits into three integrals by means of a smooth partition of unity adapted to these three open sets, and we write accordingly

$$I(|x|) = \sum_{\pm} I_\pm(|x|) + I_{\mathrm{eq}}(|x|).$$

The integral I_{eq} has a nonstationary phase and decays as $|x|^{-N}$ for any N; see Lemma 4.14. However, the integrals $I_\pm$ exactly fit into the framework of Lemma 4.15, and this yields the desired result. Note that one can absorb

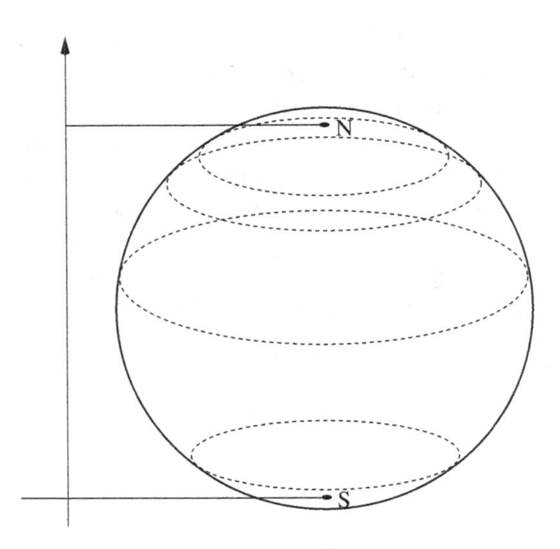

Figure 4.1. The height function on S^2 with its two critical points.

the equatorial piece $I_{\mathrm{eq}}(|x|)$ into either $e^{+i|x|}\omega_+(|x|)$ or $e^{-i|x|}\omega_-(|x|)$ without changing the stated properties of $\omega_\pm$, and (4.20) follows. □

4.3. The Fourier transform on locally compact Abelian groups

4.3.1. Generalities

Fourier transforms on $\mathbb{T}$ or $\mathbb{R}^d$ are special cases of a general construction on locally compact Abelian (LCA) groups, which we now sketch. Let G be an Abelian group as well as a locally compact Hausdorff topological space and assume that the group operations are continuous (we will write the composition on the group using an addition sign). Each such group admits a nonnegative nonzero Borel measure $\mathfrak{m}$ that is translation invariant:

$$\mathfrak{m}(E + x) = \mathfrak{m}(E) \quad \forall \,\text{Borel}\, E,\ x \in G.$$

Moreover, the measure $\mathfrak{m}$ with this property is unique up to a scalar positive multiple. We fix one such measure and define $L^p(G)$ spaces relative to it, $1 \le p < \infty$. If G is compact, one typically requires $\mathfrak{m}(G) = 1$. The measure $\mathfrak{m}$ is then called the *Haar measure*.

As usual, $C_0(G) \subset C(G)$ denotes the continuous functions that vanish at infinity, i.e., the space of all $f \in C(G)$ such that for every $\varepsilon > 0$ there exists a compact $K \subset G$ with $|f(x)| < \varepsilon$ on $G \setminus K$. Furthermore, $C_c(G)$ are the continuous functions of compact support. Both $C_0(G)$ and $C_c(G)$ are dense in $L^p(G)$ for $1 \le p < \infty$.

The convolution operation, given by

$$(f * g)(x) = \int_G f(x-y)g(y)\,\mathfrak{m}(dy) \quad \text{for a.e. } x \in G,$$

is well defined on $L^1(G)$ (by the same Fubini argument as in the Euclidean case); its inclusion turns that space into a *commutative Banach algebra*. The latter refers to any Banach space $\mathbb{A}$ over the complex field with an Abelian multiplication $x \cdot y$ on $\mathbb{A}$ that turns $\mathbb{A}$ into a ring and respects scalar multiplication: $(zx) \cdot y = z(x \cdot y) = x \cdot (zy)$ for all $x, y \in \mathbb{A}$ and $z \in \mathbb{C}$. If G is discrete with $\mathfrak{m}(\{x\}) = 1$ for all $x \in G$ then $L^1(G)$ as a Banach algebra has a unit, namely, δ_0.

The following lemma collects some basic but important properties of integration theory relative to the Haar measure.

Lemma 4.18

(i) *translation is continuous in $L^p(G)$ for $1 \le p < \infty$ and $C_0(G)$;*
(ii) *$f, g \in C_c(G) \Rightarrow f * g \in C_c(G)$;*
(iii) *for any $f \in L^1(G)$ and $g \in L^\infty(G)$, $f * g$ is bounded and uniformly continuous;*
(iv) *Young's inequality, (see Lemma 1.1(ii)) holds;*
(v) *if $1 < p < \infty$ and $f \in L^p(G)$, $g \in L^{p'}(G)$ then $f * g \in C_0(G)$;*
(vi) *if $f \in L^1(G)$ and $\varepsilon > 0$ then there exists a neighborhood U of $0 \in G$ such that $\|f - f * u\|_1 < \varepsilon$ for all $u \ge 0$, $\int u(x)\,\mathfrak{m}(dx) = 1$, $\operatorname{supp}(u) \subset U$.*

Proof We invoke the measure theory fact that $C_c(G)$ is dense in $L^p(G)$ for finite p. Otherwise, the details are similar to those in the proof of Lemma 1.1. □

Definition 4.19 A character on G is a continuous homomorphism $\gamma : G \to \mathbb{T} \simeq S^1$. These characters form a group under pointwise multiplication, called the *dual group* of G. We denote it by Γ and endow it with the topology induced by the characters. A neighborhood basis of the trivial character is given by

$$\{\gamma \in \Gamma \mid \gamma(x) \in U \;\; \forall x \in K\},$$

where U is any neighborhood of $0 \in \mathbb{T}$ and $K \subset G$ is compact. Neighborhood bases around other points are obtained by translation (this is the topology of uniform convergence on compact sets). The Fourier transform of $f \in L^1(G)$ is a function on Γ,

$$\hat{f}(\gamma) = \int_G f(x)\gamma(-x)\,\mathfrak{m}(dx) \quad \forall \gamma \in \Gamma.$$

We write $\widehat{L^1(G)} =: \mathbb{A}(G)$.

In other words, a character is a continuous map $\gamma : G \to \mathbb{C}$ with $|\gamma(x)| = 1$ for all $x \in G$ and such that $\gamma(x+y) = \gamma(x)\gamma(y)$. The trivial character $\gamma(x)$ is 1 for all $x \in G$. From the homomorphism property, $\gamma(0) = 1$ and $\overline{\gamma(x)} = \gamma(-x)$.

Exercise 4.11 Verify the following properties.
(a) For any $f, g \in L^1(G)$ one has $\widehat{f * g} = \hat{f}\hat{g}$.
(b) $\mathbb{A}(G)$ is a subalgebra of $C(\Gamma)$, and $\|\hat{f}\|_{C(\Gamma)} \le \|f\|_{L^1(G)}$ for any $f \in L^1(G)$.
(c) The algebra $\mathbb{A}(G)$ is invariant under translation and multiplication by $\eta(x)$ for any $\eta \in \Gamma$, and $(f * \eta)(x) = \eta(x)\hat{f}(\eta)$.

Examples are of course $\mathbb{T}$, $\mathbb{R}^d$ as well as $\mathbb{T}^d$. Note that for the compact case considered here the dual group is discrete: $G = \mathbb{T}^d$, $\Gamma = \mathbb{Z}^d$. Moreover, the dual of $\mathbb{Z}^d$ is $\mathbb{T}^d$ with Fourier transform

$$\hat{\mathfrak{c}}(x) = \sum_{n \in \mathbb{Z}^d} c_n\, e(n \cdot x),$$

where $\mathfrak{c} = \{c_n\}_{n \in \mathbb{Z}^d} \in \ell^1(\mathbb{Z}^d)$, $x \in \mathbb{T}^d$. Another example is $G = \mathbb{Z}_2$ with the discrete topology. The characters are the trivial character and that defined by $\eta(0) = 1$, $\eta(1) = -1$. In other words, G is self-dual. More generally, any finite cyclic group (with characters defined by the roots of unity), as well as any finite abelian group, is self-dual; the latter follows from the fact that such groups are direct sums of cyclic groups. Another basic example is given by the infinite direct products $\prod_j G_j$ of compact groups G_j with the product topology. By the Tychonoff theorem, this is a compact group. Amongst those we single out the case $\mathbb{T}^\infty := \prod_{n \in \mathbb{N}} \mathbb{T}$, since it will appear later, in the proof of Theorem 6.21. Denoting points in $\mathbb{T}^\infty$ by $\underline{\theta} = \{\theta_n\}_{n \ge 1}$, it is clear that

$$\underline{\theta} \mapsto \prod_{j=1}^{J} e(\ell_i \theta_{n_i}), \quad 1 \le n_1 < n_2 < \cdots < n_J, \quad \ell_i \in \mathbb{Z} \setminus 0, \tag{4.21}$$

are characters that form a subgroup Γ_0 of the dual group $\mathbb{Z}^\infty$ of $\mathbb{T}^\infty$. We shall now argue that these characters make up the entire dual group. This requires some simple general facts. The first is that distinct characters are orthogonal for compact groups.

Lemma 4.20 *Let G be compact. For any distinct $\gamma, \eta \in \Gamma$ one has $\int_G \gamma\, \overline{\eta}\, d\mathfrak{m} = 0$.*

Proof If γ is not the trivial character then let $x_0 \in G$ satisfy $\gamma(x_0) \neq 1$. Then

$$\int_G \gamma(x)\, \mathfrak{m}(dx) = \int_G \gamma(x + x_0)\, \mathfrak{m}(dx) = \gamma(x_0) \int_G \gamma(x)\, \mathfrak{m}(dx),$$

whence $\int_G \gamma(x)\, \mathfrak{m}(dx) = 0$. Now apply this case to $\gamma\overline{\eta}$. □

The second fact is that the dual group of a compact group is discrete. Indeed, $1 \in L^1(G)$ and $\hat{1}(\gamma)$ equals 1 for the trivial character $\gamma = 0$ and 0 otherwise. Since, however, $\hat{1} \in C(\Gamma)$ by Exercise 4.11, it follows that $\{0\}$ is open and, by translation invariance, that Γ carries the discrete topology.

Returning to the dual group of $\mathbb{T}^\infty$, we note that Trig(Γ_0), the trigonometric polynomials generated from all functions in (4.21), is a self-adjoint subalgebra of $C(\mathbb{T}^\infty)$ (i.e., it is invariant under complex conjugation), that it separates points, and that it contains all constants. By the *Stone–Weierstrass theorem*, Trig(Γ_0) is dense in $C(\mathbb{T}^\infty)$ and therefore also in Γ, since the latter carries the topology induced by $C(\mathbb{T}^\infty)$. However, as Γ is discrete, we conclude that each $\gamma \in \Gamma$ equals some element of Trig(Γ_0). By the homomorphism property, one now infers that $\gamma \in \Gamma_0$ as claimed.

An important question at this point is whether *nontrivial characters exist*, or even more pressingly, whether the subalgebra Trig(G) of $C(G)$ consisting of *trigonometric polynomials* $\sum_n a_n \gamma_n$ (a finite sum with $a_n \in \mathbb{C}$ and $\gamma_n \in \Gamma$) separates points. If this is so then, by the Stone–Weierstrass theorem, Trig(G) is in fact dense in $C(G)$ if G is compact. This is clearly a rather nontrivial question, and is at the same level of difficulty as the inversion theorem for the Fourier transform. By Exercise 4.11, the map $f \mapsto \hat{f}(\gamma)$ on $L^1(G)$ for arbitrary but fixed γ is a continuous multiplicative linear functional $L^1(G) \to \mathbb{C}$. We shall now show the converse, i.e., that all such functionals arise in this fashion. The importance of this lies in the fact that by Gelfand's theory of maximal ideals we shall be able to show that there are many such functionals, whence Γ will be seen to be sufficiently "rich" to allow for the Fourier inversion and Plancherel theorems to hold in this degree of generality. To draw an analogy with a simpler fact, consider the dual X^* of an arbitrary Banach space X. It is of course essential to know that there are "many" bounded linear functionals, in other words, that X^* is "rich enough". This is precisely the content of the Hahn–Banach theorem, which states that there are enough bounded functionals $x^* \in X^*$ to separate any point $x_0 \in X$ from a compact, convex, $K \subset X$ with $x_0 \notin K$. Here we cannot rely on the Hahn–Banach theorem, owing to the multiplicative structure that is superimposed over the linear one, but we shall find an analogue in the theory of maximal ideals (as in the case of Hahn–Banach one needs to invoke Zorn's lemma, or Hausdorff's maximality principle, in order to obtain the main existence theorem for multiplicative functionals).

Theorem 4.21 *Let $\varphi : L^1(G) \to \mathbb{C}$ be a multiplicative nonzero linear functional on $L^1(G)$, i.e., φ is a linear functional such that $\varphi(f * g) = \varphi(f)\varphi(g)$. Then there exists $\gamma \in \Gamma$ with $\varphi(f) = \hat{f}(\gamma)$ for all $f \in L^1(G)$.*

Proof We shall show later that $\|\varphi\| = 1$ for any such functional. By duality, there exists $\phi \in L^\infty(G)$ with $\|\phi\|_\infty = 1$ and

$$\varphi(f) = \int_G f(x)\phi(x)\,\mathfrak{m}(dx) \quad \forall f \in L^1(G).$$

Since, for any $f, g \in L^1(G)$,

$$\begin{aligned}\varphi(f * g) &= \int_{G\times G} f(x)g(y)\phi(x+y)\,(\mathfrak{m}\otimes\mathfrak{m})(d(x,y))\\ &= \int_{G\times G} f(x)g(y)\phi(x)\phi(y)\,(\mathfrak{m}\otimes\mathfrak{m})(d(x,y)) = \varphi(f)\varphi(g)\end{aligned}$$

and since, moreover, *tensor functions* are dense in $L^1(G \times G)$ (this can be seen via the Stone–Weierstrass theorem), it follows that

$$\phi(x+y) = \phi(x)\phi(y) \quad \text{for almost all } (x,y) \in G \times G. \tag{4.22}$$

A tensor function embraces all possible linear combination of functions of the form $f(x)g(y)$. This implies that $\varphi(f_y) = \phi(y)\varphi(f)$, where $f_y(x) := f(x-y)$. Thus $\phi(y)$ can be taken to be continuous, whence, from (4.22) and $|\phi| \le 1$, one concludes that $\phi \in \Gamma$. Indeed, from (4.22) and $\|\phi\|_\infty = 1$ we infer that $|\phi| = 1$ everywhere on G. □

We shall now construct sufficiently many multiplicative functionals, as in Theorem 4.21, and thus also sufficiently many characters.

4.3.2. Commutative Banach algebras

We will limit our treatment of this important topic to the most basic results needed for the harmonic analysis considered in this section. We have already defined a commutative Banach algebra $\mathbb{A}$ and will now introduce a multiplicative unit, denoted by $\mathfrak{e}$. Such a unit is necessarily unique. This of course causes an immediate problem for $L^1(G)$ if, for example, $G = \mathbb{T}$, since $L^1(\mathbb{T})$ does not contain a unit under convolution. However, we can fix this formally by passing to the larger algebra $\tilde{\mathbb{A}} := \mathbb{A} \times \mathbb{C}$, which has an evident linear structure and the multiplication $(a, z) \cdot (b, \zeta) := (a \cdot b + a\zeta + bz, z\zeta)$. The norm is $\|(a, z)\| := \|a\| + |z|$. Then $\mathfrak{e} = (0, 1)$ is a unit in $\tilde{\mathbb{A}}$. In the case of $L^1(G)$, this amounts to adjoining δ_0 to $L^1(G)$ and considering the algebra generated by these elements in $\mathcal{M}(G)$, the convolution algebra of measures on G.

So, henceforth, we take $\mathbb{A}$ to be commutative with unit $\mathfrak{e}$. It is then natural to study the invertible elements, which are all $a \in \mathbb{A}$, so that for some $b \in \mathbb{A}$ one has $ab = ba = \mathfrak{e}$ (where we write $ab = a \cdot b$). A few simple but important facts follow.

- The inequality $\|a\| < 1$ implies that $\mathfrak{e} - a$ is invertible, since $(\mathfrak{e} - a)^{-1} = \sum_{n=0}^{\infty} a^n$ (with $a^0 := \mathfrak{e}$).
- That a is invertible implies that $a - b$ is invertible if $\|b\|\|a^{-1}\| < 1$.
- The set $\rho(a) := \{z \in \mathbb{C} \mid z\mathfrak{e} - a \text{ invertible}\}$ is open and the *spectrum* $\sigma(a) := \mathbb{C} \setminus \rho(a)$ is compact and nonempty. In fact, $\sigma(a) \subset \{|z| \leq \|a\|\}$ follows from the fact that $\mathfrak{e} - z^{-1}a$ is invertible for any $|z| > \|a\|$. That $\sigma(a) \neq \emptyset$ follows from Liouville's theorem since otherwise $z \mapsto L((z\mathfrak{e} - a)^{-1})$ would be entire and bounded for every bounded linear functional L on $\mathbb{A}$, whence $L((z\mathfrak{e} - a)^{-1})$ would be zero for all L and all z, which is impossible.

We now turn to multiplicative linear functionals on $\mathbb{A}$. This of course refers to any linear functional φ on $\mathbb{A}$ with $\varphi(ab) = \varphi(a)\varphi(b)$. We only consider functionals φ that do not vanish identically. It is customary to call such φ *homomorphisms*.

Lemma 4.22 *Any homomorphism φ is bounded with $\|\varphi\| = 1$ and $\varphi(\mathfrak{e}) = 1$. Moreover, if a is invertible then $\varphi(a) \neq 0$.*

Proof From $\varphi(a) = \varphi(a\mathfrak{e}) = \varphi(a)\varphi(\mathfrak{e})$ we see that $\varphi(\mathfrak{e}) = 1$, since $\varphi(a) \neq 0$ for some a. Furthermore, if a is invertible then $1 = \varphi(a)\varphi(a^{-1})$ implies that $\varphi(a) \neq 0$. In particular, if $|z| > \|a\|$, where $a \in \mathbb{A}$ is arbitrary but fixed, then $\varphi(z\mathfrak{e} - a) = z - \varphi(a) \neq 0$, whence $|\varphi(a)| \leq \|a\|$. □

Let us return to the proof of Theorem 4.21, where $\|\varphi\| \leq 1$ was used. This follows from Lemma 4.22, but if $L^1(G)$ does not contain a unit at the outset then we need to adjoin the unit δ_0 in order to apply the lemma as above. It is easy to see that this construction does not affect the desired bound. Indeed, let φ be as in Theorem 4.21 and define $\tilde{\varphi}((a, z)) = \varphi(a) + z$ on the augmented algebra. Then $|\varphi(a)| = |\tilde{\varphi}((a, 0))| \leq \|a\|$. Moreover, from Lemma 4.18(vi) one obtains $\|\varphi\| = 1$.

The following is the main result on Banach algebras that we will prove here.

Theorem 4.23 *An element $a \in \mathbb{A}$ is invertible if and only if $\varphi(a) \neq 0$ for every homomorphism φ.*

Proof By Lemma 4.22 it suffices to show that for any noninvertible $a \in \mathbb{A}$ we can find some homomorphism φ such that $\varphi(a) = 0$. If there is such a φ then $I := \ker(\varphi)$ is a *proper ideal*, i.e., a subspace of $\mathbb{A}$ which is strictly smaller than $\mathbb{A}$ and for which $I\mathbb{A} = I$. Moreover, it has to be maximal since I has codimension 1. The term maximal refers here to the property that I is not contained in another proper ideal I'.

Consider the ideal $I(a) := a\mathbb{A}$ (which is proper since a is noninvertible) and all proper ideals containing it. This set is partially ordered and, by the Hausdorff maximality theorem, it possesses a maximally linearly ordered subset. Taking the union then yields a maximal proper ideal I (the fact that it is proper hinges on the fact that $\mathfrak{e}$ is not contained in any of the ideals).

Next, $\mathbb{A}/I$ with the quotient structure is another commutative Banach algebra with a unit, and the maximality of I implies that every nonzero element of $\mathbb{A}/I$ is invertible. However, since the spectrum of any element in $\mathbb{A}/I$ is nonempty we can see that $\mathbb{A}/I \simeq \mathbb{C}$. Indeed, denoting the cosets of $\mathbb{A}/I$ by $[a]$, for every such $[a]$ and for some $z \in \mathbb{C}$ one has that $z[\mathfrak{e}] - [a]$ is noninvertible and thus equal to $[0]$. Therefore the natural projection map $\mathbb{A} \to \mathbb{A}/I$ is the required homomorphism. □

As an immediate corollary, we note the following.

Corollary 4.24 *For any $a \in \mathbb{A}$ one has that*

$$\sigma(a) = \{\varphi(a) \mid \varphi \text{ is a homomorphism on } \mathbb{A}\}.$$

In particular,

$$\sigma(p(a)) = p(\sigma(a)) \quad \text{for every } p \in \mathbb{C}[z].$$

Proof The first part follows from the fact that $\varphi(z\mathfrak{e} - a) = z - \varphi(a) = 0$ for some φ if and only if $z\mathfrak{e} - a$ is noninvertible. For the second part we use $\varphi(p(a)) = p(\varphi(a))$. □

Another important relation is the following formula, which allows for a computation of the so-called spectral radius.

Proposition 4.25 *For any $a \in \mathbb{A}$,*

$$r(a) := \sup\left\{|z| \big| z \in \sigma(a)\right\} = \lim_{n\to\infty} \|a^n\|^{1/n} = \inf_{n\geq 1} \|a^n\|^{1/n}.$$

Proof The second identity is a general property of subadditive sequences, of which $\log\|a^n\|$ is an example. In particular, the limit exists. Clearly $r(a) \leq \|a\|$. By Corollary 4.24, for any $n \geq 1$, we have $(r(a))^n = r(a^n) \leq \|a^n\|$, whence $r(a) \leq \|a^n\|^{1/n}$. This gives one half of the statement. For the other half, consider the analytic function $\Phi(z) := (\mathfrak{e} - za)^{-1}$ on the set

$$\frac{1}{\rho(a)} \cup \{0\}.$$

The radius of convergence of the power series around $z = 0$ is $r(a)^{-1}$ and also equals $\lim_{n\to\infty} \|a^n\|^{-1/n}$, as claimed. □

We can now draw the conclusion given in the next theorem for the Fourier transform on an LCA group G. This theorem is a statement to the effect that there are sufficient characters in G. It will play a key role in the inversion theorem in subsection 4.3.3. Note that the difficult half of the following result is the upper bound.

Theorem 4.26 *Let $f \in L^1(G)$. Then, with $f^n := f * f * \cdots * f$ the n-fold convolution of f with itself, one has*

$$\lim_{n\to\infty} \|f^n\|^{1/n} = \sup\left\{|\hat{f}(\gamma)| \,\big|\, \gamma \in \Gamma\right\}, \tag{4.23}$$

where Γ is the dual group.

Proof As discussed above, we need to augment the convolution algebra $L^1(G)$ by adjoining to it the element δ_0. We leave the reader to check that this does not affect the statement in any way. By Proposition 4.25 the left-hand side equals $r(f)$, the spectral radius. By Corollary 4.24 one has that

$$r(f) := \sup\left\{|\varphi(f)| \,\big|\, \varphi \text{ is a homomorphism on } L^1(G)\right\}.$$

However, by Theorem 4.21 the right-hand side is the same as that of (4.23), whence the claim. □

Let us conclude with two examples.

(1) The complex-valued continuous functions $C(\mathbb{T}^d)$ form an algebra with a unit (in fact they do this on any compact Hausdorff space X). The maximal ideals are easily seen to be $\mathcal{M} = \{f \in C(\mathbb{T}^d) \mid f(x_0) = 0\}$ for some $x_0 \in \mathbb{T}^d$. Indeed, suppose that $\mathcal{M}$ is an ideal such that for every $x_0 \in \mathbb{T}^d$ one has $f(x_0) \neq 0$ for some $f \in \mathcal{M}$. Then, by multiplying by a bump function, we may conclude that for every $x_0 \in \mathbb{T}^d$ there exists a function $f \in \mathcal{M}$ such that $\operatorname{Re} f > 0$ near x_0. By compactness, we can take a finite collection of these such that the sum has a strictly positive real part. Hence $\mathcal{M}$ contains a function with strictly positive real part and which is therefore invertible, whence the claim. This shows that all the homomorphisms on $\mathbb{A}(\mathbb{T})$ are given by evaluations $f \mapsto f(x_0)$.

(2) Consider the subalgebra $\mathbb{A}(\mathbb{T}^d)$ of $C(\mathbb{T}^d)$ comprising functions with $\sum_{n\in\mathbb{Z}^d} |\hat{f}(n)| =: \|f\| < \infty$ (the Wiener algebra). One may check that $\|fg\| \le \|f\|\|g\|$. Suppose that φ is a homomorphism on $\mathbb{A}(\mathbb{T}^d)$. Clearly, $e(\theta) \in \mathbb{A}$ and $\varphi(e(\theta))\varphi(e(-\theta)) = 1$. In conjunction with $|\varphi(e(\theta))| \le 1$ this implies that $|\varphi(e(\theta))| = 1$ or, in other words, $\varphi(e(\theta)) = e(\omega)$ for some $\omega \in \mathbb{R}$. But then $\varphi(e(n\theta)) = e(n\omega)$ and $\varphi(p) = p(\omega)$ for any trigonometric polynomial p on $\mathbb{T}^d$. As these polynomials are dense in $\mathbb{A}(\mathbb{T}^d)$, it follows that all homomorphisms are given by evaluation at a point. Hence, we obtain *Wiener's theorem* directly from Theorem 4.23, as follows.

Corollary 4.27 *Suppose that $f \in C(\mathbb{T}^d)$ has an absolutely convergent Fourier series. If f does not vanish anywhere, then $1/f$ also has an absolutely convergent Fourier series.*

4.3.3. Inversion and Plancherel

We shall now apply these results on Banach algebras to the Fourier transform on LCA groups. First, we return to the topology on the dual group and note the following. For a Banach algebra $\mathbb{A}$, let $\Delta \subset \mathbb{A}^*$ (the latter being the dual Banach space of the Banach space $\mathbb{A}$) denote all complex homomorphisms, endowed with the weak-$*$ topology. By Alaoglu's theorem, the unit ball in $\mathbb{A}^*$ is weak-$*$ compact, and one can check that $\Delta \cup \{0\}$ is weak-$*$ closed, whence also compact. It follows that Δ is locally compact. In the following exercise we ask the reader to complete this topological picture (the impatient reader may take this for granted).

> **Exercise 4.12** Show that $\Delta \cup \{0\}$ is weak-$*$ closed and that Δ is locally compact as well as Hausdorff. Show that the topology introduced earlier on Γ coincides with the weak-$*$ topology. Then verify that Γ is an LCA group. Finally, show that $\mathbb{A}(G)$ as defined above is dense in $C_0(\Gamma)$. *Hint:* See Rudin [97, pp. 9,10, 262].

The key to the inversion theorem for the Fourier transform is provided by *positive definite functions*. We encountered this concept before, in the proof of Herglotz's theorem; see Problem 2.7. The following definition coincides with that of Problem 2.7 for $G = \mathbb{Z}$.

Definition 4.28 A function $\psi : G \to \mathbb{C}$ is called *positive definite* if

$$\sum_{j,k=1}^{n} z_j \overline{z_k}\, \psi(x_j - x_k) \geq 0 \tag{4.24}$$

for all $\{z_j\}_{j=1}^n \in \mathbb{C}$ and $\{x_j\}_{j=1}^n \in G$.

There are two canonical examples of such sequences.

- For $f \in L^2(G)$, set $\psi := f * \tilde{f}$ where $\tilde{f}(x) = \overline{f(-x)}$. Then

$$\sum_{j,k=1}^{n} z_j \overline{z_k}\, \psi(x_j - x_k) = \int_G \left| \sum_{j=1}^{n} z_j f(x - x_j) \right|^2 \mathfrak{m}(dx) \geq 0. \tag{4.25}$$

  Note also that $\psi \in C(G)$.
- If $\mu \in \mathcal{M}(\Gamma)$ is a complex measure then $\psi(x) := \int_\Gamma \gamma(x)\, \mu(d\gamma)$ is positive definite. Indeed,

$$\sum_{j,k=1}^{n} z_j \overline{z_k}\, \psi(x_j - x_k) = \left| \sum_{j=1}^{n} \int_\Gamma z_j \gamma(x_j)\, \mu(d\gamma) \right|^2 .$$

  Again $\psi \in C(G)$.

In analogy with Herglotz's theorem, we will now show that every positive definite function is exactly as in the second example above. This is *Bochner's theorem.*

Theorem 4.29 *Let $\psi \in C(G)$ be positive definite. Then there exists $\mu \in \mathcal{M}(\Gamma)$ with*

$$\psi(x) = \int_\Gamma \gamma(x)\,\mu(d\gamma) \tag{4.26}$$

for all $x \in G$. The measure μ with this property is unique.

Proof It follows from an application of (4.24) to sequences of length 2 that $\overline{\psi(x)} = \psi(-x)$ for all $x \in G$ and $|\psi(x)| \le \psi(0)$. We normalize in such a way that $\psi(0) = 1$. Define a functional on $L^1(G)$

$$T_\psi(f) = \int_G f(x)\psi(x)\,\mathrm{m}(dx).$$

Then by (4.24), approximating by functions in $C_c(G)$,

$$T_\psi(f * \tilde{g}) = \int_{G\times G} f(x)\overline{g(y)}\psi(x-y)\,(\mathrm{m}\otimes\mathrm{m})(d(x,y)) =: [f,g].$$

By a suitable choice of ψ one has $[f,f] \ge 0$. Thus $[f,g]$ obeys the Cauchy–Schwarz inequality: $|[f,g]|^2 \le [f,f][g,g]$ for all $f, g \in L^1(G)$. We claim that, for all $f \in L^1(G)$,

$$|T_\psi(f)|^2 \le [f,f]. \tag{4.27}$$

Fix $f \in L^1(G)$ and let u_ε be a sequence as in Lemma 4.18(vi). Then

$$|T_\psi(f * \tilde{u}_\varepsilon)|^2 \le [u_\varepsilon, u_\varepsilon][f,f] = T_\psi(u_\varepsilon * \tilde{u}_\varepsilon)[f,f]$$

and (4.27) follows on taking the limit $\varepsilon \to 0$ (since $\psi(0) = 1$). Set $h := f * \tilde{f}$ and $h_n := h_{n-1} * h$, for $n \ge 2$. Then iterating (4.27) yields

$$|T_\psi(f)|^2 \le T_\psi(h) \le \left(T_\psi(h_{2^n})\right)^{2^{-n}} \le \|h_{2^n}\|_1^{2^{-n}}$$

for all $n \ge 1$. By (4.23) we conclude that, as $n \to \infty$,

$$|T_\psi(f)| \le \|\hat{f}\|_\infty \quad \forall f \in L^1(G). \tag{4.28}$$

This is the key step in the proof. Note that it expresses that there are sufficient characters for information on functions on G to be obtained. By (4.28), T_ψ can be regarded as a bounded linear functional on $\mathbb{A}(\Gamma) \subset C_0(\Gamma)$ (see Exercise 4.12). It is well defined since $\hat{f} = 0$ with $f \in L^1(G)$ implies that

$T_\psi(f) = 0$. However, we cannot conclude yet that $f = 0$. Extend T_ψ to a functional on $C_0(\Gamma)$ with the same norm. Then, by the Riesz representation theorem, there exists $\mu \in \mathcal{M}(\Gamma)$ with $\|\mu\| \leq 1$ and

$$T_\psi(f) = \int_\Gamma \hat{f}(-\gamma)\,\mu(d\gamma) = \int_G f(x) \int_\Gamma \gamma(x)\,\mu(d\gamma)\mathfrak{m}(dx) \quad \forall f \in L^1(G).$$

Hence (4.26) holds for a.e. $x \in G$. By continuity it holds for all $x \in G$. Setting $x = 0$ therefore yields

$$1 = \psi(0) = \int_\Gamma \mu(d\gamma) = \mu(\Gamma) \leq \|\mu\| = 1,$$

whence $\mu \geq 0$. For the uniqueness, let $f \in L^1(G)$. Then

$$\int_G \psi(x) f(-x)\,\mathfrak{m}(dx) = \int_\Gamma \hat{f}(\gamma)\mu(d\gamma).$$

Since $\mathbb{A}(G)$ is dense in $C_0(\Gamma)$, we are done. □

The importance of this result for Fourier inversion can be seen by means of the first class of examples in the list after Definition 4.28. Indeed, given $f \in L^2(G)$, Theorem 4.29 implies that

$$(f * \tilde{f})(x) = \int_\Gamma \gamma(x)\,\mu(d\gamma).$$

If G is compact then Γ is discrete; multiplying the previous formula by $\eta(-x)$ and then integrating over G yields

$$|\hat{f}(\eta)|^2 = \mu(\{\eta\}) \quad \forall \eta \in \Gamma.$$

In other words, we arrive at an example of Fourier inversion:

$$(f * \tilde{f})(x) = \sum_{\gamma \in \Gamma} \gamma(x) |\hat{f}(\eta)|^2.$$

For the general case, we introduce $B(G)$ as the class of functions that can be written in the form (4.26) with arbitrary $\mu \in \mathcal{M}(\Gamma)$. Since any $\mathcal{M}(\Gamma)$ can be split into real and imaginary parts, which then can each be split into positive and negative measures as in the Jordan decomposition, we see that $B(G)$ is precisely the class of functions that are finite linear combinations of positive definite functions. Furthermore, we define $B^1(G) := L^1(G) \cap B(G)$. The general Fourier inversion theorem now reads as follows.

Theorem 4.30 *For any $f \in B^1(G)$ one has $\hat{f} \in L^1(\Gamma)$ and*

$$f(x) = \int_\Gamma \gamma(x) \hat{f}(\gamma)\,\mathfrak{n}(d\gamma) \tag{4.29}$$

for all $x \in G$. Here $\mathfrak{n}$ is the Haar measure on the LCA group Γ, normalized in a fashion that depends only on the choice of the Haar measure $\mathfrak{m}$ on G.

Proof For every $f \in B^1(G)$ denote by μ_f the unique measure associated with f as in Theorem 4.29. We claim that

$$\hat{f}\,\mu_g = \hat{g}\mu_f \quad \forall f, g \in B^1(G). \tag{4.30}$$

Indeed, with $h \in L^1(G)$ arbitrary,

$$\begin{aligned}((f * g) * h)(0) &= \int_{G\times G} f(x)g(y)h(-x-y)\,(\mathfrak{m}\otimes\mathfrak{m})(d(x,y)) \\ &= \int_{G\times G}\int_\Gamma \gamma(x-y)\,\mu_f(d\gamma)\,g(y)h(-x)\,(\mathfrak{m}\otimes\mathfrak{m})(d(x,y)) \\ &= \int_\gamma \hat{h}(\gamma)\,\hat{g}(\gamma)\,\mu_f(d\gamma).\end{aligned}$$

By symmetry in f and g the latter equals

$$\int_\gamma \hat{h}(\gamma)\,\hat{f}(\gamma)\,\mu_g(d\gamma)$$

and since $\mathbb{A}(G)$ is dense in $C_0(\Gamma)$ we obtain (4.30).

Next, we define a linear functional T on functions in $C_c(\Gamma)$. For any given $\gamma_0 \in \Gamma$ choose some $f \in L^2(G)$ with $\hat{f}(\gamma_0) \neq 0$ and note that $F := f * \tilde{f}$ satisfies $\hat{F} \geq 0$ and $\hat{F}(\gamma) > 0$ near γ_0. Taking finite sums of such functions, there exists $g \in B^1(G)$; in fact, g is positive definite, with $\hat{g} > 0$ on $K := \operatorname{supp}(\psi)$ where $\psi \in C_c(\Gamma)$. Then define

$$T(\psi) = \int_\Gamma \frac{\psi(\gamma)}{\hat{g}(\gamma)}\,\mu_g(d\gamma). \tag{4.31}$$

By (4.30) this does not depend on the specific choice of g, so T is well defined. Moreover $T(\psi) \geq 0$ if $\psi \geq 0$ since μ_g is a positive measure. It is also evident that $T \neq 0$: pick ψ and μ_f with $\int_\Gamma \psi\mu_f \neq 0$; then $T(\psi\hat{f}) = \int_\Gamma \psi\mu_f \neq 0$.

Finally, T is translation invariant, as one can check by taking $f(x) = \gamma(x_0)g(x)$ in (4.31). It follows that

$$T(\psi) = \int_\Gamma \psi(\gamma)\,\mathfrak{n}(d\gamma)$$

for some Haar measure $\mathfrak{n}$ on Γ. If $f \in B^1$ and $\psi \in C_c(\Gamma)$ then

$$\int_\Gamma \psi\mu_f = T(\psi\hat{f}) = \int_\Gamma \psi\hat{f}\mathfrak{n}(d\gamma),$$

whence $\mu_f = \hat{f}\mathfrak{n}$. Since $\mu_f \in \mathcal{M}(\Gamma)$, it follows that $\hat{f} \in L^1(\Gamma)$ and also that (4.29) holds. □

As a simple consequence of Fourier inversion, we invite the reader to check the following.

Exercise 4.13 The trigonometric polynomials are dense in $C(G)$ if G is compact. In particular, if $\hat{f} = 0$ for some $f \in L^1(G)$ then $f = 0$.

Further consequences are the Plancherel and Parseval theorems.

Theorem 4.31 *For any $f \in X := L^1(G) \cap L^2(G)$ one has $\|f\|_{L^2(G)} = \|\hat{f}\|_{L^2(\Gamma)}$. Then $\hat{X}$ is dense in $L^2(\Gamma)$, and the Fourier transform extends uniquely as an isometry from $L^2(G)$ onto $L^2(\Gamma)$. For any $f, g \in L^2(G)$ one has $\langle f, g\rangle = \langle \hat{f}, \hat{g}\rangle$, where $\langle \cdot, \cdot\rangle$ stands for the pairings in $L^2(G)$ and $L^2(\Gamma)$, respectively.*

Proof For $f \in L^2(G)$ set $h := f * \tilde{f}$. Then $h \in B^1(G)$ and, by Theorem 4.30, one has

$$\begin{aligned}\int_G |f(x)|^2\,\mathfrak{m}(dx) &= \int_G f(x)\bar{f}(-x)\,\mathfrak{m}(dx) = h(0)\\ &= \int_\Gamma \hat{h}(\gamma)\,\mathfrak{n}(d\gamma) = \int_\Gamma |\hat{f}(\gamma)|^2\,\mathfrak{n}(d\gamma)\end{aligned}$$

for all $f \in X$. To check that $\hat{X}$ is dense in $L^2(\Gamma)$, suppose that $\chi \perp \hat{X}$ in $L^2(\Gamma)$. Then, since $\hat{f}\chi \in L^1(\Gamma)$ and by the translation invariance of X, one has

$$\int_\Gamma \hat{f}(\gamma)\,\chi(\gamma)\,\gamma(-x)\,\mathfrak{n}(d\gamma) = 0 \quad \forall x \in G,$$

for all $f \in X$. Multiplying by any $g \in L^1(G)$ and integrating yields

$$\int_\Gamma \hat{f}(\gamma)\chi(\gamma)\hat{g}(\gamma)\mathfrak{n}(d\gamma) = 0.$$

Since $\mathbb{A}(\Gamma)$ is dense in $C_0(\Gamma)$ we conclude that $\hat{f}\chi = 0$ on Γ. Again, by translation invariance, this time of Γ, and the continuity of $\hat{f}$, one obtains $\chi = 0$ as desired. □

We have already encountered the compact case $G = \mathbb{T}^d$ as well as the noncompact case $G = \mathbb{R}^d$, with their respective Plancherel theorems. What is apparent in all these examples is *duality*, in the sense that taking duals twice (the character group of the character group) takes one back to the original LCA group. This is a general fact, known as the *Pontryagin duality theorem*, which says that the map taking $x \in G$ onto the character on the dual group Γ of G defined by $\gamma \mapsto \gamma(x)$ is an isomorphism between G and the character group of Γ.

We conclude this section with an example, namely the *Cantor group* $G := \mathbb{Z}_2^{\mathbb{N}}$ where $\mathbb{Z}_2$ is the group of integers mod 2. The Cantor group G endowed with the product topology is a compact group that is metrizable with metric

$$d(\underline{\omega}, \underline{\omega}') := 2^{-n}, \quad n := \min\{k \geq 1 \mid \omega_k \neq \omega'_k\},$$

where $\underline{\omega}, \underline{\omega}' \in G$ are distinct. By inspection, G is homeomorphic to the usual middle-third Cantor set. As in the case of the infinite torus $\mathbb{T}^\infty$, one may verify that the characters of G are finite products of the form $\chi = \prod \omega_{n_j}$, where $\omega_n : G \to \{1, -1\}$ is the projection of $\underline{\omega}$ onto the nth coordinate of $\underline{\omega}$, writing $\mathbb{Z}_2 = \{1, -1\}$. Looking ahead to the next chapter, these characters can be interpreted as products of finitely many *Rademacher functions*, since G can be interpreted as the probability space of the simplest Bernoulli sequence (a sequence of fair coin tosses). One then calls the characters *Walsh functions*, and the Plancherel theorem amounts to the fact that the Walsh functions are an orthogonal basis in $L^2([0, 1])$; that they are complete is not immediately obvious.

Notes

Many important topics are omitted here, such as spherical harmonics. A classic, as well as comprehensive discussion, of the Euclidean Fourier transform is found in the book by Stein and Weiss [112]. Folland [41] and especially Rudin [97] contain more information on the topic of tempered distributions. For Exercise 4.4 see Theorem 6.25 in [97]. A nice classical introduction is given by Chandrasekharan [18]. Some material in this chapter can be found in Wolff's notes [128].

Hörmander [56, Theorem 7.6.6] showed that the Fourier transform on L^p with $p > 2$ takes values in the distributions outside Lebesgue spaces. For more on the topic of Sobolev embeddings and trace estimates see for example the comprehensive treatments in Gilbarg and Trudinger [48], as well as in Evans [34], which do not invoke the Fourier transform. Stein [108] develops Sobolev spaces in the context of the Fourier transform but also uses the "fundamental theorem of calculus" approach. A standard reference for asymptotic expansions in the stationary-phase case is the classic textbook [56].

The whole of Section 4.3 (and much more) can be found in Rudin [95, 97]. It can be skipped on first reading as it is used only once in this book, namely in the proof of Rider's theorem in Chapter 6, where we need the infinite torus and its dual. However, it constitutes an essential part of modern mathematics and is most relevant to representation theory, number theory, etc. For much more on Banach algebras, for the Stone–Weierstrass theorem, and for the Hausdorff maximality theorem see [97]. The route to the inversion and Plancherel theorems via Bochner's classification of positive definite functions is standard, as is reliance on the spectral theorem for commutative Banach algebras (Gelfand theory). Harmonic analysis on LCA groups plays a fundamental role in the representation theory of these groups, just to name one area where it appears; for a recent account see Wolf's treatise [126], which presents harmonic analysis for LCA groups in complete detail. For an introduction to [126] see the final chapter of Dym and McKean [33], as well as the book by Chandrasekharan [19]. Finally, Katznelson's book [65] also contains an introduction to the Fourier transform on LCA groups.

Problems

Problem 4.1 Compute the Fourier transform of the principal value of $1/x$. In other words, determine

$$\lim_{\varepsilon\to 0}\int_{|x|>\varepsilon} e^{-2\pi i x\cdot\xi}\,\frac{dx}{x}$$

for every $\xi \in \mathbb{R}$. Conclude that (up to a normalizing constant) the Hilbert transform is an isometry on $L^2(\mathbb{R})$. This is the analogue of Lemma 3.12 for the line.

Problem 4.2 Using a Fourier transform, solve the boundary-value problem in the upper half-plane for the Laplace equation. Compare your findings with Problem 3.11. Generalize to higher dimensions.

Problem 4.3 Carry out the analogue of Problem 1.13 with $\mathbb{R}$ instead of $\mathbb{T}$ and $\mathbb{R}^d$ instead of $\mathbb{T}^d$ (the case of the Schrödinger equation was addressed in Exercise 4.10). Compute the explicit form of the heat kernel, i.e., the kernel G_t such that $G_t * u_0$ with $t > 0$ is the solution to the Cauchy problem for the heat equation $u_t - \Delta u = 0$, $u(0) = u_0$. Discuss the limit $t \to 0+$.

Problem 4.4 Prove the following generalization of Bernstein's inequality as it appears in Lemma 4.13. If $\operatorname{supp}(\hat{f}) \subset E \subset \mathbb{R}^d$, where E is measurable and f is Schwartz, say, prove that

$$\|f\|_q \le |E|^{1/p-1/q}\|f\|_p \quad \forall\, 1 \le p \le q \le \infty, \tag{4.32}$$

where $|E|$ stands for the Lebesgue measure of E. *Hint:* First handle the case $q = \infty$, $p = 2$ using Plancherel and Cauchy–Schwarz and then dualize and interpolate.

Problem 4.5 Show that if $f \in L^2(\mathbb{R})$ is supported in $[-R, R]$ then $\hat{f}$ extends to $\mathbb{C}$ as an entire function, of exponential growth $2\pi R$. The Paley–Wiener theorem states the converse: if F is an entire function of exponential growth $2\pi R$ and such that $F \in L^2(\mathbb{R})$ then $F = \hat{f}$, where $f \in L^2(\mathbb{R})$ is supported in $[-R, R]$. Prove this via a deformation of the contour, and extend it to higher dimensions.

Problem 4.6 Consider the kernel $K(x, y) = 1/(x + y)$ on $\mathbb{R}^2_+$ and define the operator T by

$$(Tf)(x) := \int_0^\infty \frac{f(y)}{x+y}\,dy \quad \forall x > 0. \tag{4.33}$$

Show that T is bounded on $L^p(\mathbb{R}_+)$ for every $1 < p < \infty$, but not for $p = 1$ or $p = \infty$. *Hint:* Substitute $y = xu$. See also the Chapter 9 problems.

Problem 4.7 The Laplace transform is defined as

$$(\mathcal{L}f)(s) := \int_0^\infty e^{-sx} f(x)\,dx, \qquad s > 0.$$

Show that $\mathcal{L}$ is bounded on $L^2(\mathbb{R}_+)$ but not on any $L^p(\mathbb{R}_+)$ with $p \neq 2$. Verify that $\|\mathcal{L}\|_{2\to 2} = \sqrt{\pi}$. Use this to show that $\|T\|_{2\to 2} = \pi$, where T is as in (4.33). *Hint:* For the L^2-boundedness, write $e^{-sx} = e^{-sx/2}x^{-1/4}e^{-sx/2}x^{1/4}$ and apply Cauchy–Schwarz. For T, consider $\mathcal{L}^2$.

Problem 4.8 With $\mathcal{L}$ the Laplace transform as in the previous problem, find an inversion formula for $\mathcal{L}$. State sufficient conditions under which your inversion formula applies.

Problem 4.9 Establish a representation as in Corollary 4.17 but for the ball instead of the sphere. That is, find the asymptotics as in that corollary for $\widehat{\chi_B}$, where $B \subset \mathbb{R}^d$ is the unit ball with $d \geq 2$. Generalize to the case of ellipsoids, i.e., for

$$\left\{ x = (x_1, \ldots, x_d) \in \mathbb{R}^d \,\middle|\, \sum_{j=1}^{d} \lambda_j^{-2} x_j^2 \leq 1 \right\}$$

where the $\lambda_j > 0$ are constants. Contrast this with the Fourier transform of the indicator function of the cube.

Problem 4.10 Let A be a positive $d \times d$ matrix. Compute the Fourier transform of the Gaussian $e^{-\langle Ax,x\rangle}$. Now do the same for $e^{-i\langle Ax,x\rangle}$ where $A \in GL(d,\mathbb{R})$ (the Fourier transform will need to be taken in the distributional sense).

Problem 4.11 Compute the inverse Fourier transform of $(4\pi^2|\xi|^2 + k^2)^{-1}$ in $\mathbb{R}^3$, where $k > 0$ is a constant. Let this function be denoted $G(x;k)$. Show that the operator

$$(R(k)f)(x) = \int_{\mathbb{R}^d} G(x-y;k)f(y)\,dy, \qquad f \in \mathcal{S}(\mathbb{R}^3),$$

is bounded on $L^2(\mathbb{R}^d)$ and that $R(k)f \in C^\infty(\mathbb{R}^3)$ as well as

$$(-\Delta + k^2)R(k)f = f \quad \forall f \in \mathcal{S}(\mathbb{R}^3).$$

Conclude that $R(k) = (-\Delta + k^2)^{-1}$ for $k > 0$, i.e., that $R(k)$ is the resolvent of the Laplacian.

Problem 4.12 Show that any $\psi : \mathbb{R} \to \mathbb{R}$ that is continuous, nonnegative, and even, and such that on $(0,\infty)$ one has $\psi' < 0$ and $\psi'' > 0$, is positive definite.

Problem 4.13 Let $f \in L^1(\mathbb{R}^d)$. Show that there exists $g \in L^1(\mathbb{R}^d)$ with $(1+\hat{f})(1+\hat{g}) = 1$ if and only if $1+\hat{f}$ does not vanish anywhere on $\mathbb{R}^d$. Give two proofs of this fact:

(a) a proof via the theory of commutative Banach algebras;
(b) A direct proof, just using basic properties of the Fourier transform in $\mathbb{R}^d$ but no Banach algebra theory. *Hint:* See for example the treatment of Wiener's Tauberian theorem in [18].

5

Introduction to probability theory

In this chapter we provide a brief introduction to some basic concepts and techniques of probability theory. This serves two main purposes: first, to develop probabilistic methods used in harmonic analysis and second to establish some level of intuition for probabilistic reasoning of the kind that has proven useful in analysis. In fact, as we shall see later, some ideas in harmonic analysis become transparent only when viewed from a probabilistic angle.

5.1. Probability spaces; independence

To begin, a probability space $(\Omega, \Sigma, \mathbb{P})$ is a measure space with $\mathbb{P}$ a positive measure such that $\mathbb{P}(\Omega) = 1$. Elements $A, B, \ldots$ of the σ-algebra Σ are called *events* and $\mathbb{P}(A)$ is the *probability of event* A. Real- or complex-valued functions that are measurable relative to such a space are called *random variables*. Of central importance is the concept of *independence*: we say that events A, B are independent if and only if $\mathbb{P}(A \cap B) = \mathbb{P}(A)\mathbb{P}(B)$. This is exactly what naive probabilistic reasoning dictates. More generally, finitely many σ-subalgebras Σ_j of Σ are called independent if, for any $A_j \in \Sigma_j$, one has

$$\mathbb{P}\Big(\bigcap_j A_j\Big) = \prod_j \mathbb{P}(A_j).$$

Finitely many random variables $\{X_j\}_j$ are called independent if and only if the σ-algebras $\{X_j^{-1}(\mathcal{B})\}_j$ are independent, where $\mathcal{B}$ is the Borel σ-algebra over the scalars. The pairwise independence of more than two variables is not the same as the type of independence defined above; a typical example of independent random variables is given by a *coin-tossing sequence*, which (at least intuitively) is a sequence obtained by repeatedly tossing a coin that comes up *heads* with

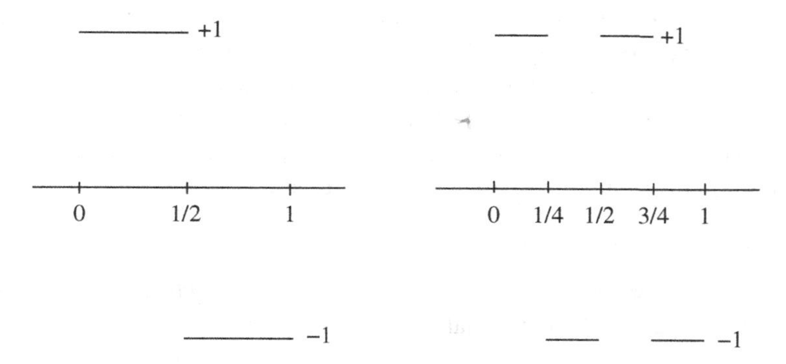

Figure 5.1. The first two Rademacher functions.

probability $p \in (0,1)$ and *tails* with probability $q = 1 - p$. This is also referred to as a *Bernoulli sequence*. A fair coin is one for which $p = q = \frac{1}{2}$. Mathematically, one can easily realize any sequence of this type of length N by means of a product space consisting of elements $\underline{\omega} = \{\omega_j\}_{j=1}^N$ with events $\omega_j \in \{0,1\}$ and

$$\Omega_N := \{0,1\}^N, \quad \mathbb{P}(\underline{\omega}) = \prod_{j=1}^{N} (p\chi_{[\omega_j=0]} + q\chi_{[\omega_j=1]}).$$

Here Σ consists of all subsets of Ω_N and χ_A equals 1 if the event A occurs and 0 otherwise. The random variable that gives the jth outcome of the sequence is simply the respective coordinate map or projection, i.e., $X_j(\underline{\omega}) := \omega_j$. By the definition of the product measure these variables are independent. Another way to encode the same random variables (for a fair coin) is by means of the *Rademacher functions*

$$r_j(t) := \operatorname{sign}(\sin(2\pi\, 2^j t)), \quad 0 < t < 1, \quad j = 0, 1, 2, \ldots$$

The corresponding space is the interval $[0,1]$ with the Lebesgue measure. The reader will have no difficulty verifying that

$$\mathbb{P}\big(\{r_{j_1} = \varepsilon_1, \ldots, r_{j_n} = \varepsilon_n\}\big) = 2^{-n},$$

for all $n \geq 1$ and where $\varepsilon_i = \pm 1$; this means precisely that the r_j are independent random variables. Intuitively, this is clear; see Figure 5.1. By construction, knowing the value of r_j does not allow one to determine the values of any other Rademacher function; what is more, knowing the values of any finite collection of these functions does not determine any of the other functions. It is moreover clear that by splitting intervals according to the ratio $p : q$ one can encode the general coin-tossing sequence. In fact, any random variable that takes only finitely many values gives rise to an infinite sequence of independent variables on $[0, 1]$ by the same construction.

From the point of view of probability theory, the actual probability space on which random variables are defined is completely immaterial. Given a finite collection of real-valued random variables $\{X_j\}_{j=1}^N$, all relevant information is contained in the (joint) *distribution function* of these variables, i.e.,

$$F_N(\lambda_1, \ldots, \lambda_N) := \mathbb{P}(\{X_1 > \lambda_1, \ldots, X_N > \lambda_N\}),$$

where $\lambda_j \in \mathbb{R}$. For example, independence means precisely that the right-hand side is the product of the individual distribution functions:

$$F_N(\lambda_1, \ldots, \lambda_N) := \prod_{j=1}^N \mathbb{P}(\{X_j > \lambda_j\}). \tag{5.1}$$

The abbreviation i.i.d., which stands for *independent identically distributed* refers to any finite or infinite collection of random variables X_j such that the distribution functions of the individual variables are identical and such that (5.1) holds for any finite subcollection of these variables. The point of focusing on the distribution functions lies with the fact that we can always associate a distribution function with a random variable. Indeed, given any non-increasing function $F : \mathbb{R} \to [0, 1]$ that is continuous from the right and such that

$$\lim_{\lambda \to -\infty} F(\lambda) = 1 \quad \text{and} \quad \lim_{\lambda \to +\infty} F(\lambda) = 0,$$

one can define a probability measure on the line $(0, \infty)$ by setting

$$\mathbb{P}((\lambda, \infty)) := F(\lambda).$$

Then the identity on the line is a random variable whose distribution relative to this probability measure equals F. Moreover, to construct any finite i.i.d. sequence with this distribution one takes $\mathbb{R}^N$ with the product measure. Finally, passing to an infinite i.i.d. sequence is more tricky and requires Kolmogorov's theorem on infinite product spaces. We shall take this theorem for granted, especially since for coin-tossing sequences the Rademacher functions allow for an explicit construction.

5.2. Sums of independent variables

Now let $\{X_j\}_{j=1}^\infty$ be an i.i.d. coin-tossing sequence with $\mathbb{P}(\{X_1 = 0\}) = p > 0$, $\mathbb{P}(\{X_1 = 1\}) = q > 0$, where $p + q = 1$. Then we would expect that $n^{-1} \sum_{j=1}^n X_j \simeq q$ for large n. For example, for a fair coin this simply says that heads should come up about half the time. Below we shall rigorously prove such a statement, known as a *law of large numbers*. Before stating this, we remark that it is traditional to call

$$\int_\Omega X \, d\mathbb{P} =: \mathbb{E}\, X$$

the *expectation* of X. It is well defined for any $X \in L^1(\Omega)$. Throughout this chapter, $\|X\|_p^p := \mathbb{E}\,|X|^p$ and $\|X\|_\infty$ are the same as in general measure theory. If F denotes the distribution function of X then this L^1 condition is the same as $\int_{\mathbb{R}} |x| \, F(dx) < \infty$ and $\mathbb{E}\, X = \int x \, F(dx)$. In particular, it has nothing to do with the underlying probability space, and we may simply write $X \in L^1$. Similarly, $X \in L^2$ means that $\int_{\mathbb{R}} x^2 \, F(dx) < \infty$, and the *variance* of $X \in L^2$ is

$$\mathbb{V}X := \mathbb{E}\,|X - \mathbb{E}\, X|^2.$$

Of course, one can speak of L^p-random variables with an obvious meaning, for any $1 \le p \le \infty$. If X, Y are independent L^2 variables then $\mathbb{E}(XY) = \mathbb{E}\, X \, \mathbb{E}\, Y$. This is clear from the definition of independent events if X, Y are indicator functions and therefore, by linearity, also if X, Y are step functions. Finally, the general case follows by approximation. There is an obvious generalization to products of any finite collection of random variables.

In analogy with measure theory, which distinguishes between convergence in measure and convergence almost everywhere, exactly the same notions in probability theory are referred to as convergence in probability and almost sure (abbreviated a.s.) convergence. We now prove the *weak law of large numbers*, admittedly under somewhat strong conditions.

Proposition 5.1 *Let $\{X_j\}_{j=1}^\infty$ be an i.i.d. sequence of L^2 variables. Then $N^{-1}\sum_{j=1}^N X_j \to \mathbb{E}\, X_1$ in probability.*

Proof We may assume that $\mathbb{E}\, X_1 = 0$. Write $S_N := \sum_{j=1}^N X_j$. We need to show that

$$\mathbb{P}(\{|S_N| > \varepsilon N\}) \to 0, \quad N \to \infty,$$

for any $\varepsilon > 0$. This follows easily using the orthogonality of the X_j:

$$\mathbb{E}\,(X_i X_j) = \mathbb{E}\, X_i \mathbb{E}\, X_j = 0 \quad \forall i \ne j.$$

Therefore

$$\begin{aligned}\mathbb{P}\big(\{|S_N| > \varepsilon N\}\big) &\le \varepsilon^{-2}N^{-2}\mathbb{E}\,|S_N|^2 \\ &= \varepsilon^{-2}N^{-2}\left(\sum_{j=1}^{N}\mathbb{E}\,X_j^2 + 2\sum_{1\le j<k\le N}\mathbb{E}\,X_jX_k\right) \qquad (5.2)\\ &= \varepsilon^{-2}N^{-1}\mathbb{E}\,X_1^2,\end{aligned}$$

which concludes the proof. □

An immediate example of the weak law is given by the expansion of a real number $x \in (0, 1)$ with respect to an integer basis $b \ge 2$. Proposition 5.1 implies that if we expand x to n digits then, for most x (up to a set of small measure), each of the b "letters" will appear with frequency about b^{-1}. The strong law will in fact say that for a.e. x, the frequency of each letter converges exactly to b^{-1}.

Exercise 5.1 This exercise presents two less immediate applications of the weak law.

(a) Derive the Weierstrass approximation theorem from Proposition 5.1, as follows. Given any $f \in C([0, 1])$, define the Bernstein polynomial

$$B_f(x) := \sum_{k=0}^{n}\binom{n}{k}x^k(1-x)^{n-k}f\left(\frac{k}{n}\right), \qquad 0 \le x \le 1.$$

Interpret this as $\mathbb{E}\,f(S_n/n)$ for a suitable i.i.d. sequence and use (5.2) to conclude your derivation.

(b) Prove the Shannon–McMillan–Breiman theorem for the i.i.d. case. Let Ω_0 be a set of cardinality r and place a measure on Ω_0 according to $\sum_{j=1}^{r} p_j = 1$ with $p_j > 0$. Define the *entropy* $H := -\sum_{j=1}^{r} p_j \log p_j$ and set $\Omega_N := \Omega_0^N$ with the product measure. Show that, for any small $\varepsilon, \delta > 0$ there exists N_0 such that, for any $N \ge N_0$, there exists a subset $\Omega_N' \subset \Omega_N$ such that

- $\mathbb{P}(\Omega_N') > 1 - \varepsilon$,
- $e^{-N(H+\delta)} < \mathbb{P}(\{\omega\}) < e^{-N(H-\delta)}$ for all $\omega \in \Omega_N'$,
- $e^{N(H-\delta)} < \#(\Omega_N') < e^{N(H+\delta)}$, where the symbol # refers to the cardinality of Ω_N'.

This is referred to as an *equipartition result* since the probabilities are evenly distributed over the set Ω_N' by the second property. Verify that H is maximized in the case where $p_j = r^{-1}$ for every j. This means that the equipartition theorem allows one to "compress" the state space Ω_N if the underlying distribution on Ω_0 is unbalanced.

One would expect that the weak law of large numbers (Proposition 5.1) should hold for L^1 variables, and this is indeed the case as we shall see later. For now, however, we wish to move on to the strong law – again assuming more than needed. We first make a simple observation about an infinite sequence of events $\{A_j\}_{j=1}^{\infty}$: the series $\sum_{j=1}^{\infty} \chi_{A_j}$ counts the number of events occurring. In particular, by the monotone convergence theorem we may write

$$\mathbb{E} \sum_{j=1}^{\infty} \chi_{A_j} = \sum_{j=1}^{\infty} \mathbb{E}\, \chi_{A_j} = \sum_{j=1}^{\infty} \mathbb{P}(A_j).$$

Thus, if the sum on the right-hand side is finite then almost surely (a.s.) $\sum_{j=1}^{\infty} \chi_{A_j}$ is finite. In other words, a.s. at most finitely many of the A_j can occur. This is the easy part of the following *Borel–Cantelli* lemma.

Lemma 5.2 *Let $\{A_j\}_{j=1}^{\infty}$ be an infinite sequence of events. Then*

$$\sum_{j=1}^{\infty} \mathbb{P}(A_j) < \infty \quad \Longrightarrow \quad \mathbb{P}\big(A_j \textit{ occurs infinitely often}\big) = 0.$$

Now assume in addition that the A_j are independent. Then

$$\sum_{j=1}^{\infty} \mathbb{P}(A_j) = \infty \quad \Longrightarrow \quad \mathbb{P}\big(A_j \textit{ occurs infinitely often}\big) = 1.$$

Proof To prove the second part, note that

$$\begin{aligned}
\mathbb{P}\Big(\bigcap_{j=k}^{N} A_j^c\Big) &= \prod_{j=k}^{N}(1 - \mathbb{P}(A_j)) \\
&\leq \prod_{j=k}^{N} \exp(-\mathbb{P}(A_j)) = \exp\Big(-\sum_{j=k}^{N} \mathbb{P}(A_j)\Big) \to 0, \quad N \to \infty.
\end{aligned}$$

This means that

$$\mathbb{P}\Big(\bigcup_{j=k}^{\infty} A_j\Big) = 1 \quad \forall k,$$

and we are done. □

The following exercise uses both statements in the Borel–Cantelli lemma.

Exercise 5.2 Let $\{X_n\}_{n=1}^{\infty}$ be an i.i.d. sequence with $\mathbb{P}(\{X_1 > \lambda\}) = e^{-\lambda}$ for all $\lambda > 0$. Show that $\limsup_{n\to\infty} (\log n)^{-1} X_n = 1$ a.s.

5.2.1. Strong law of large numbers

Now we formulate a version of the *strong law of large numbers*. Note that it applies to coin-tossing sequences.

Proposition 5.3 *Let $\{X_j\}_{j=1}^{\infty}$ be an i.i.d. sequence of L^4 variables. Then $N^{-1}\sum_{j=1}^{N} X_j \to \mathbb{E}\, X_1$ almost surely.*

Proof We can again assume that $\mathbb{E}\, X_1 = 0$. Fix $\varepsilon > 0$ and define

$$A_n := \{|S_n| > \varepsilon n\}.$$

Then

$$\mathbb{P}(A_n) \le \varepsilon^{-4} n^{-4} \mathbb{E}\, |S_n|^4 \le C n^{-2} \varepsilon^{-4}.$$

Here we have used that

$$\mathbb{E}\, |S_n|^4 = \sum_{j_1, j_2, j_3, j_4 = 1}^{n} \mathbb{E}\left(X_{j_1} X_{j_2} X_{j_3} X_{j_4}\right),$$

which can be seen to be $\le Cn^2$ by noting that the largest contribution comes from $j_1 = j_2 \ne j_3 = j_4$ and permutations thereof. But this means that $\sum_n \mathbb{P}(A_n) < \infty$, and we are done by Lemma 5.2. □

Inspection of the proofs of the law of large numbers immediately yield generalizations to sequences that are not necessarily identically distributed. Another generalization pertains to the question whether $X_1 \in L^1$ is a sufficient condition (it is, see below).

For an i.i.d. sequence with $\mathbb{P}(\{X_1 = 1\}) = \mathbb{P}(\{X_1 = -1\}) = \frac{1}{2}$ (such as the Rademacher sequence), the strong law of large numbers states that

$$\frac{1}{N}\sum_{j=1}^{N} X_j \to 0 \quad \text{a.s. for } N \to \infty.$$

In fact, the proof of Proposition 5.3 reveals that in this case

$$\mathbb{P}\big(\{|S_N| > \lambda\sqrt{N}\}\big) \le C\lambda^{-4}, \quad \lambda \ge 1.$$

In fact, repeating the argument with $\mathbb{E}\, |S_N|^{2p}$ for arbitrary positive integers p, one may improve this estimate to $\le C_p \lambda^{-p}$ for any p.

5.2.2. Sub-Gaussian bounds

By working with exponentials instead of powers, we shall now show that the above estimate can be improved dramatically.

Lemma 5.4 *Let r_j be the Rademacher sequence from Section 5.1. For any positive integer N and $\{a_j\}_{j=1}^N \subset \mathbb{C}$, one has*

$$\mathbb{P}\left(\left\{\left|\sum_{j=1}^{N} r_j a_j\right| > \lambda\left(\sum_{j=1}^{N} |a_j|^2\right)^{1/2}\right\}\right) \leq 4e^{-\lambda^2/2} \tag{5.3}$$

for all $\lambda > 0$.

Proof Assume first that $a_j \in \mathbb{R}$. Then

$$\mathbb{E}\, e^{tS_N} = \Pi_{j=1}^N \mathbb{E}\big(e^{tr_j a_j}\big) = \Pi_{j=1}^N \cosh(ta_j),$$

where we have set $S_N = \sum_{j=1}^N r_j a_j$. Now invoke the fact from calculus that

$$\cosh x \leq e^{x^2/2} \quad \forall x \in \mathbb{R}$$

to conclude that

$$\mathbb{E}\, e^{tS_N} \leq \Pi_{j=1}^N e^{t^2 a_j^2/2} = \exp\left(t^2 \sum_{j=1}^{N} \tfrac{1}{2} a_j^2\right).$$

Hence, with $\sigma^2 = \sum_{j=1}^N a_j^2$,

$$\mathbb{P}\big(\{S_N > \lambda\sigma\}\big) \leq e^{t^2\sigma^2/2} e^{-\lambda t\sigma} \leq e^{-\lambda^2/2},$$

where the final inequality follows by minimizing with respect to t, i.e., by setting $t = \lambda/\sigma$. Similarly

$$\mathbb{P}\big(\{S_N < -\lambda\sigma\}\big) \leq e^{-\lambda^2/2},$$

so that

$$\mathbb{P}\big(\{|S_N| > \lambda\sigma\}\big) \leq 2e^{-\lambda^2/2}.$$

The case where $a_j \in \mathbb{C}$ follows by means of a decomposition into real and imaginary parts.

Indeed, squaring the inequality within the set on the left-hand side of (5.3) shows that either the real or imaginary part, or possibly both, satisfy a real-valued estimate, whence the loss of a factor 2 in the upper bound in (5.3). □

This sub-Gaussian estimate on the "tails" of the distribution of the random sum $S_N := \sum_{j=1}^N r_j a_j$ implies that the expressions $(\mathbb{E}\,|S_N|^p)^{1/p}$ are all comparable. This fact, which is known as *Khinchine's inequality*, is extremely useful as we shall see later, for example in Chapter 8. The precise statement is as follows.

Lemma 5.5 *For any $1 \le p < \infty$ there exists a constant $C = C(p)$ such that*

$$C^{-1}\Big(\sum_{j=1}^{N} |a_j|^2\Big)^{p/2} \le \mathbb{E}\Big|\sum_{j=1}^{N} r_j a_j\Big|^p \le C\Big(\sum_{j=1}^{N} |a_j|^2\Big)^{p/2} \tag{5.4}$$

for any choice of positive integer N and $\{a_j\}_{j=1}^N \subset \mathbb{C}$.

Proof We start with the upper bound in (5.4). It suffices to consider the case $\sum_{j=1}^N |a_j|^2 = 1$. Setting $\sum_{j=1}^N a_j r_j = S_N$ one has

$$\begin{aligned}
\mathbb{E}\,|S_N|^p &= \int_0^\infty \mathbb{P}(\{|S_N| > \lambda\})\, p\lambda^{p-1}\, d\lambda \\
&\le \int_0^\infty 4e^{-\lambda^2/2} p\lambda^{p-1}\, d\lambda =: C(p) < \infty.
\end{aligned}$$

For the lower bound it suffices to assume that $1 \le p \le 2$; in fact, that $p = 1$. By Hölder's inequality,

$$\begin{aligned}
\mathbb{E}\,|S_N|^2 = \mathbb{E}|S_N|^{2/3}|S_N|^{4/3} &\le (\mathbb{E}\,|S_N|)^{2/3}(\mathbb{E}\,|S_N|^4)^{1/3} \\
&\le C(\mathbb{E}\,|S_N|)^{2/3}(\mathbb{E}\,|S_N|^2)^{2/3},
\end{aligned}$$

where the final inequality follows from the case $2 \le p < \infty$ just considered. This implies that

$$\mathbb{E}\,|S_N|^2 \le C(\mathbb{E}\,|S_N|)^2,$$

and we are done. □

The explicit form of the Rademacher functions allows for a different proof of Khinchine's inequality, based on expanding and integrating out $\mathbb{E}\,|S_N|^p$ when p is an even integer. We invite the reader to determine the growth (at least an upper bound on the growth) of the constant $C(p)$ in Khinchine's inequality as $p \to \infty$.

Exercise 5.3 By explicit expansion and integration for even integers p, prove that

$$\int_0^1 \Big|\sum_{j=1}^{N} a_j r_j(t)\Big|^p dt \le C(p)\Big(\sum_{j=1}^{N} |a_j|^2\Big)^{p/2}.$$

Then recover the general case of $1 \le p < \infty$. Show that $C(p) \le C_0\sqrt{p}$ with an absolute constant C_0 and that this is optimal.

It is of course natural to ask whether the estimate in Lemma 5.4 is sharp.

5.2.3. Central limit theorem

The following remarkable *central limit theorem* (CLT) says that it is indeed the case that the estimate (5.4) is sharp.

Theorem 5.6 *Let $\{X_j\}_{j=1}^{\infty}$ be i.i.d. L^2 variables with $\mathbb{E}\, X_1 = 0$, $\mathbb{E}\, X_1^2 = 1$. Then for all $\lambda_1 < \lambda_2$ one has*

$$\mathbb{P}\Big(\Big\{\lambda_1\sqrt{n} < \sum_{j=1}^{n} X_j < \lambda_2\sqrt{n}\Big\}\Big) \to (2\pi)^{-1/2}\int_{\lambda_1}^{\lambda_2} e^{-u^2/2}\,du \qquad (5.5)$$

as $n \to \infty$.

In other words, the distribution of the L^2-normalized sums $S_n/\sqrt{n}$ converges to the Gaussian normal distribution, denoted by $N(0, 1)$, which is defined as the right-hand side of (5.5). We remark that $\mathbb{E}\, X_1 = \xi$ and $\mathbb{V} X_1 = \sigma^2$ leads to the CLT with $N(\xi, \sigma)$ defined as the distribution with density

$$\frac{1}{\sqrt{2\pi}\,\sigma}\exp\Big(-\frac{(x-\xi)^2}{2\sigma^2}\Big).$$

To prove the theorem, we invoke the Fourier transform of a real-valued random variable X, defined by $\mathbb{E}\, e^{-i\xi X} = \int e^{-ix\xi}\, F(dx)$, where F is the distribution of X, also known as the *characteristic function* of X. We shall use the fact that

$$\mathbb{E}\, e^{-i\xi(X+Y)} = \mathbb{E}\, e^{-i\xi X}\, \mathbb{E}\, e^{-i\xi Y} \quad \forall \xi \in \mathbb{R}$$

if X and Y are independent.

Exercise 5.4 Let X, Y be independent random variables. Show (directly, without using a Fourier transform) that the distribution of $X + Y$ is the convolution of the distribution of X with that of Y. This also serves to explain that the characteristic function of a sum of independent variables X and Y is the product of the respective characteristic functions of X and Y.

Proof of Theorem 5.6 In order to prove (5.5) it suffices to show that, for any $\phi \in \mathcal{S}(\mathbb{R})$,

$$\mathbb{E}\,\frac{1}{2\pi}\int_{-\infty}^{\infty} e^{i\xi S_n/\sqrt{n}}\,\hat{\phi}(\xi)\,d\xi \to \frac{1}{\sqrt{2\pi}}\int_{-\infty}^{\infty} e^{-x^2/2}\phi(x)\,dx \quad \text{as } n \to \infty. \qquad (5.6)$$

In this proof the Fourier transform is defined as $\hat{\phi}(\xi) = \int_{\mathbb{R}} e^{-ix\xi}\phi(x)\,dx$ and the inversion formula is $\phi(x) = \frac{1}{2\pi}\int_{\mathbb{R}} e^{ix\xi}\hat{\phi}(\xi)\,d\xi$.

The reader is asked to provide the details of this reduction in the following exercise. Let $\mathbb{E}\, e^{i\xi X_1} =: \chi(\xi)$. Then, by independence and in fact by i.i.d.,

$$\mathbb{E}\, e^{i\xi S_n/\sqrt{n}} = \left(\chi\left(\frac{\xi}{\sqrt{n}}\right)\right)^n.$$

Using a Taylor expansion, the right-hand side equals

$$\left(1 - \frac{\xi^2}{2n} + o\big(\xi^2 n^{-1}\big)\right)^n \to e^{-\xi^2/2}$$

uniformly bounded in ξ as $n \to \infty$. Indeed, this follows from setting

$$e^{i\xi X} = 1 + i\xi X - \frac{\xi^2}{2}X^2 - \xi^2 X^2 \int_0^1 (1-t)\big(e^{it\xi X} - 1\big)\, dt,$$

taking expectations, and using the dominated convergence theorem. Therefore, the left-hand side of (5.6) converges to

$$\frac{1}{2\pi}\int_{-\infty}^{\infty} e^{-\xi^2/2}\, \hat{\phi}(\xi)\, d\xi = \frac{1}{\sqrt{2\pi}}\int_{-\infty}^{\infty} e^{-x^2/2}\phi(x)\, dx$$

as $n \to \infty$; the final expression follows by moving the Fourier transform from ϕ to the Gaussian; see (4.5). □

Exercise 5.5 Show that (5.6) implies (5.5).

Needless to say, the CLT has been generalized in many different directions, such as by removing the assumption that the variables are identically distributed or by relaxing the independence assumption. Other important questions concern the rate of convergence (as given by the Berry–Esseen theorem) in the CLT, as well as the effect of letting λ_1, λ_2 vary with n. We shall not pursue these matters here. We do, however, ask the reader to derive one important, as well as simple, extension of Theorem 5.6 to higher dimensions.

Exercise 5.6 Let X_j be L^2 i.i.d. $\mathbb{R}^d$-valued variables with $\mathbb{E}\, X_1 = 0$ and $\mathbb{E}\,(X_1^j X_1^k) =: \Sigma^{jk}$ for all $1 \le j, k \le d$. Show that Σ is a nonnegative matrix (called the *covariance matrix*); assuming that it is positive, prove that $n^{-1/2} S_n$ converges in distribution (the analogue of (5.5)) to the Gaussian distribution in $\mathbb{R}^d$ with density

$$(2\pi)^{-d/2}(\det \Sigma)^{-1/2} \exp\big(-\tfrac{1}{2}\langle \Sigma^{-1}x, x\rangle\big). \tag{5.7}$$

5.2.4. Another look at the law of large numbers

We now return to the law of large numbers.

Exercise 5.7 Prove the weak law of large numbers for L^1 variables. *Hint:* Truncate X_j to $\{|X_j| < \delta n\}$ where $\delta > 0$ needs to be adjusted to the size of ε in the bound on $\mathbb{P}(\{|S_n - n\mathbb{E}\, X_1| > \varepsilon n\})$.

The proof of the strong law in L^1 is more challenging and is based on the following estimate by Kolmogorov.

Lemma 5.7 *Let $\{X_j\}$ be independent L^2 variables with $\mathbb{E}\,X_j = 0$. Then*

$$\mathbb{P}\left(\left\{\max_{1\le \ell\le n} |S_\ell| > \varepsilon\right\}\right) \le \varepsilon^{-2}\sum_{j=1}^{n} \mathbb{E}\,X_j^2 \tag{5.8}$$

for all $\varepsilon > 0$.

Proof We take the variables to be real-valued for simplicity. Let

$$A_\ell := \left\{\max_{1\le j<\ell} |S_j| \le \varepsilon,\ |S_\ell| > \varepsilon\right\}$$

for $1 \le \ell \le n$. Then

$$\begin{aligned}\mathbb{P}\left(\left\{\max_{1\le \ell\le n} |S_\ell| > \varepsilon\right\}\right) &= \sum_{\ell=1}^{n} \mathbb{P}(A_\ell) \le \varepsilon^{-2}\sum_{\ell=1}^{n}\int_{A_\ell} S_\ell^2\, d\mathbb{P} \\ &\le \varepsilon^{-2}\sum_{\ell=1}^{n}\int_{A_\ell} S_n^2\, d\mathbb{P} \le \varepsilon^{-2}\mathbb{E}\,|S_n|^2,\end{aligned}$$

as claimed. To pass to the second line we used

$$\begin{aligned}\mathbb{E}\left(\chi_{A_\ell} S_n^2\right) &= \mathbb{E}\left(\chi_{A_\ell} S_\ell^2\right) + 2\mathbb{E}\left(\chi_{A_\ell} S_\ell(S_n - S_\ell)\right) + \mathbb{E}\left(\chi_{A_\ell}(S_n - S_\ell)^2\right) \\ &\ge \mathbb{E}\left(\chi_{A_\ell} S_\ell^2\right) + 2\mathbb{E}\left(\chi_{A_\ell} S_\ell\right)\mathbb{E}\,(S_n - S_\ell) = \mathbb{E}\left(\chi_{A_\ell} S_\ell^2\right),\end{aligned}$$

since $A_\ell S_\ell$ depends only on the variables X_j for $1 \le j \le \ell$; $S_n - S_\ell$ is independent of these variables. □

Exercise 5.8 Generalize Lemma 5.7 to Hilbert-space-valued variables. To be precise, let the X_n be independent and let them take values in some Hilbert space $\mathcal{H}$. Assume that $\mathbb{E}\,\|X_n\|^2 < \infty$ for each $n \ge 1$ and also that $\mathbb{E}\,X_n = 0$. Prove that

$$\mathbb{P}\left(\left\{\max_{1\le \ell\le n} \|S_\ell\| > \varepsilon\right\}\right) \le \varepsilon^{-2}\sum_{\ell=1}^{n} \mathbb{E}\,\|X_\ell\|^2. \tag{5.9}$$

As an immediate consequence one has the following special case of *Kolmogorov's three-series theorem.*

Corollary 5.8 *Let $\{X_n\}_{n=1}^\infty$ be independent L^2 variables with both $\sum_{n=1}^\infty \mathbb{E}\,X_n$ and $\sum_{n=1}^\infty \mathbb{V}X_n$ convergent. Then $\sum_{n=1}^\infty X_n$ converges almost surely.*

Proof We may assume that $\mathbb{E}\,X_n = 0$ for all n. Then, for any $\varepsilon > 0$, by Lemma 5.7 we have

$$\mathbb{P}\left(\left\{\sup_{N\le n\le m}\left|\sum_{j=n}^{m} X_j\right| > \varepsilon\right\}\right) \le \varepsilon^{-2}\sum_{n=N}^{\infty}\mathbb{V}X_n,$$

which implies that

$$\mathbb{P}\left(\left\{\lim_{N\to\infty}\sup_{N\le n\le m}\left|\sum_{j=n}^{m} X_j\right| > 0\right\}\right) = 0,$$

and we are done. □

A standard example is of course $\sum_{n=1}^{\infty} \pm n^{-\sigma}$, where the signs are i.i.d. with equal probabilities. By Corollary 5.8 this series converges if $\sigma > \frac{1}{2}$. In fact, by the aforementioned three-series theorem, this is also necessary. It is interesting to note at this point that a series $\sum_{n=1}^{\infty} X_n$ of independent random variables either *converges a.s.* or *diverges a.s.* In other words, it is not possible that the series converges with some positive probability other than 1. This is due to another theorem of Kolmogorov, namely his zero–one law for tail events. The latter are defined as follows. Given an infinite sequence X_n defining a series, *a tail event* A is defined as one that is measurable with respect to the intersection $\bigcap_{N=1}^{\infty} \Sigma_N$ of the σ-algebras Σ_N generated by $\{X_n\}_{n=N}^{\infty}$, where $N \ge 1$ is arbitrary. The theorem states that either $\mathbb{P}(A) = 0$ or $\mathbb{P}(A) = 1$. The idea here is simply that A needs to be independent of itself, which yields $\mathbb{P}(A) = \mathbb{P}(A)^2$, whence the claim. This independence is intuitively clear and can be proved as follows. By general measure theory (see the notes at the end of this chapter), for every $\varepsilon > 0$ there exists an A_ε that is measurable with respect to finitely many of the X_n and which is such that $\mathbb{P}(A\Delta A_\varepsilon) < \varepsilon$ (the symmetric difference). However, since by definition A is measurable with respect to the σ-algebra generated by the remaining variables, it follows that A is indeed independent of itself.

5.2.5. The law of the iterated logarithm

Another application of Lemma 5.7 yields the following statement concerning the largest possible deviation of a random sum (*large-deviation estimates*).

Exercise 5.9 Let X_j be i.i.d. L^2 variables with $\mathbb{E}\,X_1 = 0$ and $\mathbb{E}\,X_1^2 = 1$. Then for any $\sigma > \frac{1}{2}$ and with $S_n = \sum_{j=1}^{n} X_j$ as before,

$$\mathbb{P}\left(\left\{\limsup_{n\to\infty} n^{-1/2}(\log n)^{-\sigma}\,|S_n| > 0\right\}\right) = 0. \tag{5.10}$$

In particular, conclude that the strong law of large numbers holds in this context. *Hint:* Partition n into dyadic sizes, i.e., set $n \simeq 2^k$, and apply Kolmogorov's estimate followed by the Borel–Cantelli lemma 5.2.

The estimate (5.10) is not optimal. In fact, from the central limit theorem one would expect that

$$\mathbb{P}(\{n^{-1/2}(2\log\log n)^{-1/2}|S_n| > A\}) \le C(\log n)^{-A^2}.$$

Letting $n = 2^k$ and summing over k reveals a transition at $A = 1$. In fact, the *law of the iterated logarithm* states that

$$\limsup_{n\to\infty} \frac{|S_n|}{\sqrt{2n\log\log n}} = 1 \quad \text{a.s.} \tag{5.11}$$

We shall return to this law later in the chapter. As a final application of Lemma 5.7, we now deduce the strong law from it.

Theorem 5.9 *The strong law of large numbers holds for i.i.d. L^1 variables.*

Proof We make the following preliminary claim: if an arbitrary sequence $\{X_n\}_{n=1}^{\infty}$ of independent L^2 random variables satisfies $\mathbb{E}\,X_n = 0$ and $\sum_{n=1}^{\infty} n^{-2}\mathbb{V}X_n < \infty$ then these variables obey the strong law of large numbers. To verify this, fix $\varepsilon > 0$ and, for any $k \ge 0$, set

$$A_k := \big\{|S_n| > \varepsilon n \ \text{ for some } 2^k \le n < 2^{k+1}\big\}.$$

Then

$$\mathbb{P}(A_k) \le C\varepsilon^{-2}2^{-2k}\sum_{n=1}^{2^{k+1}}\mathbb{E}\,X_n^2,$$

whence

$$\sum_{k=0}^{\infty}\mathbb{P}(A_k) \le C\varepsilon^{-2}\sum_{k=0}^{\infty}2^{-2k}\sum_{n=1}^{2^{k+1}}\mathbb{E}\,X_n^2 \le C\varepsilon^{-2}\sum_{n=1}^{\infty}\frac{\mathbb{V}X_n}{n^2} < \infty.$$

The claim now follows from the Borel–Cantelli lemma. Next, define $Y_n := X_n\,\chi_{[|X_n|<n]}$. Then

$$\sum_{n=1}^{\infty}\mathbb{P}(\{X_n \ne Y_n\}) \le \sum_{n=1}^{\infty}\mathbb{P}(\{|X_1| > n\}) \le \mathbb{E}\,|X_1|,$$

whence a.s. $X_n = Y_n$ for all but finitely many choices of n. Moreover,

$$\frac{1}{N}\sum_{n=1}^{N}\mathbb{E}\,Y_n \to 0 \quad \text{as } N \to \infty$$

and

$$\sum_{n=1}^{\infty} \frac{\mathbb{V} Y_n}{n^2} \le C \sum_{n=1}^{\infty} \sum_{k=0}^{n-1} \frac{(k+1)^2 \mathbb{P}(\{k \le |X_1| < k+1\})}{n^2}$$

$$\le C \sum_{k=0}^{\infty} (k+1) \mathbb{P}(\{k \le |X_1| < k+1\}) \le C \mathbb{E}\, |X_1|,$$

and so the theorem follows by an appeal to the preliminary claim above. □

5.2.6. Stopping times

We conclude this section by generalizing an idea from Lemma 5.7.

Definition 5.10 Let $\{\mathcal{F}_n\}_{n=1}^{\infty}$ be an increasing sequence of σ-algebras (a *filtration*). Then we say that an integer-valued random variable $T \ge 1$ is a *stopping time* relative to this filtration if and only if $\{T = j\} \in \mathcal{F}_j$ for each $j \ge 1$.

For example, suppose that $\{X_n\}_{n=1}^{\infty}$ are random variables and, for each $n \ge 1$, $\mathcal{F}_n$ is generated by $\{X_j\}_{j=1}^{n}$. An example of a stopping time is then given by the first instance, i.e., the smallest $n \ge 1$, at which $S_n \ge c$, say, where c is any constant. It is understood that we consider only instances where this happens a.s. (one can also allow cases in which $T = \infty$, but we avoid that here). The largest $n \le 1000$, say, at which this happens is an example of a variable that is *not* a stopping time. A standard result in this context is the following result, known as *Wald's identity*.

Proposition 5.11 *Suppose that $\{X_n\}_{n=1}^{\infty}$ is an i.i.d. L^1 sequence and let T be an associated stopping time with $\mathbb{E}\, T < \infty$. Then*

$$\mathbb{E}\, S_T = \mathbb{E}\, T\, \mathbb{E}\, X_1.$$

Now suppose that $\mathbb{E}\, X_1 = 0$ and that X_1 is an L^2 variable. Then

$$\mathbb{E}\, S_T^2 = \mathbb{E}\, T\, \mathbb{E}\, X_1^2. \tag{5.12}$$

Proof First, assume that $X_1 \ge 0$. Then, by the monotone convergence theorem,

$$\mathbb{E}\, S_T = \mathbb{E} \sum_{k=1}^{\infty} X_k \chi_{[k \le T]} = \sum_{k=1}^{\infty} \mathbb{E}\, X_1\, \mathbb{P}(\{T \ge k\}) = \mathbb{E}\, X_1\, \mathbb{E}\, T.$$

For the second equality we used $\chi_{[k \le T]} = 1 - \chi_{[k > T]}$ and that $\{[k > T]\}$ is independent of X_k since T is a stopping time. The general case follows by writing $X_1 = X_1^+ - X_1^-$.

For the second moment statement (5.12), consider $S_{T\wedge n}$ for any $n \geq 1$. Then, for any $n \geq 1$,

$$S_{T\wedge n}^2 = \left(\sum_{k=1}^{n} X_k \ \chi_{[k\leq T]}\right)^2$$

$$= \left(\sum_{k=1}^{n-1} X_k \chi_{[k\leq T]}\right)^2 + 2X_n \chi_{[n\leq T]} \sum_{k=1}^{n-1} X_k \chi_{[k\leq T]} + X_n^2 \chi_{[n\leq T]},$$

where the first sum in the second line vanishes for $n = 1$. Now note that

$$\chi_{[n\leq T]} \sum_{k=1}^{n-1} X_k \chi_{[k\leq T]} \quad \text{and} \quad X_n$$

are independent. Therefore, for each $n \geq 1$,

$$\mathbb{E}\, S_{T\wedge n}^2 = \mathbb{E}\, S_{T\wedge(n-1)}^2 + \mathbb{E}\, X_1^2\, \mathbb{P}(\{n \leq T\}),$$

whence, upon summing over $n \geq 1$, one concludes that

$$\lim_{n\to\infty} \mathbb{E}\, S_{T\wedge n}^2 = \mathbb{E}\, X_1^2\, \mathbb{E}\, T.$$

It remains to show that the left-hand side of the previous equation is $\mathbb{E}\, S_T^2$. On the one hand, we have $T \wedge n \to T$ almost surely. On the other hand, by the preceding discussion,

$$\mathbb{E}\,(S_{T\wedge n} - S_{T\wedge m})^2 = \mathbb{E}\, X_1^2\, \mathbb{E}\,(T \wedge n - T \wedge m)$$

for any $m < n$. Thus, $S_{T\wedge n}$ is Cauchy in L^2 and we are done. □

By means of a *random walk* on the discrete line we will now show that $\mathbb{E}\, T < \infty$ cannot be dropped in Wald's identity. On the discrete lattice $\mathbb{Z}^d$, $d \geq 1$, the simplest example of a random walk is the sum $S_n = \sum_{j=1}^{n} X_j$ for $n \geq 1$, $S_0 := 0$, where the X_j are i.i.d. and

$$\mathbb{P}(\{X_1 = e_j\}) = 2^{-d} \quad \forall 1 \leq j \leq 2^d;$$

the e_j are the nearest neighbors $(\pm 1, 0, 0, \ldots, 0)$ etc. of the origin. It is easy to formulate a long list of natural questions about these simple random walks; for example, we have the following. What is the probability that the walk will return to the origin? How long does it take on average for it to come back? What is the average distance from the origin after N steps (this is easy: $\mathbb{E}\, |S_N|^2 = N$)? What is the probability that the walk will arrive at a given region? In addition, one can of course modify the random walk by choosing a different lattice or a different law.

Exercise 5.10 By considering the Rademacher functions, and with $T = n$ the first time at which $S_n > 0$, show that one cannot remove the assumption $\mathbb{E}\, T < \infty$ from (5.12). *Hint:* Find $\mathbb{P}(\{T = 2n\})$ by counting the paths in the plane of length $2n$, obtained by taking steps, in either the direction $(1, 1)$ or the direction $(1, -1)$, which remain above the x-axis until the final step, when the x-axis is reached. This leads to the Catalan numbers C_n.

Stopping times play an important role in the theory of martingales, to which we now turn.

5.3. Conditional expectations; martingales

While sums of independent random variables are of basic importance, they are too restrictive for many purposes. For example, given such a sum $S_n = \sum_{j=1}^{n} X_j$ it is natural to consider objects such as $\varphi(S_n)$, where φ is increasing and convex; an example is the exponential function. Another object of interest is $\max_{1\le k\le n} |S_k|$. In either case, these types of new random variable lose the property of being sums of independent variables.

This section introduces a structure that is at the same time more robust but yet rigid enough to allow for general assertions to hold, such as convergence theorems. This structure is that of a *martingale* or, more generally, *sub- or supermartingales*. To define these objects, we need the notion of *conditional expectation*.

5.3.1. Conditional expectation

For two events A, B the conditional probability of A given B is defined as

$$\mathbb{P}(A|B) := \frac{\mathbb{P}(A \cap B)}{\mathbb{P}(B)}$$

(assuming of course that $\mathbb{P}(B) \neq 0$), which is precisely what intuition dictates. If A and B are independent then $\mathbb{P}(A|B) = \mathbb{P}(A)$, which means that the knowledge that B has occurred does not affect the occurrence of A. At the other extreme, if $A = B$ then one has maximal information and $\mathbb{P}(A|B) = 1$.

Passing from events to random variables, we see that if $X = \sum_j c_j \chi_{A_j}$ takes finitely many values, then the conditional expectation is given by

$$\mathbb{E}[X \mid B] := \sum_j c_j \frac{\chi_{A_j \cap B}}{\mathbb{P}(B)},$$

which has the property that

$$\int_B \mathbb{E}[X \mid B]\, d\mathbb{P} = \int_B X\, d\mathbb{P}.$$

Generalizing from this, we define the conditional expectation of any L^1 variable, relative to a sub σ-algebra $\mathcal{F} \subset \Sigma$ of the underlying σ-algebra Σ in the probability space, as follows:

$$\int_B \mathbb{E}[X \mid \mathcal{F}]\, d\mathbb{P} = \int_B X\, d\mathbb{P} \quad \forall B \in \mathcal{F}. \tag{5.13}$$

We need to show that the conditional expectation exists; this could be done easily using the Radon–Nikodym theorem. However, we shall derive it here via orthogonal projections in a Hilbert space. This is very natural, as the following example shows. Consider $f \in L^1([0, 1])$ and let $\mathcal{F}$ be generated by a partition of $[0, 1]$ into finitely many intervals

$$I_j := [x_j, x_{j+1}), \quad 0 \leq j < N, \quad x_0 := 0 < x_1 < \cdots < x_N := 1. \tag{5.14}$$

Then, if a crossed integral sign $⨍$ denotes the average, one has

$$\mathbb{E}[f \mid \mathcal{F}] = \sum_{j=1}^{N} \chi_{I_j} ⨍_{I_j} f(y)\, dy. \tag{5.15}$$

If $f \in L^2$ then the right-hand side is nothing other than the orthogonal projection of f onto the subspace of functions in $L^2([0, 1])$ that are measurable with respect to $\mathcal{F}$.

Proposition 5.12 *For any L^1 random variable X and sub σ-algebra $\mathcal{F}$ the conditional expectation $\mathbb{E}[X \mid \mathcal{F}]$ exists and is uniquely characterized by* (5.13). *If $X \geq 0$, but we do not necessarily have $X \in L^1$, then $\mathbb{E}[X \mid \mathcal{F}]$ again exists and is uniquely characterized by* (5.13).

Proof If X is $L^2(\Omega)$ then denote by Q the orthogonal projection onto the subspace in $L^2(\Omega)$ of all $\mathcal{F}$-measurable functions. Set $\mathbb{E}[X \mid \mathcal{F}] := Q(X)$. Since Q is self-adjoint, for any $B \in \mathcal{F}$ we have

$$\langle Q(X) | \chi_B \rangle = \langle X | Q(\chi_B) \rangle = \langle X | \chi_B \rangle,$$

which is (5.13). Now let $Y \in L^2$ be $\mathcal{F}$-measurable and assume that

$$\langle Y | \chi_B \rangle = 0 \quad \forall B \in \mathcal{F}.$$

Then $\langle Y | Z \rangle = 0$ for all Z in L^2 that are $\mathcal{F}$-measurable, and so $Y = 0$. Hence $Q(X) = \mathbb{E}[X | \mathcal{F}]$ is uniquely characterized by (5.13). We also see that $X \geq 0$ almost surely implies that $\mathbb{E}[X | \mathcal{F}] \geq 0$ and $|\mathbb{E}[X | \mathcal{F}]| \leq \mathbb{E}[|X| \mid \mathcal{F}]$.

If $X \in L^1$ then write $X_n := |X| \wedge n$ for every $n \geq 1$. By the preceding argument,

$$\mathbb{E}\left|\mathbb{E}[X_n | \mathcal{F}] - \mathbb{E}[X_m | \mathcal{F}]\right| \leq \mathbb{E}|X_n - X_m| \to 0$$

as $n, m \to \infty$. Letting $\lim_{n\to\infty} \mathbb{E}[X_n|\mathcal{F}] = \mathbb{E}[X|\mathcal{F}]$ in an L^1 sense completes the proof. The uniqueness over L^1 is also elementary and is left to the reader.

The final claim is obtained by the same limit procedure but using the monotone convergence theorem. □

Exercise 5.11 Suppose that X, Y are two L^1 variables such that

$$\mathbb{E}[X|Y] = Y, \quad \mathbb{E}[Y|X] = X.$$

Here $\mathbb{E}[X|Y] = \mathbb{E}[X|\mathcal{F}]$, where $\mathcal{F}$ is the σ-algebra generated by Y. Show that $X = Y$ almost surely. Is either of the above relations by itself enough to reach the same conclusion? What can you conclude from

$$\mathbb{E}[X|Y] = X, \quad \mathbb{E}[Y|X] = Y?$$

Conditional expectations have a number of simple properties, which we now collect in the following result. All variables will be assumed to be L^1. As always, it is helpful to keep (5.15) in mind.

Lemma 5.13 *The conditional expectation $\mathbb{E}[X \mid \mathcal{F}]$ enjoys the following properties:*

(i) *if X is $\mathcal{F}$-measurable then $\mathbb{E}[X \mid \mathcal{F}] = X$;*
(ii) *if X is $\mathcal{F}$-independent then $\mathbb{E}[X \mid \mathcal{F}] = \mathbb{E}\, X$;*
(iii) *if Y is $\mathcal{F}$-measurable and bounded (or of a definite sign if X is not L^1) then $\mathbb{E}[XY|\mathcal{F}] = Y\mathbb{E}[X|\mathcal{F}]$;*
(iv) *if $X \geq 0$ then $\mathbb{E}[X|\mathcal{F}] \geq 0$;*
(v) $|\mathbb{E}[X|\mathcal{F}]| \leq \mathbb{E}[|X| \mid \mathcal{F}]$;
(vi) $\mathbb{E}(\mathbb{E}[X|\mathcal{F}]) = \mathbb{E}\, X$;
(vii) *$\mathbb{E}[\cdot \mid \mathcal{F}]$ is linear;*
(viii) *if $\varphi \geq 0$ is convex on the line then $\varphi(\mathbb{E}[X|\mathcal{F}]) \leq \mathbb{E}[\varphi(X)|\mathcal{F}]$;*
(ix) *$\|\mathbb{E}[X|\mathcal{F}]\|_p \leq \|X\|_p$ for every $1 \leq p < \infty$.*

Proof These follow in a routine fashion from (5.13) and the existence and uniqueness statement, Proposition 5.12. Claim (viii) uses Jensen's inequality. Note that $\varphi(X)$ is not necessarily in L^1, but since it is nonnegative we are on safe ground. Part (ix) follows from (vi) and (viii). □

Before defining martingales, we will record the behavior of conditional expectations under the taking of limits.

Lemma 5.14 *One has the following analogues of the monontone and dominated convergence theorems, as well as of Fatou's lemma, respectively:*

(i) *Let $\{X_n\}_{n\geq 1}$ be an increasing sequence of nonnegative variables. Then*

$$\lim_{n\to\infty} \mathbb{E}\,[X_n|\mathcal{F}] = \mathbb{E}\big[\lim_{n\to\infty} X_n\big|\mathcal{F}\big].$$

(ii) *Let $\{X_n\}_{n=1}^{\infty}$ be dominated by a fixed L^1 variable Y and assume that $X_n \to X$ almost surely. Then*

$$\lim_{n\to\infty} \mathbb{E}\,[X_n|\mathcal{F}] = \mathbb{E}\,[X|\mathcal{F}].$$

(iii) *Let $\{X_n\}_{n=1}^{\infty}$ be a sequence of nonnegative variables. Then*

$$\mathbb{E}\big[\liminf_{n\to\infty} X_n\big|\mathcal{F}\big] \leq \liminf_{n\to\infty} \mathbb{E}\,[X_n|\mathcal{F}].$$

The following *Marcinkiewicz–Zygmund* inequality shows how the concept of conditional expectation may be applied.

Proposition 5.15 *Let $\{X_j\}_{j=1}^n$ be independent L^1 variables with $\mathbb{E}\,X_j = 0$. Then for every $1 \leq p < \infty$ one has*

$$A_p \left\| \Big(\sum_{j=1}^n |X_j|^2\Big)^{1/2} \right\|_p \leq \left\| \sum_{j=1}^n X_j \right\|_p \leq B_p \left\| \Big(\sum_{j=1}^n |X_j|^2\Big)^{1/2} \right\|_p, \tag{5.16}$$

where A_p, B_p depend only on p.

Proof Fix some $1 \leq p < \infty$. It is a simple exercise based on independence to show that

$$\sum_{j=1}^n X_j \in L^p \iff \Big(\sum_{j=1}^n |X_j|^2\Big)^{1/2} \in L^p \iff X_j \in L^p\ \forall 1 \leq j \leq n,$$

which has the meaning that all X_j are in L^p. First, suppose that the distributions of X_j and $-X_j$ are identical; i.e., the X_j are *symmetric*. Let r_j be a Rademacher function, viewed as being independent of $\{X_j\}_{j=1}^n$. Then we have

$$\mathbb{E}\left|\sum_{j=1}^n r_j X_j\right|^p = \mathbb{E}\left|\sum_{j=1}^n X_j\right|^p,$$

where the expectation on the left is with respect to both the r_j and the X_j whereas on the right-hand side it is only with respect to the X_j. Hence, (5.16) follows immediately from Kinchine's inequality Lemma 5.5.

If the X_j are not symmetric then we use the device of *symmetrization*: with $\{X'_j\}_{j=1}^n$ another copy of $\{X_j\}_{j=1}^n$, but independent of it, we set $\tilde{X}_j := X_j - X'_j$, which is clearly symmetric. Then

$$\mathbb{E}\left|\sum_{j=1}^n \tilde{X}_j\right|^p \leq 2^p\,\mathbb{E}\left|\sum_{j=1}^n X_j\right|^p.$$

To obtain a lower bound, simply note that

$$\mathbb{E}\left[\sum_{j=1}^{n} \tilde{X}_j \mid X_1, \ldots, X_n\right] = \sum_{j=1}^{n} X_j,$$

whence

$$\mathbb{E}\left|\sum_{j=1}^{n} X_j\right|^p \leq \mathbb{E}\left|\sum_{j=1}^{n} \tilde{X}_j\right|^p.$$

The same argument proves that

$$\mathbb{E}\left|\sum_{j=1}^{n} r_j X_j\right|^p \leq \mathbb{E}\left|\sum_{j=1}^{n} r_j \tilde{X}_j\right|^p \leq 2^p\, \mathbb{E}\left|\sum_{j=1}^{n} r_j X_j\right|^p.$$

Combining these statements with Lemma 5.5 finishes the proof. □

Later, Proposition 5.15 will serve to motivate Littlewood–Paley theory; see Chapter 8. For now, we formulate a simple corollary that states an optimal L^p estimate on sums of independent variables.

Corollary 5.16 *Let X_j be independent $L^{p\vee 2}$ variables, $1 \leq p < \infty$, with $\mathbb{E}\, X_j = 0$ for all j as well as $\sup_j \|X_j\|_{p\vee 2} < \infty$. Set $S_n := \sum_{j=1}^{n} X_j$. Then one has $\mathbb{E}\,|S_n|^p = O\big(n^{p/2}\big)$ as $n \to \infty$.*

Proof Assume first that $p \geq 2$. Then, by Hölder's inequality,

$$\left(\sum_{j=1}^{n} X_j^2\right)^{1/2} \leq n^{1/2-1/p}\left(\sum_{j=1}^{n} |X_j|^p\right)^{1/p}.$$

Taking L^p norms and appealing to Proposition 5.15 finishes the proof. If $1 \leq p < 2$ then one may estimate directly via the Hölder inequality, obtaining

$$\|S_n\|_p \leq \|S_n\|_2 \leq C n^{1/2}$$

and we are done. □

5.3.2. Martingales

All variables will be assumed to be either nonnegative or integrable, i.e., L^1.

Definition 5.17 Let $\{\mathcal{F}_n\}_{n=0}^{\infty}$ be an increasing sequence of σ-algebras (a filtration) and assume that $\{X_n\}_{n=0}^{\infty}$ is a sequence of variables such that X_n is measurable with respect to $\mathcal{F}_n$. We say that $\{X_n, \mathcal{F}_n\}_{n=0}^{\infty}$ (or for brevity $\{X_n\}_{n=0}^{\infty}$) is a martingale provided that

$$\mathbb{E}\,[X_{n+1}|\mathcal{F}_n] = X_n \quad \forall n \geq 0. \tag{5.17}$$

If in (5.17) the equals sign is replaced by the symbol $\geq$ then this object is known as a *submartingale*; if the equals sign is replaced by the symbol $\leq$ then we have a *supermartingale*.

A trivial but nonetheless helpful example is given by constant random variables. For such a sequence of numbers to form a martingale, it is necessary and sufficient that they form a constant sequence; for a submartingale this is the case if and only if the sequence increases whereas for a supermartingale the sequence must decrease. Evidently, one can turn submartingales into supermartingales by changing the signs.

If $f \in L^1([0, 1])$, and $\mathcal{F}_n$ is induced by a partition as in (5.14) in such a fashion that the partitions become finer as n increases, then the projections in (5.15) form a martingale. In particular, by the differentiation theorem we see that $\mathbb{E}[f|\mathcal{F}_n]$ converges almost everywhere, as well as in the sense of $L^1([0, 1])$, as $n \to \infty$.

If u is convex on the line, and if the X_j are independent real-valued Gaussians with mean zero, then

$$Y_n := u\Big(\sum_{j=0}^{n} X_j\Big)$$

is a submartingale with respect to the natural filtration. In dimensions $d \geq 2$, if the X_j are independent mean-zero Gaussian d-dimensional vectors with covariance matrix $\sigma\mathrm{Id}$, $\sigma > 0$, and if u is *subharmonic* in $\mathbb{R}^d$ then, by the same procedure, we obtain a submartingale (hence the "sub"). Subharmonic functions in $\mathbb{R}^d$ are defined in the same way as they were for $d = 2$ in Chapter 3. The proof of this claim proceeds via the sub-mean-value property of subharmonic functions. A martingale is obtained by taking u to be harmonic.

By means of simple random walks, one obtains the following discrete analogue of the previous example. Let $u : \mathbb{Z}^d \to \mathbb{R}$ be a subharmonic function, i.e., with e_j the nearest neighbors of 0 as in the discussion before Exercise 5.10:

$$\frac{1}{2d}\sum_{j=1}^{2d} u(x + e_j) \geq u(x) \quad \forall x \in \mathbb{Z}^d.$$

Suppose that S_n is a simple random walk on $\mathbb{Z}^d$. Then $X_n := u(S_n)$ is a submartingale.

Sums S_n of independent variables with the natural filtration form a martingale, but so do sums

$$S_{k,n} := \sum_{1\leq i_1 < i_2 < \cdots < i_k \leq n} X_{i_1} \cdots X_{i_k}, \tag{5.18}$$

for fixed $k \geq 1$ and $n \geq 1$; they do not have the sum-of-independent-variables structure if $k > 1$. What is more, by Jensen's inequality, as in Lemma 5.13(viii),

$\{\varphi(S_{k,n})\}_{n\geq 1}$ forms a submartingale for any convex $\varphi \geq 0$, as does $|S_{k,n}|$. We now show that super- and submartingales are also stable under the action of a stopping time.

Lemma 5.18 *Let T be a stopping time adapted to a filtration $\{\mathcal{F}_n\}_{n=0}^{\infty}$, i.e., T takes its values in $\mathbb{Z}_0^+ \cup \{+\infty\}$ and $\{T \leq j\} \in \mathcal{F}_j$ for every $j \geq 0$. Then if $\{X_n\}_{n=0}^{\infty}$ is a martingale, so is $\{X_{T\wedge n}\}_{n=0}^{\infty}$. The same applies to super- and submartingales.*

Proof We leave it to the reader to check the measurability and integrability requirements for $X_{T\wedge n}$. Now suppose that $\{X_n\}_{n=0}^{\infty}$ is a submartingale. The proof hinges on the following general construction: suppose that $Y_n \geq 0$ are bounded and measurable with respect to $\mathcal{F}_{n-1}$ for every $n \geq 1$. Then

$$Z_n := \sum_{k=1}^{n} Y_k(X_k - X_{k-1}), \quad n \geq 1, \quad Z_0 := 0, \tag{5.19}$$

is a submartingale. Indeed, Z_n is $\mathcal{F}_n$-measurable and, for each $n \geq 0$,

$$\mathbb{E}\,[Z_{n+1}|\mathcal{F}_n] = Y_{n+1}\mathbb{E}\,[X_{n+1} - X_n|\mathcal{F}_n] + Z_n \geq Z_n,$$

as claimed. Now set $Y_k := \chi_{[k\leq T]}$ for each $k \geq 1$; this is measurable with respect to $\mathcal{F}_{k-1}$. Then Z_n as in (5.19) satisfies $Z_n + X_0 = X_{T\wedge n}$ and we are done. □

5.3.3. Doob convergence theorem

We now state and prove the basic Doob convergence theorem for martingales.

Theorem 5.19 *Let $\{X_n\}_{n=0}^{\infty}$ be a supermartingale that is L^1-bounded, i.e., such that* $\sup_{n\geq 0} \mathbb{E}\,|X_n| < \infty$. *Then $X_n \to X_\infty$ a.s., where X_∞ is real-valued. In particular, nonnegative supermartingales converge a.s. to a finite limit.*

Proof The final statement of the theorem follows from the first since $\mathbb{E}\,|X_n| = \mathbb{E}\,X_n \leq \mathbb{E}\,X_0$. For the first we follow the usual path of bounding the number of *upcrossings* of X_n. To define this, fix any $a < b$ and set $T_0(a,b) := 0$ and, for every $k \geq 0$, set

$$S_{k+1}(a,b) := \inf\{n \geq T_k(a,b) \mid X_n < a\},$$

$$T_{k+1}(a,b) := \inf\{m \geq S_{k+1}(a,b) \mid X_m > b\},$$

where $\inf \emptyset := \infty$. These are clearly stopping times (possibly infinite-valued, but this makes no difference) which satisfy $0 = T_0 \leq S_1 < T_1 < S_2 < T_2 < \ldots$. Then define

$$N_n(a,b) := \sup\{k \geq 0 \mid T_k \leq n\}, \quad N(a,b) = \sup_n N_n(a,b),$$

where $N(a,b)$ is the total number of upcrossings. The main claim now is the following. For each $n \geq 1$,

$$\mathbb{E}\, N_n(a,b) \leq (b-a)^{-1}\, \mathbb{E}\,(X_n - a)^-, \tag{5.20}$$

where $f^- := -f \wedge 0$. If this claim holds then it is easy to complete the proof. Indeed, by monotone convergence,

$$\mathbb{E}\, N(a,b) = \sup_n \mathbb{E}\, N_n(a,b) \leq (b-a)^{-1} \Big(\sup_n \mathbb{E}\, |X_n| + |a| \Big) < \infty,$$

whence $N(a,b) < \infty$ almost surely. It follows that a.s. the total number of upcrossings for any (rational) pair $a < b$ is finite, whence the convergence follows. From Fatou's lemma it further follows that $\mathbb{E}\, |X_\infty| < \infty$ and so X_∞ is finite a.s.

It remains to prove the bound (5.20). For this, dropping the arguments a, b from the stopping times, let

$$Y_n := \sum_{k=1}^{\infty} \chi_{[S_k < n \leq T_k]}, \qquad n \geq 1.$$

Then Y_n is $\mathcal{F}_{n-1}$-measurable and

$$Z_n := \sum_{k=1}^{n} Y_k (X_k - X_{k-1}), \quad n \geq 1, \quad Z_0 := 0,$$

is a supermartingale by the same argument as in the proof of Lemma 5.18. Now

$$\begin{aligned} Z_n &= \sum_{j=1}^{N_n} (X_{T_j} - X_{S_j}) + (X_n - X_{S_{N_n+1}})\, \chi_{[S_{N_n+1} \leq n]} \\ &\geq (b-a) N_n + (X_n - a)\, \chi_{[X_n \leq a]} \\ &\geq (b-a) N_n - (X_n - a)^-. \end{aligned}$$

In conclusion,

$$(b-a)\, \mathbb{E}\, N_n - \mathbb{E}\,(X_n - a)^- \leq \mathbb{E}\, Z_n \leq \mathbb{E}\, Z_0 = 0,$$

and we are done. □

Note that the L^1 condition is the same as the condition $\sup_{n \geq 0} \mathbb{E}\, X_n^- < \infty$. For sequences of numbers, Theorem 5.19 simply means that decreasing sequences that are bounded from below (in particular, if they are nonnegative) converge.

It is now natural to find conditions under which the convergence in Theorem 5.19 occurs in the sense of L^p for $1 \leq p < \infty$. As usual, $p = 2$ is special.

Exercise 5.12 Show that any L^2 martingale $\{X_n\}_{n=0}^\infty$ with respect to some filtration has the property that the increments $X_{n+1} - X_n$ are pairwise orthogonal. Show that $\sup_n \|X_n\|_2 < \infty$ is equivalent to

$$\sum_{n=1}^{\infty} \mathbb{E}\,|X_n - X_{n-1}|^2 < \infty$$

and that X_n converges in L^2 in this case.

Treating general p requires the following maximal function estimate.

Proposition 5.20 *Let $\{X_n\}_{n=0}^\infty$ be a submartingale, and let $M_n := \max_{1\le k\le n} X_k$. Then*

$$\lambda\mathbb{P}(\{M_n \ge \lambda\}) \le \mathbb{E}\big(X_n\,\chi_{[M_n\ge\lambda]}\big) \le \mathbb{E}\,X_n^+ \tag{5.21}$$

for every $\lambda > 0$ and $n \ge 0$.

Proof First we claim that the estimate $\mathbb{E}\,X_n \ge \mathbb{E}\,X_{T\wedge n}$ for all $n \ge 0$ holds for any stopping time T. This is natural in view of the monotonicity $\mathbb{E}\,X_n \ge \mathbb{E}\,X_k$ for all $0 \le k \le n$. To prove the claim, let $Y_k := \chi_{[T<k\le n]}$, which is $\mathcal{F}_{k-1}$-measurable. Then this ensures as before that

$$Z_n := \sum_{k=1}^{n} Y_k(X_k - X_{k-1}), \qquad n \ge 1, \;\; Z_0 := 0,$$

is a submartingale. However, $Z_n = X_n - X_{T\wedge n}$ whence

$$\mathbb{E}\,Z_n = \mathbb{E}\,X_n - \mathbb{E}\,X_{T\wedge n} \ge 0,$$

as claimed. Now define $T := \inf\{n \ge 0 \mid X_n \ge \lambda\}$. Then $\{T \le n\}$ is the same as $\{M_n \ge \lambda\}$, whence

$$\begin{aligned}\mathbb{E}\,X_n \ge \mathbb{E}\,X_{T\wedge n} &= \mathbb{E}\,(X_n\,\chi_{[T>n]}) + \mathbb{E}\,(X_T\,\chi_{[T\le n]})\\ &\ge \mathbb{E}\,(X_n\,\chi_{[T>n]}) + \lambda\mathbb{P}(\{T \le n\})\end{aligned}$$

and we are done. □

For sequences of numbers, Proposition 5.20 amounts to the fact that finite increasing sequences are dominated by the last term. For finite $p > 1$ one now has the following L^p bound on the maximal function, due to Doob.

Corollary 5.21 *Let $\{X_n\}_{n=0}^\infty$ be a martingale. Define $X_n^* := \sup_{0\le k\le n} |X_k|$ and suppose that $1 < p < \infty$. Then*

$$\|X_n^*\|_p \le \frac{p}{p-1}\|X_n\|_p.$$

Proof Apply Proposition 5.20 to the submartingale $\{|X_n|\}_{n=0}^{\infty}$:

$$\begin{aligned}\mathbb{E}\,|X_n^*|^p &= \int_0^\infty p\,\mathbb{P}(\{X_n^* > \lambda\})\,\lambda^{p-1}\,d\lambda \\ &\le \int_0^\infty p\,\mathbb{E}\left(|X_n|\,\chi_{[X_n^*>\lambda]}\right)\lambda^{p-2}\,d\lambda \\ &\le \frac{p}{p-1}\|X_n\|_p\|X_n^*\|_p^{p-1},\end{aligned}$$

where one integrates over λ and then applies Hölder's inequality to pass to the final estimate. □

Now assume that $\{X_n\}_{n=0}^{\infty}$ is an L^p-bounded martingale in the sense that

$$\sup_{n\ge 0}\|X_n\|_p < \infty \tag{5.22}$$

with $1 < p < \infty$. Then this martingale is also L^1-bounded, and Theorem 5.19 implies that $X_n \to X_\infty$ almost surely. Furthermore, $X^* := \sup_{n\ge 0} X_n^*$ satisfies $\|X^*\|_p < \infty$ by the monontone convergence theorem and Corollary 5.21. But then one can apply the dominated convergence theorem to conclude that $X_n \to X_\infty$ in the sense of L^p. This establishes the implication (i) $\Longrightarrow$ (ii) in the following theorem; the other implications are fairly immediate and are left to the reader.

Theorem 5.22 *Let $X := \{X_n\}_{n=0}^{\infty}$ be a martingale and $1 < p < \infty$. Then the following are equivalent:*

(i) *X is L^p bounded as in (5.22);*
(ii) *X converges a.s. as well as in the L^p sense;*
(iii) *there exists $X_\infty \in L^p$ with $X_n = \mathbb{E}[X_\infty|\mathcal{F}_n]$ for all $n \ge 0$.*

Property (iii) here in some sense reduces us to the projection example stated after Definition 5.17.

Exercise 5.13 Show that Theorem 5.22 fails for $p = 1$, i.e., show that an L^1-bounded martingale need not converge in L^1. However, prove the analogue of Theorem 5.22 for martingales that are uniformly integrable; see Problem 2.1.

5.3.4. Applications

As a first application of martingale ideas, we now establish the upper bound of the law of the iterated logarithm. In order not to become distracted from the main ideas by technicalities, we restrict ourselves to sums of Rademacher variables.

Proposition 5.23 *Let $S_n = \sum_{j=1}^n r_j$, where the r_j form a Rademacher sequence. Then*

$$\limsup_{n\to\infty} \frac{|S_n|}{\sqrt{2n\log\log n}} \le 1 \quad a.s. \tag{5.23}$$

Proof One has, with $n_k := \lfloor c^k \rfloor$ and $\sigma_k := \mathbb{E}\,|S_{n_k}|^2$, where $c > 1$ is arbitrary but fixed,

$$\mathbb{P}\Big(\Big\{\max_{n_{k-1}<n\le n_k} |S_n| > \lambda\Big\}\Big) = \mathbb{P}\Big(\Big\{\max_{n_{k-1}<n\le n_k} e^{t|S_n|} > e^{t\lambda}\Big\}\Big)$$
$$\le e^{-t\lambda}\,\mathbb{E}\,e^{t|S_{n_k}|} \le 2e^{-t\lambda}e^{t^2\sigma_k^2/2} \le 2e^{-\lambda^2/2\sigma_k^2}.$$

To obtain the first inequality we used the fact that both $|S_n|$ and $e^{t|S_n|}$ for $t > 0$ are submartingales, followed by Proposition 5.20. For the second inequality we argued as in Lemma 5.4. Now set $\lambda = \lambda_k := (1+\varepsilon)\sqrt{2\sigma_k^2 \log\log n_k}$, with $\varepsilon > 0$ fixed, to conclude that

$$\mathbb{P}\Big(\Big\{\max_{n_{k-1}<n\le n_k} |S_n| > \lambda_k\Big\}\Big) \le 2(\log n_k)^{-(1+\varepsilon)^2},$$

which is summable in k. Apply Borel–Cantelli and let $\varepsilon \to 0$ and $c \to 1$. □

As seen in (5.11), equality may hold in (5.23). The details are sketched in the following exercise.

Exercise 5.14 Show that

$$\limsup_{n\to\infty} \frac{S_n}{\sqrt{2n\log\log n}} = 1 \quad \text{a.s.}, \tag{5.24}$$

by following these steps.

(a) Let $a \ge 2$, $n_k := a^k$, $k \ge 1$. Set $A_k := \{S_{n_k} - S_{n_{k-1}} > \mu_k\}$ with a suitable choice of μ_k. Use a CLT to obtain a lower bound for $\mathbb{P}(A_k)$ and conclude that $\sum_k \mathbb{P}(A_k) = \infty$. Note that Theorem 5.6 does not apply since $\mu_k/\sqrt{n_k} \to \infty$; however, since S_n is a Bernoulli variable, the needed CLT can be obtained from Stirling's formula since this divergence is very slow (alternatively, look up the deMoivre–Laplace CLT).

(b) Conclude that A_k occurs infinitely often, by the independence version of Borel–Cantelli. By intersecting A_k with the event $\{|S_{n_{k-1}}| < \mu_k'\}$ for appropriate μ_k' and letting $a \to \infty$, infer that (5.24) holds.

As is to be expected, we can derive the strong law of large numbers from martingale convergence theory. A standard tool in deriving such implications is *Kronecker's lemma*.

Lemma 5.24 *Let B be some Banach space and assume that $\{a_n\}_{n=1}^{\infty} \subset B$ is a sequence of vectors such that $\sum_{n=1}^{\infty} a_n$ converges in B. If $b_n > 0$ is increasing and tends to ∞ then $b_n^{-1} \sum_{k=1}^{n} a_k b_k \to 0$ in B.*

Proof With $A_n := \sum_{k=n}^{\infty} a_k$,

$$\sum_{k=1}^{n} a_k b_k = \sum_{k=1}^{n} (A_k - A_{k+1}) b_k = A_1 b_1 - A_{n+1} b_n + \sum_{k=2}^{n} (b_k - b_{k-1}) A_k,$$

which implies the lemma. □

The following result is an example of a strong law derived via martingales.

Theorem 5.25 *Assume that $\{X_j\}_{j=1}^{\infty}$ are independent L^{2p} variables with $1 < p < \infty$ arbitrary but fixed, $\mathbb{E}\, X_j = 0$, and $\sup_j \mathbb{E}\, |X_j|^{2p} < \infty$. Then, with $S_{n,2}$ as in* (5.18), *one has*

$$\binom{n}{2}^{-1} S_{n,2} \to 0 \quad \textit{a.s. and in the } L^p \textit{ sense}$$

as $n \to \infty$.

Proof With the usual sum $S_n = \sum_{j=1}^{n} X_j$, and $b_n := \binom{n}{2}$, define

$$T_n := \sum_{m=2}^{n} X_m S_{m-1} b_m^{-1}.$$

Then $\{T_n\}_{n=2}^{\infty}$ is a martingale with respect to the natural filtration. By Corollary 5.16,

$$\sup_{n \geq 2} \|T_n\|_p \leq \sum_{m=2}^{\infty} b_m^{-1} \|X_m\|_{2p} \|S_{m-1}\|_{2p} \leq C \sum_{m=2}^{\infty} m^{-3/2} < \infty,$$

whence Theorem 5.22 implies that $T_n \to T_\infty$ a.s. and in L^p. We infer from Lemma 5.24 that

$$b_n^{-1} S_{n,2} = b_n^{-1} \sum_{m=2}^{n} X_m S_{m-1} \to 0$$

almost surely and in the L^p sense. □

Exercise 5.15 Establish the analogue of the previous theorem for $S_{n,k}, k \geq 3$.

Notes

For a brief and intuitive introduction to probability theory, see Sinai [101]. Everything presented in this chapter can be found in Chow and Teicher [20] and most of it also in

Durrett [32]. These two references go far beyond the material of this chapter, Durrett being technically less intense than Chow and Teicher. For the measure theory fact alluded to in the discussion of the zero–one law, see Halmos [51]. A far-reaching generalization of Khinchine's inequality is due to J. P. Kahane. It applies to Banach-space-valued sequences rather than scalar sequences and reads as follows: there exists a constant $C(p)$ for any $1 \le p < \infty$ such that for arbitrary sequences $\{x_n\}_{n=1}^N \subset X$, where X is a normed linear space, one has

$$\mathbb{E}\left\|\sum_{n=1}^N r_n x_n\right\| \le \left(\mathbb{E}\left\|\sum_{n=1}^N r_n x_n\right\|^p\right)^{1/p} \le C(p)\,\mathbb{E}\left\|\sum_{n=1}^N r_n x_n\right\|,$$

where the r_n are the Rademacher variables. This inequality was the starting point for far-reaching investigations in Banach space geometry involving the notions of *type* and *co-type*; see for example Wojtaszczyk [125].

Problems

Problem 5.1 Let X be an L^1 variable. Show that the family

$$\big\{\mathbb{E}[X|\mathcal{F}] \mid \mathcal{F} \text{ is a sub } \sigma\text{-algebra}\big\}$$

is uniformly integrable.

Problem 5.2 Suppose that $\{X_n\}$ are i.i.d. and satisfy the condition that $n^{-1}\sum_{j=1}^n X_j \to Y$ converges almost surely. Show that X_1 is an L^1 variable and that $Y = \mathbb{E}\,X_1$. This shows that Theorem 5.9 is sharp. Does convergence take place in L^1?

Problem 5.3

(a) If $\{X_n\}$ are i.i.d. and satisfy the condition that $n^{-1}\sum_{j=1}^n X_j$ converges almost surely, then show that

$$\sum_{n=1}^\infty \mathbb{P}(\{|X_n| > \varepsilon n\}) < \infty \quad \forall \varepsilon > 0.$$

(b) Show that if a real sequence $\{\sigma_n\}$ satisfies $\sum_{n=1}^\infty n^{-2}\sigma_n^2 = \infty$ then there exists $\{X_n\}$, where the X_n are independent, with mean zero and with $\mathbb{V}X_n = \sigma_n^2$, and are such that the strong law of large numbers fails.

Problem 5.4 Let the random variables $\{X_n\}$ be independent with $\mathbb{E}\,X_n = 0$ and $\sum_{n=1}^\infty n^{-2}\,\mathbb{V}\,X_n < \infty$. We have shown that $n^{-1}\sum_{j=1}^n X_j \to 0$ a.s. Show that this convergence also holds in the sense of L^2. Find conditions that ensure that this holds for other L^p norms, such as L^4.

Problem 5.5 This problem generalizes Lemma 5.4 to martingales with bounded increments. Let $\{X_n\}_{n=0}^\infty$ with $X_0 = 0$ be a martingale with respect to some filtration, and assume that

$$\|X_n - X_{n-1}\|_\infty \le d_n, \quad n \ge 1.$$

Show that

$$\mathbb{P}\left(\left\{\sup_{0\le k\le N} X_k > \lambda \left(\sum_{n=1}^{N} d_n^2\right)^{1/2}\right\}\right) \le e^{-\lambda^2/2}$$

for any $N \ge 1$ and $\lambda > 0$. *Hint:* Use the same idea as in the proof of Lemma 5.4.

Problem 5.6 Use the previous problem to obtain a more general version of the upper bound in the law of the iterated logarithm, cf. Proposition 5.23.

Problem 5.7 Let $\{X_n\}_{n\ge 0}$ be a (super)martingale with respect to some filtration, and let T be a stopping time that is a.s. finite. Show that $\mathbb{E}\, X_T = \mathbb{E}\, X_0$ (or $\mathbb{E}\, X_T \le \mathbb{E}\, X_0$) under either of the following two conditions:

- $\sup_n \|X_n\|_\infty < \infty$;
- $\mathbb{E}\, T < \infty$ and $\sup_{n\ge 1} \|X_n - X_{n-1}\|_\infty < \infty$.

Problem 5.8 Consider the random sum $S = \sum_{n=1}^{\infty} 3^{-n} X_n$ with i.i.d. random variables X_n with law $\mathbb{P}(\{X_n = 0\}) = \mathbb{P}(\{X_n = 2\}) = \frac{1}{2}$. Show that the distribution μ of S is singular with respect to Lebesgue measure and that $\mu(\mathcal{C}) = 1$, where $\mathcal{C}$ is the middle-third Cantor set in $[0, 1]$. The function $\mu([0, \lambda))$ is the Cantor staircase function. Show that the Fourier transform $\hat{\mu}(k)$ does not go to zero as $|k| \to \infty$, in contrast with the Riemann–Lebesgue lemma, Corollary 1.8. Prove that $\hat{\mu}(3k) = \hat{\mu}(k)$ for any integer k. *Hint:* Compute the characteristic function of $2\pi S$.

6

Fourier series and randomness

In this chapter we apply some probabilistic tools developed in Chapter 5 to Fourier series. Our main goal here is to be able to answer, using probabilistic arguments, questions which in and of themselves are deterministic. For example, in the following section we address the problem of a.e. convergence of the partial sums of the Fourier series of $L^1(\mathbb{T})$ functions. By means of a combination of deterministic and probabilistic ideas we shall show that L^1 functions exist for which the partial sums of their Fourier series diverge almost everywhere. Section 6.2 is more probabilistic from the start, as it deals with the problem of deciding the convergence of Fourier series with random and independent coefficients. This is a classic topic, going back to the work of Paley and Zygmund in the 1930s; they introduced randomness into Fourier series as a tool for answering deterministic questions.

6.1. Fourier series on $L^1(\mathbb{T})$: pointwise questions

This section is devoted to the question of the almost everywhere convergence of Fourier series for $L^1(\mathbb{T})$ functions. From the previous chapters we know that this is not an elementary question, since the partial sums S_N are given by convolutions, with the Dirichlet kernels D_N, which *do not* form an approximate identity. Note also that we have already encountered an a.e. convergence question that is not associated with an approximate identity, namely, that for the Hilbert transform; see Proposition 3.18. However, the needed uniform control was provided by the weak-L^1 bound on the maximal function $\tilde{u}^*$; see Theorem 3.15.

By a natural analogy, one sees that a similar bound on the maximal function associated with $S_N f$, i.e., $\sup_N |S_N f|$, would lead to an a.e. convergence result.

It was realized by Kolmogorov that essentially no bound on this maximal function is possible for $f \in L^1(\mathbb{T})$. This has to do with the oscillatory nature of the Dirichlet kernel D_N and the growth of $\|D_N\|_{L^1}$. In fact, by a careful choice of $f \in L^1$ we may arrange the convolution $(D_N * f)(x)$ so that only the positive peaks of D_N contribute at many points x.

Calderón (see Zygmund [133]) and Stein [107] developed the idea that a weak-type bound on

$$\sup_N |(S_N f)(x)|$$

is not only sufficient but also *necessary* for the a.e. convergence $S_N f$. In view of this fact we will conclude indirectly that $S_N f$ fails to converge a.e. for some $f \in L^1(\mathbb{T})$. Kolmogorov's original argument involved the construction of a function $f \in L^1(\mathbb{T})$ for which a.e. convergence failed. What we will do here is very close to his original approach even though it is somewhat more streamlined.

In the second volume of this book we will take up the more challenging question of the a.e. convergence of $S_N f$ for $f \in L^2(\mathbb{T})$ or $f \in L^p(\mathbb{T})$ for $1 < p < \infty$. This is the subject of Carleson's theorem, and its proof introduces some essentially new and far-reaching ideas. As well as presenting Kolomogorov's theorem, this chapter also serves to illustrate the use of probabilistic ideas in a deterministic problem.

6.1.1. Maximal functions and weak-L^1 bounds

We begin with a probabilistic construction that is a version of the Borel–Cantelli lemma, Lemma 5.2.

Lemma 6.1 *Let $\{E_n\}_{n=1}^\infty$ be a sequence of measurable subsets of $\mathbb{T}$ such that*

$$\sum_{n=1}^{\infty} |E_n| = \infty.$$

Then there exists a sequence $\{x_n\}_{n=1}^\infty \in \mathbb{T}$ such that

$$\sum_{n=1}^{\infty} \chi_{E_n}(x + x_n) = \infty \tag{6.1}$$

for almost every $x \in \mathbb{T}$.

Proof View $\Omega := \prod_{n=1}^\infty \mathbb{T}$ as a probability space equipped with the infinite-product measure. Given $x \in \mathbb{T}$, let $A_x \subset \Omega$ be the event characterized by (6.1). We claim that

$$\mathbb{P}(A_x) = 1 \quad \forall x \in \mathbb{T}. \tag{6.2}$$

By Fubini's theorem it then follows that, for a.e. $\{x_n\}_{n=1}^{\infty} \in \Omega$, the event (6.1) holds for a.e. $x \in \mathbb{T}$. Hence, fix an arbitrary $x \in \mathbb{T}$. Then

$$\begin{aligned} A_x &= \big\{\{x_n\}_{n=1}^{\infty} \mid x \text{ belongs to infinitely many } E_n(\cdot + x_n)\big\} \\ &= \left\{\{x_n\}_{n=1}^{\infty} \;\middle|\; x \in \bigcap_{N=1}^{\infty} \bigcup_{m=N}^{\infty} (x_m + E_m)\right\} = \bigcap_{N=1}^{\infty} A_N, \end{aligned}$$

where

$$\begin{aligned} A_N^c &:= \left\{\{x_n\}_{n=1}^{\infty} \;\middle|\; x \in \bigcap_{m=N}^{\infty} (x_m + E_m^c)\right\} \\ &= \big\{\{x_n\}_{n=1}^{\infty} \mid x_m \in x - E_m^c \quad \forall m \geq N\big\}. \end{aligned}$$

By definition of the product measure on Ω it follows that

$$\mathbb{P}(A_N^c) = \prod_{m=N}^{\infty} (1 - |E_m|) = 0,$$

by (6.1). Hence (6.2) holds, and the lemma follows. □

Lemma 6.1 will play an important role in the proof of the following theorem, which establishes the necessity of a weak-type maximal function bound for a.e. convergence. In addition to the previous probabilistic lemma, we shall also require the basic *Kolmogorov zero-one law*, which we stated in Chapter 5.

In our case, the Carleson maximal operator

$$\mathcal{C} f(x) := \sup_N |S_N f(x)| \tag{6.3}$$

is the main object that needs to be bounded. However, we shall formulate the following result in much greater generality. In fact, the reader will easily see that it does not depend on any special properties of $\mathbb{T}$, although we restrict ourselves to that case for simplicity.

Theorem 6.2 *Let T_n be a sequence of linear operators, bounded on $L^1(\mathbb{T})$ and translation invariant. Define*

$$\mathcal{M} f(x) := \sup_{n \geq 1} |T_n f(x)|$$

and assume that $\|\mathcal{M} f\|_\infty < \infty$ for every trigonometric polynomial f on $\mathbb{T}$.

If, for any $f \in L^1(\mathbb{T})$,

$$|\{x \in \mathbb{T} \mid \mathcal{M} f(x) < \infty\}| > 0$$

then there exists a constant A such that

$$|\{x \in \mathbb{T} \mid \mathcal{M}f(x) > \lambda\}| \le \frac{A}{\lambda}\|f\|_1$$

for any $f \in L^1(\mathbb{T})$ and $\lambda > 0$.

Proof We will prove this by contradiction. Thus, assume that there exists a sequence $\{f_j\}_{j=1}^{\infty} \subset L^1(\mathbb{T})$, with $\|f_j\|_1 = 1$ for all $j \ge 1$ as well as $\lambda_j > 0$, such that

$$E_j := \{x \in \mathbb{T} \mid \mathcal{M}f_j(x) > \lambda_j\}$$

satisfies

$$|E_j| > \frac{2^j}{\lambda_j}$$

for each $j \ge 1$. By the definition of $\mathcal{M}$ we then also have

$$\lim_{m\to\infty} |\{x \in \mathbb{T} \mid \sup_{1\le k\le m} |T_k f_j(x)| > \lambda_j\}| > \frac{2^j}{\lambda_j}$$

for each $j \ge 1$. Hence there are $M_j < \infty$ with the property that

$$\left|\left\{x \in \mathbb{T} \,\middle|\, \sup_{1\le k\le M_j} |T_k f_j(x)| > \lambda_j\right\}\right| > \frac{2^j}{\lambda_j}$$

for each $j \ge 1$. Let σ_N denote the Nth Cesàro sum, i.e., $\sigma_N f = K_N * f$ where K_N is the Fejér kernel. Since each T_j is bounded on L^1, we conclude that

$$\lim_{N\to\infty}\left|\left\{x \in \mathbb{T} \,\middle|\, \sup_{1\le k\le M_j} |T_k\sigma_N f_j(x)| > \lambda_j\right\}\right| > \frac{2^j}{\lambda_j}.$$

Hence, we may assume from now on that each f_j is a trigonometric polynomial. Let m_j be a positive integer with the property that

$$m_j \le \frac{\lambda_j}{2^j} < m_j + 1.$$

Then

$$\sum_{j=1}^{\infty} m_j |E_j| = \infty$$

by construction. Counting each set E_j with multiplicity m_j, the previous lemma implies that there exists a sequence of points $x_{j,\ell}$, $j \ge 1$, $1 \le \ell \le m_j$, such that

$$\sum_{j=1}^{\infty}\sum_{\ell=1}^{m_j}\chi_{E_j}(x-x_{j,\ell})=\infty \tag{6.4}$$

for almost every $x\in\mathbb{T}$. Let

$$\delta_j := \frac{1}{j^2 m_j}$$

and define

$$f(x):=\sum_{j=1}^{\infty}\sum_{\ell=1}^{m_j}\pm\delta_j f_j(x-x_{j,\ell}),$$

where the signs $\pm$ are to be chosen randomly. First note that, irrespectively of the choice of these signs,

$$\|f\|_1\le\sum_{j=1}^{\infty}m_j\delta_j<\infty.$$

Now we choose the signs in such a way that

$$\mathcal{M}f(x)=\infty$$

for almost every $x\in\mathbb{T}$. For this purpose, select $x\in\mathbb{T}$ such that $x\in E_j+x_{j,\ell}$ for infinitely many j and $\ell=\ell(j)$ (just pick any one such $\ell(j)$ if there are more than one). Denote the set of these j by $\mathcal{J}$. Since the T_n are translation invariant, so is $\mathcal{M}$. Hence

$$\mathcal{M}f_j(x-x_{j,\ell})>\lambda_j$$

for all $j\in\mathcal{J}$. We conclude that for these j there is a positive integer $n(j,x)$ such that

$$|T_{n(j,x)}f_j(x-x_{j,\ell})|>\lambda_j.$$

At the cost of having to remove another set of measure zero we may assume that

$$T_nf(x)=\sum_{j=1}^{\infty}\sum_{\ell=1}^{m_j}\pm\delta_jT_nf_j(x-x_{j,\ell})$$

for all positive integers n. In particular, we have that, for all $j\in\mathcal{J}$,

$$T_{n(j,x)}f(x)=\sum_{k=1}^{\infty}\sum_{\ell=1}^{m_k}\pm\delta_kT_{n(j,x)}f_k(x-x_{k,\ell}),$$

which implies that

$$\mathbb{P}\big(\big\{|T_{n(j,x)}f(x)| > \delta_j\lambda_j\big\}\big) \geq \tfrac{1}{2},$$

where the probability measure is with respect to the choice of signs $\pm$. Since $\delta_j\lambda_j \to \infty$, we obtain that

$$\mathbb{P}\big(\big\{\mathcal{M}f(x) = \infty\big\}\big) \geq \tfrac{1}{2}.$$

We claim that the event corresponding to the left-hand side is a tail event. Indeed, this holds since

$$\mathcal{M}f(x) \leq \sum_{j=1}^{\infty} \mathcal{M}f_j(x)$$

and each summand on the right-hand side here is finite (f_j is a trigonometric polynomial and we are assuming that $\mathcal{M}$ is uniformly bounded on trigonometric polynomials). By Kolmogorov's zero–one law we therefore have

$$\mathbb{P}\big(\big\{\mathcal{M}f(x) = \infty\big\}\big) = 1.$$

It follows from Fubini's theorem that almost surely (with respect to a random choice of $\pm$)

$$\mathcal{M}f(x) = \infty \quad \text{for a.e. } x.$$

This would contradict our main hypothesis, and so we are done. □

Lemma 6.1 reduces Kolmogorov's theorem on the failure of the a.e. convergence of Fourier series of L^1 functions to disproving a weak-L^1 bound for the Carleson maximal operator $\mathcal{C}$. More precisely, we arrive at the following corollary. It will be technically more convenient later on to formulate this result in terms of measures rather than on $L^1(\mathbb{T})$.

Corollary 6.3 *Suppose that $\{S_N f\}_{N=1}^{\infty}$ converges a.e. for every $f \in L^1(\mathbb{T})$. Then there exists a constant A such that*

$$|\{x \in \mathbb{T} \mid \mathcal{C}\mu(x) > \lambda\}| \leq \frac{A}{\lambda}\|\mu\| \tag{6.5}$$

for any complex Borel measure μ on $\mathbb{T}$ and $\lambda > 0$, where $\mathcal{C}$ is as in (6.3).

Proof By Theorem 6.2, our assumption implies that there exists a constant A such that

$$|\{x \in \mathbb{T} \mid \mathcal{C}f(x) > \lambda\}| \leq \frac{A}{\lambda}\|f\|_1$$

for all $\lambda > 0$. This is the same as

$$\left|\left\{x \in \mathbb{T} \,\middle|\, \max_{1 \le n \le N} |S_n f(x)| > \lambda\right\}\right| \le \frac{A}{\lambda}\|f\|_1$$

for all $N \ge 1$ and $\lambda > 0$. If μ is a complex measure then we set $f = V_m * \mu$, where V_m is de la Vallée Poussin's kernel; see (1.22). It follows that, for all $N \ge 1$,

$$\left|\left\{x \in \mathbb{T} \,\middle|\, \max_{1 \le n \le N} |S_n[V_m * \mu](x)| > \lambda\right\}\right| \le \frac{A}{\lambda}\|\mu\|$$

for all $m \ge 1$ and $\lambda > 0$. For $m \ge N$ the kernel V_m can be removed from the left-hand side, and so we obtain our desired conclusion. □

6.1.2. Resonant measures for the Dirichlet kernel

The idea behind Kolmogorov's theorem is to find a measure μ that violates (6.5). This measure is chosen to create "resonances", i.e., such that the peaks of the Dirichlet kernel all appear with the same sign. More precisely, for every positive integer N we choose

$$\mu_N := \frac{1}{N}\sum_{j=1}^{N} \delta_{x_{j,N}}, \tag{6.6}$$

where the $x_{j,N}$ are close to j/N. Then

$$(S_n\mu_N)(x) = \frac{1}{N}\sum_{j=1}^{N} \frac{\sin((2n+1)\pi(x - x_{j,N}))}{\sin(\pi(x - x_{j,N}))}. \tag{6.7}$$

If x is fixed, we will then argue that there exists n such that the summands on the right-hand side all have the same sign (for this we need to make a careful choice of the $x_{j,N}$). Thus, the size of the entire sum is given by

$$\sum_{j=1}^{N} \frac{1}{j} \simeq \log N,$$

because of the denominators in (6.7); this clearly contradicts (6.5).

The choice of the points $x_{j,N}$ is based on the following lemma, due to Kronecker.

Lemma 6.4 *Assume that* $(\theta_1, \ldots, \theta_d) \in \mathbb{T}^d$ *is* incommensurate, *i.e., for any* $(n_1, \ldots, n_d) \in \mathbb{Z}^d \setminus \{0\}$ *one has that*

$$n_1\theta_1 + \cdots + n_d\theta_d \notin \mathbb{Z}.$$

Then the orbit

$$\left\{(n\theta_1, \ldots, n\theta_d) \bmod \mathbb{Z}^d \,\middle|\, n \in \mathbb{Z}\right\} \subset \mathbb{T}^d \tag{6.8}$$

is dense in $\mathbb{T}^d$.

Proof It will suffice to show that, for any smooth function f on $\mathbb{T}^d$,

$$\frac{1}{N}\sum_{n=1}^{N} f(n\theta_1, \ldots, n\theta_d) \to \int_{\mathbb{T}^d} f(x)\,dx. \tag{6.9}$$

Indeed, if the orbit (6.8) is not dense then we could find $f \geq 0$ such that the set $\{\mathbb{T}^d \mid f > 0\}$ does not intersect it. Clearly, this would contradict (6.9). To prove (6.9), expand f in a Fourier series; see Proposition 1.15. Then

$$\begin{aligned}\frac{1}{N}\sum_{n=1}^{N} f(n\theta_1, \ldots, n\theta_d) &= \frac{1}{N}\sum_{n=1}^{N}\sum_{\nu\in\mathbb{Z}^d} \hat{f}(\nu)e^{2\pi i n\theta\cdot\nu} \\ &= \hat{f}(0) + \sum_{\nu\in\mathbb{Z}^d\setminus\{0\}} \hat{f}(\nu)\frac{1}{N}\left(\frac{1 - e^{2\pi i(N+1)\theta\cdot\nu}}{1 - e^{2\pi i\theta\cdot\nu}}\right),\end{aligned}$$

where the ratio on the right-hand side is well defined, by our assumption. Clearly, $\hat{f}(\nu)$ is rapidly decaying in $|\nu|$ since $f \in C^\infty(\mathbb{T}^d)$. Thus, since

$$\left|\frac{1}{N}\left(\frac{1 - e^{2\pi i(N+1)\theta\cdot\nu}}{1 - e^{2\pi i\theta\cdot\nu}}\right)\right| \leq 1$$

for all $N \geq 1$, $\nu \neq 0$, it follows that

$$\lim_{N\to\infty} \sum_{\nu\in\mathbb{Z}^d\setminus\{0\}} \hat{f}(\nu)\frac{1}{N}\left(\frac{1 - e^{2\pi i(N+1)\theta\cdot\nu}}{1 - e^{2\pi i\theta\cdot\nu}}\right) = 0,$$

and we are done. □

We can now carry out our construction of the μ_N (see (6.6)).

Lemma 6.5 *There exists a sequence* $\{\mu_N\}_{N\geq 1}$ *of probability measures on* $\mathbb{T}$ *with the property that, for each* $N \geq 1$,

$$\limsup_{n\to\infty} (\log N)^{-1}|S_n\mu_N(x)| > 0$$

for a.e. $x \in \mathbb{T}$.

Proof For every $N \geq 1$ and $1 \leq j \leq N$ choose $x_{j,N} \in \mathbb{T}$ that satisfy

$$\left|x_{j,N} - \frac{j}{N}\right| \leq \frac{1}{N^2}$$

and such that $\{x_{j,N}\}_{j=1}^{N} \in \mathbb{T}^N$ is an incommensurate vector. This can be done since commensurate vectors have measure zero. Clearly, the set of $x \in \mathbb{T}$ such that $\{2(x - x_{j,N})\}_{j=1}^{N} \in \mathbb{T}^N$ is a commensurate vector is at most countable. It follows that, for a.e. $x \in \mathbb{T}$,

$$\left\{\{2n(x - x_{j,N})\}_{j=1}^{N} \bmod \mathbb{Z}^N \,\middle|\, n \in \mathbb{Z}\right\}$$

is dense in $\mathbb{T}^N$. Hence, for a.e. $x \in \mathbb{T}$,

$$\left\{\{(2n+1)(x - x_{j,N})\}_{j=1}^{N} \bmod \mathbb{Z}^N \,\middle|\, n \in \mathbb{Z}\right\}$$

is also dense in $\mathbb{T}^N$. It follows that for a.e. $x \in \mathbb{T}$ there are infinitely many choices of $n \geq 1$ such that

$$\sin((2n+1)\pi(x - x_{j,N})) \geq \tfrac{1}{2}$$

for all $1 \leq j \leq N$. In particular, for those n the sum in (6.7) satisfies

$$|(S_n \mu_N)(x)| \geq \frac{1}{2N} \sum_{j=1}^{N} \frac{1}{|\sin(\pi(x - x_{j,N}))|} \geq C \frac{1}{N} \sum_{j=1}^{N} \frac{1}{j/N} \geq C \log N,$$

as desired. □

Finally, combining Lemma 6.5 with Corollary 6.3 yields Kolmogorov's theorem. The second statement in the following theorem follows from Theorem 6.2.

Theorem 6.6 *There exists $f \in L^1(\mathbb{T})$ such that $S_n f$ does not converge almost everywhere. Furthermore, there exists $f \in L^1(\mathbb{T})$ such that*

$$\{x \in \mathbb{T} \,|\, \mathcal{C}f(x) < \infty\}$$

has measure zero.

It is known that this statement also holds for everywhere divergence. In fact, Konyagin [70] showed the following stronger result. Let $\phi : [0, \infty) \to [0, \infty)$ be nondecreasing and let it satisfy $\phi(u) = o(u\sqrt{\log u}/\sqrt{\log \log u})$ as $u \to \infty$. Then there exists $f \in L^1(\mathbb{T})$ such that

$$\int_{\mathbb{T}} \phi(|f(x)|)\, dx < \infty$$

and $\limsup_{m \to \infty} |S_m f(x)| = \infty$, for every $x \in \mathbb{T}$.

6.2. Random Fourier series: the basics

We now turn our attention to Fourier series of the form

$$\sum_{n=0}^{\infty} X_n e(n\theta), \tag{6.10}$$

where $\{X_n\}_{n=0}^{\infty}$ are independent complex-valued random variables that satisfy $a_n^2 := \mathbb{E}\,|X_n|^2 < \infty$ and $\mathbb{E}\,X_n = 0$. Typical examples are $X_n = \pm a_n$, where the signs are symmetric and random (the Rademacher series) and $X_n = a_n e^{2\pi i \omega_n}$, where ω_n is uniformly distributed in $[0, 1]$ (the Steinhaus series). Finally, one can take $X_n = \gamma_n a_n$, where the γ_n are standard normal variables (the Gaussian series), or some combination or variant of these examples. The most basic questions about the random series (6.10) concern its convergence: whether this is a.e., L^p, uniform, etc. Note that the events characterized by these properties are tail events. In other words, they are either a.s. true or a.s. false.

From the Kolmogorov's estimate (5.9) we immediately recognize the convergence or divergence of $\sum_n \mathbb{E}\,|X_n|^2$ as a decisive property.

6.2.1. The L^2 theory of random Fourier series

The most basic result on (6.10) reads as follows.

Theorem 6.7 *Under the stated conditions,* (6.10) *displays the following dichotomy: if $\sum_n a_n^2 < \infty$ then a.s.* (6.10) *converges a.e. and in the sense of $L^2(\mathbb{T})$. If $\sup_n a_n^{-p}\,\mathbb{E}\,|X_n|^p < \infty$ for some finite $p > 2$ then a.s.* (6.10) *converges in $L^p(\mathbb{T})$.*

If $\sum_n a_n^2 = \infty$, and $\mathbb{E}\,|X_n|^4 \le M a_n^4$ for all $n \ge 0$ with some finite constant M, then a.s. (6.10) *diverges a.e. and is not the Fourier series of any measure.*

In particular, if $\sum_n a_n^2 < \infty$ then the Rademacher, Steinhaus, or Gaussian Fourier series all converge a.s. in all L^p norms with finite p, as well as a.e. In Exercise 6.2 the reader will be asked to show for these three classes that a.s. $e^{k|S|^2} \in L^1(\mathbb{T})$ for any $k > 0$, where S is a Fourier series. Proposition 6.9 below demonstrates that in general we have $S \notin L^\infty(\mathbb{T})$ without making any further assumptions apart from the ℓ^2 condition.

Theorem 6.7 also shows that there are no subtleties hidden in the summation over $n \ge 0$ in (6.10). This is not immediately obvious because we have a Hilbert transform. However, by this theorem matters are essentially reduced to L^2 and orthogonality, whence it is customary (for convenience) to sum over $n \ge 0$ rather than over all integers.

In the proof of the theorem we will use a generalized triangle inequality, which is given in the following exercise.

Exercise 6.1

(a) Show that, for any $\infty \geq p \geq q \geq 1$ and on any measure space (X, μ),

$$\Big\| \, \|\{f_j\}_{j\in\mathbb{Z}}\|_{\ell^q(\mathbb{Z})} \Big\|_{L^p(\mu)} \leq \Big\| \{\|f_j\|_{L^p(\mu)}\}_{j\in\mathbb{Z}} \Big\|_{\ell^q(\mathbb{Z})}.$$

Also show that for $\infty \geq q \geq p \geq 1$ one has the reverse inequality.

(b) Show that (4.12) fails for every $1 \leq p < 2$.

Also needed for the proof is the second-moment method in probability. It can only be applied if $\mathbb{E}\, X^2$ is comparable with $(\mathbb{E}\, X)^2$. Note that this excludes highly localized examples such as $X = \delta^{-1}\chi_{[0<x<\delta]}$ as a random variable on $[0, 1]$ when $\delta \to 0$.

Lemma 6.8 *Let $X \geq 0$ be an L^1 random variable. Then, for any $0 < \varepsilon < 1$,*

$$\mathbb{P}(\{X > \varepsilon\mathbb{E}\, X\}) \geq (1-\varepsilon)^2 \frac{(\mathbb{E}\, X)^2}{\mathbb{E}\, X^2}. \tag{6.11}$$

Proof Fix ε and denote the event appearing on the left-hand side by A. Then by inspection and the Cauchy–Schwarz inequality,

$$\mathbb{E}\, X = \mathbb{E}\,(\chi_A X) + \mathbb{E}\,(\chi_{A^c} X) \leq \mathbb{P}(A)^{1/2}(\mathbb{E}\, X^2)^{1/2} + \varepsilon\mathbb{E}\, X,$$

whence the lemma. □

Proof of Theorem 6.7 Assume that $\sum_{n=0}^{\infty} a_n^2 < \infty$ and write

$$S_N(\theta) := \sum_{n=0}^{N} X_n e(n\theta).$$

Then by Lemma 5.7, for any $\varepsilon > 0$ and $\theta \in \mathbb{T}$,

$$\mathbb{P}\left(\left\{\max_{N<n\leq M} |S_n(\theta) - S_N(\theta)| > \varepsilon\right\}\right) \leq \varepsilon^{-2} \sum_{n>N} |a_n|^2$$

for all integers $M > N > 1$. As usual, this implies that for every θ the series $\{S_n(\theta)\}_n$ converges a.s. But then a.s. one has a.e. convergence, by Fubini's theorem. Next, we apply Kolmogorov's estimate in the Hilbert space L^2, see (5.9):

$$\mathbb{P}\left(\left\{\max_{N<n\leq M} \|S_n - S_N\|_2 > \varepsilon\right\}\right) \leq \varepsilon^{-2} \sum_{n>N} |a_n|^2$$

for every $\varepsilon > 0$. One concludes that $\{S_n\}$ converges a.s. in the sense of $L^2(\mathbb{T})$.

For $p > 2$ we use Proposition 5.15 and the generalized triangle inequality as stated in Exercise 6.1(a) above. In fact, for any $\theta \in \mathbb{T}$,

$$\sup_N \mathbb{E}\,|S_N(\theta)|^p \le C \sup_N \mathbb{E}\left(\sum_{n=0}^{N} |X_n|^2\right)^{p/2}$$

$$\le C \left\|\left\{\left(\mathbb{E}\,|X_n|^p\right)^{1/p}\right\}\right\|_{\ell^2}^p \le C\left(\sum_{n=0}^{\infty} |a_n|^2\right)^{p/2}.$$

Integrating over θ shows that

$$\sup_N \mathbb{E}\,\|S_N\|_{L^p(\mathbb{T})}^p < \infty$$

whence by Fatou's lemma one has $\lim_{n\to\infty} S_n =: S \in L^p(\mathbb{T})$ a.s. and $\mathbb{E}\,\|S\|_p^p < \infty$.

Next, suppose that $\sum_n a_n^2 = \infty$. Then apply Lemma 6.8 with $X = |S_n(\theta)|^2$ for $\theta \in \mathbb{T}$ and $\varepsilon = \frac{1}{2}$:

$$\mathbb{P}\left(\left\{|S_n(\theta)|^2 > \frac{1}{2}\,\mathbb{E}\,|S_n|^2\right\}\right) \ge \frac{1}{4}\frac{(\mathbb{E}\,|S_n(\theta)|^2)^2}{\mathbb{E}\,|S_n(\theta)|^4}.$$

By our fourth-moment assumption (see Theorem 6.7), we can easily verify that

$$\mathbb{E}\,|S_N(\theta)|^4 \le C\left(\sum_{n=0}^{N} a_n^2\right)^2 = C\left(\mathbb{E}\,|S_N(\theta)|^2\right)^2$$

for all $N \ge 1$. In summary,

$$\mathbb{P}\left(\left\{|S_n(\theta)|^2 > \frac{1}{2}\,\mathbb{E}\,|S_n|^2\right\}\right) \ge c_0 \quad \forall\, n \ge 1,$$

where $c_0 > 0$ is some constant. This implies (via the zero–one law) that a.s.

$$\limsup_{n\to\infty} |S_n(\theta)| = \infty.$$

Hence, a.s. the random series diverges a.e., as claimed. By an analogous argument one finds that the Fejér means associated with (6.10),

$$\sigma_N(\theta) := \sum_{n=0}^{N}\left(1 - \frac{n}{N+1}\right) X_n e(n\theta),$$

a.s. diverge a.e. as $N \to \infty$.

Now assume that (6.10) is the Fourier series for some measure $\mu \in \mathcal{M}(\mathbb{T})$. In that case the Fejér means $(\sigma_n\,\mu)(\theta)$ converge for a.e. θ to, say, μ' (such a limit exists a.e.); this is a general property of radially bounded approximate identities; see Exercise 2.6 and the following exercise. However, as we have just seen, these Fejér means diverge almost surely almost everywhere. □

Exercise 6.2 Show that for the convergence part of Theorem 6.7 one has the following property: almost surely, $e^{\tau|S|^2} \in L^1(\mathbb{T})$ for any $\tau > 0$ where S is a Rademacher, Steinhaus, or Gaussian series as in (6.10). Moreover, give a proof, of the fact that in the divergent case (6.10) is not the Fourier series of a measure, that *does not* use the result found in Exercise 2.6. Rather, obtain a contradiction from Lemma 6.8.

Another way to view Theorem 6.7 is as follows. For any $f \in L^2(\mathbb{T})$ consider the random L^2 function

$$F_\omega(\theta) := \sum_{n\in\mathbb{Z}} \omega_n \hat{f}(n)\, e(n\theta), \tag{6.12}$$

with i.i.d. symmetric signs $\omega_n = \pm 1$. Then, by Theorem 6.7, we have a.s. that $f_\omega \in L^p(\mathbb{T})$ for any finite p. In other words, the randomness serves to regularize a function f that is no better than $L^2(\mathbb{T})$ to begin with. However, if $f \in L^q(\mathbb{T}) \setminus L^2(\mathbb{T})$ for any $1 \le q < 2$ then the random signs have no effect: a.s. F_ω does not exist, even as a measure. Remarkably, the random signs destroy the L^q convergence which holds for (6.12) provided that $\omega_n = 1$ for all $n \in \mathbb{Z}$ and $1 < q < 2$.

6.2.2. The L^∞ case

The following proposition shows that in general $f \in L^2(\mathbb{T})$ cannot be "upgraded" to $L^\infty(\mathbb{T})$ by means of randomization.

Proposition 6.9 *Let* $\{c_n\} \in \ell^2(\mathbb{Z}^+) \setminus \ell^1(\mathbb{Z}^+)$ *be a sequence of reals. Then* $\sum_{n=1}^\infty \pm\, c_n e(3^n\theta) \in L^2(\mathbb{T}) \setminus L^\infty(\mathbb{T})$ *for* every *choice of signs.*

Proof Define the measures

$$\mu_N(d\theta) := \prod_{n=1}^{N} \big(1 + \epsilon_n \cos(2\pi 3^n\theta)\big)\, d\theta, \tag{6.13}$$

where $\epsilon_n = \pm 1$ are chosen such that $\epsilon_n c_n = |c_n|$ for all $n \ge 1$. By inspection, the μ_N are positive measures of mass 1 and

$$\widehat{\mu_N}(3^n) = \tfrac{1}{2}\epsilon_n \quad \forall\, 1 \le n \le N, \tag{6.14}$$

since the representation of an integer as a finite sum $\sum \pm 3^n$ is unique. Thus

$$\int_{\mathbb{T}} \sum_{n=1}^{N} c_n e(3^n\theta)\, \mu_N(d\theta) = \tfrac{1}{2} \sum_{n=1}^{N} |c_n|,$$

whence

$$\sum_{n=1}^{N} |c_n| \leq 2 \left\| \sum_{n=1}^{N} \pm c_n\, e(3^n\theta) \right\|_{\infty}$$

for any choice of signs, and any $N \geq 1$. Letting $N \to \infty$ yields the desired conclusion. □

The construction carried out in the above proof is an example of a *Sidon set*, and the measure μ_N is known as a *Riesz measure*. We shall investigate these notions in more detail in the following section. While Proposition 6.9 may seem surprising at first sight, it relies on a very natural mechanism: that the characters $e(3^n\theta)$ have many features in common with an i.i.d. sequence, as can be seen by comparison with the Rademacher sequences. To be more specific, for any given n consider the set

$$I_n := \left\{\theta \in \mathbb{T} \,\middle|\, \mathrm{Re}\left(c_n e(3^n\theta)\right) > \tfrac{1}{2}|c_n|\right\} \subset \mathbb{T}.$$

It is not hard to see that owing to the tripling of frequency as $n \to n+1$ one retains a large portion of I_{n+1} within I_n, etc. In this way, $\bigcap_n I_n \neq \emptyset$ as desired. By the same logic, the randomness of the signs does not have any effect here, and this explains the deterministic nature of the result.

Since Theorem 6.7 is sharp in the sense that the ℓ^2 condition does not imply the boundedness or continuity of (6.10), it is natural to ask for a condition that will ensure this. As one would expect, it suffices to make "small" (in some sense) changes to the ℓ^2 condition (due to the sub-Gaussian bound (5.5) or the central limit theorem). We now make this precise. Recall that a finite sequence $\{x_j\}$ in some metric space (X, d) is called an ϵ-net if for every $x \in X$ there is some element of the sequence, say x_k, with $d(x, x_k) < \epsilon$.

Lemma 6.10 *Let $\{r_j\}_1^\infty$ be the Rademacher sequence. Then, for any $\{a_j\} \in \mathbb{C}$,*

$$\mathbb{P}\left(\left\{\sup_{0\leq\theta\leq 1} \left|\sum_{j=1}^{N} r_j\, a_j e(j\theta)\right| > 6\left(\sum_{j=1}^{N} |a_j|^2\right)^{1/2} \sqrt{\log N}\right\}\right) \leq N^{-2}$$

for all integers $N \geq N_0$, where N_0 is some absolute constant.

Proof By Lemma 5.5 one has

$$\mathbb{P}\left(\left\{\left|\sum_{j=1}^{N} r_j\, a_j e(j\theta)\right| > 3\left(\sum_{j=1}^{N}|a_j|^2\right)^{1/2}\sqrt{\log N}\right\}\right) \le N^{-4} \tag{6.15}$$

for any fixed $\theta \in \mathbb{T}$ and large N. Bernstein's inequality, Proposition 1.11, implies that

$$\left\|\sum_{j=1}^{N} r_j\, a_j e(j\theta)\right\|_\infty \le 2\max_k \left|\sum_{j=1}^{N} r_j\, a_j e(j\theta_k)\right|,$$

where $\{\theta_k\}_k$ is a $(CN)^{-1}$-net, C being some large constant. Summing the probabilities in (6.15) over this net concludes the proof. □

As a corollary we can give the following sufficient condition, which ensures continuity when the signs are random.

Corollary 6.11 *For any $\varepsilon > 0$ one has the following property: the randomization* (6.12) *of an arbitrary $f \in H^\varepsilon(\mathbb{T})$ belongs a.s. to $C(\mathbb{T})$.*

Proof For any $f \in L^2$ define

$$(P_j f)(\theta) := \sum_{|n|\simeq 2^j} \hat{f}(n)\, e(n\theta), \quad j \ge 0.$$

Fix $\varepsilon > 0$ and $f \in H^\varepsilon(\mathbb{T})$. Denote the randomization of f by F_ω. Then, by Lemma 6.10, one has a.s.

$$\|P_j F_\omega\|_\infty \le Cj\|P_j F_\omega\|_2 = Cj\|P_j f\|_2$$

for all but finitely many j, say for $j \ge j_0$ (the latter being random). But this means that a.s.

$$\sum_{j\ge j_0} \|P_j F_\omega\|_\infty \le C\sum_{j\ge j_0} j\|P_j f\|_2 \le C\|f\|_{H^\varepsilon},$$

by the Cauchy–Schwarz inequality. Invoking the a.e. convergence of Theorem 6.7 now concludes the proof. □

Exercise 6.3 Using Lemma 6.10, prove a weaker form of Proposition 1.14 in which $C^{1/2}(\mathbb{T})$ is replaced by $C^\alpha(\mathbb{T})$ with arbitrary $\alpha < \frac{1}{2}$. To be more specific, substitute a random construction for the Rudin–Shapiro polynomials in the proof of that proposition.

To conclude this introductory section, we would like to point out a specific distinction between the integers in general and the powers of 3, as it emerges from Proposition 6.9 and Corollary 6.11. For any $\Lambda \subset \mathbb{Z}$ and Banach space $B \subset L^2(\mathbb{T})$ we denote by $B_\Lambda(\mathbb{T})$ those functions in B with Fourier support in Λ. For example, the $C_\Lambda(\mathbb{T})$ are those $f \in C(\mathbb{T})$ such that $\hat{f}(n) = 0$ for all $n \in \mathbb{Z} \setminus \Lambda$. Moreover, we denote by $B_\Lambda^\varepsilon(\mathbb{T})$ those functions in $L^2(\mathbb{T})$ for which the random series

$$\sum_{n \in \Lambda} \varepsilon_n \hat{f}(n) e(n\theta) \in B \quad \text{a.s.},$$

where the $\varepsilon_n = \pm 1$ are i.i.d. and symmetric. On the one hand, by Proposition 6.9, if $\Lambda = \{3^n \mid n \in \mathbb{N}\}$ then $C_\Lambda^\varepsilon(\mathbb{T}) = C_\Lambda(\mathbb{T})$. On the other hand, it is clear from Corollary 6.11 that this fails for $\Lambda = \mathbb{Z}$ or if Λ equals any infinite arithmetic progression.

In the following section we shall give a complete characterization of those $\Lambda \subset \mathbb{Z}$ for which the equality $C_\Lambda(\mathbb{T}) = C_\Lambda^\varepsilon(\mathbb{T})$ holds.

6.3. Sidon sets

Definition 6.12 An infinite subset $\Lambda \subset \mathbb{Z}$ is called a *Sidon set* provided that for every trigonometric polynomial $p = \sum_{\lambda \in \Lambda} a_\lambda e_\lambda$ one has $\|\{a_\lambda\}\|_{\ell^1(\Lambda)} \le C(\Lambda)\|p\|_{C(\mathbb{T})}$, with a constant $C(\Lambda)$ that does not depend on the trigonometric polynomial. The smallest such constant is called the Sidon constant of Λ and is denoted $K = K(\Lambda)$.

In other words, $\Lambda \subset \mathbb{Z}$ is Sidon if and only if $\mathbb{A}_\Lambda(\mathbb{T}) = C_\Lambda(\mathbb{T})$. Indeed, since $\|f\|_{C(\mathbb{T})} \le \|f\|_{\mathbb{A}(\mathbb{T})}$ the open-mapping theorem implies that $\|f\|_{\mathbb{A}(\mathbb{T})} \le K\|f\|_{C(\mathbb{T})}$ in that case, which agrees with Definition 6.12. Note that this definition carries over verbatim to any compact Abelian group in place of $\mathbb{T}$ (see Section 4.3 for the Fourier transform in that degree of generality). In particular we may take $\mathbb{T}^d$ instead of $\mathbb{T}$, but we will restrict ourselves to the latter case for the sake of simplicity. Inspection of the proof of Proposition 6.9 reveals that $\Lambda := \{3^n \mid n \ge 0\}$ is Sidon. Indeed, we constructed a measure $\mu \ge 0$ on $\mathbb{T}$ with $\mu(\mathbb{T}) = 1$ and such that $\hat{\mu}(\lambda)$ equals a prescribed bounded sequence, namely $\pm\frac{1}{2}$. This was accomplished by taking the weak-$*$ limit of the Riesz measures μ_N; see (6.14). Given any $f \in C_\Lambda(\mathbb{T})$ we can thus construct a measure μ of bounded mass with the property that $\hat{f}\hat{\mu} = \frac{1}{2}|\hat{f}|$, whence Λ is Sidon:

$$\tfrac{1}{2} \sum_{n \in \Lambda} |\hat{f}(n)| = \sum_{n \in \mathbb{Z}} \widehat{f * \mu}(n) = (f * \mu)(0) \le \|f * \mu\|_\infty \le \|f\|_{C(\mathbb{T})} \|\mu\|.$$

6.3.1. Interpolating measures

Both the idea of an interpolating measure and the actual construction of such a measure are of a more general nature than we have considered so far; they have little to do with the specific choice of the powers of 3. We begin with a characterization of Sidon sets via interpolating measures.

Proposition 6.13 *For any $\Lambda \subset \mathbb{Z}$ the following are equivalent:*

(i) *Λ is Sidon;*

(ii) *for every $\{b_\ell\}_{\ell\in\Lambda} \in \ell^\infty(\Lambda)$ there exists a measure $\mu \in \mathcal{M}(\mathbb{T})$ with $\hat{\mu}(\ell) = b_\ell$ for all $\ell \in \Lambda$;*

(iii) *for every complex sequence $\{b_\ell\}_{\ell\in\Lambda}$ with $|b_\ell| = 1$ for all $\ell \in \Lambda$ there exists a measure $\mu \in \mathcal{M}(\mathbb{T})$ for which*

$$\sup_{\ell\in\Lambda} |\hat{\mu}(\ell) - b_\ell| < 1. \tag{6.16}$$

Proof If Λ is Sidon and $\{b_\ell\}_{\ell\in\Lambda}$ is in the unit ball of $\ell^\infty(\Lambda)$ then define a functional on $C_\Lambda(\mathbb{T})$ by

$$L(f) := \sum_{\ell\in\Lambda} b_\ell \, \hat{f}(\ell).$$

By construction $\|L\| \le K$, where K is the Sidon constant. Extend L by the Hahn–Banach theorem to $C(\mathbb{T})$ with the same norm, and denote the extension by $\tilde{L}$. By the Riesz representation theorem, there exists $\mu \in \mathcal{M}(\mathbb{T})$ with $\|\mu\| \le K$ and

$$\tilde{L}(f) = \int_{\mathbb{T}} f(\theta)\,\mu(d\theta) \quad \forall f \in C(\mathbb{T}),$$

whence

$$b_\ell = \tilde{L}(e(-\ell\cdot)) = \hat{\mu}(\ell).$$

as desired. The implication (ii) $\Longrightarrow$ (iii) is trivial. To prove (iii) $\Longrightarrow$ (i), let $f \in C_\Lambda(\mathbb{T})$ be a trigonometric polynomial and let b_ℓ satisfy $b_\ell \hat{f}(\ell) = |\hat{f}(\ell)|$, $|b_\ell| = 1$ for all $\ell \in \Lambda$. There exists $\mu \in \mathcal{M}(\mathbb{T})$ and $\delta > 0$ such that

$$|\hat{\mu}(\ell) - b_\ell| < 1 - \delta \quad \forall \ell \in \Lambda.$$

Then

$$\begin{aligned}\sum_{\ell\in\Lambda} |\hat{f}(\ell)| &= \sum_{\ell\in\mathbb{Z}} \widehat{\mu * f}(\ell) + \sum_{\ell\in\Lambda} (b_\ell - \hat{\mu}(\ell))\hat{f}(\ell)\\ &\le \|\mu * f\|_\infty + (1-\delta)\sum_{\ell\in\Lambda} |\hat{f}(\ell)|,\end{aligned}$$

which implies that

$$\sum_{\ell\in\Lambda} |\hat{f}(\ell)| \le C\|f\|_{C(\mathbb{T})}, \quad C := \delta^{-1}\|\mu\|.$$

If $f \in C_\Lambda(\mathbb{T})$ then this same bound shows that we can uniformly approximate f by trigonometric polynomials obtained by replacing Λ with finite subsets of itself. □

Exercise 6.4 Show the following: $\Lambda \subset \mathbb{Z}$ is Sidon if and only if for every sequence $\{b_\ell\}_{\ell\in\Lambda}$ with $b_\ell = \pm 1$ for all $\ell \in \Lambda$ and every $\delta > 0$ there exists a measure $\mu \in \mathcal{M}(\mathbb{T})$ with

$$\sup_{\ell\in\Lambda} |\hat{\mu}(\ell) - b_\ell| < \delta. \tag{6.17}$$

Before generalizing the concept of a Riesz measure, we observe that Sidon sets cannot be too close to an *arithmetic progression*, since arithmetic sequences are far from lacunary whereas a Sidon set should behave in some ways like a lacunary sequence. While this is a completely deterministic statement, we shall use a random construction in the proof.

Proposition 6.14 *Let $\Lambda \subset \mathbb{Z}$ be a Sidon set with Sidon constant K. Then, for any arithmetic progression $\mathcal{A}$ of length N, one has*

$$\#(\Lambda \cap \mathcal{A}) \le CK^2 \log N,$$

where C is an absolute constant.

Proof Suppose that $\mathcal{A} = \{pn + m \mid 1 \le n \le N\}$ with integer m and integer $p \ge 1$. We may assume that $\mathcal{A} = \{n \mid 1 \le n \le N\}$: indeed, if

$$\Lambda' := \mathcal{A} \cap \Lambda = \{pn + m \mid n \in \Lambda''\}$$

then the polynomials

$$p(\theta) = \sum_{n\in\Lambda''} a_n e((pn+m)\theta), \quad \tilde{p}(\theta) = \sum_{n\in\Lambda''} a_n e(n\theta)$$

are equivalent when regarded as Sidon sets. Suppose that $\#(\mathcal{A} \cap \Lambda) = M$. Apply Lemma 6.10 with $a_j = \chi_{\mathcal{A}\cap\Lambda}(j)$ to conclude from the Sidon property that, for some choice of signs $\pm$,

$$M \le K \left\| \sum_{j\in\mathcal{A}\cap\Lambda} \pm a_j e(j\cdot) \right\|_\infty \le CK\sqrt{M \log N},$$

whence the claim. □

6.3.2. Riesz measures

We now turn to the construction of Riesz measures.

Definition 6.15 For any $\Lambda \subset \mathbb{Z}$, denote by $R_s(\Lambda, n)$ the number of representations of $n \in \mathbb{Z}$ in the form

$$n = \pm \ell_1 \pm \ell_2 \pm \cdots \pm \ell_s, \quad \ell_j \in \Lambda, \ \ell_1 < \ell_2 < \cdots < \ell_s.$$

In the proof of Proposition 6.9 (see (6.14)) we noted that for $\Lambda = \{3^n \mid n \geq 0\}$ one has $R_s(\Lambda, n) \leq 1$ for all $n \in \mathbb{Z}$. This was originally used in the construction of the Riesz measures. However, less is needed as we shall now demonstrate.

Theorem 6.16 *Suppose that $\Lambda \subset \mathbb{Z}$ satisfies $R_s(\Lambda, 0) \leq B^s$ for all $s \geq 1$, where $1 \leq B < \infty$ is fixed. If $\varphi = \pm 1$ on $\Lambda \cup (-\Lambda)$ then for every $\varepsilon > 0$ there exists a measure $\nu \in \mathcal{M}(\mathbb{T})$ with*

$$\begin{aligned} |\hat{\nu}(\ell) - \varphi(\ell)| &< \varepsilon \quad \forall \ell \in \Lambda \cup (-\Lambda), \\ |\hat{\nu}(\ell)| &< \varepsilon \quad \forall \ell \in \mathbb{Z} \setminus \big(\Lambda \cup (-\Lambda)\big). \end{aligned} \tag{6.18}$$

Let $\rho > 1$ be arbitrary but fixed, and suppose that $\Lambda := \{n_j\}_{j=1}^{\infty}$ is an infinite increasing sequence of integers with $n_{j+1} > \rho n_j$ for all $j \geq 1$. Then we call Λ *lacunary*. A standard corollary of Theorem 6.16 is that *lacunary sequences* are Sidon.

Corollary 6.17 *Any lacunary Λ is Sidon. The same applies to any finite union of lacunary sequences.*

Proof This is done by thinning the sequence Λ. More precisely, write $\Lambda = \bigcup_k \Lambda_k$ as a finite disjoint union with Λ_k a lacunary sequence with $\rho \geq 3$. Fix $\varphi : \Lambda \to \{-1, 1\}$. Then on each Λ_j the condition of Theorem 6.16 holds, and so we obtain measures μ_j which satisfy (6.18) on Λ_j. Taking ε sufficiently small, we see that $\sum_j \mu_j$ has the desired properties. □

Exercise 6.5 Show that

$$\Lambda := \left\{3^{4\times 2^m} + 3^{2^m + j} \,\middle|\, 0 \leq j \leq 2^m - 1, \ m \geq 0\right\}$$

is Sidon but not a finite union of lacunary sets.

Proof of Theorem 6.16 We may assume that $0 \notin \Lambda$ and $\Lambda \cap (-\Lambda) = \emptyset$. We now make the following claim:

$$\sum_{s=1}^{\infty} (2B)^{-s} R_s(\Lambda, n) \leq 2 \quad \forall n \in \mathbb{Z}. \tag{6.19}$$

Let $\beta := (2B)^{-1}$ and define

$$P_N(x) := \prod_{k=1}^{N}(1 + 2\beta\cos(2\pi n_k x)),$$

where $\{n_k\}_{k\geq 1}$ is some enumeration of Λ. By construction, $P_N \geq 0$ and

$$P_N(x) = 1 + \sum_{n\in\Lambda} C_N(n)\, e(nx),$$

$$0 \leq C_N(n) \leq \sum_{s=1}^{N} \beta^s R_s(\Lambda, n).$$

In particular,

$$0 \leq C_N(0) \leq \sum_{s=1}^{N} \beta^s R_s(\Lambda, 0) \leq \sum_{s=1}^{\infty}(\beta B)^s \leq 1.$$

Thus $P_N \geq 0$ implies that $\| P_N \|_1 = 1 + C_N(0) \leq 2$, whence also

$$C_N(n) = |\widehat{P_N}(n)| \leq 2 \quad \forall n \neq 0.$$

Passing to the limit $N \to \infty$ yields

$$\lim_{N\to\infty} C_N(n) = \sum_{s=1}^{\infty} \beta^s R_s(\Lambda, n),$$

and the claim (6.19) holds. Changing B we may therefore assume that $R_s(\Lambda, n) \leq B^s$ for all $s \geq 1$ and each $n \in \mathbb{Z}$.

Let φ be as in the statement of the theorem. We now partition $\Lambda = \Lambda_1 \cup \Lambda_2$ into disjoint sets, defined as

$$\Lambda_1 := \{n \in \Lambda \mid \varphi(n) = \varphi(-n)\},$$

$$\Lambda_2 := \{n \in \Lambda \mid \varphi(n) = -\varphi(-n)\}.$$

Set $\gamma := (AB^2)^{-1}$, with $A \geq 2$ to be determined, and

$$g(n) := \begin{cases} \gamma\varphi(n) & \text{if } n \in \Lambda_1, \\ i\gamma\varphi(n) & \text{if } n \in \Lambda_2. \end{cases}$$

We define $f_\ell(x) := 1 + g(\ell)e(\ell x) + \overline{g(\ell)}e(-\ell x)$ and form the Riesz products

$$P_{N,j}(x) := \prod_{\ell\in\Lambda_j} f_\ell(x), \quad j = 1, 2,$$

where ℓ runs through the first N elements of Λ_j relative to some fixed enumeration. Passing to the weak-$*$ limit of the measures $P_{N,j}$ yields measures

$\mu_j \geq 0$ satisfying the following properties:

$$\|\mu_j\| \leq 1 + \sum_{s=2}^{\infty} \gamma^s R_s(\Lambda, 0),$$

$$|\widehat{\mu_j}(\pm n) - \varphi(n)| \leq \sum_{s=2}^{\infty} \gamma^s R_s(\Lambda, n) \quad \forall n \in \Lambda_j,$$

$$|\widehat{\mu_j}(n)| \leq \sum_{s=2}^{\infty} \gamma^s R_s(\Lambda, n) \quad \forall n \in \mathbb{Z} \setminus (\Lambda_j \cup (-\Lambda_j) \cup \{0\}).$$

However,

$$\sum_{s=2}^{\infty} \gamma^s R_s(\Lambda, n) \leq \sum_{s=2}^{\infty} (\gamma B)^s \leq \frac{(\gamma B)^2}{1 - \gamma B} < \frac{1}{A(A-1)B^2}.$$

Taking A large enough, depending on ε, one may check from the preceding that $\nu := \gamma^{-1}(\mu_1 - i\mu_2)$ has the desired properties. □

Exercise 6.6 Show the following: if there exists $n_0 \in \Lambda$ such that $R_s(\Lambda, n_0) \leq B^s$ for all $s \geq 1$ then $R_s(\Lambda, 0) \leq 3B^{s+1}$ for all $s \geq 1$.

6.3.3. The $\Lambda(p)$ property

We shall now obtain a far-reaching generalization of Proposition 6.9. We say that a Banach space $B \subset L^1(\mathbb{T})$ that contains all possible characters is *homogeneous* if both B and the norm on B are invariant under translations and if translation is continuous in B. The spaces in which we are most interested are $C_\Lambda(\mathbb{T})$ and $L^p_\Lambda(\mathbb{T})$. The continuity assumption ensures that the trigonometric polynomials are dense in B (by convolution with the Fejér kernel). The following proposition is based on the facts that $\|f * \mu\|_B \leq \|f\|_B \|\mu\|$ for any $\mu \in \mathcal{M}(\mathbb{T})$ and B is an arbitrary homogeneous Banach space.

Proposition 6.18 *Let $\Lambda \subset \mathbb{Z}$ be a Sidon set. Then, for any homogeneous Banach space B, one has*

$$C(B, \Lambda)^{-1} \|f\|_{B_\Lambda(\mathbb{T})} \leq \Big\| \sum_{n \in \Lambda} \varepsilon_n \hat{f}(n) e(n\cdot) \Big\|_{B_\Lambda(\mathbb{T})} \leq C(B, \Lambda) \|f\|_{B_\Lambda(\mathbb{T})}, \quad (6.20)$$

for any choice of the $\varepsilon_n = \pm 1$, where the constant $C(B, \Lambda)$ depends only on the space B and Λ. The same holds for arbitrary ζ_n in place of ε_n, where $|\zeta_n| = 1$.

Proof For arbitrary ζ_n as above we can find $\mu \in \mathcal{M}(\mathbb{T})$ with $\|\mu\| \leq C(\Lambda)$ and $\hat{\mu}(n) = \zeta_n$ for $\forall n \in \Lambda$. Convolving an arbitrary trigonometric polynomial in $B_\Lambda(\mathbb{T})$ with μ concludes the proof. □

Combining Proposition 6.18 with Khinchine's inequality now implies the following basic property.

Corollary 6.19 *Let $\Lambda \subset \mathbb{Z}$ be Sidon. Then, for any $2 \le p < \infty$, one has that*

$$\|f\|_p \le C(p)\,\|f\|_2 \tag{6.21}$$

for any $f \in L^2_\Lambda(\mathbb{T})$ and that $C(p) \le C\sqrt{p}$. Moreover, $\|f\|_2 \le C\|f\|_1$ for all $f \in L^1_\Lambda(\mathbb{T})$.

Proof Set $B = L^p(\mathbb{T})$ with $2 \le p < \infty$ and apply Proposition 6.18 with random signs ± 1. Raising (6.20) to the power p, taking expectations, and applying Lemma 5.5 yields the desired conclusion. For the L^1 estimate we use the same Hölder estimate as in Lemma 5.5,

$$\|f\|_2 \le \|f\|_4^{2/3}\,\|f\|_1^{1/3} \le C\|f\|_2^{2/3}\,\|f\|_1^{1/3},$$

and we are done. The estimate on $C(p)$ is the same as for the Kinchine inequality; see Exercise 5.3. □

Sets $\Lambda \subset \mathbb{Z}$ with the property that (6.21) holds for all $f \in L^2_\Lambda(\mathbb{T})$ with some constant $C(\Lambda, p)$ are called $\Lambda(p)$-sets. We will not study these sets systematically here. We simply remark that it was shown by Rudin that for any $2 \le p < \infty$ this is a *strictly weaker property* than the property of being Sidon; see Problem 6.8(d). Therefore, Proposition 6.18 with $B = L^p(\mathbb{T})$ and $2 \le p < \infty$ fixed *does not characterize Sidon sets*. However, we shall now show that Proposition 6.18 with $B = C(\mathbb{T})$ *does characterize* Sidon sets.

Definition 6.20 For the remainder of this section, for any $f \in L^2(\mathbb{T})$ we define the randomizations

$$f_\omega(x) := \sum_{n \in \mathbb{Z}} e(\omega_n)\hat{f}(n)\,e(nx),$$

$$f_\epsilon(x) := \sum_{n \in \mathbb{Z}} \epsilon_n \hat{f}(n)\,e(nx),$$

where the ω_n are i.i.d. uniformly distributed in $[0, 1]$ and the ϵ_n form a Rademacher sequence. For any $\Lambda \subset \mathbb{Z}$ set

$$\begin{aligned} C^\epsilon_\Lambda(\mathbb{T}) &:= \{f \in L^2_\Lambda(\mathbb{T}) \mid f_\epsilon \in C(\mathbb{T}) \text{ a.s.},\ \ \mathbb{E}\,\|f_\epsilon\|_\infty < \infty\}, \\ C^\omega_\Lambda(\mathbb{T}) &:= \{f \in L^2_\Lambda(\mathbb{T}) \mid f_\omega \in C(\mathbb{T}) \text{ a.s.},\ \ \mathbb{E}\,\|f_\omega\|_\infty < \infty\}, \end{aligned} \tag{6.22}$$

with norms given by the respective expectations on the right-hand sides.

6.3.4. Rider's characterization of Sidonicity

It follows from Definition 6.20 that $C^{\omega}_{\Lambda}(\mathbb{T}) = C^{\epsilon}_{\Lambda}(\mathbb{T}) = C_{\Lambda}(\mathbb{T})$ with equivalent norms if Λ is Sidon. The following remarkable theorem due to Rider states the converse direction, thus providing a probabilistic characterization of a deterministic object. The proof is an example where Fourier analysis on a compact group (in fact, the infinite torus) enters into a probabilistic argument.

Theorem 6.21 *If $\Lambda \subset \mathbb{Z}$ satisfies $C^{\omega}_{\Lambda}(\mathbb{T}) = C_{\Lambda}(\mathbb{T})$ with equivalent norms then Λ is Sidon.*

Before proving Theorem 6.21 we point out the following nontrivial result, which follows easily from the theorem. The difficulty here lies with the fact that if $f \in C(\mathbb{T})$ then it does not follow that $\sum_{n\in E} \hat{f}(n)\, e(nx) \in C(\mathbb{T})$, where $E \subset \mathbb{Z}$ (an example is provided by the Hilbert transform). However, this is *true* for a.s. continuous functions.

Corollary 6.22 *The union of two Sidon sets in $\mathbb{Z}$ is another Sidon set.*

Proof Let $\Lambda_1, \Lambda_2 \subset \mathbb{Z}$ be Sidon sets with respective Sidon constants K_1, K_2. Since subsets of Sidon sets are again Sidon, we may assume that $\Lambda_1 \cap \Lambda_2 = \emptyset$. Denote $\Lambda := \Lambda_1 \cup \Lambda_2$. For any random sequence $\{\omega_n\}_{n\in\Lambda}$ as in Definition 6.20 define

$$\widetilde{\omega}_n := \omega_n + \tfrac{1}{2} \pmod 1 \quad \forall n \in \Lambda_2,$$

with $\widetilde{\omega}_n := \omega_n$ otherwise. Denote this random sequence by $\tilde{\omega}$.

If $f \in C^{\omega}_{\Lambda}$ then by definition $f_{\tilde{\omega}}$ has the same distribution as f_{ω}, whence

$$\mathbb{E}\,\|f_{\omega} + f_{\tilde{\omega}}\|_{\infty} \le 2\,\mathbb{E}\,\|f_{\omega}\|_{\infty}.$$

Since

$$f^{(1)}_{\omega}(x) := \tfrac{1}{2}(f_{\omega} + f_{\tilde{\omega}})(x) = \sum_{n\in\Lambda_1} \widehat{f_{\omega}}(n)\, e(nx),$$

one has

$$\|f^{(1)}_{\omega}\|_{\mathbb{A}(\mathbb{T})} \le K_1 \|f^{(1)}_{\omega}\|_{\infty} \le K_1\, \mathbb{E}\,\|f_{\omega}\|_{\infty}.$$

Similarly, the Fourier restriction of f_{ω} to Λ_2, denoted $f^{(2)}_{\omega}$, satisfies

$$\|f^{(2)}_{\omega}\|_{\mathbb{A}(\mathbb{T})} \le K_2 \|f^{(2)}_{\omega}\|_{\infty} \le K_2\, \mathbb{E}\,\|f_{\omega}\|_{\infty}.$$

Therefore, for every $f \in C^{\omega}_{\Lambda}(\mathbb{T})$ one has

$$\|f\|_{\mathbb{A}(\mathbb{T})} \le (K_1 + K_2)\,\mathbb{E}\,\|f_{\omega}\|_{\infty}$$

and Theorem 6.21 implies that Λ is Sidon, as claimed. ∎

In the following proof we shall apply the machinery (and some of the notation) of Section 4.3 to the infinite torus.

Proof of Theorem 6.21 By inspection, the underlying probability space Ω is the infinite torus $\mathbb{T}^\infty = \prod_\Lambda \mathbb{T}$ (we may of course assume that $\#\Lambda = \infty$), which carries the structure of a compact group. The normalized Haar measure on Ω is denoted by $\mathbb{P}$. By assumption there exists a constant B such that $\|f\|_\infty \le B|||f|||$, where $|||f||| := \mathbb{E}\,\|f_\omega\|_\infty$ for all trigonometric polynomials f with Fourier support in Λ. On this subspace, the functional $L(f) := \sum_{n\in\Lambda} \hat{f}(n)$ satisfies

$$|L(f)| = |f(0)| \le \|f\|_\infty \le B|||f|||.$$

Extending it by the Hahn–Banach theorem to the whole of $C^\omega_{\mathbb{Z}}(\mathbb{T})$ yields a linear functional $\tilde{L}$ with the same norm. Since $C^\omega_{\mathbb{Z}}(\mathbb{T}) = L^1(\Omega; C(\mathbb{T}))$ has dual space $L^\infty(\Omega; \mathcal{M}(\mathbb{T}))$ there exists $\mu_\omega \in L^\infty(\Omega; \mathcal{M}(\mathbb{T}))$ such that

$$\tilde{L}(f) = \int_\Omega \int_\mathbb{T} f_\omega(-\theta)\, \mu_\omega(d\theta)\, \mathbb{P}(d\omega) \quad \forall f \in C^\omega_{\mathbb{Z}}(\mathbb{T}), \qquad (6.23)$$
$$\|\mu_\omega\|_{\mathcal{M}(\mathbb{T})} \le B \text{ a.s. in } \omega.$$

We chose a minus sign for the argument in $f_\omega(-\theta)$ since it will turn out to be convenient to do so. Another convention that we shall follow in this proof is to denote the characters in the dual group of $\mathbb{T}$, which is isomorphic to $\mathbb{Z}$, as γ. In other words, $\gamma(x) = e(nx)$ for a unique $n \in \mathbb{Z}$. The random variables ω_n in f_ω can then be viewed as the projections $\varphi_\gamma : \mathbb{T}^\infty \to \mathbb{T}$ onto a single component of $\mathbb{T}^\infty$ (in other words we are identifying γ with n). Therefore, (6.23) implies that

$$1 = L(\gamma) = \int_\Omega \int_\mathbb{T} \gamma(-x)\varphi_\gamma(-\omega)\, \mu_\omega(dx)\, \mathbb{P}(d\omega) \quad \forall \gamma \in \Lambda.$$

Considering φ_γ as a character on $\Omega = \mathbb{T}^\infty$, we may rewrite this as

$$\widehat{g_\gamma}(\varphi_\gamma) = 1 \quad \forall \gamma \in \Lambda, \qquad g_\gamma(\omega) := \widehat{\mu_\omega}(\gamma). \qquad (6.24)$$

Define

$$\nu_\omega := \int_\Omega \mu_{\omega-\omega'} * \mu_{\omega'}\, \mathbb{P}(d\omega'),$$

where $*$ indicates convolution on $\mathbb{T}$. Then $\|\nu_\omega\| \le B^2$ a.s. on Ω, and we have

$$\widehat{\nu_\omega}(\gamma) = \int_\Omega \widehat{\mu_{\omega-\omega'}}(\gamma)\widehat{\mu_{\omega'}}(\gamma)\, \mathbb{P}(d\omega') = (g_\gamma \# g_\gamma)(\omega),$$

where # indicates convolution on Ω.

Now set $h_\gamma(\omega) := \widehat{\nu_\omega}(\gamma)$. Then $h_\gamma = g_\gamma \# g_\gamma$ implies that, with Ω^* the dual group of Ω,

$$\|h_\gamma\|_{\mathbb{A}(\Omega)} = \|\widehat{h_\gamma}\|_{L^1(\Omega^*)} = \|\widehat{g_\gamma}\|^2_{L^2(\Omega^*)} = \|g_\gamma\|^2_{L^2(\Omega)} \le \|g_\gamma\|^2_{L^\infty(\Omega)} \le B^2,$$

where we have used the Plancherel theorem to obtain the third equality. Furthermore,

$$\widehat{h_\gamma}(\varphi_\gamma) = [\widehat{g_\gamma}(\varphi_\gamma)]^2 = 1 \quad \forall \gamma \in \Lambda. \tag{6.25}$$

We now carry out a Riesz measure construction on Ω, using the fact that the projections $\{\varphi_\gamma\}_{\gamma\in\mathbb{Z}}$ form an independent set. This means that there are no nontrivial relations of the form $\sum_j n_j \varphi_{\gamma_j} = 0$, where the sum is finite, in the dual group Ω^* (with the composition written additively). In analogy with the Riesz measures in the proof of Theorem 6.16, we set

$$P_N = \prod_{j=1}^{N} \left(1 + \beta \varphi_{\gamma_j} + \beta \overline{\varphi_{\gamma_j}}\right),$$

where $\{\gamma_j\}_{j\ge 1}$ is some enumeration of Λ and $\beta > 0$ is small. Then $P_N \ge 0$ on $\mathbb{T}^\infty$ and, passing to the weak-$*$ limit, we obtain after division by β a measure λ on $\mathbb{T}^\infty$ with, relative to the Fourier transform on Ω,

$$\hat{\lambda}(\varphi_\gamma) = 1 \quad \forall \gamma \in \Lambda,$$
$$|\hat{\lambda}(\alpha)| < \delta \quad \forall \alpha \in \Omega^* \setminus \Lambda,$$

where we have identified the characters φ_γ, $\gamma \in \Lambda$ with the set Λ in the second line. Here $\delta > 0$ is some small parameter that will be fixed later.
Define

$$\theta_\omega := \int_\Omega \nu_{\omega-\omega'}\, \lambda(d\omega').$$

Then $\theta_\omega \in \mathcal{M}(\mathbb{T})$, $\|\theta_\omega\| \le B^2\|\lambda\|$ for almost every ω, and

$$\widehat{\theta_\omega}(\gamma) = \int_\Omega \widehat{\nu_{\omega-\omega'}}(\gamma)\, \lambda(d\omega') = (h_\gamma \# \lambda)(\omega),$$

which is continuous in ω since $h_\gamma \in \mathbb{A}(\Omega)$. Let $k_\gamma(\omega) = \widehat{\theta_\omega}(\gamma)$. Then

$$k_\gamma = \sum_{\rho\in\Lambda} \widehat{h_\gamma}(\varphi_\rho)\varphi_\rho + \sum_{\alpha\in\Omega^*\setminus\Lambda} \widehat{h_\gamma}(\alpha)\hat{\lambda}(\alpha)\,\alpha \quad \forall \gamma \in \mathbb{Z}. \tag{6.26}$$

Denoting the second sum by η_γ, we have $\|\eta_\gamma\|_{\mathbb{A}(\Omega)} < \delta B^2$.

In the final step of the argument, define a map $\mathbb{T} \to \Omega$ by

$$x \mapsto \bar{x} := \{\gamma(x)\}_{\gamma\in\Lambda}.$$

Let $\delta_x \in \mathcal{M}(\mathbb{T})$ be the δ-measure at $x \in \mathbb{T}$, and set

$$\sigma_\omega := \int_{\mathbb{T}} \theta_{\omega+\bar{x}} * \delta_x \, dx.$$

Then $\sigma_\omega \in \mathcal{M}(\mathbb{T})$ and, for $\gamma \in \mathbb{Z}$,

$$\widehat{\sigma_\omega}(\gamma) = \int_{\mathbb{T}} \widehat{\theta_{\omega+\bar{x}}}\, \widehat{\delta_x} \, dx = \int_{\mathbb{T}} k_\gamma(\omega+\bar{x})\, \overline{\gamma(x)}\, dx. \tag{6.27}$$

Moreover,

$$\int_{\mathbb{T}} \varphi_\rho(\omega+\bar{x})\overline{\gamma(x)}\, dx = \varphi_\rho(\omega) \int_{\mathbb{T}} \rho(x)\, \overline{\gamma(x)}\, dx = \varphi_\rho(\omega),$$

if $\gamma = \rho$; it vanishes otherwise. Combining the above relation with (6.25) and (6.26) in order to evaluate (6.27) yields the following:

$$\widehat{\sigma_\omega}(\gamma) = \begin{cases} \varphi_\gamma(\omega) + \tau_\gamma(\omega), & \gamma \in \Lambda, \\ \tau_\gamma(\omega), & \gamma \in \mathbb{Z} \setminus \Lambda, \end{cases}$$

where

$$\tau_\gamma(\omega) = \int_{\mathbb{T}} \eta_\gamma(\omega+\bar{x})\, \overline{\gamma(x)}\, dx.$$

Thus, for all $\gamma \in \mathbb{Z}$,

$$\|\tau_\gamma\|_\infty \le \|\tau_\gamma\|_{\mathbb{A}} < \delta B^2.$$

Choosing $\delta > 0$ small enough we conclude from Proposition 6.13 that Λ is a Sidon set, and we are done. □

We will now show that one can pass from the random case with i.i.d. variables uniformly distributed on $\mathbb{T}$ (a Steinhaus sequence) to the ± 1-symmetric Bernoulli sequence.

Corollary 6.23 *The spaces $C^{\epsilon}_{\Lambda}(\mathbb{T})$ and $C^{\omega}_{\Lambda}(\mathbb{T})$ are identical and have comparable norms. Therefore, if $\Lambda \subset \mathbb{Z}$ satisfies $C^{\epsilon}_{\Lambda}(\mathbb{T}) = C_\Lambda(\mathbb{T})$ with equivalent norms then Λ is Sidon.*

Proof This is a standard argument, based on the contraction method. Given any $\Lambda' \subset \Lambda$ and a Bernoulli sequence ϵ, let ϵ' be obtained from ϵ by reversing the sign of ϵ_n if and only if $n \in \Lambda \setminus \Lambda'$. Since $f_{\epsilon'}$ has the same distribution as f_ϵ, taking the expectation of $\|\frac{1}{2}(f_\epsilon + f_{\epsilon'})\|_\infty$ implies that

$$\mathbb{E} \left\| \sum_{n\in\Lambda'} \epsilon_n \hat{f}(n) e(n\theta) \right\|_\infty \le \mathbb{E} \left\| \sum_{n\in\Lambda} \epsilon_n \hat{f}(n) e(n\theta) \right\|_\infty . \tag{6.28}$$

Next, we claim that for any real sequence $\{\rho_n\}_{n\in\Lambda}$ with $|\rho_n| \le 1$ one has

$$\mathbb{E}\left\|\sum_{n\in\Lambda} \epsilon_n \rho_n \hat{f}(n)e(n\theta)\right\|_\infty \le \mathbb{E}\left\|\sum_{n\in\Lambda} \epsilon_n \hat{f}(n)e(n\theta)\right\|_\infty; \qquad (6.29)$$

this follows by expanding each ρ_n as $\pm\sum_{k=1}^\infty \rho_{n,k}2^{-k}$, where $\rho_{n,k} = 0$ or $\rho_{n,k} = 1$, and then invoking (6.28). Finally, if ζ_n is complex with $|\zeta_n| = 1$ then by splitting it into real and imaginary parts we infer from (6.29) that

$$\mathbb{E}\left\|\sum_{n\in\Lambda} \epsilon_n \zeta_n \hat{f}(n)e(n\theta)\right\|_\infty \le 2\,\mathbb{E}\left\|\sum_{n\in\Lambda} \epsilon_n \hat{f}(n)e(n\theta)\right\|_\infty. \qquad (6.30)$$

Now let $\{\zeta_n\}_{n\in\Lambda}$ be i.i.d. and uniformly distributed on $\mathbb{T}$ as well as independent of the Bernoulli sequence ϵ. Then we note that $\{\epsilon_n\zeta_n\}_{n\in\Lambda}$ is (probabilistically speaking) just another version of the Steinhaus sequence, whence we conclude from (6.30) that

$$\mathbb{E}\left\|\sum_{n\in\Lambda} \zeta_n \hat{f}(n)e(n\theta)\right\|_\infty \le 2\,\mathbb{E}\left\|\sum_{n\in\Lambda} \epsilon_n \hat{f}(n)e(n\theta)\right\|_\infty. \qquad (6.31)$$

Therefore $C^\epsilon_\Lambda(\mathbb{T}) \subset C^\omega_\Lambda(\mathbb{T})$ as a continuous inclusion of Banach spaces. Next we argue that these Banach spaces are isometric and that both sides in (6.31) are comparable. To see this one again switches to $\epsilon_n\zeta_n$, with independent Bernoulli and Steinhaus sequences. Then, by the same argument as before,

$$\mathbb{E}_\zeta\left\|\sum_{n\in\Lambda} \zeta_n \hat{f}(n)e(n\theta)\right\|_\infty = \mathbb{E}_\zeta\,\mathbb{E}_\epsilon\left\|\sum_{n\in\Lambda} \epsilon_n\zeta_n \hat{f}(n)e(n\theta)\right\|_\infty.$$

Hence we can fix a unimodular sequence $\zeta^{(0)}$ such that

$$\mathbb{E}_\zeta\left\|\sum_{n\in\Lambda} \zeta_n \hat{f}(n)e(n\theta)\right\|_\infty \ge \mathbb{E}_\epsilon\left\|\sum_{n\in\Lambda} \epsilon_n\zeta_n^{(0)} \hat{f}(n)e(n\theta)\right\|_\infty.$$

Now we may again contract, by multiplying with $\overline{\zeta_n^{(0)}}$, whence

$$2\,\mathbb{E}_\zeta\left\|\sum_{n\in\Lambda} \zeta_n \hat{f}(n)e(n\theta)\right\|_\infty \ge \mathbb{E}_\epsilon\left\|\sum_{n\in\Lambda} \epsilon_n \hat{f}(n)e(n\theta)\right\|_\infty$$

and we are done. The final statement follows from Theorem 6.21. □

Notes

The original Kolmogorov reference is [69]. Calderón initially observed the connection between a.e. convergence of $S_n f$ for $f \in L^2$ and a weak-type bound on the associated maximal function; see Zygmund [133]. Stein [107] devised a general theory expounding this connection, which applies outside the realm of Fourier series. Much more on pointwise and a.e. convergence questions, as well as on Sidon sets and Riesz products, can be found in [133] and Katznelson [65], as well as in the treatise by Rudin [95]. Kahane [63] is a standard reference on random Fourier (and Taylor) series. For a seminal paper on random Fourier series see Paley, Wiener, and Zygmund [90]. Marcus and Pisier [79] settled the problem of characterizing the space $C^{\epsilon}(\mathbb{T}) = C^{\omega}(\mathbb{T})$ (the equality had been shown earlier by Billard; see Kahane [63]). This was accomplished by showing the equality of these spaces with that obtained by Gaussian randomization and then using Dudley–Fernique theory for the continuity of sample paths of Gaussian processes. The necessary and sufficient condition on the coefficients of $f \in L^2(\mathbb{T})$ found by Marcus and Pisier, for the randomized Fourier series of f to converge uniformly, is quite complicated and is expressed as the finiteness of an entropy integral, as in the Dudley–Fernique theory.

Section 6.3 follows Rider's papers [92, 93]; see also Rudin [95, Chapter 5]. In particular, the remarkable Theorem 6.21 is from [93], where it is proved in general for compact groups. Corollary 6.22 was first obtained by Drury [30] using different methods. Sidon sets, as well as other "thin" sets of integers remain an active research area in harmonic analysis, with connections to Banach space theory, probability, and number theory; see for example Li, Queffélec, and Rodríguez-Piazza [76] and Lefèvre and Rodríguez-Piazza [74]. Bourgain showed that for every $2 \le p < q < \infty$ there is a $\Lambda(p)$ set that is not a $\Lambda(q)$ set, a problem raised by Rudin [94]. See also Problem 6.8 below.

Problems

Problem 6.1 Consider the following variants of Theorem 6.2.

(a) Let $\{\mu_n\}_{n=1}^{\infty}$ be a sequence of positive measures on $\mathbb{R}^d$ supported in a common compact set. Define

$$\mathcal{M}f(x) := \sup_{n \ge 1} |(f * \mu_n)(x)|.$$

Let $1 \le p < \infty$ and assume that, for each $f \in L^p(\mathbb{R}^d)$,

$$\mathcal{M}f(x) < \infty \quad \text{on a set of positive measure.}$$

Then show that $f \mapsto \mathcal{M}f$ is weak-type (p, p).

(b) Now suppose that the μ_n are *complex* measures of the form $\mu_n(dx) = K_n(x)\,dx$, but again with common compact support. Show that the conclusion of part (a) holds, but only for $1 \le p \le 2$.

Problem 6.2 If ω is an irrational number, show that

$$\left\| \frac{1}{N} \sum_{n=1}^{N} f(\cdot + n\omega) - \int_{\mathbb{T}} f(\theta)\, d\theta \right\|_{L^2} \to 0$$

for any $f \in L^2(\mathbb{T})$. In particular, if $f \in L^2$ is such that $f(x + \omega) = f(x)$ for a.e. x then f = constant.

Problem 6.3 Let $\{x_n\}_{n=1}^{\infty}$ be an infinite sequence of real numbers. Show that the following three conditions are equivalent.

(a) For any $f \in C(\mathbb{T})$,

$$\frac{1}{N}\sum_{n=1}^{N} f(x_n) \to \int_{\mathbb{T}} f(x)\,dx.$$

(b) One has $N^{-1}\sum_{n=1}^{N} e(kx_n) \to 0$ for all $k \in \mathbb{Z}^+$.
(c) One has $\lim_{N\to\infty} \sup_{I\subset\mathbb{T}} \left|\frac{1}{N}\#\{1 \le j \le N \mid x_j \in I \bmod 1\} - |I|\right| = 0$.

If these conditions hold we say that $\{x_n\}_{n=1}^{\infty}$ is *uniformly distributed* modulo 1 (abbreviated as u.d.). The quantity

$$D(\{x_j\}_{j=1}^{N}) := \sup_{I\subset\mathbb{T}} \left|\#\{1 \le j \le N \mid x_j \in I \bmod 1\} - N|I|\right|$$

is called the *discrepancy* of the finite sequence $\{x_j\}_{j=1}^{N}$.

Problem 6.4 Using Problem 1.4 with a suitable choice of T, prove the following. If $\{x_n\}_{n=1}^{\infty}$ is a sequence for which $\{x_{n+k} - x_n\}_{n=1}^{\infty}$ is u.d. mod 1 for some $k \in \mathbb{Z}^+$ then $\{x_n\}_1^{\infty}$ is also u.d. mod 1. In particular, show that $\{n^d\omega\}_{n=1}^{\infty}$ is u.d. mod 1 for any irrational ω and $d \in \mathbb{Z}^+$.

Problem 6.5 Let $p \ge 2$ be a positive integer. Show that for a.e. $x \in \mathbb{T}$ the sequence $\{p^n x\}_{n=1}^{\infty}$ is u.d. mod 1. Can you characterize those x that have this property? *Hint:* Consider expansions to the power p.

Problem 6.6 Let $\{\theta_j\}_{j=1}^{N} \subset \mathbb{T}$ be N distinct points, and consider

$$P(z) = \prod_{j=1}^{N}(z - e(\theta_j)).$$

Assume that $\sup_{|z|=1} |P(z)| \le e^{\tau}$. We may assume that $1 \le \tau \le N\log 2$. Prove the following bound on the discrepancy of $\{\theta_j\}_{j=1}^{N}$ as defined in Problem 6.3:

$$\Delta_N := D(\{\theta_j\}_{j=1}^{N}) \le C\sqrt{\tau N}.$$

Conclude that

$$\left|\{\theta \in \mathbb{T} \,|\, |P(e(\theta))| < e^{-AN}\}\right| \le Ce^{-AN/\Delta_N} \le Ce^{-A\sqrt{N}/\tau}$$

for any $A \ge 0$. The constants C are absolute but may differ. *Hint:* Consider the function $F(\theta) = \mu([\theta_0, \theta)) - N(\theta - \theta_0)$ with $\mu = \sum_{j=1}^{N} \delta_{\theta_j}$, where $\theta_0 \in (-\frac{1}{2}, \frac{1}{2})$ is chosen such that $\int_{-1/2}^{1/2} F(\theta)\,d\theta = 0$. Relate F to $u(\theta) := \log|P(e(\theta))|$ via a Hilbert transform. Then write $F = K * \mu$ with a suitable kernel K. See Bourgain, Goldstein, and Schlag [11, Lemma 2.3] for more details.

Problem 6.7 Let $\{r_j\}_1^{\infty}$ be the Rademacher sequence. Then, for any trigonometric polynomials P_j of degree at most N, show that

$$\mathbb{P}\left(\left\{\sup_{0\le\theta\le1}\left|\sum_{j=1}^{N} r_j\, P_j(\theta)\right| > 6\left(\sum_{j=1}^{N}\|P_j\|_\infty^2\right)^{1/2}\sqrt{\log N}\right\}\right) \le N^{-2}$$

for all integers $N \ge N_0$, where N_0 is some absolute constant.

Problem 6.8 Let $2 < p < \infty$. Then a $\Lambda(p)$-set is a set of integers $E \subset \mathbb{Z}$ such that, for some constant $B = B(E, p)$,

$$\|f\|_p \le B\|f\|_2 \quad \forall f \in L^2_E(\mathbb{T}).$$

(a) Assume E is $\Lambda(p)$. Show that for any arithmetic progression $\mathcal{A}$ of length $N \ge 1$ one has

$$\#(E \cap \mathcal{A}) \le CB^2 N^{2/p}, \tag{6.32}$$

where C is some absolute constant.

(b) Suppose that $E \subset \mathbb{Z}$ satisfies

$$\sup_{\ell \in \mathbb{Z}} \#\{(n, m) \in E^2 \,|\, n + m = \ell\} \le A,$$

$$\sup_{\ell \in \mathbb{Z}} \#\{(n, m) \in E^2 \,|\, n - m = \ell\} \le B.$$

Show that $\|f\|_4 \le \min(A^{1/4}, B^{1/4})\|f\|_2$ for any $f \in L^2_E(\mathbb{T})$.

(c) Use part (b) to show that the bound in (6.32) is sharp for $p = 4$, possibly up to a logarithm (use the squares of the integers as E; this requires a little number theory).

(d) Give an example of a $\Lambda(4)$-set that is neither Sidon nor a $\Lambda(6)$-set.

Problem 6.9 Suppose that $\{x_n\}_{n\ge1}$ and $\{\varphi_n\}_{n\ge1}$ are two real sequences. Assume that

$$\sum_{n\ge1} x_n^2 = \infty.$$

Prove that $\sum_{n=1}^{\infty} x_n^2 \cos^2(nx + \varphi_n) = \infty$ for a.e. $x \in \mathbb{R}$. Use this to give a self-contained proof of the fact that the Rademacher series $\sum_{n\ge1} \epsilon_n x_n \cos(nx + \varphi_n)$ diverges a.s. for a.e. $x \in \mathbb{R}$.

Problem 6.10 Let $1 \le p < \infty$ and suppose that there exists a constant $C(p)$ such that

$$\sup_{\varepsilon_n = \pm1} \left\| \sum_{-N}^{N} \varepsilon_n \hat{f}(n) e(n\theta) \right\|_p \le C(p)\|f\|_p \quad \forall f \in L^p([0, 1]),\ \forall N \ge 1. \tag{6.33}$$

Show that then necessarily $p = 2$. Next show that (6.33) is equivalent to the property that $\sum_{-\infty}^{\infty} \varepsilon_n \hat{f}(n) e(n\cdot)$ converges in $L^p([0, 1])$ for each choice of sign $\varepsilon_n = \pm1$ and $f \in L^p([0, 1])$. This latter property is called *unconditional convergence*, and this problem therefore amounts to proving that the exponential system is unconditional only for $p = 2$. In Section 8.4 we shall see that the Haar functions are unconditional on $L^p([0, 1])$ for $1 < p < \infty$. *Hint:* Use Khinchine's inequality.

7

Calderón–Zygmund theory of singular integrals

7.1. Calderón–Zygmund kernels

In this chapter we take up the problem of developing a real-variable theory of the Hilbert transform. In fact, we will subsume the Hilbert transform into a class of operators defined in any dimension. We shall make no reference to harmonic functions, as in Chapter 3 but, rather, rely on the properties of the kernel alone. For simplicity, we shall mostly restrict ourselves to a translation-invariant setting. We begin with the class of kernels in which operators are defined by convolution. As for the Hilbert transform, we include a cancellation condition; see part (iii) in the definition below. For example, this condition guarantees that the principal value as in (7.1) exists; contrast this with the case $K(x) = |x|^{-d}$. In addition, this condition will be convenient in proving the L^2-boundedness of the associated operators.

This being said, it is often preferable to assume the L^2-boundedness of the Calderón–Zygmund operator instead of the cancellation condition. The logic is then to develop L^2-boundedness theory separately from L^p theory, with the goal of finding necessary and sufficient conditions for L^2-boundedness; L^p theory requires only condition (ii) in the definition. In many ways (ii), which is called the *Hörmander condition*, is therefore the most important.

7.1.1. The basic definition

Definition 7.1 Let $K : \mathbb{R}^d \backslash \{0\} \to \mathbb{C}$ satisfy, for some constant B,

(i) $|K(x)| \leq B|x|^{-d}$ for all $x \in \mathbb{R}^d$,
(ii) $\int_{[|x|>2|y|]} |K(x) - K(x-y)|\,dx \leq B$ for all $y \neq 0$,
(iii) $\int_{[r<|x|<s]} K(x)\,dx = 0$ for all $0 < r < s < \infty$.

Then K is called a *Calderón–Zygmund* kernel.

With such a kernel we associate a translation-invariant operator, by means of its principal value. Thus, the *singular integral operator* (or Calderón–Zygmund operator) with kernel K is defined as

$$Tf(x) := \lim_{\varepsilon \to 0} \int_{[|x-y|>\varepsilon]} K(x-y) f(y)\, dy \tag{7.1}$$

for all $f \in \mathcal{S}(\mathbb{R}^d)$.

Exercise 7.1

(a) Check that the limit exists in (7.1) for all $f \in \mathcal{S}(\mathbb{R}^d)$. How much regularity on f suffices for this property (you may take f to be compactly supported if you wish)?

(b) Check that the Hilbert transform on the line with kernel x^{-1} is a singular integral operator.

The following simple condition guarantees Definition 7.1(ii).

Lemma 7.2 *Suppose that $|\nabla K(x)| \le B|x|^{-d-1}$ for all $x \neq 0$ and some constant B. Then*

$$\int_{[|x|>2|y|]} |K(x) - K(x-y)|\, dx \le CB, \tag{7.2}$$

with $C = C(d)$.

Proof Fix $x, y \in \mathbb{R}^d$ with $|x| > 2|y|$. Connect x and $x - y$ by the line segment $x - ty$, $0 \le t \le 1$. This line segment lies entirely inside the ball $B(x, |x|/2)$. Hence

$$\begin{aligned} |K(x) - K(x-y)| &= \left| -\int_0^1 \nabla K(x - ty) y\, dt \right| \\ &\le \int_0^1 |\nabla K(x-ty)||y|\, dt \le B 2^{d+1} |x|^{-d-1} |y|. \end{aligned}$$

Inserting this bound into the left-hand side of (7.2) yields the desired bound. □

For the remainder of this chapter, we shall refer to kernels as in Definition 7.1 that satisfy the gradient condition $|\nabla K(x)| \le B|x|^{-d-1}$ as *strong Calderón–Zygmund kernels*. The associated operators are *strong Calderón–Zygmund operators*.

Exercise 7.2 Check that, for any fixed $0 < \alpha \leq 1$ and all $x \neq 0$,

$$\sup_{|y|<|x|/2} \frac{|K(x) - K(x-y)|}{|y|^\alpha} \leq B|x|^{-d-\alpha}$$

implies (7.2).

7.1.2. The L^2-boundedness of T

We now prove the L^2-boundedness of T, by computing $\hat{K}$ and applying Plancherel's theorem. This is an instance where the full force of Definition 7.1 comes into play. We shall use all three conditions, in particular the cancellation condition (iii).

Proposition 7.3 *Let K be as in Definition 7.1. Then $\|T\|_{2\to 2} \leq CB$ with $C = C(d)$.*

Proof Fix $0 < r < s < \infty$ and consider

$$(T_{r,s} f)(x) = \int_{\mathbb{R}^d} K(y)\chi_{[r<|y|<s]}(y) f(x-y)\, dy.$$

Let

$$m_{r,s}(\xi) := \int_{\mathbb{R}^d} e^{-2\pi i x\cdot\xi} \chi_{[r<|x|<s]} K(x)\, dx$$

be the Fourier transform of the restricted kernel. By Plancherel's theorem it suffices to prove that

$$\sup_{0<r<s} \|m_{r,s}\|_\infty \leq CB. \tag{7.3}$$

Indeed, if (7.3) holds then

$$\|T_{r,s}\|_{2\to 2} = \|m_{r,s}\|_\infty \leq CB$$

uniformly in r, s. Moreover, for any $f \in \mathcal{S}(\mathbb{R}^d)$ one has

$$Tf(x) = \lim_{\substack{r\to 0\\ s\to\infty}} (T_{r,s} f)(x)$$

pointwise in $x \in \mathbb{R}^d$. Fatou's lemma therefore implies that $\|Tf\|_2 \leq CB\|f\|_2$ for any $f \in \mathcal{S}(\mathbb{R}^d)$. To verify (7.3) we split the integration in the Fourier transform into the regions $|x| < |\xi|^{-1}$ and $|x| \geq |\xi|^{-1}$. In the former case,

$$\begin{aligned}\left|\int_{[r<|x|<|\xi|^{-1}]} e^{-2\pi i x\cdot\xi} K(x)\,dx\right| &= \left|\int_{[r<|x|<|\xi|^{-1}]} (e^{-2\pi i x\cdot\xi} - 1)K(x)\,dx\right| \\ &\leq \int_{[|x|<|\xi|^{-1}]} 2\pi|x||\xi||K(x)|\,dx \\ &\leq 2\pi|\xi| \int_{[|x|<|\xi|^{-1}]} B|x|^{-d+1}\,dx \\ &\leq CB|\xi||\xi|^{-1} \leq CB,\end{aligned}$$

as desired. Note that we used the cancellation condition (iii) to obtain the first equality. To deal with the case $|x| > |\xi|^{-1}$ one uses the cancellation in $e^{-2\pi i x\cdot\xi}$, which in turn requires the smoothness of K, i.e., condition (ii) (but compare Lemma 7.2). First, observe that

$$\int_{[s>|x|>|\xi|^{-1}]} K(x)e^{-2\pi i x\cdot\xi}\,dx = -\int_{[s>|x|>|\xi|^{-1}]} K(x)e^{-2\pi i\left(x+\xi/(2|\xi|^2)\right)\cdot\xi}\,dx \tag{7.4}$$

$$= -\int_{[s>|x-\xi/2|\xi|^2|>|\xi|^{-1}]} K\!\left(x - \frac{\xi}{2|\xi|^2}\right) e^{-2\pi i x\cdot\xi}\,dx. \tag{7.5}$$

Denoting the expression on the left-hand side of (7.4) by F one thus has

$$2F = \int_{[s>|x|>|\xi|^{-1}]} \left(K(x) - K\left(x - \frac{\xi}{2|\xi|^2}\right)\right) e^{-2\pi i x\cdot\xi}\,dx + O(1). \tag{7.6}$$

The $O(1)$ term stands for a term bounded by CB. Its origin is of course the difference between the regions of integration in (7.4) and (7.5). We leave it to the reader to check that condition (i) implies that this error term is really no larger than CB. Estimating the integral in (7.6) by means of (ii) now yields

$$|2F| \leq \int_{[|x|>|\xi|^{-1}]} \left|K(x) - K\left(x - \frac{\xi}{2|\xi|^2}\right)\right| dx + CB \leq CB,$$

as claimed. We have shown (7.3) and the proposition follows. □

Exercise 7.3 Supply the details concerning the $O(1)$ term in the previous proof.

7.1.3. Calderón–Zygmund decomposition, weak-L^1 bound

Our next goal is to show that singular integrals are bounded as operators from $L^1(\mathbb{R}^d)$ into weak-L^1. The proof of this fact is based on the fundamental *Calderón–Zygmund decomposition*, which we have already encountered in Lemma 2.17. Note that the following formulation of the crucial weak-L^1 bound for singular integrals makes assumptions that are more general than those of Definition 7.1.

Proposition 7.4 *Suppose that T is a linear operator bounded on $L^2(\mathbb{R}^d)$ such that*

$$(Tf)(x) = \int_{\mathbb{R}^d} K(x-y)f(y)\,dy \quad \forall f \in L^2_{\text{comp}}(\mathbb{R}^d)$$

and all $x \notin \operatorname{supp}(f)$. Assume that K satisfies condition (ii) of Definition 7.1 but not necessarily any of the other conditions. Then for every $f \in \mathcal{S}(\mathbb{R}^d)$ there is the weak-L^1 bound

$$\left|\{x \in \mathbb{R}^d \mid |Tf(x)| > \lambda\}\right| \le \frac{CB}{\lambda}\|f\|_1 \quad \forall \lambda > 0,$$

where $C = C(d)$.

Proof Dividing by B if necessary, we may assume that $B = 1$. Now fix $f \in \mathcal{S}(\mathbb{R}^d)$ and let $\lambda > 0$ be arbitrary. By Lemma 2.17 one can write $f = g + b$ with this value of λ. We now set

$$f_1 = g + \sum_{Q \in \mathcal{B}} \chi_Q \frac{1}{|Q|}\int_Q f(x)\,dx,$$

$$f_2 = b - \sum_{Q \in \mathcal{B}} \chi_Q \frac{1}{|Q|}\int_Q f(x)\,dx = \sum_{Q \in \mathcal{B}} f_Q,$$

where

$$f_Q := \chi_Q\left(f - \frac{1}{|Q|}\int_Q f(x)\,dx\right).$$

Note that

$$f = f_1 + f_2, \quad \|f_1\|_\infty \le 2^d\lambda$$
$$\|f_2\|_1 \le 2\|f\|_1, \quad \|f_1\|_1 \le \|f\|_1,$$

and

$$\int_Q f_Q(x)\,dx = 0$$

for all $Q \in \mathcal{B}$. We now proceed as follows:

$$\begin{aligned}
&\left|\{x \in \mathbb{R}^d \mid |(Tf)(x)| > \lambda\}\right| \\
&\le \left|\left\{x \in \mathbb{R}^d \,\middle|\, |(Tf_1)(x)| > \frac{\lambda}{2}\right\}\right| + \left|\left\{x \in \mathbb{R}^d \,\middle|\, |(Tf_2)(x)| > \frac{\lambda}{2}\right\}\right| \\
&\le \frac{C}{\lambda^2}\|Tf_1\|_2^2 + \left|\left\{x \in \mathbb{R}^d \,\middle|\, |(Tf_2)(x)| > \frac{\lambda}{2}\right\}\right|. \qquad (7.7)
\end{aligned}$$

The first term in (7.7) is controlled by Proposition 7.3:

$$\frac{C}{\lambda^2}\|Tf_1\|_2^2 \le \frac{C}{\lambda^2}\|f_1\|_2^2 \le \frac{C}{\lambda^2}\|f_1\|_\infty\|f_1\|_1 \le \frac{C}{\lambda}\|f\|_1.$$

To estimate the second term in (7.7) we define, for any $Q \in \mathcal{B}$, the cube Q^* to be the dilate of Q by the fixed factor $2\sqrt{d}$ (i.e., Q^* has the same center as Q but side length equal to $2\sqrt{d}$ times that of Q). Thus

$$\begin{aligned}
&\left|\left\{x \in \mathbb{R}^d \,\middle|\, |(Tf_2)(x)| > \frac{\lambda}{2}\right\}\right| \\
&\quad\le \left|\cup_{\mathcal{B}} Q^*\right| + \left|\left\{x \in \mathbb{R}^d \backslash \cup_{\mathcal{B}} Q^* \,\middle|\, |(Tf_2)(x)| > \frac{\lambda}{2}\right\}\right| \\
&\quad\le C\sum_Q |Q| + \frac{2}{\lambda}\int_{\mathbb{R}^d\backslash\cup Q^*} |(Tf_2)(x)|\,dx \\
&\quad\le \frac{C}{\lambda}\|f\|_1 + \frac{2}{\lambda}\sum_{Q\in\mathcal{B}}\int_{\mathbb{R}^d\backslash Q^*} |(Tf_Q)(x)|\,dx.
\end{aligned}$$

Perhaps the most crucial point of this proof is the fact that f_Q has mean zero, which allows one to exploit the smoothness of the kernel K. More precisely, for any $x \in \mathbb{R}^d\backslash Q^*$,

$$\begin{aligned}
(Tf_Q)(x) &= \int_Q K(x-y) f_Q(y)\,dy \\
&= \int_Q (K(x-y) - K(x-y_Q)) f_Q(y)\,dy, \qquad (7.8)
\end{aligned}$$

where y_Q denotes the center of Q. Thus

$$\begin{aligned}
\int_{\mathbb{R}^d\backslash Q^*} |(Tf_Q)(x)|\,dx &\le \int_{\mathbb{R}^d\backslash Q^*}\int_Q \left|K(x-y) - K(x-y_Q)\right||f_Q(y)|\,dy\,dx \\
&\le \int_Q |f_Q(y)|\,dy \le 2\int_Q |f(y)|\,dy.
\end{aligned}$$

To pass to the second inequality sign we used condition (ii) in Definition 7.1 and our choice of Q^*. Hence the second term under the integral sign in the second line of (7.8) is no larger than

$$\frac{C}{\lambda}\sum_Q \int_Q |f(y)|dy \le \frac{C}{\lambda}\|f\|_1,$$

and we are done. □

The assumption $f \in \mathcal{S}(\mathbb{R}^d)$ in Proposition 7.4 was for convenience only. It ensured that one could define Tf by means of the principal value (7.1). However, observe that Proposition 7.3 allows one to extend T to a bounded operator on L^2. This in turn implies that the weak-L^1 bound in Proposition 7.4

holds for all $f \in L^1 \cap L^2(\mathbb{R}^d)$. Indeed, inspection of the proof of Proposition 7.4 reveals that, apart from the L^2-boundedness of T, see (7.7), the definition of T in terms of K was only used in (7.8) where x and y were assumed to be sufficiently separated that the integrals would be absolutely convergent. The following result is the fundamental *Calderón–Zygmund theorem.*

7.1.4. The L^p-boundedness of T

Theorem 7.5 *Let T be a singular integral operator as in Definition 7.1. Then for every $1 < p < \infty$ one can extend T to a bounded operator on $L^p(\mathbb{R}^d)$ with the bound $\|T\|_{p\to p} \leq CB$, where $C = C(p, d)$.*

Proof By Propositions 7.3 and 7.4 (and the remark above), one obtains this statement for the range $1 < p \leq 2$ from the Marcinkiewicz interpolation theorem 1.17. The range $2 \leq p < \infty$ now follows by duality. Indeed, for $f, g \in \mathcal{S}(\mathbb{R}^d)$, one has

$$\langle Tf, g\rangle = \langle f, T^*g\rangle \quad \text{where } T^*g(x) = \lim_{\varepsilon\to 0}\int_{|x-y|>\varepsilon} K^*(x-y)g(y)\,dy$$

and $K^*(x) := \overline{K(-x)}$. Since K^* verifies conditions (i)–(iii) in Definition 7.1, we are done. □

Remark It is important to realize that the cancellation condition (iii) was used only to prove the L^2-boundedness of T; it did not appear in the proof of Proposition 7.4 explicitly. Therefore, T is bounded on $L^p(\mathbb{R}^d)$ provided that it is bounded for $p = 2$ and that condition (ii) of Definition 7.1 holds.

7.1.5. Boundedness on Hölder spaces

In light of Theorem 7.5 it is natural to ask whether a Calderón–Zygmund operator is bounded on Hölder spaces $C^\alpha(\mathbb{R}^d)$ where $0 < \alpha < 1$. As we shall now see, this turns out to be true and more elementary than the L^p case. Recall that the norm in C^α is given by

$$\|f\|_\alpha := \|f\|_\infty + [f]_\alpha, \quad [f]_\alpha := \sup_{x\neq y}\frac{|f(x)-f(y)|}{|x-y|^\alpha}.$$

Since C^α functions do not necessarily decay at infinity, some care is needed when applying a singular integral operator to them. In the following result we will therefore assume compact support.

Theorem 7.6 *Let K be a strong Calderón–Zygmund kernel, and fix some $0 < \alpha < 1$. Then for every $f \in C^\alpha$ with* $\operatorname{supp}(f) \subset B(0,1)$ *one has*

$$\|Tf\|_\alpha \leq C(\alpha, d)\|f\|_\alpha. \tag{7.9}$$

Proof Let $\|f\|_\alpha \le 1$ and also let $B = 1$, where B is the constant in the estimates on K. Then, if on the one hand $|x| \le 2$,

$$|Tf(x)| = \left|\int K(x-y)f(y)\,dy\right| \le \int_{[|x-y|\le 3]} |K(x-y)||f(y)-f(x)|\,dy$$
$$\le \int_{[|x-y|\le 3]} |x-y|^{-d}|x-y|^{\alpha}\,dy \le C.$$

On the other hand, if $|x| > 2$ then for all $|y| < 1$ we have $|K(x-y)| \le 1$, whence

$$|Tf(x)| = \left|\int K(x-y)f(y)\,dy\right| \le \int |K(x-y)||f(y)|\,dy \le C.$$

Next, we let $|x-x'| = \delta < 1$ and write

$$\begin{aligned}|Tf(x)-Tf(x')| &\le \left|\int_{[|y|<3\delta]} K(y)\big(f(x-y)-f(x'-y)\big)\,dy\right| \\ &\quad + \left|\int_{[|y|\ge 3\delta]} K(y)\big(f(x-y)-f(x'-y)\big)\,dy\right| \\ &= I_1 + I_2.\end{aligned} \tag{7.10}$$

Then, on the one hand,

$$\begin{aligned}I_1 &\le \int_{[|y|<3\delta]} |K(y)||f(x-y)-f(x)|\,dy \\ &\quad + \int_{[|y|<3\delta]} |K(y)||f(x'-y)-f(x')|\,dy \le \int_{[|y|<3\delta]} |y|^{-d}|y|^{\alpha}\,dy \le C\delta^{\alpha}\end{aligned} \tag{7.11}$$

and on the other hand, setting $K_\delta(x) := \chi_{[|x|>3\delta]}K(x)$, we have

$$I_2 \le \int_{\mathbb{R}^d} \big|K_\delta(x-y) - K_\delta(x'-y)\big|\;|f(y)-f(x)|\,dy. \tag{7.12}$$

In order for the integrand not to vanish we need $|x-y| > 2\delta$ and $|x'-y| > 2\delta$. But then the previous line is estimated by

$$\int_{[|x-y|>2\delta]} \frac{\delta|x-y|^{\alpha}}{|x-y|^{d+1}}\,dy \le C\delta^{\alpha}, \tag{7.13}$$

as desired. □

An alternative proof, relying on the Fourier transform, is presented in Chapter 8. For future reference we also note the following corollary of this proof. If $f \in \mathcal{S}(\mathbb{R}^d)$ and $0 < \alpha < 1$ then

$$[Tf]_\alpha \le C(\alpha, d)\,[f]_\alpha. \tag{7.14}$$

7.2. The Laplacian: Riesz transforms and fractional integration

We now present some basic examples of singular integrals. Consider the Poisson equation $\Delta u = f$ in $\mathbb{R}^d$, where $f \in \mathcal{S}(\mathbb{R}^d)$, $d \geq 2$. In the next exercise we ask the reader to show that, in dimensions $d \geq 3$,

$$u(x) = C_d \int_{\mathbb{R}^d} |x-y|^{2-d} f(y)\,dy \tag{7.15}$$

with some dimensional constant C_d is a solution to this equation. In fact, one can show that (7.15) gives the unique solution that vanishes at infinity. In two dimensions the formula reads

$$u(x) = C_2 \int_{\mathbb{R}^2} \log(|x-y|) f(y)\,dy. \tag{7.16}$$

The kernels $|x|^{2-d}$ and $\log|x|$ are called *Newton potentials*. It turns out that the second derivatives $\partial^2 u/(\partial x_i \partial x_j)$ with u defined by the above convolutions can be expressed as Calderón–Zygmund operators acting on f (the so-called "double Riesz transforms").

Exercise 7.4

(a) With $f \in \mathcal{S}$ and a suitable constant C_d, show that (7.15) satisfies

$$\Delta u = f.$$

Hint: In (7.15), pull the Laplacian inside the integral but apply it to f and introduce a cutoff to the set $\{|x-y| > \varepsilon\}$. Apply Green's formula, and let $\varepsilon \to 0$ to recover $f(x)$.

(b) With u as in (7.15) show that, for $f \in \mathcal{S}(\mathbb{R}^d)$ with $d \geq 3$,

$$\frac{\partial^2 u}{\partial x_i \partial x_j}(x) = C_d \int_{\mathbb{R}^d} K_{ij}(x-y) f(y)\,dy + \frac{1}{d}\delta_{ij} f(x)$$

in the principal-value sense, where

$$K_{ij}(x) = \begin{cases} \dfrac{x_i x_j}{|x|^{d+2}} & \text{if } i \neq j, \\[2ex] \dfrac{x_i^2 - d^{-1}|x|^2}{|x|^{d+2}} & \text{if } i = j. \end{cases}$$

Find the analogue for $d = 2$.

(c) Verify that the K_{ij} as above are singular integral kernels. Also show that $K_i(x) = x_i/|x|^{d+1}$ is a singular integral kernel.

The operators R_i and R_{ij} defined in terms of the kernels K_i and K_{ij}, respectively, are called the Riesz transform or the double Riesz transform, respectively. By Exercise 7.4,

$$R_{ij}(\Delta\varphi) = \frac{\partial^2 \varphi}{\partial x_i \partial x_j} \tag{7.17}$$

for any $\varphi \in \mathcal{S}(\mathbb{R}^d)$. The following estimate is a basic L^p estimate for elliptic equations.

Corollary 7.7 *Let $u \in \mathcal{S}(\mathbb{R}^d)$. Then*

$$\sup_{1\le i,j\le d} \left\| \frac{\partial^2 u}{\partial x_i \partial x_j} \right\|_{L^p(\mathbb{R}^d)} \le C_{p,d} \|\Delta u\|_{L^p(\mathbb{R}^d)} \tag{7.18}$$

for any $1 < p < \infty$. Here $C_{p,d}$ depends only on p and d. Similarly, for any $0 < \alpha < 1$ one has

$$\sup_{1\le i,j\le d} \left[\frac{\partial^2 u}{\partial x_i \partial x_j} \right]_\alpha \le C_{\alpha,d} [\Delta u]_\alpha. \tag{7.19}$$

Proof This is an immediate consequence of the representation formula (7.17) and the Calderón–Zygmund theorem 7.5, as well as of Theorem 7.6 (or (7.14)). □

The inequality (7.19) is the simplest example of a *Schauder estimate*. Corollary 7.7 is remarkable since it states that the Hessian is bounded by its trace. This cannot hold in a pointwise sense, however, and this means that the corollary fails at $p = \infty$ (or $\alpha = 0$) and by duality therefore also at $p = 1$.

Exercise 7.5 This exercise explains in more detail the aforementioned failure of Corollary 7.7 at $p = 1$ or $p = \infty$. Take u to be the Newton potential. Then $\Delta u = \delta_0$, which, at least heuristically, belongs to $L^1(\mathbb{R}^d)$. However, one may check that $\partial^2 u/(\partial x_i \partial x_j) \notin L^1(B(0,1))$ for any i, j; see the kernels K_{ij} from Exercise 7.4. In order to transfer δ_0 to L^1, we convolve the Newton potentials with an approximate identity (this is called *mollifying*); in addition, we truncate to a bounded set.

(a) Let $\varphi \ge 0$ be in $C_0^\infty(\mathbb{R}^d)$ with $\int_{\mathbb{R}^d} \varphi \, dx = 1$. Set $\varphi_\varepsilon(x) := \varepsilon^{-d}\varphi(x/\varepsilon)$ for any $0 < \varepsilon < 1$. Clearly, $\{\varphi_\varepsilon\}_{0<\varepsilon<1}$ form an approximate identity provided that the latter are defined as in Definition 1.3 but on $\mathbb{R}^d$ rather than $\mathbb{T}$. Moreover, let $\chi \in C_0^\infty(\mathbb{R}^d)$ be arbitrary with $\chi(0) \neq 0$. Verify that, with $\Gamma_d(x) = |x|^{2-d}$ for $n \ge 3$ and $\Gamma_2(x) = \log|x|$, $u_\varepsilon(x) := (\varphi_\varepsilon * \Gamma_d)(x)\chi(x)$ one has the following properties:

$$\sup_{\varepsilon>0} \|\Delta u_\varepsilon\|_{L^1} < \infty$$

and

$$\limsup_{\varepsilon\to 0} \left\| \frac{\partial^2 u_\varepsilon}{\partial x_i \partial x_j} \right\|_{L^1} = \infty$$

for any $1 \le i, j \le d$. Thus Corollary 7.7 fails on $L^1(\mathbb{R}^d)$.

(b) Now show that Corollary 7.7 also fails on L^∞.

The convolution in (7.15) is an example of *fractional integration*. More generally, for any $0 < \alpha < d$ define

$$(I_\alpha f)(x) := \int_{\mathbb{R}^d} |x-y|^{\alpha-d} f(y)\,dy, \quad f \in \mathcal{S}(\mathbb{R}^d). \tag{7.20}$$

We cannot use Young's inequality (Lemma 1.1(ii)) since although the kernel is not in any $L^r(\mathbb{R}^d)$ space it is clearly in weak-L^r, with $r = d/(d-\alpha)$. The following proposition shows that nevertheless one still has (up to an endpoint) the same conclusion as Young's inequality would imply.

Proposition 7.8 *For any $1 < p < q < \infty$ and any $0 < \alpha < d$ there is a bound*

$$\|I_\alpha f\|_q \le C(p,q,d)\|f\|_p \quad \textit{provided that } \frac{1}{q} = \frac{1}{p} - \frac{\alpha}{d}$$

for any $f \in \mathcal{S}(\mathbb{R}^d)$.

Proof By Marcinkiewicz's interpolation theorem it suffices for this bound to hold for weak L^q. Thus, we need to show that

$$\left|\left\{x \in \mathbb{R}^d \mid |I_\alpha f(x)| > \lambda\right\}\right| \le C\lambda^{-q}\|f\|_p^q. \tag{7.21}$$

Fix some f with $\|f\|_p = 1$ and an arbitary $\lambda > 0$. With $r > 0$ to be determined, break up the kernel as follows:

$$k_\alpha(x) := |x|^{\alpha-d} = k_\alpha(x)\chi_{[|x|<r]} + k_\alpha(x)\chi_{[|x|>r]} =: k_\alpha^{(1)}(x) + k_\alpha^{(2)}(x).$$

Then

$$\left\|k_\alpha^{(2)} * f\right\|_\infty \le \left\|k_\alpha^{(2)}\right\|_{p'}\|f\|_p = Cr^{\alpha-d/p}.$$

Determine r such that the right-hand side equals $\lambda/2$. Then

$$\begin{aligned}\left|\left\{x \in \mathbb{R}^d \mid |I_\alpha f(x)| > \lambda\right\}\right| &\le \left|\left\{x \in \mathbb{R}^d \mid \left|\left(k_\alpha^{(1)} * f\right)(x)\right| > \lambda/2\right\}\right| \\ &\le C\lambda^{-p}\left\|k_\alpha^{(1)} * f\right\|_p^p \le C\lambda^{-p}\left\|k_\alpha^{(1)}\right\|_1^p \\ &\le C\lambda^{-p}r^{\alpha p} = C\lambda^{-q},\end{aligned}$$

which is (7.21). □

The condition on p and q is of course dictated by scaling. By Exercise 4.5 one has

$$(-\Delta)^{s/2} f = C(s,d)\, I_s f \quad \forall f \in \mathcal{S}(\mathbb{R}^d). \tag{7.22}$$

Thus, Proposition 7.8 implies the following *Sobolev embedding estimate*.

Corollary 7.9 *Let $0 \le s < d/2$. Then for all*

$$\frac{1}{2} - \frac{1}{q} = \frac{s}{d}$$

one has

$$\|f\|_{L^q(\mathbb{R}^d)} \le C(s,d)\|f\|_{\dot{H}^s(\mathbb{R}^d)} \quad \forall f \in \mathcal{S}(\mathbb{R}^d). \tag{7.23}$$

Proof Fix $f \in \mathcal{S}(\mathbb{R}^d)$ and set $g := (-\Delta)^{s/2} f$. Then by (7.22) one has

$$f = C(s,d) I_s g$$

and the estimate (7.23) follows from Proposition 7.8. □

7.3. Almost everywhere convergence; homogeneous kernels

We now address the question whether a singular integral operator can be defined by means of formula (7.1) even if we have $f \in L^p(\mathbb{R}^d)$ rather than $f \in \mathcal{S}(\mathbb{R}^d)$, say. The answer to this question is an analogue of Proposition 3.18 and should be understood as follows: for $1 < p < \infty$ we defined T abstractly as an operator on $L^p(\mathbb{R}^d)$ by extension from $\mathcal{S}(\mathbb{R}^d)$ via the a priori bounds $\|Tf\|_p \le C_{p,d}\|f\|_p$ for all $f \in \mathcal{S}(\mathbb{R}^d)$ (the latter space is dense in $L^p(\mathbb{R}^d)$, cf. Proposition 1.5). We now ask whether the principal value (7.1) converges almost everywhere to this extension T for any $f \in L^p(\mathbb{R}^d)$. On the basis of our previous experience with such questions, we introduce the associated maximal operator T_*, by

$$T_* f(x) = \sup_{\varepsilon > 0} \left| \int_{[|y|>\varepsilon]} K(y) f(x-y)\, dy \right|$$

and seek weak-L^1 bounds for it. As in previous cases where we studied a.e. convergence, this will suffice for us to conclude the a.e. convergence on L^p. In what follows we shall need the Hardy–Littlewood maximal function

$$Mf(x) = \sup_{x \in B} \frac{1}{|B|} \int_B |f(y)|\, dy,$$

where the supremum runs over all balls B containing x. The operator M satisfies the same type of bounds as those stated in Proposition 2.9.

We prove the desired bounds on T_* only for a subclass of kernels, namely those that are homogeneous. This class includes the Riesz transforms mentioned after Exercise 7.4. More precisely, let

$$K(x) := \frac{\Omega(x/|x|)}{|x|^d} \quad \text{for } x \neq 0, \tag{7.24}$$

where $\Omega : S^{d-1} \to \mathbb{C}$, $\Omega \in C^1(S^{d-1})$ and $\int_{S^{d-1}} \Omega(x)\,\sigma(dx) = 0$ with σ the surface measure on S^{d-1}. Observe that the homogeneity $K(tx) = t^{-d} K(x)$ holds for all $t > 0$.

Exercise 7.6 Check that any such K satisfies the conditions in Definition 7.1.

7.3.1. A maximal function bound

Proposition 7.10 *Suppose K is of the form* (7.24). *Then T_* satisfies*

$$(T_*f)(x) \leq C_d(M(Tf)(x) + Mf(x))$$

with some absolute constant C_d. In particular, T_* is bounded on $L^p(\mathbb{R}^d)$ for $1 < p < \infty$. Furthermore, T_* is also weak-L^1-bounded.

Proof Let $\tilde{K}(x) := K(x)\chi_{[|x|\geq 1]}$ and, more generally, let

$$\tilde{K}_\varepsilon(x) = \varepsilon^{-d}\tilde{K}\left(\frac{x}{\varepsilon}\right) = K(x)\chi_{[|x|\geq\varepsilon]}.$$

Pick a smooth bump function $\varphi \in C_0^\infty(\mathbb{R}^d)$, $\varphi \geq 0$, $\int_{\mathbb{R}^d} \varphi\, dx = 1$. Set

$$\Phi := \varphi * K - \tilde{K}$$

and observe that $\varphi * K$ is well defined in the principal-value sense. For any function F on $\mathbb{R}^d$ let $F_\varepsilon(x) := \varepsilon^{-d}F(x/\varepsilon)$ be its L^1-normalized rescaling. Then $K_\varepsilon = K$ and $\tilde{K}_\varepsilon = (\tilde{K})_\varepsilon$, and thus

$$\Phi_\varepsilon = (\varphi * K)_\varepsilon - \tilde{K}_\varepsilon = \varphi_\varepsilon * K_\varepsilon - \tilde{K}_\varepsilon = \varphi_\varepsilon * K - \tilde{K}_\varepsilon.$$

Hence, for any $f \in \mathcal{S}(\mathbb{R}^d)$, one has

$$K_\varepsilon * f = \varphi_\varepsilon * (K * f) - \Phi_\varepsilon * f.$$

We now invoke an analogue of Lemma 2.11, for radially bounded approximate identities in $\mathbb{R}^d$. This of course requires that $\{\Phi_\varepsilon\}_{\varepsilon>0}$ be bounded above by such a radially bounded approximate identity (up to a multiplicative constant), the verification of which we leave to the reader (the case of φ_ε is obvious). Indeed, one checks that

$$|\Phi(x)| \leq C\min\left(1, |x|^{-d-1}\right),$$

which implies the desired property. Therefore

$$T_*f \leq C(M(Tf) + Mf),$$

as claimed. The boundedness of T_* on $L^p(\mathbb{R}^d)$ for $1 < p < \infty$ now follows from that of T and M. The proof of the weak-L^1-boundedness of T_* is a variation of that for the same property of T, found in Proposition 7.4. □

Exercise 7.7 Show that $\{\Phi_\varepsilon\}_{\varepsilon>0}$ as it appears in the proof of Proposition 7.10 forms a radially bounded approximate identity.

7.3.2. Almost everywhere convergence

Corollary 7.11 *Let K be a homogeneous singular integral kernel as in* (7.24). *Then, for any $f \in L^p(\mathbb{R}^d)$, $1 \le p < \infty$, the limit in* (7.1) *exists almost everywhere.*

Proof Let

$$\Lambda(f)(x) = \left| \limsup_{\varepsilon \to 0} (T_\varepsilon f)(x) - \liminf_{\varepsilon \to 0} (T_\varepsilon f)(x) \right|$$

Observe that $\Lambda(f) \le 2T_* f$. Fix $f \in L^p$ and let $g \in \mathcal{S}(\mathbb{R}^d)$ be such that

$$\|f - g\|_p < \delta$$

for a given small $\delta > 0$. Then $\Lambda(f) = \Lambda(f - g)$ and therefore

$$\|\Lambda f\|_p \le C \|\Lambda(f - g)\|_p < C\delta$$

if $1 < p < \infty$. Hence, $\Lambda f = 0$. The case $p = 1$ is similar. □

Operators of the form (7.24) arise very naturally when one considers *multiplier operators T_m* that act as $T_m f := (m\hat{f})^\vee$, where m is homogeneous of degree zero and belongs to $C^\infty(\mathbb{R}^d \setminus \{0\})$ (for simplicity we assume infinite regularity)[1]. Indeed, with $f \in \mathcal{S}(\mathbb{R}^d)$ one has, over tempered distributions,

$$Tf = K * f, \quad K = \check{m},$$

where $\check{m} \in \mathcal{S}'(\mathbb{R}^d)$ is the distributional inverse Fourier transform. From the dilation law (4.2) one sees that K is homogeneous of degree $-d$, and this means that, for any $\lambda > 0$,

$$\begin{aligned} \langle \widehat{m(\lambda \cdot)}, \varphi \rangle &= \langle m, \lambda^{-d} \hat{\varphi}(\lambda^{-1} \cdot) \rangle = \langle m, \widehat{\varphi(\lambda \cdot)} \rangle \\ &= \langle \hat{m}, \varphi(\lambda \cdot) \rangle = \langle \lambda^{-d} \hat{m}(\lambda^{-1} \cdot), \varphi \rangle \end{aligned}$$

for all $\varphi \in \mathcal{S}(\mathbb{R}^d)$. Observe that constants are homogeneous of degree zero, whence $\hat{1} = \delta_0$ is homogeneous of degree $-d$.

This might seem less strange if one remembers that as $\varepsilon \to 0$ one has

$$\frac{1}{|B(0, \varepsilon)|} \chi_{B(0,\varepsilon)} \to \delta_0 \quad \text{in } \mathcal{S}'(\mathbb{R}^d),$$

since the left-hand side scales as a function that is homogeneous of degree $-d$.

[1] The inverse hat symbol indicates an inverse Fourier transform.

7.3.3. A characterization of homogeneous kernels

We now prove the following general statement, which is very natural in view of the fact that the representation of $\hat{m}$ needs to be invariant under the addition of constants to m.

Lemma 7.12 *Let $m \in C^\infty(\mathbb{R}^d \setminus \{0\})$ be homogeneous of degree zero. Let $\langle m \rangle_{S^{d-1}}$ denote the mean of m on S^{d-1}. Then $\hat{m} \in \mathcal{S}'(\mathbb{R}^d)$ satisfies*

$$\hat{m} = \langle m \rangle_{S^{d-1}}\, \delta_0 + \text{P.V.}\, K_\Omega$$

where K_Ω is of the form (7.24) with $\langle \Omega \rangle_{S^{d-1}} = 0$ and $\Omega \in C^\infty(S^{d-1})$.

Proof We may assume that m has mean zero on the sphere, i.e.,

$$\int_{S^{d-1}} m(\omega)\, \sigma(d\omega) = 0.$$

Observe that $u_j := \partial_j^d m \in \mathcal{S}'(\mathbb{R}^d)$ is homogeneous of degree $-d$ and has mean zero on S^{d-1}. This implies that the principal value of u_j is well defined in the usual integral sense and, moreover, one has

$$u_j = \text{P.V.}\, u_j + \sum_\alpha c_\alpha\, \partial^\alpha \delta_0,$$

by Exercise 4.4, where the sum is finite. In fact, by homogeneity, the sum is necessarily of the form $c_j\, \delta_0$. Passing to the Fourier transforms in $\mathcal{S}'$ one obtains

$$\widehat{u_j} = \widehat{\text{P.V.}\, u_j} + c_j.$$

Next, one needs to verify that $\widehat{\text{P.V.}\, u_j}(x)$ is a smooth function on $\mathbb{R}^d \setminus \{0\}$ that is homogeneous of degree zero. Indeed, with $\chi(\xi)$ a smooth, radial, compactly supported function that equals 1 on a neighborhood of the origin, one has for any $x \neq 0$

$$\begin{aligned}\widehat{\text{P.V.}\, u_j}(x) = &\int_{\mathbb{R}^d} \partial_j^d m(\xi)\chi(\xi)(e^{-2\pi i x\cdot\xi} - 1)\, d\xi \\ &+ \sum_{j=1}^d \frac{x_j}{2\pi i |x|^2} \int_{\mathbb{R}^d} \partial_j\big(\partial_j^d m(\xi)(1 - \chi(\xi))\big) e^{-2\pi i x\cdot\xi}\, d\xi,\end{aligned}$$

where $p_j = \partial/\partial\xi_j$. Both integrals are absolutely convergent, and the first clearly defines a smooth function in x. We may repeat integration by parts of the second integral any number of times, which implies that the second integral also defines a smooth function. The homogeneity of degree zero follows from the homogeneity of $\text{P.V.}\, u_j$. We may therefore write

$$\hat{m}(x) = \frac{\Omega(x/|x|)}{|x|^d}, \qquad x \neq 0,$$

where $\Omega \in C^\infty(S^{d-1})$. Let $\varphi \in \mathcal{S}(\mathbb{R}^d)$ be radial. Then

$$\langle \hat{m}, \varphi \rangle = \langle m, \hat{\varphi} \rangle = 0,$$

since m has mean zero on spheres. Thus, Ω also has mean zero on S^{d-1}. By construction, $\hat{m} - \text{P.V. } K_\Omega$ is supported at $\{0\}$. By Exercise 4.4 we conclude that this difference is therefore a polynomial in the derivatives of δ_0. By homogeneity it must be a constant. By the vanishing of the means, this constant is zero and we are done. □

As a consequence we see that singular integral operators of the type T_m form an algebra, T_m being invertible if and only if $m \neq 0$ on S^{d-1}. In Chapter 8 we shall see that one can still obtain the L^p-boundedness of T_m without the strong assumption that m is homogeneous of degree zero. Instead, one requires that $|\xi|^k |\partial^\alpha m(\xi)|$ is bounded for all $|\alpha| = k$ for finitely many k. While these conditions are satisfied if m is homogeneous of degree zero, they are much weaker than that property.

7.4. Bounded mean oscillation space

We now wish to *quantify* the failure of the boundedness of singular integral operators on L^∞. It turns out that the correct extension of L^∞, which contains $T(L^\infty)$ for any singular integral operator T, is the space of functions having *bounded mean oscillation* on $\mathbb{R}^d$, denoted $\mathrm{BMO}(\mathbb{R}^d)$. In what follows,

$$⨍_Q f(y)\,dy := |Q|^{-1} \int_Q f(y)\,dy.$$

Definition 7.13 Let $f \in L^1_{\mathrm{loc}}(\mathbb{R}^d)$. Define

$$f^\sharp(x) := \sup_{Q \ni x} ⨍_Q |f(y) - f_Q|\,dy, \tag{7.25}$$

where the supremum runs over cubes Q and f_Q is the mean of f over Q. We have $f \in \mathrm{BMO}(\mathbb{R}^d)$ if and only if $f^\sharp \in L^\infty(\mathbb{R}^d)$. Then $\|f\|_{\mathrm{BMO}} := \|f^\sharp\|_\infty$. If Q_0 is a fixed cube, $\mathrm{BMO}(Q_0)$ is obtained by defining $f^\sharp$ in terms of all subcubes of Q_0.

7.4.1. The most basic example: the (discrete) logarithm

Of course one may replace cubes here with balls without changing anything. Moreover, $\|\cdot\|_{\mathrm{BMO}}$ is a norm only after the constants have been factored out. Clearly $L^\infty \subset \mathrm{BMO}$ but BMO is strictly larger. In addition BMO scales like L^∞, i.e., $\|f(\lambda\cdot)\|_{\mathrm{BMO}} = \|f\|_{\mathrm{BMO}}$. Suppose that $Q_2 \subset Q_1$ are two cubes, with Q_2 half the side length of Q_1. Then

$$|f_{Q_1} - f_{Q_2}| \le \frac{|Q_1|}{|Q_2|} \fint_{Q_1} |f(y) - f_{Q_1}|\, dy \le 2^d \|f\|_{\mathrm{BMO}}. \tag{7.26}$$

Iterating this relation leads to the following. Suppose that $Q_n \subset Q_{n-1} \subset \cdots Q_1 \subset Q_0$ is a sequence of nested cubes such that the side lengths decrease by a factor 2 at each step. Then

$$|f_{Q_n} - f_{Q_0}| \le n\, 2^d \|f\|_{\mathrm{BMO}}. \tag{7.27}$$

The important feature here is that, while the side length of Q_n is smaller by a factor 2^{-n} than Q_0, the averages f_{Qn} increase only linearly in n. As an example, consider the function g defined on the interval $[0, 1]$ as follows:

$$g = \sum_{n=0}^{\infty} a_n \chi_{[0,2^{-n}]} \tag{7.28}$$

with $1 \le a_n$ for all $n \ge 0$.

Exercise 7.8

(a) Show that $g \in \mathrm{BMO}([0, 1])$ if and only if $a_n = O(1)$. In that case verify that $g(x) \simeq |\log x|$ for $0 < x < \frac{1}{2}$ (the function g is referred to as a *discrete logarithm*).

(b) Show that $\chi_{[|x|<1]} \log |x| \in \mathrm{BMO}([-1, 1])$ but

$$\chi_{[|x|<1]}\, \mathrm{sign}(x) \log |x| \notin \mathrm{BMO}([-1, 1]).$$

(c) Show that, for any f as in Definition 7.13,

$$f^{\sharp\sharp}(x) := \sup_{x\in Q} \inf_{c} \fint_{Q} |f(y) - c|\, dy$$

satisfies $f^{\sharp\sharp} \le f^{\sharp} \le 2 f^{\sharp\sharp}$.

We shall not distinguish between $f^{\sharp}$ and $f^{\sharp\sharp}$.

7.4.2. Singular integrals on L^∞

It is easy to show that singular integrals take L^∞ to the BMO space.

Theorem 7.14 *Let T be a singular integral operator as in Definition 7.1. Then*

$$\|Tf\|_{\mathrm{BMO}} \le CB\|f\|_\infty \quad \forall f \in L^2(\mathbb{R}^d) \cap L^\infty(\mathbb{R}^d).$$

We are assuming here that $f \in L^2(\mathbb{R}^d)$, so that Tf is well defined.

Proof We may assume that $B = 1$ in Definition 7.1. Fix some f as above and a ball $B_0 := B(x_0, R) \subset \mathbb{R}^d$, and define

$$c_{B_0} := \int_{[|y-x_0|>2R]} K(x_0 - y) f(y)\, dy.$$

Then, with $B_0^* := B(x_0, 2R)$, one can obtain the following bound:

$$\begin{aligned}
\int_{B_0} |(Tf)(x) - c_{B_0}|dx &\le \int_{B_0} \int_{[|y-x_0|>2R]} |K(x-y) - K(x_0 - y)||f(y)|\, dy\, dx \\
&\quad + \int_{B_0} T\big(\chi_{B_0^*} f\big)(x)\, dx \\
&\le C|B_0|\|f\|_\infty + |B_0|^{1/2}\|T\|_{2\to 2}\, \|\chi_{B_0^*} f\|_2 \\
&\le C|B_0|\|f\|_\infty,
\end{aligned}$$

and we are done. □

In fact, it turns out that BMO is the smallest space that contains $T(L^\infty(\mathbb{R}^d))$ for every singular integral operator T as in Definition 7.1; see the notes at the end of this chapter.

7.4.3. An interpolation result

If BMO is intended as a useful substitute to $L^\infty(\mathbb{R}^d)$ then we would like to be able to use it in interpolation theory. This is indeed possible, as the following theorem shows.

Theorem 7.15 *Let T be a linear operator bounded on $L^{p_0}(\mathbb{R}^d)$ for some $1 \le p_0 < \infty$ and bounded from $L^\infty \to$ BMO. Then T is bounded on $L^p(\mathbb{R}^d)$ for any $p_0 < p < \infty$.*

The proof of this result requires some machinery, more specifically a comparison between the Hardy–Littlewood maximal function and the sharp maximal function (7.25). For this purpose it is convenient to use the *dyadic maximal function*, which we denote by M_{dyad} and which is defined by the following:

$$(M_{\text{dyad}} f)(x) := \sup_{x \ni Q} \fint_Q |f(y)|\, dy,$$

where the supremum now runs over dyadic cubes defined as the collection

$$\mathcal{Q}_{\text{dyad}} := \left\{2^k[0,1)^d + 2^k\mathbb{Z}^d \mid k \in \mathbb{Z}\right\};$$

see (2.28). Note that any two dyadic cubes are either disjoint, or one contains the other. Suppose $Q_0 \in \mathcal{Q}_{\text{dyad}}$, and let $f \in L^1(Q_0)$. Then, for any

$$\lambda > \fint_{Q_0} |f(y)|\, dy, \tag{7.29}$$

we can perform a Calderón–Zygmund decomposition as in Lemma 2.17 to conclude that

$$\big\{x \in Q_0 \,\big|\, (M_{\mathrm{dyad}} f)(x) > \lambda\big\} = \bigcup_{Q \in \mathcal{B}} Q,$$

whence in particular the measure of the left-hand side is $< \lambda^{-1}\|f\|_{L^1(Q_0)}$. In other words, we have reestablished the weak-L^1 bound for the maximal function; see Exercise 2.12.

The condition (7.29) serves as a starting condition for the stopping-time construction in Lemma 2.17. The other values of λ are of no interest, as then

$$\lambda^{-1}\|f\|_{L^1(Q_0)} \ge |Q_0|,$$

so the weak-L^1 bound becomes trivial. Of course, the construction does not necessarily need to be localized to any Q_0 but can also be carried out on $\mathbb{R}^d$ globally.

7.4.4. The Hardy–Littlewood and sharp maximal functions

We now come to the aforementioned comparison between $M_{\mathrm{dyad}} f$ and $f^\sharp$. While it is clear that $f^\sharp \le 2Mf$, where Mf is the usual Hardy–Littlewood maximal function, we require a lower bound. In essence, the following theorem says that, on L^p with finite p, the sharp function contains no new information.

Theorem 7.16 *Let $1 \le p_0 \le p < \infty$ and suppose that $f \in L^{p_0}(\mathbb{R}^d)$. Then*

$$\int_{\mathbb{R}^d} (M_{\mathrm{dyad}} f)^p(x)\, dx \le C(p,d) \int_{\mathbb{R}^d} \big(f^\sharp(x)\big)^p\, dx. \tag{7.30}$$

Proof For simplicity we write M instead of M_{dyad}. The proof is an immediate consequence of the following estimate, which is an example of a *good-λ inequality*. For all $\gamma > 0$ and $\lambda > 0$ one has

$$\begin{aligned} &\big|\big\{x \in \mathbb{R}^d \,\big|\, Mf(x) > 2\lambda,\ f^\sharp(x) \le \gamma\lambda\big\}\big| \\ &\qquad \le 2^d \gamma \big|\big\{x \in \mathbb{R}^d \,\big|\, Mf(x) > \lambda\big\}\big|. \end{aligned} \tag{7.31}$$

To prove (7.31), we write the set on the second line as $\bigcup_{Q \in \mathcal{B}} Q$ with dyadic Q by means of Lemma 2.17. We use $f \in L^{p_0}$ here since this ensures that we may start the stopping-time argument at any level $\lambda > 0$ by taking the side length of the first dyadic cube to be very large. By construction these Q are the *maximal dyadic cubes* contained in $\{Mf > \lambda\}$. Indeed, if $\tilde{Q} \supset Q$ denotes the unique dyadic cube having twice the side length of Q (the parent cube) then, by the definition of $\mathcal{B}$, one has

$$⨍_{\tilde{Q}} |f(y)|\, dy \le \lambda \quad \text{and} \quad \lambda < ⨍_{Q} |f(y)|\, dy \le 2^d \lambda.$$

Now fix one such Q and its parent $\tilde{Q}$. Then, by the preceding, if $x \in Q$,

$$(Mf)(x) > 2\lambda \quad \Longrightarrow \quad (M(f - \chi_Q f_{\tilde{Q}}))(x) > \lambda.$$

Therefore

$$\begin{aligned}\big|\{x \in Q \mid (Mf)(x) > 2\lambda\}\big| &\le \big|\{x \in Q \mid (M(f - \chi_Q f_{\tilde{Q}}))(x) > \lambda\}\big| \\ &\le \lambda^{-1} \int_Q |f - f_{\tilde{Q}}|(y)\, dy \le \gamma |\tilde{Q}|.\end{aligned}$$

To pass to the final bound, we used the fact that if there exists some $x_0 \in Q$ such that $f^\sharp(x_0) \le \gamma\lambda$ (which we may assume) then

$$\int_Q |f - f_{\tilde{Q}}|(y)\, dy \le \int_{\tilde{Q}} |f - f_{\tilde{Q}}|(y)\, dy \le \gamma\lambda |\tilde{Q}|.$$

Summing over all $Q \in \mathcal{B}$ establishes the claim (7.31). Now

$$\begin{aligned}&\int_0^{\lambda_0} |\{Mf(x) > 2\lambda\}|\, p\lambda^{p-1}\, d\lambda \\ &\quad \le \int_0^{\lambda_0} \big|\{Mf(x) > 2\lambda,\ f^\sharp(x) \le \gamma\lambda\}\big|\, p\lambda^{p-1}\, d\lambda \\ &\qquad + \int_0^{\lambda_0} |\{f^\sharp(x) > \gamma\lambda\}|\, p\lambda^{p-1}\, d\lambda \\ &\quad \le 2^{d+p}\gamma \int_0^{\lambda_0} |\{Mf(x) > 2\lambda\}|\, p\lambda^{p-1}\, d\lambda + \gamma^{-p} \|f^\sharp\|_p^p.\end{aligned}$$

Choosing $\gamma = 2^{-d-p-1}$ and letting $\lambda_0 \to \infty$ concludes the proof. □

It is now easy to prove the interpolation result.

Proof of Theorem 7.15 Consider $S : f \mapsto (Tf)^\sharp$. While S is not linear, it is sublinear and this is sufficient for Marcinkiewicz interpolation. By definition S is bounded on L^{p_0} and L^∞, whence it is also bounded on $L^p(\mathbb{R}^d)$ for any $p_0 < p < \infty$. Now take $f \in L^p \cap L^{p_0}$. Then $Tf \in L^{p_0}$ and we conclude from Theorem 7.16 and the fact that $Tf \le M_{\mathrm{dyad}}(Tf)$ a.e. that T is bounded on L^p. □

7.4.5. Exponential bounds: the John–Nirenberg inequality

We will now take a closer look at the property (7.27). In fact, as is suggested by Exercise 7.8, we shall now show that any BMO function is exponentially integrable. The precise statement is given by the *John–Nirenberg* inequality, Theorem 7.17 below. The logic behind the exponential decay given below in (7.32) is revealed by the behavior of the averages in (7.27): *they increase on an arithmetic scale while the magnitudes of the corresponding supports decrease on a geometric scale.* The reader should note that the following proof is based exactly on this principle.

Theorem 7.17 *Let $f \in \mathrm{BMO}(\mathbb{R}^d)$. Then, for any cube Q, one has*

$$\left|\left\{x \in Q \,\middle|\, |f(x) - f_Q| > \lambda\right\}\right| \leq C|Q| \exp\Big(-c\frac{\lambda}{\|f\|_{\mathrm{BMO}}}\Big) \tag{7.32}$$

for any $\lambda > 0$. The constants c, C are absolute.

Proof We may take $Q = [0,1)^d$. Assume that $\|f\|_{\mathrm{BMO}} \leq 1$ with $f_Q = 0$, and let $\lambda > 10$. Perform a Calderón–Zygmund decomposition of f on the cube Q at a level $\mu \geq 2$ that we shall fix later. This yields $f = g + b$, $|g| \leq 2^d\mu$, and

$$b = \sum_{Q' \in \mathcal{B}} \chi_{Q'}(f - f_{Q'}), \quad \sum_{Q' \in \mathcal{B}} |Q'| \leq \mu^{-1}.$$

Define

$$E(\lambda) := \sup_{\|\varphi\|_{\mathrm{BMO}(Q)} \leq 1} \left|\left\{x \in Q \,\middle|\, |\varphi(x) - \varphi_Q| > \lambda\right\}\right|.$$

Then, assuming that f attains this supremum up to a factor 2.

$$\begin{aligned} E(\lambda) &\leq 2\left|\left\{x \in Q \,\middle|\, |b(x)| > \lambda - 2^d\mu\right\}\right| \\ &\leq 2\sum_{Q' \in \mathcal{B}} \left|\left\{x \in Q' \,\middle|\, |f(x) - f_{Q'}| > \lambda - 2^d\mu\right\}\right| \\ &\leq 2\sum_{Q' \in \mathcal{B}} E(\lambda - 2^d\mu)|Q'| \leq 2\mu^{-1}E(\lambda - 2^d\mu). \end{aligned} \tag{7.33}$$

Iterating this relation implies that

$$E(\lambda) \leq 2^n\mu^{-n}E(\lambda - 2^d\mu n).$$

Setting for example $\mu = 3$ leads to the desired bound. □

Exercise 7.9 Show that one can characterize BMO also by the following condition, where $1 \leq r < \infty$ is fixed but arbitrary:

$$\sup_Q \left(\fint_Q |f(y) - f_Q|^r\,dy\right)^{1/r} < \infty. \tag{7.34}$$

In fact, show that if the left-hand side is ≤ 1 then $\|f\|_{\mathrm{BMO}} \leq C(d,r)$. Conversely, by Hölder's inequality, $\|f\|_{\mathrm{BMO}} \leq 1$ implies that the left-hand side of (7.34) is ≤ 1.

7.4.6. Commutators of singular integral operators with BMO functions

A remarkable characterization of BMO was found by Coifman, Rochberg, and Weiss. They considered the commutator of a Calderón–Zygmund operator T and an operator b corresponding to multiplication by a function, in other words, $[T, b] = Tb - bT$. While it is clear that $[T, b]$ is bounded on L^p for $1 < p < \infty$ if $b \in L^\infty$, these authors discovered that this property remains true for $b \in \mathrm{BMO}(\mathbb{R}^d)$. What is more, if this property holds for a single Riesz transform (or a single nonzero operator of the form (7.24)) then necessarily $b \in \mathrm{BMO}$. We shall now prove one direction of this statement, the *Coifman–Rochberg–Weiss theorem.*

Theorem 7.18 *Let T be a linear operator as in Proposition 7.4 but with the added property that, for some fixed $0 < \delta \le 1$, one has*

$$|K(x-y) - K(x_0 - y)| \le A\frac{|x-x_0|^\delta}{|y-x_0|^{d+\delta}} \quad \forall |y-x_0| > 2|x-x_0|, \tag{7.35}$$

as well as $\|T\|_{2\to 2} \le A$. Then for any $b \in \mathrm{BMO}(\mathbb{R}^d)$ one has

$$\|[T, b]\|_{p\to p} \le C(d, p)A\|b\|_{\mathrm{BMO}}$$

for all $1 < p < \infty$.

Proof Since (7.35) implies the Hörmander condition (see Definition 7.1(ii)), we see that any operator T as in the theorem is a Calderón–Zygmund operator. We can also assume that $\|b\|_{\mathrm{BMO}} \le 1$ and $A = 1$. Fix $f \in \mathcal{S}(\mathbb{R}^d)$, say. We shall show that, for any $1 < r < \infty$,

$$([T, b]f)^\sharp \le C(r, d)(M_r(Tf) + M_r f), \tag{7.36}$$

where

$$(M_r f)(x) := \sup_{Q \ni x} \left(\fint_Q |f(y)|^r \, dy \right)^{1/r}.$$

Letting $1 < r < p < \infty$ and noting that M_r is L^p-bounded, we conclude from (7.36) that $\|([T, b]f)^\sharp\|_p \le C(d, p)\|f\|_p$, and Theorem 7.16 then implies the desired estimate (the p_0 condition in that theorem clearly holds, by Theorem 7.17 and $f \in \mathcal{S}(\mathbb{R}^d)$).

To establish (7.36) fix any cube Q, which by scaling and translation invariance may be taken to be the unit cube centered at 0. Denote by $\tilde{Q}$ the cube of side length $2\sqrt{d}$, and write

$$[T, b]f = T\big((b - b_{\tilde{Q}})f_1\big) + T\big((b - b_{\tilde{Q}})f_2\big) - (b - b_{\tilde{Q}})Tf =: g_1 + g_2 + g_3,$$

where $f_1 := \chi_{\tilde{Q}} f$, $f_2 = f - f_1$. First, by Exercise 7.9,

$$\fint_Q |g_3(x)|\,dx \le \Big(\fint_Q |b - b_Q|^{r'}\,dx\Big)^{1/r'} \Big(\fint_Q |Tf|^r(x)\,dx\Big)^{1/r} \le C \inf_Q M_r(Tf).$$

Second, choose some $1 < s < r$. Then by the boundedness of T on L^s one obtains

$$\begin{aligned}\fint_Q |g_1(x)|\,dx &\le \Big(\fint_Q |g_1(x)|^s\,dx\Big)^{1/s} \le \Big(\fint_{\tilde{Q}} |b - b_Q|^s |f(x)|^s\,dx\Big)^{1/s}\\ &\le C \inf_Q M_r f,\end{aligned}$$

where we have used (7.26) to compare b_Q with $b_{\tilde{Q}}$. Finally, set

$$c_Q := T\big((b - b_{\tilde{Q}}) f_2\big)(0).$$

Then by (7.35) one has the estimate

$$\begin{aligned}&\fint_Q |g_2(x) - c_Q|\,dx\\ &\quad\le C \int_Q \int_{\tilde{Q}^c} |K(x-y) - K(0-y)|\,|b(y) - b_{\tilde{Q}}|\,|f(y)|\,dy\,dx\\ &\quad\le C \sum_{\ell \ge 0} \int_{2^{\ell+1}\tilde{Q} \setminus 2^\ell \tilde{Q}} 2^{-\ell(d+\delta)} |b(y) - b_Q|\,|f(y)|\,dy\\ &\quad\le C \sum_{\ell \ge 0} 2^{-\ell\delta} \Big(\fint_{2^{\ell+1}\tilde{Q}} |b(y) - b_Q|^{r'}\,dy\Big)^{1/r'} \Big(\fint_{2^{\ell+1}\tilde{Q}} |f(y)|^r\,dy\Big)^{1/r}\\ &\quad\le C \sum_{\ell \ge 0} \ell\, 2^{-\ell\delta} \inf_Q M_r f \le C \inf_Q M_r f,\end{aligned}$$

where we have used (7.27) to pass to the final estimate. □

7.5. Singular integrals and A_p weights

We shall now show that singular integrals are bounded on $L^p(\mathbb{R}^d, w(x)\,dx)$ with $1 < p < \infty$ and $w \in A_p$. The latter class was defined in Chapter 2, where it was shown to be both necessary and sufficient for the boundedness of the Hardy–Littlewood maximal function. In establishing the strong L^p bound in that case *we did not make explicit use* of the following fundamental property of the A_p class, which amongst other things ensures the "openness" of the A_p condition. This property is that if $w \in A_p$ with $p > 1$ then $w \in A_q$ for some

$q < p$. For the theory of singular integrals, however, it is most convenient to invoke this property.

7.5.1. The reverse Hölder inequality

We begin with the fundamental *reverse Hölder inequality*, which follows by an iterative application of a Calderón–Zygmund decomposition.

Lemma 7.19 *Let $w \in A_p$ for some $1 \le p < \infty$. Then there exist constants C and $\delta > 0$, that depend on p and the A_p constant of w, such that for any cube Q one has*

$$\left(\fint_Q w^{1+\delta}\right)^{1/(1+\delta)} \le C \fint_Q w. \tag{7.37}$$

Proof Without loss of generality, we take Q to be the unit cube $[0,1)^d$. We perform a Calderón–Zygmund decomposition at increasing heights $\{\lambda_k\}_{k=0}^{\infty}$, starting from $w(Q) =: \lambda_0$. For each λ_k we obtain a sequence of cubes $Q_{k,j}$ with the properties

$$w(x) \le \lambda_k \quad \text{if } x \notin \Omega_k = \bigcup_j Q_{k,j},$$

$$\lambda_k < \fint_{Q_{k,j}} w < 2^d \lambda_k.$$

Then Ω_k is a decreasing sequence of sets. Writing $Q_{k,j_0} \cap \Omega_{k+1} = \bigcup_i Q_{k+1,i}$, we conclude that

$$\begin{aligned} |Q_{k,j_0} \cap \Omega_{k+1}| &= \sum_i |Q_{k+1,i}| \le \lambda_{k+1}^{-1} \sum_i \int_{Q_{k+1,i}} w \\ &\le \lambda_{k+1}^{-1} \int_{Q_{k,j_0}} w \le \frac{2^d \lambda_k}{\lambda_{k+1}} |Q_{k,j_0}|. \end{aligned} \tag{7.38}$$

With $\alpha < 1$ fixed, we define $2^d \lambda_k / \lambda_{k+1} = \alpha$, whence

$$\lambda_k = (2^d \alpha^{-1})^k w(Q) \quad \forall k \ge 0.$$

By (2.18) there exists $\beta < 1$ such that

$$w(Q_{k,j_0} \cap \Omega_{k+1}) \le \beta w(Q_{k,j_0}).$$

Summing over all Q_{k,j_0} in Ω_k yields $w(\Omega_{k+1}) \le \beta w(\Omega_k)$. Since $\bigcap_k \Omega_k$ has measure zero, using $|\Omega_k| \le \alpha^k |\Omega_0|$ we can finally make the following estimation:

$$\begin{aligned}\fint_Q w^{1+\delta} &= \int_{Q\setminus\Omega_0} w^{1+\delta} + \sum_{k=0}^{\infty}\int_{\Omega_k\setminus\Omega_{k+1}} w^{1+\delta}\\ &\le \lambda_0^\delta w(Q) + \sum_{k=0}^{\infty}\lambda_{k+1}^\delta w(\Omega_k)\\ &= \lambda_0^\delta w(Q) + \sum_{k=0}^{\infty}(2^d\alpha^{-1})^{(k+1)\delta}\lambda_0^\delta\,\beta^k w(\Omega_0).\end{aligned}$$

We now choose $\delta > 0$ small enough that $(2^d\alpha^{-1})^\delta\beta < 1$. Then the entire right-hand side is bounded by a constant times the first term, as desired. □

7.5.2. The A_∞ condition

The reverse Hölder inequality implies the following regularity statement, which is more precise than (2.18) on small sets.

Corollary 7.20 *If $w \in A_p$ for some $1 \le p < \infty$ then there exist constants $\delta > 0$ and C such that, for any cube Q and $S \subset Q$, one has*

$$\frac{w(S)}{w(Q)} \le C\left(\frac{|S|}{|Q|}\right)^\delta. \tag{7.39}$$

Proof Taking $Q = [0,1)^d$ without loss of generality then, by Hölder and reverse Hölder inequalities,

$$w(S) = \int_Q \chi_S w \le \left(\int_Q w^{1+\delta}\right)^{1/(1+\delta)} |S|^{\delta/(1+\delta)} \le Cw(Q)|S|^{\delta/(1+\delta)}$$

as claimed. □

The class of weights satisfying (7.39) is referred to as the A_∞ class; this is justified by Problem 7.10. Next, we ask the reader to establish the weighted analogue of Theorem 7.16 as an exercise.

Exercise 7.10 Show that, for any $w \in A_p$, one has

$$\int_{\mathbb{R}^d} (M_{\text{dyad}} f)^p(x)\, w(x)dx \le C(p,d)\int_{\mathbb{R}^d} \left(f^\sharp(x)\right)^p w(x)\,dx \tag{7.40}$$

for every f for which $M_{\text{dyad}} f \in L^{p_0}(w)$. *Hint:* Proceed as in the proof of Theorem 7.16, and use (7.39) to switch from Lebesgue measure to $w\,dx$.

7.5.3. The singular integral bound

We can now establish the main result concerning singular integrals and A_p weights.

Theorem 7.21 *If T is a strong Calderón–Zygmund operator then for all $1 < p < \infty$ one has*

$$\int_{\mathbb{R}^d} |(Tf)(x)|^p w(x)\,dx \le C \int_{\mathbb{R}^d} |f(x)|^p w(x)\,dx$$

for all $f \in \mathcal{S}(\mathbb{R}^d)$.

Proof We first make the following claim. For any $s > 1$ one has, pointwise,

$$(Tf)^\sharp \le C_s (M(|f|^s))^{1/s}$$

where M is the Hardy–Littlewood maximal operator. Assuming the claim, we can easily finish the proof using (7.40) and Theorem 2.19. Indeed, one then has

$$\begin{aligned}\int_{\mathbb{R}^d} |(Tf)(x)|^p\, w(x)\,dx &\le \int_{\mathbb{R}^d} (M_{\mathrm{dyad}}(Tf))^p(x)\, w(x)\,dx \\ &\le C \int_{\mathbb{R}^d} |(Tf)^\sharp(x)|^p\, w(x)\,dx \\ &\le C \int_{\mathbb{R}^d} |M(|f|^s)(x)|^{p/s}\, w(x)\,dx \\ &\le C \int_{\mathbb{R}^d} |f(x)|^p\, w(x)\,dx.\end{aligned}$$

It therefore remains to verify the claim. By translation and scaling invariance it suffices to check that

$$\inf_{c\in\mathbb{R}} \int_{Q_0} |(Tf)(x) - c|\,dx \le C_s (M(|f|^s))^{1/s}(0),$$

where $Q_0 = \left[\frac{-1}{2}, \frac{1}{2}\right)^d$. Split f as $f_0 + g_0$, where $f_0 = f\chi_{Q_1}$ with $Q_1 = [-1, 1)^d$. Then, on the one hand,

$$\int_{Q_0} |(Tf_0)(x)|\,dx \le \left(\int_{Q_0} |(Tf_0)(x)|^s\,dx\right)^{1/s} \le C\|f_0\|_{L^s}.$$

On the other hand, we write $\Omega_\ell := 2^{\ell+1}Q_0 \setminus 2^\ell Q_0$ and

$$g_0 = \sum_{\ell=1}^{\infty} \chi_{\Omega_\ell} f =: \sum_{\ell=1}^{\infty} f_\ell$$

and set

$$c_0 := (Tg_0)(0) = \sum_{\ell=1}^{\infty} (Tf_\ell)(0).$$

Then for each $\ell \geq 1$ one has

$$\begin{aligned}\int_{Q_0} |(Tf_\ell)(x) - (Tf_\ell)(0)|\, dx &\leq C \int_{\Omega_\ell} \int_{Q_0} \big|K(x-y) - K(0-y)\big|\, |f_\ell(y)|\, dxdy \\ &\leq C \int_{\Omega_\ell} \int_{Q_0} \frac{|x|}{|x-y|^{d+1}} |f_\ell(y)|\, dxdy \\ &\leq C2^{-\ell} \fint_{2^{\ell+1}Q_0} |f(y)|\, dy \\ &\leq C2^{-\ell} (M(|f|^s))^{1/s}(0)\end{aligned}$$

which suffices. □

7.6. A glimpse of H^1–BMO duality and further remarks

We would like to determine a duality relation analogous to the basic relation $(L^1(\mathbb{R}^d))^* = L^\infty(\mathbb{R}^d)$. This turns out to be $(H^1(\mathbb{R}^d))^* = \mathrm{BMO}(\mathbb{R}^d)$, where $H^1(\mathbb{R}^d)$ is the completion of $\mathcal{S}(\mathbb{R}^d)$ under the norm

$$\|f\|_{H^1} := \|f\|_1 + \sum_{j=1}^{d} \|R_j f\|_1; \tag{7.41}$$

here the R_j are Riesz transforms. Dually, one can then write

$$\mathrm{BMO} = L^\infty + \sum_{j=1}^{d} R_j L^\infty, \tag{7.42}$$

where the action of the Riesz transforms on L^∞ needs to be defined suitably. In particular, this shows that BMO is the smallest space into which general singular integral operators map L^∞. We shall prove the H^1–BMO duality in a *dyadic setting* in Section 8.4. There we shall also derive an atomic decomposition of dyadic H^1.

Another important role played by BMO functions is as logarithms of A_p weights. This is explored in Problem 7.8.

A more elementary, but very fundamental, question turns out to be the following. Is there a proof of the L^2-boundedness of Calderón–Zygmund operators (or the Hilbert transform, for that matter) that does not rely on the Fourier transform? In other words, we wish to find a proof of this L^2-boundedness that relies exclusively on the properties of the kernel itself. We shall answer this question in the affirmative in the larger context of Calderón–Zygmund operators, introducing the important method of *almost orthogonality* in the process; see Chapter 9.

Finally, it is natural to seek a Calderón–Zygmund theory for kernels that are not translation invariant. This, too, has far-reaching consequences especially with respect to L^2-boundedness. By considerations similar to those in the convolution case, and under suitable conditions on the kernels, L^p theory reduces to the $p = 2$ case; see Problem 7.3. Establishing the required boundedness at $p = 2$ is much more delicate, and the standard tool for this purpose is the $T(1)$ theorem of David and Journé. We will present this theorem for the case of the line in Section 9.4.

Notes

For most of the basic material of this chapter, excluding BMO, see Stein's 1970 book [108]. A brief account of singular integral theory from about the same time can be found in Zygmund's lecture notes [131]. The foundational paper by Calderón and Zygmund is [15], and a survey of the earlier theory, including pseudodifferential operators, was given by Calderón in [13].

For a thorough development of H^1 and BMO see Stein [111], which moreover gives a systematic account of many aspects of Calderón–Zygmund theory that appeared from 1970 to roughly 1992. The classic reference for H^1 and the duality with BMO is the paper by Fefferman and Stein [39], which contains (7.41). Uchiyama [124] gives a constructive proof of the decomposition of any BMO function as a sum $f_0 + \sum_{j=1}^d R_j f_j$, where all $f_j \in L^\infty$ and the R_j are the Riesz transforms; see (7.42). Coifman and Meyer [24] offer another perspective of Calderón–Zygmund theory with many applications and connections with wavelets. They also focused on the multilinear aspects of Calderón–Zygmund theory, which are explored in considerable detail in the second volume of this book. The proof of the Coifman–Rochberg–Weiss commutator estimate is from Janson [61] (credited to Strömberg); see also Torchinsky [123]. For the nontranslation-invariant theory, and in particular the $T(1)$ theorem, see Stein [111] as well as Duoandikoetxea [31]. The latter two references also contain a discussion of the method of rotations, which can be used to reduce higher-dimensional results to one dimensional. As mentioned above, Section 9.4 will present a proof of the David–Journé $T(1)$ theorem using Haar functions.

Problems

Problem 7.1 Let $\{a_{ij}(x)\}_{i,j=1}^d$ be continuous in some domain $\Omega \subset \mathbb{R}^d$, $d \geq 2$, and suppose that these matrices are elliptic in the sense that, for all $x \in \Omega$,

$$\theta|\xi|^2 \leq \sum_{i,j=1}^{d} a_{ij}(x)\xi^i\xi^j \leq \Theta|\xi|^2 \quad \forall \xi \in \mathbb{R}^d$$

for some fixed $0 < \theta < \Theta$. Assume that $u \in C^2(\Omega)$ satisfies

$$\sum_{i,j=1}^{d} a_{ij}(x) \frac{\partial}{\partial x_i} \frac{\partial}{\partial x_j} u(x) = f(x).$$

Then show that, for any compact $K \subset \Omega$ and any $1 < p < \infty$, one has the bound

$$\| D^2 u \|_{L^p(K)} \leq C(\|u\|_{L^p(\Omega)} + \|f\|_{L^p(\Omega)}),$$

where C depends only on p, K, Ω, and the a_{ij}.

Problem 7.2 This problem shows how symmetries limit the boundedness properties of operators that commute with these symmetries. We shall consider translation and dilation symmetries, respectively.

(a) Show that for *any* translation-invariant nonzero operator $T : L^p(\mathbb{R}^d) \to L^q(\mathbb{R}^d)$ one necessarily has $q \geq p$.
(b) Show that a homogeneous kernel such as that in (7.24) can be bounded from

$$L^p(\mathbb{R}^d) \to L^q(\mathbb{R}^d)$$

only if $p = q$.

Problem 7.3 Let T be a bounded linear operator on $L^2(\mathbb{R}^d)$ such that, for some function $K(x, y)$ that is measurable and locally bounded on $\mathbb{R}^d \times \mathbb{R}^d \setminus \Delta$, with $\Delta = \{x = y\}$ the diagonal, one has for all $f \in L^2_{\text{comp}}(\mathbb{R}^d)$

$$(Tf)(x) = \int_{\mathbb{R}^d} K(x, y)\, f(y)\, dy, \quad x \notin \text{supp}(f).$$

Furthermore, assume the Hörmander conditions

$$\begin{aligned} &\int_{[|x-y|>2|y-y'|]} |K(x, y) - K(x, y')|\, dx \leq A \quad \forall y \neq y', \\ &\int_{[|x-y|>2|x-x'|]} |K(x, y) - K(x', y)|\, dy \leq A \quad \forall\, x \neq x'. \end{aligned} \tag{7.43}$$

Then T is bounded from L^1 to weak-L^1 and is bounded strongly on L^p for every $1 < p < \infty$.

Problem 7.4 This is the continuum analogue of Problem 3.9. Let m be a function having bounded variation on $\mathbb{R}$. Show that $T_m f = (m\hat{f})^\vee$ is bounded on $L^p(\mathbb{R})$ for any $1 < p < \infty$. Find a suitable generalization to higher dimensions, starting with $\mathbb{R}^2$.

Problem 7.5 Let $f \in \text{BMO}(\mathbb{R}^d)$. Show that, for any $\sigma > d$ and $1 \leq r < \infty$, one has

$$\int_{\mathbb{R}^d} \frac{|f(x) - f_{Q_0}|^r}{1 + |x|^\sigma}\, dx \leq C(d, \sigma, r) \|f\|^r_{\text{BMO}},$$

where Q_0 is any cube of side length 1. What does one obtain by rescaling this inequality?

Problem 7.6 Let $f \in L^1(\mathbb{R}^d)$ have compact support, say $B := B(0, 1)$. Let M be the Hardy–Littlewood maximal operator. Show that $Mf \in L^1(B)$ if and only if $f \log(2 + |f|) \in L^1(B)$. Now redo Problem 3.12 using this fact and the F. and M. Riesz theorem. *Hint:* Prove the weak-L^1 bound on M by means of the Calderón–Zygmund decomposition. In particular, show that in order to bound $|\{Mf > \lambda\}|$ it is enough to bound the L^1 norm of f on $\{|f| > c\lambda\}$, where $c > 0$ is some constant. See Stein [108, p. 23] for more details. Compare Exercise 2.10.

Problem 7.7 Let A, B be reflexive Banach spaces (a reader not familiar with reflexivity may take A and B to be Hilbert spaces). Let $K(x, y)$ be defined, measurable, and locally bounded on $\mathbb{R}^d \times \mathbb{R}^d \setminus \Delta$, where $\Delta = \{x = y\}$ is the diagonal, and let K take its values in $\mathcal{L}(A, B)$, the bounded linear operators from A to B. Define an operator T on $C^0_{\mathrm{comp}}(\mathbb{R}^d; A)$, the continuous compactly supported functions taking their values in A, such that

$$(Tf)(x) = \int_{\mathbb{R}^d} K(x, y) f(y)\, dy, \qquad x \notin \mathrm{supp}(f),$$

for all $f \in C^0_{\mathrm{comp}}(\mathbb{R}^d; A)$. Assume that, for some finite constant C_0,

$$\int_{[|x-y|>2|x-x'|]} \|K(x, y) - K(x', y)\|_{\mathcal{L}(A,B)}\, dy \leq C_0 \quad \forall x \neq x',$$

$$\int_{[|x-y|>2|y-y'|]} \|K(x, y) - K(x, y')\|_{\mathcal{L}(A,B)}\, dx \leq C_0 \quad \forall y \neq y'.$$

Further, assume that $T : L^r(\mathbb{R}^d; A) \to L^r(\mathbb{R}^d; B)$, for some $1 < r < \infty$, with norm bounded by C_0. Show that then $T : L^p(\mathbb{R}^d; A) \to L^p(\mathbb{R}^d; B)$, for all $1 < p < \infty$, with norm bounded by $C_0\, C_1$ where $C_1 = C_1(d, r)$.

Problem 7.8 Show that $w \in A_2$ implies that $\log w \in \mathrm{BMO}$ and that, conversely, if $\varphi \in \mathrm{BMO}$ then $e^{\tau\varphi} \in A_2$ for small $|\tau|$. Generalize to A_p.

Problem 7.9 Investigate the endpoint $p = d/\alpha$ of Proposition 7.8. More precisely, find examples that show that fractional integration in that case is not bounded as an operator from L^p into L^∞. Verify that your examples are mapped into $\mathrm{BMO} \setminus L^\infty(\mathbb{R}^d)$. For a positive result that proves the boundedness of fractional integration in the $L^p \to \mathrm{BMO}$ sense, see Adams [1].

Problem 7.10 Show that $\bigcup_{1 \leq p < \infty} A_p = A_\infty$. *Hint:* Show that w^{-1} satisfies a reverse Hölder inequality with measure $w\, dx$.

Problem 7.11 Establish a weighted version of Theorem 7.18. *Hint:* See Segovia and Torrea [98].

8

Littlewood–Paley theory

8.1. The Mikhlin multiplier theorem

A central concern of harmonic analysis is the study of multiplier operators, i.e., operators T defined by a relation of the form

$$(T_m f)(x) = \int_{\mathbb{R}^d} e^{2\pi i x\cdot\xi} m(\xi) \hat{f}(\xi)\, d\xi, \tag{8.1}$$

where $m : \mathbb{R}^d \to \mathbb{C}$ is bounded. By Plancherel's theorem, $\|T_m\|_{2\to 2} \le \|m\|_\infty$ but in fact it is easy to see that one has equality here. Over distributions (see Definition 4.5) we may write $T_m f = K * f$ for any $f \in \mathcal{S}(\mathbb{R}^d)$ with $K := \check{m} \in \mathcal{S}'$.[1] Then T_m is bounded on L^1 if and only if the associated kernel is in L^1, in which case

$$\|T_m\|_{1\to 1} = \|K\|_1.$$

There are many cases, though, where $K \notin L^1(\mathbb{R}^d)$ but T is bounded on L^p for some (or all) $1 < p < \infty$. The Hilbert transform is one such example. We shall now discuss a basic result in the field which describes a large class of multipliers m that give rise to L^p-bounded operators for $1 < p < \infty$.

In Chapter 7 we studied the case where m is homogeneous of degree zero and we showed that the T_m are Calderón–Zygmund operators with homogeneous kernel and thus L^p-bounded for $1 < p < \infty$. In the following theorem, known as the Mikhlin multiplier theorem, we shall allow more general functions m, which satisfy finitely many derivative conditions *of the zero-degree type*. The proof of this theorem requires a standard partition of unity over a geometric scale.

[1] As earlier, the inverse hat indicates an inverse Fourier transform.

8.1.1. A partition of unity over a geometric scale

Lemma 8.1 *There exists $\psi \in C^\infty(\mathbb{R}^d)$ with the property that* $\mathrm{supp}(\psi) \subset \mathbb{R}^d \setminus \{0\}$ *is compact and such that*

$$\sum_{j=-\infty}^{\infty} \psi(2^{-j}x) = 1 \quad \forall x \neq 0. \tag{8.2}$$

For any given $x \neq 0$, at most two terms in this sum are nonzero. Moreover, ψ can be chosen to be a radial nonnegative function.

Proof Let $\chi \in C^\infty(\mathbb{R}^d)$ satisfy $\chi(x) = 1$ for all $|x| \leq 1$ and $\chi(x) = 0$ for $|x| \geq 2$. Set $\psi(x) := \chi(x) - \chi(2x)$. For any positive N one has

$$\sum_{j=-N}^{N} \psi(2^{-j}x) = \chi(2^{-N}x) - \chi(2^{N+1}x).$$

If $x \neq 0$ is given then we can take N sufficiently large that $\chi(2^{-N}x) = 1$ and $\chi(2^{N+1}x) = 0$. This implies (8.2), and the other properties are immediate as well. □

8.1.2. The multiplier theorem

We can now state the main result of this section, the multiplier Theorem 8.2. A typical example to which this theorem applies is the Riesz transform R_j, which is given by the multiplier $m(\xi) = c\xi_j|\xi|^{-1}$ with a suitable constant c. This has its only singularity at the origin, and we shall impose conditions in Theorem 8.2 that guarantee the same property for general multipliers $m(\xi)$.

As a rule, one analyses the boundedness properties of the associated operator T_m by breaking up m according to a geometric length scale relative to the origin:

$$m = \sum_j m_j, \qquad m_j(\xi) = m(\xi)\psi\left(2^{-j}\xi\right),$$

where ψ is as in Lemma 8.1. As a first step, one notes easily that T_{m_j} is L^p bounded via Young's inequality *for all* $1 \leq p \leq \infty$. The real difficulty is the summation over j, which requires Calderón–Zygmund theory. Note that in the process of summation one necessarily loses the validity at $p = 1$ and $p = \infty$.

Theorem 8.2 *Let $m : \mathbb{R}^d \setminus \{0\} \to \mathbb{C}$ satisfy, for any multi-index γ of length $|\gamma| \leq d + 2$,*

$$|\partial^\gamma m(\xi)| \leq B|\xi|^{-|\gamma|}$$

for all $\xi \neq 0$. *Then, for any* $1 < p < \infty$, *there is a constant* $C = C(d, p)$ *such that*

$$\|(m\hat{f})^\vee\|_p \leq CB\|f\|_p$$

for all $f \in \mathcal{S}$.

Proof Let ψ give rise to a dyadic partition of unity as in Lemma 8.1. Define, for any $j \in \mathbb{Z}$,

$$m_j(\xi) = \psi(2^{-j}\xi)m(\xi)$$

and set $K_j = \check{m}_j$. Now fix some large positive integer N and set

$$K(x) = \sum_{j=-N}^{N} K_j(x).$$

We claim that under our smoothness assumption on m one has the pointwise estimates

$$|K(x)| \leq CB|x|^{-d}, \quad |\nabla K(x)| \leq CB|x|^{-d-1}, \tag{8.3}$$

where $C = C(d)$. One then applies the Calderon–Zygmund theorem and lets N tend to ∞.

We will verify the first inequality in (8.3). The second is similar and the proof will only be sketched. By assumption, $\|D^\gamma m_j\|_\infty \leq CB2^{-j|\gamma|}$ and thus

$$\|D^\gamma m_j\|_1 \leq CB2^{-j|\gamma|}2^{jd}$$

for any multi-index $|\gamma| \leq d+2$. Similarly,

$$\|D^\gamma(\xi_i m_j)\|_1 \leq CB2^{-j(|\gamma|-1)}2^{jd}$$

for the same γ. Hence,

$$\|x^\gamma \check{m}_j(x)\|_\infty \leq CB2^{j(d-|\gamma|)}$$

and

$$\|x^\gamma D\check{m}_j(x)\|_\infty \leq CB2^{j(d+1-|\gamma|)}.$$

Since $|x|^k \leq C(k, d)\sum_{|\gamma|=k}|x^\gamma|$, one concludes that

$$|\check{m}_j(x)| \leq CB2^{j(d-k)}|x|^{-k} \tag{8.4}$$

and

$$|D\check{m}_j(x)| \leq CB2^{j(d+1-k)}|x|^{-k} \tag{8.5}$$

for any $0 \le k \le d+2$ and all $j \in \mathbb{Z}$, $x \in \mathbb{R}^d \setminus \{0\}$. We shall use this with $k=0$ and $k = d+2$. Indeed,

$$\begin{aligned}|K(x)| &\le \sum_j |\check{m}_j(x)| \le \sum_{2^j \le |x|^{-1}} |\check{m}_j(x)| + \sum_{2^j > |x|^{-1}} |\check{m}_j(x)| \\ &\le CB \sum_{2^j \le |x|^{-1}} 2^{jd} + CB \sum_{2^j > |x|^{-1}} 2^{jd}(2^j|x|)^{-(d+2)} \\ &\le CB|x|^{-d} + CB|x|^2|x|^{-d-2} = CB|x|^{-d}\,,\end{aligned}$$

as claimed. To obtain the second inequality in (8.3) one uses (8.5) instead of (8.4). Otherwise the argument is unchanged. Thus we have verified that K satisfies the conditions (i) and (ii) of Definition 7.1; see Lemma 7.2. Furthermore $\|m\|_\infty \le B$, so that $\|(m\hat{f})^\vee\|_2 \le B\|f\|_2$. By Theorem 7.5 and the remark following it one concludes that

$$\|(m\hat{f})^\vee\|_p \le C(p,d)\|f\|_p$$

for all $f \in \mathcal{S}$ and $1 < p < \infty$, as claimed. □

The number of derivatives entering into the hypothesis can be decreased; see Problem 8.1. Theorem 8.2 allows one to give proofs of Corollary 7.7 without going through the procedure in Exercise 7.4. Indeed, one has

$$\widehat{\frac{\partial^2 u}{\partial x_i \partial x_j}}(\xi) = \frac{\xi_i\xi_j}{|\xi|^2}\widehat{\Delta u}(\xi).$$

Since $m(\xi) = \xi_i\xi_j/|\xi|^2$ satisfies the conditions of Theorem 8.2 (this is obvious, as m is homogeneous of degree zero and smooth away from $\xi = 0$), we are done. Observe that this example also shows that Theorem 8.2 fails if $p=1$ or $p = \infty$.

Theorem 8.2 is formulated as an a priori inequality for $f \in \mathcal{S}(\mathbb{R}^d)$. In this case $m\hat{f} \in \mathcal{S}'$, so that $(m\hat{f})^\vee \in \mathcal{S}'$. The question arises whether $(m\hat{f})^\vee$ is meaningful for $f \in L^p(\mathbb{R}^d)$. If $1 \le p \le 2$ then $\hat{f} \in L^2 + L^\infty(\mathbb{R}^d)$ (in fact $\hat{f} \in L^{p'}$ by the Hausdorff–Young inequality), so that $m\hat{f} \in \mathcal{S}'$, and therefore $(m\hat{f})^\vee \in \mathcal{S}'$, is well defined. If $p > 2$, however, it is known that there are functions $f \in L^p(\mathbb{R}^d)$ such that $\hat{f} \in \mathcal{S}'$ has positive order (see the notes to Chapter 4). For such f it is in general not possible to define $(m\hat{f})^\vee$ in $\mathcal{S}'$.

8.2. Littlewood–Paley square-function estimate

We now present an important application of Mikhlin's theorem to Littlewood–Paley theory. With ψ as in Lemma 8.1 we define $P_j f = (\psi_j \hat{f})^\vee$, where

$\psi_j(\xi) = \psi(2^{-j}\xi)$ (it is customary to assume that $\psi \ge 0$). Then, by Plancherel's theorem we have

$$C^{-1}\|f\|_2^2 \le \sum_{j\in\mathbb{Z}} \|P_j f\|_2^2 \le \|f\|_2^2 \tag{8.6}$$

for any $f \in L^2(\mathbb{R}^d)$. Observe that the middle expression is equal to $\|Sf\|_2^2$, with

$$Sf = \Big(\sum_j |P_j f|^2\Big)^{1/2}.$$

The function Sf is called the Littlewood–Paley *square function*. It is a basic result of harmonic analysis that (8.6) generalizes to

$$C^{-1}\|f\|_p \le \|Sf\|_p \le C\|f\|_p \tag{8.7}$$

for any $f \in L^p(\mathbb{R}^d)$, provided that $1 < p < \infty$ and with $C = C(p,d)$. We now offer some heuristic explanations as to why such a result might hold.

A natural point of departure is to view sequences of complex exponentials on $\mathbb{T}$ with geometric (or "lacunary") sequences of frequencies, such as $\{e(2^n\theta)\}_{n\ge1}$, from a probabilistic angle. Indeed, we can think of these functions as being close to independent random variables on the interval $[0,1]$: on any interval $[k2^{-n},(k+1)2^{-n}) \subset [0,1)$, with integer k, on which $\sin(2^n\pi\theta)$ has a fixed sign, the function $\sin(2^{n+1}\pi\theta)$ is equally likely to be positive or negative. Clearly, this behavior is close to that of the Rademacher functions r_n defined in Chapter 5. This suggests that Proposition 5.15 should hold if X_j is replaced with the random variable $a_j e(2^j\theta)$ on the interval $[0,1]$. For the same reason it is quite natural to view $P_j f$ and $P_k f$ as independent variables if $|k-j| > 10$, say; but then (8.7) is natural in view of Proposition 5.15.

Another, rather elementary, way to look at this heuristics is as follows. Consider a lacunary trigonometric polynomial of the form

$$T(\theta) = \sum_{n=1}^{M} c_n e(2^n\theta),$$

where $\theta \in \mathbb{T}$. Let $p \ge 1$ be an integer and estimate by Plancherel's theorem that

$$\int_0^1 \Big|\sum_{n=1}^{M} c_n\, e(2^n\theta)\Big|^{2p} d\theta = C(p)\int_0^1 \Big|\sum_{\substack{n_1,\dots,n_p=1\\ n_1+\dots+n_p=m}}^{M} c_{n_1}\cdots c_{n_p} e(2^m\theta)\Big|^2 d\theta$$

$$\le C(p)\sum_{n_1,\dots,n_p=1}^{M} |c_{n_1}\cdots c_{n_p}|^2 = C(p)\Big(\sum_n |c_n|^2\Big)^p.$$

We have used here the fact that, for every positive integer N, the number of ways in which it can be written in the form

$$N = 2^{n_1} + \cdots + 2^{n_p}, \quad n_j \geq 0,$$

is bounded by a constant that depends only on p. In other words, given a sequence $\{c_k\} \in \ell^2(\mathbb{Z})$ such that $c_k = 0$ unless $|k|$ is of the form 2^n with $n \geq 0$, we have the property

$$C(p)^{-1}\Big(\sum_k |c_k|^2\Big)^{1/2} \leq \Big\|\sum_k c_k e(k\theta)\Big\|_p \leq C(p)\Big(\sum_k |c_k|^2\Big)^{1/2} \tag{8.8}$$

for every $p \geq 2$ (by Hölder's inequality one can lower this to $p \geq 1$; see the final inequality in the proof of Lemma 5.5). Clearly, (8.8) is remarkable as it says that any $L^2(\mathbb{T})$ function f such that $\hat{f}(k) = 0$ unless $|k|$ is a power of 2 has finite L^p norm for every $p < \infty$ (but of course not for $p = \infty$). We remark in passing that this actually applies to any lacunary or gap Fourier series, but this generalization is harder to establish (the standard argument involves Riesz products).

The rough idea behind (8.7) can be stated as follows. If we consider an arbitrary trigonometric polynomial $\sum_n a_n e(n\theta)$, then (8.8) is false. However, we may try to salvage something of (8.8) by setting

$$c_k(\theta) := \sum_{2^{k-1} < |n| \leq 2^k} a_n e(n\theta), \quad \forall k \geq 1.$$

We might then expect that the square function

$$Sf(\theta) := \Big(\sum_k |c_k(\theta)|^2\Big)^{1/2}$$

satisfies $\|Sf\|_p \simeq \|f\|_p$.

It should therefore come as no surprise to the reader that some probabilistic elements enter into the proof of (8.7). In fact, we shall now derive (8.7) by means of a standard randomization technique involving a fair-coin-tossing sequence or, equivalently, the Rademacher functions. More specifically, let $\{r_j\}$ be a sequence of independent random variables with $\mathbb{P}(\{r_j = 1\}) = \mathbb{P}(\{r_j = -1\}) = \frac{1}{2}$ for all j. Khinchine's inequality, see Lemma 5.5, will play a crucial role in the proof of the following *Littlewood–Paley theorem*.

Theorem 8.3 *For any $1 < p < \infty$ there is a constant $C = C(p, d)$ such that*

$$C^{-1}\|f\|_p \leq \|Sf\|_p \leq C\|f\|_p$$

for any $f \in \mathcal{S}$.

Proof Let $\{r_j\}$ be as above. The proof rests on the fact that

$$m(\xi) := \sum_{j=-N}^{N} r_j \psi_j(\xi)$$

satisfies the conditions of the Mikhlin multiplier theorem uniformly in N and uniformly in the realization of the random variables $\{r_j\}$. Indeed, for any γ,

$$|D^\gamma m(\xi)| \le \sum_{j=-N}^{N} |D^\gamma \psi_j(\xi)| \le C \sum_{j=-N}^{N} |\xi|^{-|\gamma|} \big|(D^\gamma \psi)(2^{-j}\xi)\big| \le C|\xi|^{-|\gamma|}.$$

To pass to the final inequality we used the fact that only an absolutely bounded number of terms is nonzero, in the sum in the middle expression, for any $\xi \neq 0$. Hence, in view of Lemma 5.5,

$$\int_{\mathbb{R}^d} |(Sf)(x)|^p \, dx \le C \limsup_{N\to\infty} \mathbb{E} \int_{\mathbb{R}^d} \left| \sum_{j=-N}^{N} r_j (P_j f)(x) \right|^p dx \le C\|f\|_p^p,$$

as desired.

To prove the lower bound we use duality. Choose a function $\tilde{\psi}$ such that $\tilde{\psi} = 1$ on $\operatorname{supp}(\psi)$ and $\tilde{\psi}$ is compactly supported with $\operatorname{supp}(\tilde{\psi}) \subset \mathbb{R}^d \setminus \{0\}$. Defining $\tilde{P}_j$ in the same way as P_j but with $\tilde{\psi}$ instead of ψ yields $\{\tilde{P}_j\}$ satisfying $\tilde{P}_j P_j = P_j$. Therefore, for any $f, g \in \mathcal{S}$ and any $1 < p < \infty$,

$$|\langle f, g\rangle| = \left|\sum_j \langle P_j f, \tilde{P}_j g\rangle\right| \le \int_{\mathbb{R}^d} \Big(\sum_j |(P_j f)(x)|^2\Big)^{1/2} \Big(\sum_k \big|(\tilde{P}_k g)(x)\big|^2\Big)^{1/2} dx$$

$$\le \|Sf\|_p \, \|\tilde{S}g\|_{p'} \le C\|Sf\|_p \, \|g\|_{p'}.$$

For the final bound we used the fact that the argument for the upper bound applies equally well to $\tilde{S}$. Thus $\|f\|_p \le C\|Sf\|_p$, as claimed. □

It is desirable to formulate Theorem 8.3 on $L^p(\mathbb{R}^d)$ rather than on $\mathcal{S}(\mathbb{R}^d)$. This is done in the following corollary. Observe that Sf is defined pointwise if $f \in \mathcal{S}'(\mathbb{R}^d)$, since $P_j f$ is a smooth function.

Corollary 8.4

(i) *Let $1 < p < \infty$. Then, for any $f \in L^p(\mathbb{R}^d)$, one has $Sf \in L^p$ and*

$$C_{p,d}^{-1}\|f\|_p \le \|Sf\|_p \le C_{p,d}\|f\|_p.$$

(ii) *Suppose that $f \in \mathcal{S}'$ and that $Sf \in L^p(\mathbb{R}^d)$ with some $1 < p < \infty$. Then $f = g + P$, where P is a polynomial and $g \in L^p(\mathbb{R}^d)$. Moreover, $Sf = Sg$ and*

$$C_{p,d}^{-1}\|Sf\|_p \le \|g\|_p \le C_{p,d}\|Sf\|_p.$$

Proof To prove part (i), let $f_k \in \mathcal{S}$ such that $\|f_k - f\|_p \to 0$ as $k \to \infty$. We claim that

$$\lim_{k\to\infty} \|Sf_k - Sf\|_p = 0. \tag{8.9}$$

If (8.9) holds then, passing to the limit $k \to \infty$ in

$$C_{p,d}^{-1}\|f_k\|_p \le \|S(f_k)\|_p \le C_{p,d}\|f_k\|_p,$$

implies part (i). To prove (8.9) one applies Fatou's lemma repeatedly. Fix $x \in \mathbb{R}^d$. Then

$$\begin{aligned}|Sf_k(x) - Sf(x)| &= \Big|\big\|\{P_j f_k(x)\}_{j=-\infty}^{\infty}\big\|_{\ell^2} - \big\|\{P_j f(x)\}_{j=-\infty}^{\infty}\big\|_{\ell^2}\Big| \\ &\le \big\|\{P_j f_k(x) - P_j f(x)\}_{j=-\infty}^{\infty}\big\|_{\ell^2} \\ &= S(f_k - f)(x) \le \liminf_{m\to\infty} S(f_k - f_m)(x).\end{aligned}$$

Therefore

$$\|Sf_k - Sf\|_p \le \liminf_{m\to\infty} \|S(f_k - f_m)\|_p \le C_{p,d} \liminf_{m\to\infty} \|f_k - f_m\|_p.$$

The claim (8.9) now follows on letting $k \to \infty$.

For the second part we argue as in the proof of the lower bound in Theorem 8.3. Let $f \in \mathcal{S}'$ and $h \in \mathcal{S}$ with $\mathrm{supp}(\hat{h}) \subset \mathbb{R}^d \setminus \{0\}$. Then

$$|\langle f, \bar{h}\rangle| \le C_p \|Sf\|_p \,\|\tilde{S}h\|_{p'} \le C_p \|Sf\|_p \,\|h\|_{p'},$$

where $\tilde{S}$ is the modified square function from the proof of Theorem 8.3. By the Hahn–Banach theorem there exists $g \in L^p$, such that $\langle f, \bar{h}\rangle = \langle g, h\rangle$ for all h as above, satisfying $\|g\|_p \le C_p\|Sf\|_p$. By Exercise 4.4 one has $f = g + P$. Moreover, $Sf = Sg$ and

$$\|Sf\|_p = \|Sg\|_p \le C\|g\|_p$$

by part (i). $\square$

The following exercise shows that the Littlewood–Paley theorem fails at the endpoints $p = 1$ and $p = \infty$.

Exercise 8.1

(a) Show that Theorem 8.3 fails at $p = 1$. Intuitively, one takes $f = \delta_0$. For this case verify that

$$(Sf)(x) \simeq |x|^{-d}$$

such that $Sf \notin L^1(\mathbb{R}^d)$. Next, transfer this logic to $L^1(\mathbb{R}^d)$ by means of approximate identities.

(b) Show that Theorem 8.3 fails for $p = \infty$.

It is natural to ask at this point whether the Littlewood–Paley theorem 8.3 holds for square functions defined in terms of *sharp cutoffs* rather than smooth cutoffs as above. More precisely, suppose that we set

$$(S_{\text{new}} f)^2 = \sum_{j \in \mathbb{Z}} |(\chi_{[2^{j-1} \le |\xi| < 2^j]} \hat{f}(\xi))^\vee|^2.$$

Is it true that

$$C_p^{-1} \|f\|_p \le \|S_{\text{new}} f\|_p \le C_p \|f\|_p \tag{8.10}$$

for $1 < p < \infty$? Owing to the boundedness of the Hilbert transform the answer is "yes" in dimension 1 but it is a rather deep fact that it is "no" in dimensions $d \ge 2$; see the notes and problems to this chapter.

As a first application of (8.7) we will now finish the proof of (4.9).

Proposition 8.5 *For any* $0 \le s < d/2$ *and with*

$$\frac{1}{2} - \frac{1}{p} = \frac{s}{d}$$

one has

$$\|f\|_{L^p(\mathbb{R}^d)} \le C(s,d) \, \|f\|_{\dot{H}^s(\mathbb{R}^d)}$$

for all $f \in \mathcal{S}(\mathbb{R}^d)$.

Proof In Chapter 4 we settled the case where $\hat{f}$ is localized to a dyadic shell $\{|\xi| \simeq 2^j\}$. Thus, it remains to prove (4.12). In fact, we shall now prove that, for any finite $p \ge 2$,

$$\|f\|_p \le C \Big(\sum_{j \in \mathbb{Z}} \|P_j f\|_p^2 \Big)^{1/2}$$

where the constant C depends only on p and d. By (8.7),

$$\begin{aligned} \|f\|_p &\le C(p,d) \, \|Sf\|_p = C(p,d) \Big\| \, \|\{P_j f\}_{j \in \mathbb{Z}}\|_{\ell^2(\mathbb{Z})} \Big\|_p \\ &\le C(p,d) \Big\| \{\|P_j f\|_p\}_{j \in \mathbb{Z}} \Big\|_{\ell^2(\mathbb{Z})}, \end{aligned}$$

where we used the generalized triangle inequality in the final step; see Exercise 6.1. □

As evidenced by the proof of Proposition 8.5, the result in Exercise 6.1 is a useful (and sharp) tool when combined with (8.7). In particular, it may be used to compare Sobolev spaces with Besov spaces; see the problem section for this chapter. As a final general comment about Littlewood–Paley theory, we ask the reader to verify that there can be no version of (8.7) in which the square function is replaced by some ℓ^q-based norm.

Exercise 8.2 Suppose that for some $1 \le q \le \infty$ and $1 \le p \le \infty$ one has

$$\big\| \|\{P_j f\}_{j\in\mathbb{Z}}\|_{\ell^q(\mathbb{Z})} \big\|_{L^p(\mathbb{R}^d)} \simeq \|f\|_p \quad \forall f \in \mathcal{S}(\mathbb{R}^d),$$

with constants that do not depend on f. Show that $q = 2$.

8.3. Calderón–Zygmund decomposition, Hölder spaces, and Schauder estimates

We shall now revisit the question of the boundedness of singular integrals on Hölder spaces C^α for $0 < \alpha < 1$. While the proof given here is considerably longer than that in Chapter 7, it is somewhat more efficient since it applies to kernels that satisfy Definition 7.1 without any further smoothness requirements such as $|\nabla K(x)| \le C|x|^{-d-1}$, which we required for Theorem 7.6.

8.3.1. A Besov characterization of Hölder spaces

We begin with a characterization of $C^\alpha(\mathbb{R}^d)$ for $0 < \alpha < 1$ in terms of the projections P_j (see the start of Section 8.2). The proof of the following lemma is reminiscent of the proof of Bernstein's theorem 1.13.

Lemma 8.6 *Let $|f| \le 1$. Then $f \in C^\alpha(\mathbb{R}^d)$ for $0 < \alpha < 1$ if and only if*

$$\sup_{j\in\mathbb{Z}} 2^{j\alpha} \|P_j f\|_\infty \le A. \tag{8.11}$$

Moreover, the smallest A for which (8.11) *holds is comparable with $[f]_\alpha$.*

Proof Set $\check{\psi}_j(x) = 2^{jd}\check{\psi}(2^j x) =: \varphi_j(x)$. Hence $\|\varphi_j\|_1 = \|\check{\psi}\|_1$ for all $j \in \mathbb{Z}$, and

$$\int_{\mathbb{R}^d} \varphi_j(x) x^\gamma \, dx = 0$$

for all multi-indices γ. First assume that $f \in C^\alpha$. Then

$$\begin{aligned} |P_j f(x)| &\le \int_{\mathbb{R}^d} |f(x-y) - f(x)|\, |\varphi_j(y)| \, dy \\ &\le \int_{\mathbb{R}^d} [f]_\alpha |y|^\alpha |\varphi_j(y)| \, dy = 2^{-j\alpha} [f]_\alpha \int_{\mathbb{R}^d} |y|^\alpha |\check{\psi}(y)| \, dy. \end{aligned}$$

Hence $A \le C[f]_\alpha$, as claimed. Conversely, define

$$g_\ell(x) = \sum_{-\ell \le j \le \ell} (P_j f)(x)$$

for any integer $\ell \ge 0$. We need to show that, for all $y \in \mathbb{R}^d$,

$$\sup_\ell |g_\ell(x-y) - g_\ell(x)| \le CA|y|^\alpha$$

with some constant $C = C(d)$. Fix $y \neq 0$ and obtain the estimate

$$\left| \sum_{|y|^{-1} < 2^j \le 2^\ell} (P_j f)(x) \right| \le \sum_{2^j > |y|^{-1}} A 2^{-j\alpha} \le C A |y|^\alpha . \tag{8.12}$$

Second, observe that

$$\begin{aligned} \left| P_j f(x - y) - P_j f(x) \right| &\le \| \nabla P_j f \|_\infty |y| \le C 2^j \| P_j f \|_\infty |y| \\ &\le C 2^{j(1-\alpha)} A |y|, \end{aligned} \tag{8.13}$$

where we invoked Bernstein's inequality to pass to the second inequality. Combining (8.12) and (8.13) yields

$$\begin{aligned} &|g_\ell(x - y) - g_\ell(x)| \\ &\le \sum_{2^{-\ell} \le 2^j \le |y|^{-1}} |P_j f(x - y) - P_j f(x)| + \sum_{|y|^{-1} < 2^j \le 2^\ell} 2 \| P_j f \|_\infty \\ &\le \sum_{2^j \le |y|^{-1}} C A 2^{j(1-\alpha)} |y| + \sum_{|y|^{-1} < 2^j} 2 A 2^{-j\alpha} \le C A |y|^\alpha \end{aligned} \tag{8.14}$$

uniformly in $\ell \ge 1$.

So $\{g_\ell - g_\ell(0)\}_{\ell=1}^\infty$ are uniformly bounded on $C^\alpha(K)$ for any compact K. By the Arzela–Ascoli theorem one concludes that

$$g_\ell - g_\ell(0) \to g$$

uniformly on any compact set and therefore that $[g]_\alpha \le CA$ by (8.14) (up to passing to a subsequence $\{\ell_i\}$). It remains to show that f has the same property. This follows from the claim that $f = g +$ constant. To verify this property, note that $g_\ell - g_\ell(0) \to g$ in $\mathcal{S}'$. Thus, we also have $\hat{g}_\ell - \delta_0 g_\ell(0) \to \hat{g}$ in $\mathcal{S}'$, which is the same as

$$\sum_{-\ell \le j \le \ell} \psi(2^{-j}\xi) \hat{f}(\xi) - \delta_0 g_\ell(0) \quad \to \quad \hat{g} \text{ in } \mathcal{S}'.$$

So if $h \in \mathcal{S}$ with $\operatorname{supp}(h) \subset \mathbb{R}^d \setminus \{0\}$ then $\langle \hat{f} - \hat{g}, h \rangle = 0$, i.e., we have $\operatorname{supp}(\hat{f} - \hat{g}) = \{0\}$. By Exercise 4.4, therefore,

$$(f - g)(x) = \sum_{|\gamma| \le M} C_\gamma x^\gamma . \tag{8.15}$$

However, since $g(0) = 0$,

$$|(f - g)(x)| \le \| f \|_\infty + |g(x) - g(0)| \le 1 + C A |x|^\alpha .$$

Since $\alpha < 1$, comparing this bound with (8.15) shows that the polynomial in (8.15) is of degree zero. Thus, $f - g =$ constant, as claimed. □

8.3.2. Singular integrals on C^α

Next we need the following result, which gives an example of a situation where T maps into $L^1(\mathbb{R}^d)$. We will postpone the proof.

Proposition 8.7 *Let K be a singular integral kernel as in Definition 7.1. For any $\eta \in \mathcal{S}(\mathbb{R}^d)$ with $\int_{\mathbb{R}^d} \eta(x)\,dx = 0$ one has*

$$\|T\eta\|_1 \le CB,$$

where the constant C depends on η.

We can now formulate and prove the C^α bound for singular integrals.

Theorem 8.8 *Let K be as in Definition 7.1 and let $0 < \alpha < 1$. Then, for any $f \in L^2 \cap C^\alpha(\mathbb{R}^d)$, one has $Tf \in C^\alpha(\mathbb{R}^d)$ and $\left[Tf\right]_\alpha \le C_\alpha B[f]_\alpha$ with $C_\alpha = C(\alpha, d)$.*

Proof We will use Lemma 8.6. In order to do so, notice first that

$$\|Tf\|_\infty \le CB([f]_\alpha + \|f\|_2),$$

which we leave to the reader to check. Therefore it suffices to show that

$$\sup_j \left(2^{j\alpha}\|P_j Tf\|_\infty\right) \le CB[f]_\alpha. \tag{8.16}$$

Let $\tilde{P}_j$ be defined as follows:

$$\tilde{P}_j u = \left(\tilde{\psi}(2^{-j}\cdot)\hat{u}(\xi)\right)^\vee,$$

where $\tilde{\psi} \in C_0^\infty(\mathbb{R}^d \setminus \{0\})$ and $\tilde{\psi}\psi = \psi$. Thus $\tilde{P}_j P_j = P_j$. Hence

$$\begin{aligned}
\|P_j Tf\|_\infty &= \|\tilde{P}_j P_j Tf\|_\infty = \|\tilde{P}_j T P_j f\|_\infty \\
&\le \|\tilde{P}_j T\|_{\infty\to\infty}\,\|P_j f\|_\infty \\
&\le \|\tilde{P}_j T\|_{\infty\to\infty}\, C[f]_\alpha 2^{-j\alpha}.
\end{aligned}$$

It remains to show that $\sup_j \|\tilde{P}_j T\|_{\infty\to\infty} \le CB$; see (8.16). Clearly, the kernel of $\tilde{P}_j T$ is $2^{jd}\check{\tilde{\psi}}(2^j\cdot) * K$, so that

$$\|\tilde{P}_j T\|_{\infty\to\infty} \le \left\|2^{jd}\check{\tilde{\psi}}(2^j\cdot) * K\right\|_1 = \left\|\check{\tilde{\psi}} * 2^{-jd}K(2^{-j}\cdot)\right\|_1 \le CB, \tag{8.17}$$

by Proposition 8.7. Indeed, in that proposition set $\eta = \check{\tilde{\psi}}$ such that

$$\int_{\mathbb{R}^d} \eta(x)\,dx = \tilde{\psi}(0) = 0,$$

as required. Further, apply Proposition 8.7 with the rescaled kernel $2^{-jd}K(2^{-j}\cdot)$. As this kernel satisfies the conditions in Definition 7.1 *uniformly* in j, one obtains (8.17) and the theorem is proved. □

8.3.3. An atomic decomposition in $\mathcal{S}$

It remains to show Proposition 8.7. We will build up from the case of an "atom" f, as defined by the following lemma.

Lemma 8.9 *Let $f \in L^\infty(\mathbb{R}^d)$ with $\int f(x)\,dx = 0$, $\mathrm{supp}(f) \subset B(0,R)$, and $\|f\|_\infty \le R^{-d}$. Then*

$$\|Tf\|_1 \le CB.$$

Proof By the Cauchy–Schwarz inequality,

$$\begin{aligned}\int_{[|x|\le 2R]} \big|(Tf)(x)\big|\,dx &\le \|Tf\|_2 C R^{d/2} \le CB\|f\|_2 R^{d/2}\\ &\le CBR^{-d}R^{d/2}R^{d/2} = CB.\end{aligned}$$

Furthermore,

$$\begin{aligned}\int_{[|x|>2R]} |(Tf)(x)|\,dx &\le \int_{\mathbb{R}^d}\int_{[|x|>2|y|]} |K(x-y)-K(x)|\,dx\,|f(y)|\,dy\\ &\le B\|f\|_1 \le CB,\end{aligned}$$

as desired. □

In passing, we point out a simple relation between these atoms and BMO. Let $\varphi \in \mathrm{BMO}$ and $a = f$ as defined in Lemma 8.9. Then

$$\begin{aligned}|\langle a,\varphi\rangle| &= \left|\int_B a(x)(\varphi(x)-\varphi_B)\,dx\right| \le \|a\|_\infty \int_B |\varphi(x)-\varphi_B|\,dx\\ &\le ⨍_B |\varphi(x)-\varphi_B|\,dx \le \|\varphi\|_{\mathrm{BMO}}.\end{aligned}$$

Moreover, it is easy to see from this that $\sup_a |\langle a,\varphi\rangle| = \|\varphi\|_{\mathrm{BMO}}$. It is therefore natural to define the *atomic space* $H^1_{\mathrm{atom}}(\mathbb{R}^d)$ as all sums $f = \sum_j c_j a_j$ with $\sum_j |c_j| < \infty$ and to define the norm $\|f\|_{H^1}$ as the smallest possible value of the sum $\sum_j |c_j|$ for a given f.

It is a remarkable fact that this atomic space coincides with H^1 as defined by (7.41). In the following section we shall prove a result of this nature in a one-dimensional dyadic setting. In the next lemma we show that one can write a general Schwartz function with mean zero as a superposition of infinitely many "atoms".

Lemma 8.10 *Let $\eta \in \mathcal{S}(\mathbb{R}^d)$, $\int_{\mathbb{R}^d} \eta(x)\,dx = 0$. Then one can write*

$$\eta = \sum_{\ell=1}^{\infty} c_\ell a_\ell,$$

with $\int_{\mathbb{R}^d} a_\ell(x)\,dx = 0$, $\|a_\ell\|_\infty \le \ell^{-d}$, $\mathrm{supp}(a_\ell) \subset B(0,\ell)$ *for all* $\ell \ge 1$, *and*

$$\sum_{\ell=1}^{\infty} |c_\ell| \le C,$$

where the constant C *depends on* η.

Proof In this proof we let

$$\langle g \rangle_S := \frac{1}{|S|}\int_S g(x)\,dx$$

for any $g \in L^1(\mathbb{R}^d)$ and $S \subset \mathbb{R}^d$ with $0 < |S| < \infty$. Moreover, $B_\ell := B(0,\ell)$ for $\ell \ge 1$ and $\chi_\ell = \chi_{B_\ell}$ (the indicator of B_ℓ). Define

$$f_1 := (\eta - \langle \eta \rangle_{B_1})\chi_1, \qquad \eta_1 := \eta - f_1$$

and set inductively

$$\begin{cases} f_{\ell+1} := (\eta_\ell - \langle \eta_\ell \rangle_{B_{\ell+1}})\,\chi_{\ell+1}, \\ \eta_{\ell+1} := \eta_\ell - f_{\ell+1} \end{cases} \tag{8.18}$$

for $\ell \ge 1$ (one can also take $\ell = 0$ and $\eta_0 := \eta$). Observe that

$$\eta = \eta_1 + f_1 = f_1 + f_2 + \eta_2 = \cdots = \sum_{\ell=1}^{M} f_\ell + \eta_{M+1}. \tag{8.19}$$

We need to show that we can pass to the limit $M \to \infty$ and that

$$a_\ell := f_\ell \frac{\ell^{-d}}{\|f_\ell\|_\infty}, \qquad \text{for} \quad \ell \ge 1, \tag{8.20}$$

have the desired properties. By construction, for all $\ell \ge 1$, we have

$$\int_{\mathbb{R}^d} a_\ell(x)\,dx = 0, \qquad \mathrm{supp}(a_\ell) \subset B_\ell,$$

and $\|a_\ell\|_\infty \le \ell^{-d}$. It remains to show that

$$c_\ell := \ell^d \|f_\ell\|_\infty \tag{8.21}$$

satisfies $\sum_{\ell=1}^{\infty} c_\ell < \infty$ and that $\|\eta_{M+1}\|_\infty \to 0$; see (8.19). Clearly,

$$\eta_1 = \begin{cases} \langle \eta \rangle_{B_1} & \text{on } B_1, \\ \eta & \text{on } \mathbb{R}^d \setminus B_1. \end{cases}$$

Hence

$$\eta_2 = \begin{cases} \langle \eta_1 \rangle_{B_2} & \text{on } B_2, \\ \eta_1 = \eta & \text{on } \mathbb{R}^d \setminus B_2. \end{cases}$$

By induction, for $\ell \geq 1$, one checks that

$$\eta_{\ell+1} = \begin{cases} \langle \eta_\ell \rangle_{B_{\ell+1}} & \text{on } B_{\ell+1}, \\ \eta & \text{on } \mathbb{R}^d \backslash B_{\ell+1}. \end{cases} \tag{8.22}$$

Moreover, induction shows that

$$\int_{\mathbb{R}^d} \eta_\ell(x)\, dx = 0 \tag{8.23}$$

for all $\ell \geq 0$. Indeed, this is assumed for $\eta_0 = \eta$. Since $\int f_\ell(x)\, dx = 0$ for $\ell \geq 1$ by construction, one may now proceed inductively via the formula

$$\eta_{\ell+1} = \eta_\ell - f_{\ell+1}.$$

Property (8.23) implies that

$$\begin{aligned} \left| \langle \eta_\ell \rangle_{B_{\ell+1}} \right| &\leq \frac{1}{|B_{\ell+1}|} \int_{\mathbb{R}^d \backslash B_{\ell+1}} |\eta_\ell(x)|\, dx \\ &\leq \frac{1}{|B_{\ell+1}|} \int_{\mathbb{R}^d \backslash B_{\ell+1}} |\eta(x)|\, dx \leq C \ell^{-20d}, \end{aligned} \tag{8.24}$$

since $\eta_\ell = \eta$ on $\mathbb{R}^d \backslash B_\ell$, see (8.22), and since η has rapid decay. Equation (8.24) implies that

$$\|\eta_{\ell+1}\|_\infty \leq C \ell^{-20d},$$

see (8.22), and also that

$$\|f_{\ell+1}\|_\infty \leq C \ell^{-20d};$$

see (8.18). We now conclude from (8.19)–(8.21) that

$$\eta = \sum_{\ell=1}^{\infty} f_\ell = \sum_{\ell=1}^{\infty} c_\ell a_\ell$$

with, see (8.21) and (8.20),

$$\sum_{\ell=1}^{\infty} |c_\ell| = \sum_{\ell=1}^{\infty} \ell^d \|f_\ell\|_\infty \leq \sum_{\ell=1}^{\infty} C \ell^{-19d} < \infty,$$

and the lemma follows. □

Proof of Proposition 8.7 Let η be as in the statement of the proposition. By Lemma 8.10,

$$\eta = \sum_{\ell=1}^{\infty} c_\ell a_\ell$$

with c_ℓ and a_ℓ as stated there. By Lemma 8.9

$$\|T(c_\ell a_\ell)\|_1 \le |c_\ell|\,\|T a_\ell\|_1 \le CB|c_\ell|$$

for $\ell \ge 1$. Hence

$$\sum_{\ell=1}^{\infty} \|T(c_\ell a_\ell)\|_1 \le C_\eta B,$$

and this easily implies that $\|T\eta\|_1 \le C_\eta B$, as claimed. □

8.3.4. The Schauder estimate

A typical application of Theorem 8.8 is to the so-called *Schauder estimate*. We have already encountered this, in the previous chapter, but record it here again.

Corollary 8.11 *Let $f \in C_0^{2,\alpha}(\mathbb{R}^d)$. Then*

$$\sup_{1\le i,j\le d} \left[\frac{\partial^2 f}{\partial x_i \partial x_j}\right]_\alpha \le C(\alpha, d)[\Delta f]_\alpha \tag{8.25}$$

for any $0 < \alpha < 1$.

Proof As in the L^p case, see Corollary 7.7, this follows from the fact that

$$\frac{\partial^2 f}{\partial x_i \partial x_j} = R_{ij}(\Delta f) \quad \text{for } 1 \le i, j \le d,$$

where R_{ij} are double Riesz transforms. Now apply Theorem 8.8 to the singular integral operators R_{ij}. □

For an extension of this corollary to the case of variable coefficients, see Problem 8.6 below.

Exercise 8.3 Show that Theorem 8.8 fails at $\alpha = 0$ and $\alpha = 1$.

As a final application of Lemma 8.6 we remark that it easily implies Sobolev's embedding $\dot{W}^{1,p}(\mathbb{R}^d) \hookrightarrow C^\alpha(\mathbb{R}^d)$ in the range $d < p < \infty$ with $\alpha = 1 - d/p$. See Problem 8.9.

8.4. The Haar functions; dyadic harmonic analysis

We now turn to the Haar functions on $[0, 1]$ (see Figure 8.1). We shall first show that they, together with the function 1, span $L^p([0, 1])$ for any $1 \le p < \infty$. To be precise, they form a *Schauder basis* in $L^p([0, 1])$ for finite p. In contrast with an exponential system, they are in fact an *unconditional basis* if $1 < p < \infty$, cf. Problem 6.10. An elegant approach to the unconditionality of the Haar

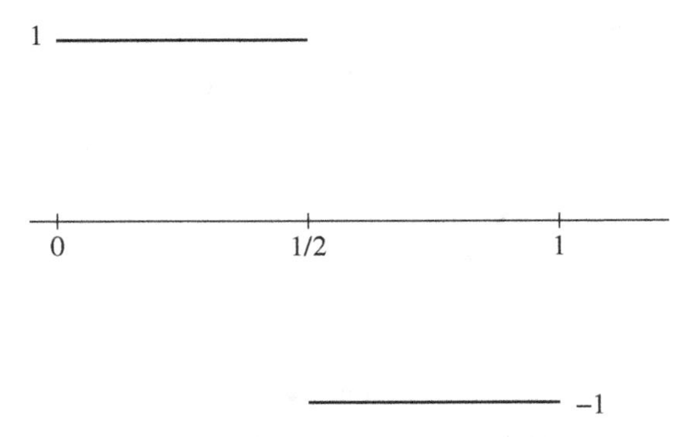

Figure 8.1. The Haar function $h_{[0,1]}$.

system proceeds via Burkholder's inequality, and we shall follow this route as well as a route based on Calderón–Zygmund theory. A square function appears naturally in this context, and we shall establish the same type of L^p estimate for it as in the Littlewood–Paley case. The failure of this square-function bound at $p = 1$ suggests an exploration of the subspace of $L^1([0, 1])$ on which the square function *is bounded*. This subspace is precisely the (dyadic) *Hardy space* $\mathcal{H}^1([0, 1])$. We shall show that its dual is (dyadic) BMO and that $\mathcal{H}^1$ admits an atomic decomposition. While less applicable than its continuum analogue, the theory of dyadic $\mathcal{H}^1$ and BMO exhibits the basic ideas more easily. Indeed, the deepest theorems established here, which are $\mathcal{H}^1$–BMO duality and atomic decomposition, respectively, rely on simple stopping-time arguments involving the binary tree.

8.4.1. Basic definitions and properties

Definition 8.12 We set

$$\mathcal{A}_n := \left\{ [(k-1)2^{-n}, k2^{-n}) \,\middle|\, 1 \le k \le 2^n \right\}$$

for each $n \ge 0$ and $\mathcal{D}_n := \bigcup_{m=0}^{n} \mathcal{A}_m$ for each $0 \le n \le \infty$. Let $I \in \mathcal{D}_\infty$ and bisect it: thus we set $I = I_1 \cup I_2$, where I_1, I_2 are the two halves of I relative to the midpoint. Define

$$h_I := (\chi_{I_1} - \chi_{I_2})|I|^{-1/2}$$

to be the *Haar function* associated with I. The Haar system is $\{h_I\}_{I \in \mathcal{D}_\infty}$.

In what follows we shall write $\mathcal{D}$ instead of $\mathcal{D}_\infty$ to denote all dyadic intervals in $[0, 1]$. The dyadic intervals are most conveniently identified with the nodes of a binary tree, as in Figure 8.2; the figure exhibits the first five levels of this tree.

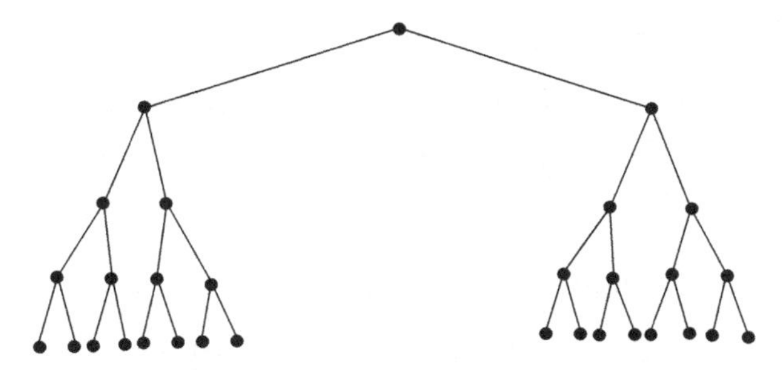

Figure 8.2

The root is $[0, 1]$ and moving down along a branch of the tree corresponds to continued bisection. In this fashion each infinite branch of the tree is precisely a unique real number in $[0, 1]$. Each row of the tree is given by the class $\mathcal{A}_n$. Let $\Sigma_n := \sigma(\mathcal{A}_n)$ denote the σ-algebra generated by $\mathcal{A}_n$, and define

$$\mathbb{E}_n(f) := \mathbb{E}[f|\Sigma_n], \quad f \in L^1([0, 1]),\ n \geq 0,$$

where $\mathbb{E}[\cdot|\cdot]$ is the conditional expectation operator; see Chapter 5. Clearly,

$$\mathbb{E}_n(f) = \sum_{I \in \mathcal{A}_n} \chi_I \fint_I f. \tag{8.26}$$

Lemma 8.13 *One has the following properties:*

$$\lim_{n\to\infty} \|\mathbb{E}_n(f) - f\|_\infty = 0 \quad \forall f \in C([0, 1]), \tag{8.27}$$

$$\lim_{n\to\infty} \|\mathbb{E}_n(f) - f\|_p = 0 \quad \forall f \in L^p([0, 1]),\ 1 \leq p < \infty.$$

Proof The first statement follows from the uniform continuity of f, whereas the second is obtained from the fact that the continuous functions are dense in L^p and the bound

$$\sup_n \|\mathbb{E}_n\|_{p\to p} \leq 1$$

for all $1 \leq p \leq \infty$. □

The connection with the Haar system is given by the following result.

Lemma 8.14 *For each $n \geq 1$, one has*

$$\mathbb{E}_n(f) = \int_0^1 f + \sum_{I \in \mathcal{D}_{n-1}} \langle f, h_I \rangle h_I \tag{8.28}$$

as well as a local version,

$$\fint_J f = \int_0^1 f + \sum_{\substack{I \supsetneq J \\ I \in \mathcal{D}}} \langle f, h_I \rangle h_I \quad \forall x \in J \tag{8.29}$$

and any $J \in \mathcal{D}$.

Proof One verifies inductively that

$$\mathbb{E}_n(f) = \mathbb{E}_{n-1}(f) + \sum_{I \in \mathcal{A}_{n-1}} \langle f, h_I \rangle h_I \tag{8.30}$$

for each $n \geq 1$. Indeed, for the case $n = 1$,

$$\begin{aligned}
&\chi_{[0,1/2)} \fint_{[0,1/2)} f + \chi_{[1/2,1)} \fint_{[1/2,1)} f \\
&= \int_0^1 f + \left(\int_0^{1/2} f - \int_{1/2}^1 f \right) \left(\chi_{[0,1/2)} - \chi_{[1/2,1)} \right) \\
&= \mathbb{E}_0(f) + \langle f, h_{[0,1)} \rangle h_{[0,1)}.
\end{aligned}$$

Rescaling this identity from $[0, 1]$ to any $I \in \mathcal{D}$ yields

$$\chi_{I_1} \fint_{I_1} f + \chi_{I_2} \fint_{I_2} f = \chi_I \fint_I f + \langle f, h_I \rangle h_I,$$

where I_1, I_2 are the two halves of I. This is (8.30), the iteration of which completes the proof. □

The previous two lemmas imply the following result.

Corollary 8.15 *The functions $\{1\} \cup \{h_I\}_{I \in \mathcal{D}}$ are an orthonormal basis in $L^2([0, 1])$.*

Proof It remains only to note that $\langle h_I, h_J \rangle = 0$ if $I \neq J$. This is due to the fact that either distinct dyadic intervals are disjoint or one contains the other, say $I \subsetneq J$. In the latter case, h_I has mean zero on any interval on which h_J is constant. □

We number the Haar functions by their position in the binary tree, counting along each row from left to right and starting from the root. In the same way we label the dyadic intervals I_ℓ. We also set $h_0 := 1$ on $[0, 1]$. It follows from (8.29) that, for each $\ell \geq 1$, there exists a σ-algebra $\mathcal{P}_\ell$ with the property that

$$P_\ell(f) := \sum_{k=0}^{\ell} \langle f, h_k \rangle h_k = \mathbb{E}[f | \mathcal{P}_\ell] \quad \forall \ell \geq 0.$$

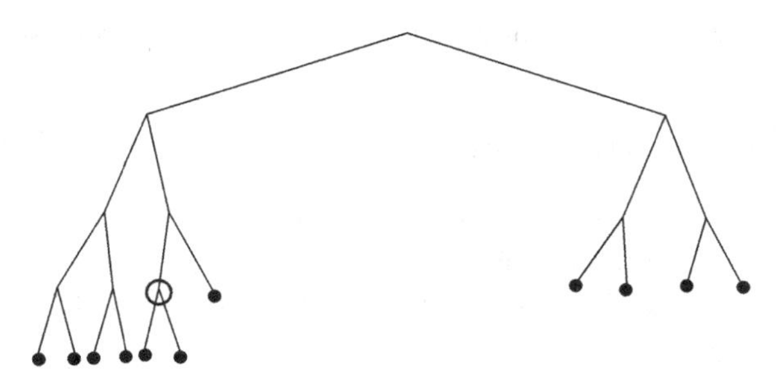

Figure 8.3. The projection P_{10} associated with h_{10}.

For $\ell = 10$ this is depicted in Figure 8.3, where the intervals generating $\mathcal{P}_{10}$ are the solid nodes. In other words, $\mathcal{P}_\ell$ is generated by the dyadic intervals on the same row of the binary tree as I_ℓ and strictly to the right of it, together with the children of I_ℓ and those of all other dyadic intervals in the same row and to the left of I_ℓ.

Thus one has

$$\sup_\ell \|P_\ell\|_{p\to p} \le 1, \quad \|P_\ell f - f\|_p \to 0,$$

as $\ell \to \infty$ any $f \in L^p([0,1])$, $1 \le p < \infty$. These properties characterize a *Schauder basis*, whence the Haar system $\{h_j\}_{j=0}^\infty$ forms just such a basis. Note that the exponential system $\{e(n\cdot)\}_{n\in\mathbb{Z}}$ is a Schauder basis in $L^p([0,1])$ for $1 < p < \infty$ but not for $p = 1$; cf. Chapter 3.

8.4.2. Multiplier operators for the Haar expansion

In further contrast with the exponential system we shall now show that the Haar system is an *unconditional* basis for $L^p([0,1])$ if $1 < p < \infty$; cf. Problem 6.10. This can be done in a number of ways, and we shall present two distinct methods leading to the required L^p bounds. The first is based on a Calderón–Zygmund decomposition and is easier conceptually.

Theorem 8.16 *Let $\{x_I\}_{I\in\mathcal{D}}$ be an arbitrary sequence of scalars such that $|x_I| \le 1$ for all $I \in \mathcal{D}$. Then the multiplier operator T, defined on functions $f \in L^1$ with finite Haar expansion, i.e.,*

$$Tf := \sum_{I\in\mathcal{D}} x_I \langle f, h_I\rangle h_I, \tag{8.31}$$

is bounded from L^1 to weak-L^1, as well as from L^p to L^p for every $1 < p < \infty$. The associated operator norms can be bounded uniformly in $\{x_I\}_{I\in\mathcal{D}}$.

Proof The bound on L^2 is immediate and, by interpolation and duality, it suffices to establish that

$$\left|\left\{x \in [0,1] \,\middle|\, |(Tf)(x)| > \lambda\right\}\right| \leq C\lambda^{-1}\|f\|_1 \quad \forall\lambda > 0 \tag{8.32}$$

and all $f \in L^1([0,1])$ with C some absolute constant. Assume that $\|f\|_1 = 1$ and let $\lambda \geq 2$. Perform a Calderón–Zygmund decomposition at level λ. The lower bound on λ is needed in order to start the decomposition. Then $f = g + b$, with

$$|g| \leq 2\lambda, \quad \|g\|_1 \leq 1,$$
$$b = \textstyle\sum_{I\in\mathcal{B}} \chi_I(f - f_I), \quad f_I = \fint_I f, \quad \textstyle\sum_{I\in\mathcal{B}} |I| \leq C\lambda^{-1}.$$

Thus

$$\begin{aligned}\left|\left\{x \in [0,1] \,\middle|\, |(Tf)(x)| > \lambda\right\}\right| &\leq \left|\left\{x \in [0,1] \,\middle|\, |(Tg)(x)| > \lambda/2\right\}\right| \\ &\quad + \left|\left\{x \in [0,1] \,\middle|\, |(Tb)(x)| > \lambda/2\right\}\right| \\ &=: A + B.\end{aligned}$$

On the one hand

$$A \leq C\lambda^{-2}\|g\|_2^2 \leq C\lambda^{-1}\|g\|_1 \leq C\lambda^{-1}$$

and, on the other hand, since $T\big((f - f_I)\chi_I\big)$ is supported in I, one has

$$B \leq \sum_{I\in\mathcal{B}} |I| \leq C\lambda^{-1},$$

and we are done. □

The previous result, amongst other things, implies the unconditionality of the Haar expansion in L^p provided that $1 < p < \infty$. Indeed, if $f = \sum_{I\in\mathcal{D}}\langle f, h_I\rangle h_I$ converges in L^p then, by Theorem 8.16, the series $\sum_{I\in\mathcal{D}} \varepsilon_I \langle f, h_I\rangle h_I$ converges in L^p for an arbitrary choice of signs $\varepsilon_I = \pm 1$ as well, provided that $1 < p < \infty$.

Inspection of the proof of Theorem 8.16 reveals the following more general fact.

Corollary 8.17 *Let T be a linear operator on $L^2([0,1])$. Assume that Th_I is supported in I for every $I \in \mathcal{D}$. Then T is bounded from $L^1([0,1])$ to weak-$L^1([0,1])$ and therefore on $L^p([0,1])$ for $1 < p \leq 2$.*

This will be useful in Section 9.4 in connection with paraproducts. In fact, one needs even less than in the corollary: it suffices for Th_I to be supported in I^*, where I^* is a dilate of I by a fixed constant. Of course that constant then enters into the operator norms.

8.4.3. Burkholder's inequality

An alternative and, to some extent, a more elementary and more efficient approach, is given by the following result. The proof is based on a convexity argument and does not require any interpolation. However, the construction is quite tricky.

Theorem 8.18 *Let $1 < p < \infty$, and suppose that $\{a_k\}_{k=0}^{\infty}$, $\{b_k\}_{k=0}^{\infty}$ are complex numbers with $|b_k| \leq |a_k|$ for all $k \geq 0$. Then*

$$\left\| \sum_{k=0}^{n} b_k h_k \right\|_p \leq (p^* - 1) \left\| \sum_{k=0}^{n} a_k h_k \right\|_p, \tag{8.33}$$

where $p^* := p \vee p'$.

Proof By duality we may assume that $2 < p < \infty$. Then $p^* = p$, and we define a function $v : \mathbb{C} \times \mathbb{C} \to \mathbb{R}$ via

$$v(x, y) := |y|^p - (p-1)^p |x|^p.$$

Furthermore, let $\alpha_p := p(1 - 1/p)^{p-1}$ and define another function $u : \mathbb{C} \times \mathbb{C} \to \mathbb{R}$ via

$$u(x, y) := \alpha_p (|x| + |y|)^{p-1} (|y| - (p-1)|x|).$$

We now claim that the following properties hold:

(1) $v(x, y) \leq u(x, y)$;

(2) $u(x, y) = u(-x, -y)$;

(3) $u(0, 0) = 0$;

(4) $u(x + a, y + b) + u(x - a, y - b) \leq 2u(x, y)$, provided that $|b| \leq |a|$.

While (2) and (3) are obvious, (1) and (4) are not. Assuming these properties for now, we will apply them in the following way. Denote

$$f_n = \sum_{k=0}^{n} a_k h_k, \quad g_n = \sum_{k=0}^{n} b_k h_k.$$

Then

$$\begin{aligned} \|g_n\|_p^p - (p-1)^p \|f_n\|_p^p &= \int_0^1 v(f_n(t), g_n(t))\, dt \leq \int_0^1 u(f_n(t), g_n(t))\, dt \\ &= \int_{h_n = 0} u(f_{n-1}(t), g_{n-1}(t))\, dt \\ &\quad + \int_{h_n > 0} u(f_{n-1}(t) + a_n, g_{n-1}(t) + b_n)\, dt \\ &\quad + \int_{h_n < 0} u(f_{n-1}(t) - a_n, g_{n-1}(t) - b_n)\, dt. \end{aligned}$$

We now extend the final two integrals to the supports of h_n and apply the convexity property (iv) to give a further estimate:

$$
\begin{aligned}
\|g_n\|_p^p - (p-1)^p \|f_n\|_p^p &= \int_{h_n=0} u(f_{n-1}(t), g_{n-1}(t))\,dt \\
&\quad + \tfrac{1}{2} \int_{h_n \neq 0} \big(u(f_{n-1}(t) + a_n, g_{n-1}(t) + b_n) \\
&\qquad\qquad + u(f_{n-1}(t) - a_n, g_{n-1}(t) - b_n)\big)\,dt \\
&\le \int_0^1 u(f_{n-1}(t), g_{n-1}(t))\,dt \le u(a_0, b_0) \\
&= \tfrac{1}{2}(u(a_0, b_0) + u(-a_0, -b_0)) \le u(0,0) = 0.
\end{aligned}
$$

It remains to prove (1) and (4) from above. For (1) we assume that $|x| + |y| = 1$, and set $|x| = s$. Then (1) amounts to showing that

$$0 \le \alpha_p(1 - sp) + (p-1)^p s^p - (1-s)^p \quad \forall 0 \le s \le 1.$$

The right-hand side vanishes identically for $p = 2$. However, denoting it by $\varphi_p(s)$ for $p > 2$, one may verify by elementary calculus that $\varphi_p \ge 0$ with a unique zero at $s = 1/p$ that also satisfies $\varphi_p'(1/p) = 0$, $\varphi_p''(1/p) > 0$. For property (4) we need to check the concavity of the function

$$G(t) = u(x + ta, y + tb) + u(x - ta, y - tb), \quad -1 \le t \le 1.$$

This amounts to verifying that $G''(t) \le 0$, which reduces further to $G''(0) \le 0$. □

Exercise 8.4 Supply the details for properties (1) and (4) in the previous proof. *Hint:* See Wojtaszczyk [125, p. 65].

Although these two proofs of the L^p bounds look very different, technically speaking, there are some similarities. Indeed, convexity enters both proofs. In the second it is explicit, whereas in the first it is implicit, being part of the interpolation theorem required to pass to the L^p estimates.

8.4.4. The square function

Next, we define the square function associated with the Haar expansion.

Definition 8.19 For any $f \in L^1([0,1])$ with $\int_0^1 f = 0$, we define

$$Sf = \left(\sum_I |a_I|^2 h_I^2 \right)^{1/2}$$

for $f = \sum_I a_I h_I$.

Note that $h_I^2 = |I|^{-1}\chi_I$ is L^1-normalized. Alternatively, we may write $f = \sum_n f_n$ where $f_n = \sum_{I \in \mathcal{A}_n} a_I h_I$ is the sum over the nth row of the binary tree. Then $Sf = \sqrt{\sum_n f_n^2}$.

The analogue of the Littlewood–Paley theorem now reads as follows. The heuristics based on estimates for independent random variables that explained Theorem 8.3 applies here as well.

Theorem 8.20 *For any $f \in L^p([0,1])$ with $1 < p < \infty$ and $\int_0^1 f = 0$, one has*

$$C(p)^{-1}\|f\|_p \le \|Sf\|_p \le C(p)\|f\|_p, \tag{8.34}$$

where $C(p)$ is some constant.

Proof This follows immediately from Khinchine's inequality and Theorem 8.18. Indeed, by the latter theorem one has

$$(p^* - 1)^{-1}\|f\|_p \le \left\| \sum_I r_I(t) a_I h_I \right\|_p \le (p^* - 1)\|f\|_p.$$

Raising this to the power p and averaging over t yields

$$(p^* - 1)^{-p}\|f\|_p^p \le \int_0^1 \left\| \sum_I r_I(t) a_I h_I \right\|_p^p dt \le (p^* - 1)^p \|f\|_p^p.$$

By Khinchine's inequality the expression in the middle satisfies

$$\begin{aligned} C(p)^{-1}\left(\sum_I |a_I|^2 h_I^2\right)^{p/2} &\le \int_0^1 \left\| \sum_I r_I(t) a_I h_I \right\|_p^p dt \\ &\le C(p)\left(\sum_I |a_I|^2 h_I^2\right)^{p/2}, \end{aligned}$$

whence the required result. □

We can now see an analogy between the Littlewood–Paley and Haar decompositions. The sum over the rows in the binary trees, i.e.,

$$\sum_{I \in \mathcal{A}_n} a_I h_I$$

plays the role of a Littlewood–Paley projection at scale 2^{-n}.

Exercise 8.5 Verify that Theorem 8.20 fails for $p = 1$ and $p = \infty$. Also show that if $1 < p < \infty$ and $f \in L^1$ satisfies $Sf \in L^p$ then $f \in L^p$.

8.4.5. Dyadic $\mathcal{H}^1$ and BMO spaces

For $p = 1$ we shall find the (dyadic) Hardy space $\mathcal{H}^1$ to be a natural substitute for L^1 in Theorem 8.20.

Definition 8.21 The dyadic Hardy space $\mathcal{H}^1([0, 1])$ is defined as

$$\mathcal{H}^1([0,1]) = \left\{ f \in L^1([0,1]) \;\middle|\; \int_0^1 f = 0,\, Sf \in L^1([0,1]) \right\},$$

and the dyadic BMO space on $[0, 1]$ is the space of all $f \in L^2([0, 1])$ with $\int_0^1 f = 0$ and

$$\|f\|_{\mathrm{BMO}} := \sup_{I \in \mathcal{D}} \left(\fint_I |f(x) - f_I|^2 \, dx \right)^{1/2} < \infty, \quad f_I := \fint_I f.$$

Since there is no pointwise estimate $|f| \le CSf$, it is not clear that $\mathcal{H}^1$ is complete as a normed vector space relative to the norm $\|f\|_{\mathcal{H}^1} := \|Sf\|_1$. However, this is indeed the case but we shall not make any reference to it yet. After we have established the atomic decomposition in $\mathcal{H}^1$ it will follow easily that $\|f\|_1 \le C\|Sf\|_1$ for mean-zero functions, see Exercise 8.6, which settles this issue.

We leave it to the reader to check that dyadic BMO is strictly larger than BMO([0, 1]) as defined in Chapter 7. For the remainder of this section, BMO denotes the dyadic space. Nevertheless, it satisfies many properties similar to BMO([0, 1]), such as the John–Nirenberg inequality (with the same proof). For now, we remark that $f \in L^2([0, 1])$ with mean zero satisfies $f \in \mathrm{BMO}$ if and only if we have the Carleson condition

$$\sup_{I \in \mathcal{D}} |I|^{-1} \sum_{\substack{J \subseteq I \\ J \in \mathcal{D}}} |a_J|^2 = \sup_{I \in \mathcal{D}} \fint_I |f(x) - f_I|^2 \, dx = \|f\|_{\mathrm{BMO}}^2 < \infty, \tag{8.35}$$

where $f = \sum_{I \in \mathcal{D}} a_I h_I$. In view of the John–Nirenberg estimate we note that we could have defined BMO by

$$\|f\|_{\mathrm{BMO}} := \sup_{I \in \mathcal{D}} \fint_I |f(x) - f_I| \, dx < \infty,$$

since these norms are comparable by multiplicative constants. However, we prefer the L^2-based definition since it shows that BMO is invariant under a change of signs in the Haar expansions. Note that $\mathcal{H}^1$ enjoys this property by definition. Furthermore, both the $\mathcal{H}^1$ and BMO norms are *contractive* in the sense that $\|f\|_{\mathcal{H}^1} \le \|g\|_{\mathcal{H}^1}$ and $\|f\|_{\mathrm{BMO}} \le \|g\|_{\mathrm{BMO}}$ for any f, g that are finite linear combinations of Haar functions with the property that the coefficients of g dominate those of f in absolute value.

Our next goal is to prove Fefferman's theorem, i.e., the duality between dyadic $\mathcal{H}^1$ and BMO.

8.4.6. The first duality estimate

Our first step towards this characterization is the following estimate. We restrict ourselves to finite linear combinations of Haar functions in $\mathcal{H}^1$ in order to avoid complications arising from integrating a general $\mathcal{H}^1$ function against a function in BMO.

Theorem 8.22 *For any f that is a finite linear combination of Haar functions and any $h \in$ BMO, one has*

$$|\langle f, h\rangle| \leq 2\sqrt{2}\|f\|_{\mathcal{H}^1}\|h\|_{\text{BMO}}. \tag{8.36}$$

Proof Let $h = \sum_J b_J h_J$ and $f = \sum_J a_J h_J$ (a finite sum), and assume without loss of generality that a_J, b_J are real-valued. Then

$$\langle f, h\rangle = \sum_J a_J b_J, \tag{8.37}$$

which is a finite sum. By the aforementioned contractivity of the BMO norm we may also assume that h has a finite Haar expansion. The difficulty with (8.36) lies in the fact that we cannot simply force a square function Sf to appear in (8.37) using the Cauchy–Schwarz inequality, say. However, we will be able to do this if we first carry out a suitable localization procedure via the following stopping-time argument. Define a localized square function

$$S(h|I) = \left(\sum_{J\subset I} b_J^2 h_J^2\right)^{1/2}.$$

For any $x \in (0, 1)$ let $I(x)$ be the largest interval $I \in \mathcal{D}$ with the property that

$$\begin{aligned} S^2(h|I)(x) &\leq 2\sup_{I\ni x} \fint_I S^2(h|I)(y)\,dy = 2\sup_{I\ni x}\fint_I |h(y) - h_I|^2\,dy \\ &= 2\sup_{I\ni x} |I|^{-1}\sum_{\substack{L\subset I\\ L\in\mathcal{D}}} b_L^2. \end{aligned} \tag{8.38}$$

This is well defined. Indeed, going down to the final level of the finite binary tree of h, the level containing the smallest dyadic intervals I, we obtain a sum $S(h|I)$ that consists of only a single term. In that case we have equality:

$$S^2(h|I)(x) = \fint_I S^2(h|I)(y)\,dy \quad \forall x \in I.$$

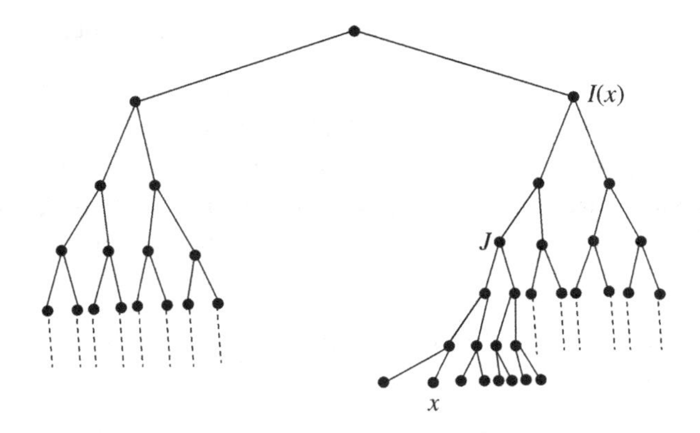

Figure 8.4. The stopping time defining $I(x)$.

Thus

$$S(h|I(x))(x) \leq \sqrt{2}\|h\|_{\mathrm{BMO}} \quad \forall x \in [0,1],$$

and it suffices to prove the estimate

$$|\langle f, h\rangle| \leq 2\int_0^1 S(f)(x)S(h|I(x))(x)\,dx. \tag{8.39}$$

This in turn will be seen to be a consequence of the following claim:

$$|\{x \in J \mid I(x) \supset J\}| \geq \tfrac{1}{2}|J| \quad \forall J \in \mathcal{D}, \tag{8.40}$$

which rests on the maximality of $I(x)$. Figure 8.4 illustrates this claim. The tree rooted at J depicts all branches $x \in J$. We claim that for at least half these branches the node $I(x)$ lies above J or equals J.

To verify this property, fix some $J \in \mathcal{D}$ and define

$$A := \big\{x \in J \mid I(x) \text{ is strictly contained in } J\big\}.$$

Then, assuming as we may that $S(h|J) \neq 0$, one has

$$\begin{aligned}\int_J S^2(h|J)(x)\,dx &\geq \int_A S^2(h|J)(x)\,dx \\ &> 2\int_A |J|^{-1}\sum_{\substack{L\subset J\\ L\in\mathcal{D}}} b_L^2 \\ &= 2|A|\,|J|^{-1}\int_J S^2(h|J)(x)\,dx,\end{aligned}$$

whence the claim. By means of (8.40) we can now conclude as follows:

$$\begin{aligned}
|\langle f, h\rangle| &\leq \sum_{J\in\mathcal{D}} |a_J b_J| \\
&\leq 2\int_0^1 \sum_{J\in\mathcal{D}} |a_J b_J|\,|J|^{-1}\chi_J\,\chi_{[x\,|\,I(x)\supset J]}\,dx \\
&\leq 2\int_0^1 \sum_{J\subset I(x)} |a_J||b_J|\,h_J^2(x) \\
&\leq 2\int_0^1 \Big(\sum_J a_J^2 h_J^2\Big)^{1/2}\Big(\sum_{J\subset I(x)} b_J^2 h_J^2\Big)^{1/2}\,dx \\
&= 2\int_0^1 (Sf)(x)\,S(h|I(x))(x)\,dx,
\end{aligned}$$

as desired. □

Define the (dyadic) sharp function as

$$h^\sharp(x) := \sup_{I\ni x}\Big(|I|^{-1}\sum_{J\subset I}\langle h, h_J\rangle^2\Big)^{1/2}$$

for any $h \in L^2([0,1])$ with $\int_0^1 h = 0$. Then $\|h\|_{\mathrm{BMO}} = \|h^\sharp\|_\infty$. The key inequality underlying the proof of Theorem 8.22 now reads as follows:

$$|\langle f, h\rangle| \leq 2\sqrt{2}\int_0^1 (Sf)(x)\,h^\sharp(x)\,dx.$$

Next, we show that the BMO space "norms" $\mathcal{H}^1$: this means that there are enough BMO functions to detect the size of $\mathcal{H}^1$ functions.

Proposition 8.23 *Given $f \in \mathcal{H}^1([0,1])$ with finite Haar expansion there exists $h \in \mathrm{BMO}$ with $\|h\|_{\mathrm{BMO}} = 1$ and*

$$|\langle f, h\rangle| \geq c\|f\|_{\mathcal{H}^1},$$

where $c > 0$ is an absolute constant.

Proof Let $f = \sum_I a_I h_I$, and set $f_t = \sum_I r_I(t)\,a_I h_I$, where the r_I are the Rademacher functions indexed by the dyadic intervals. Then from Khinchine's inequality one has

$$\int_0^1\int_0^1 |f_t(x)|\,dt\,dx \geq c\int_0^1 (Sf)(x)\,dx.$$

Choose $t_0 \in [0, 1]$ with $\|f_{t_0}\|_1 \geq c\|Sf\|_1$ and $g \in L^\infty([0, 1])$, $\|g\|_\infty = 1$, such that

$$\langle f_{t_0}, g\rangle = \|f_{t_0}\|_1.$$

Set

$$h = \sum_{I\in\mathcal{D}} r_I(t_0)c_I h_I \qquad \text{where } g = \sum_I c_I h_I.$$

On the one hand $\|h\|_{\mathrm{BMO}} = \|g - \langle g\rangle\|_{\mathrm{BMO}} \leq 2\|g\|_\infty = 2$, and on the other hand

$$\langle f, h\rangle = \sum_I r_I(t_0)a_I c_I = \langle f_{t_0}, g\rangle \geq c\|f\|_{\mathcal{H}^1}.$$

Replacing h by ah, where $a \geq \frac{1}{2}$ is some constant, concludes the proof. □

In summary, we have shown that $\mathrm{BMO} \hookrightarrow (\mathcal{H}^1)^*$. Next, we show that this is surjective.

8.4.7. The second duality estimate

Finally, we will establish that every bounded functional on $\mathcal{H}^1([0, 1])$ is given by integration against some $h \in \mathrm{BMO}([0, 1])$. This will then complete the characterization of the dual of $\mathcal{H}^1$. It is enough to consider finite linear combinations of Haar functions, since they are dense in $\mathcal{H}^1$.

Proposition 8.24 *For all $L \in (\mathcal{H}^1([0, 1]))^*$ there exists a unique $h \in \mathrm{BMO}$ with $\|h\|_{\mathrm{BMO}} \simeq \|L\|$ and such that $L(f) = \langle f, h\rangle$ for all $f \in \mathcal{H}^1([0, 1])$ with a finite Haar expansion.*

Proof The basic idea is to embed $\mathcal{H}^1([0, 1]) \hookrightarrow L^1([0, 1]; \ell^2)$ isometrically. Indeed, any $f \in \mathcal{H}^1([0, 1])$ is of the form $f = \sum_{n=0}^\infty f_n$ with

$$f_n := \sum_{I\in\mathcal{A}_n} \langle f, h_I\rangle h_I.$$

Then $Sf = \left(\sum_{n=0}^\infty f_n^2\right)^{1/2} \in L^1([0, 1])$ with $\|f\|_{\mathcal{H}^1} = \|Sf\|_1$. By the Hahn–Banach theorem, we may extend L to $\tilde{L} \in (L^1([0, 1]; \ell^2))^* = L^\infty([0, 1]; \ell^2)$ with the same norm. Therefore, there exists $\{\varphi_n\}_{n=0}^\infty \in L^\infty([0, 1]; \ell^2)$ with

$$\|L\| = \sup_{x\in(0,1)} \left(\sum_{n=0}^\infty \varphi_n^2(x)\right)^{1/2}, \qquad \tilde{L}(\{g_n\}_{n=0}^\infty) = \int_0^1 \sum_{n=0}^\infty g_n\varphi_n$$

for all $\{g_n\}_{n=0}^{\infty} \in L^1([0,1];\ell^2)$. Thus

$$L(f) = \sum_{n=0}^{\infty} \int_0^1 \sum_{I \in \mathcal{A}_n} \langle f, h_I \rangle h_I(x)\, \varphi_n(x)\, dx = \langle f, h \rangle, \tag{8.41}$$

where

$$h = \sum_{n=0}^{\infty} \sum_{I \in \mathcal{A}_n} \langle h_I, \varphi_n \rangle h_I.$$

To verify that $h \in \mathrm{BMO}([0,1])$ we compute as follows:

$$\begin{aligned}\sup_{J \in \mathcal{D}} |J|^{-1} \sum_{n=0}^{\infty} \sum_{\substack{I \subset J \\ I \in \mathcal{A}_n}} \left| \langle h_I, \varphi_n \rangle \right|^2 &\le \sup_{J \in \mathcal{D}} |J|^{-1} \int_J \sum_{n=0}^{\infty} |\varphi_n(x)|^2\, dx \\ &\le \left\| \{\varphi\}_{n=0}^{\infty} \right\|_{L^\infty([0,1];\ell^2)},\end{aligned}$$

whence $\|h\|_{\mathrm{BMO}} \le \|L\|$. The reverse direction (with some multiplicative constant) was established in Proposition 8.23, and we are done. □

8.4.8. The atomic decomposition in dyadic $\mathcal{H}^1$

To conclude our discussion of the Haar system, we now show that any $f \in \mathcal{H}^1$ admits an atomic decomposition. The version we present here is not as strong as that alluded to in Section 8.3, but it is sufficient for most purposes. The main difference is that we base our notion of an atom on L^2 rather than on L^∞.

Definition 8.25 We say that a is an $\mathcal{H}^1$ atom if and only if for some $I \in \mathcal{D}$ we have $\mathrm{supp}(a) \subset I$ and $\|a\|_2 |I|^{1/2} \le 1$ as well as $\int_0^1 a = 0$.

First we check that every atom is in $\mathcal{H}^1$.

Lemma 8.26 *If a is an atom then one has $\|a\|_{\mathcal{H}^1} \le 1$ as well as $\|a\|_1 \le 1$.*

Proof We have $a = \sum_{J \subset I} c_J h_J$, where I is as in the definition. Then

$$\int (Sa)^2(x)\, dx = \int_I (Sa)^2(x)\, dx = \sum_{J \subset I} c_J^2 = \|a\|_2^2 \le |I|^{-1}.$$

Then

$$\|Sa\|_1 \le \|Sa\|_2 |I|^{1/2} \le 1,$$

as claimed. Finally, $\|a\|_1 \le \|a\|_2 |I|^{1/2} \le 1$. □

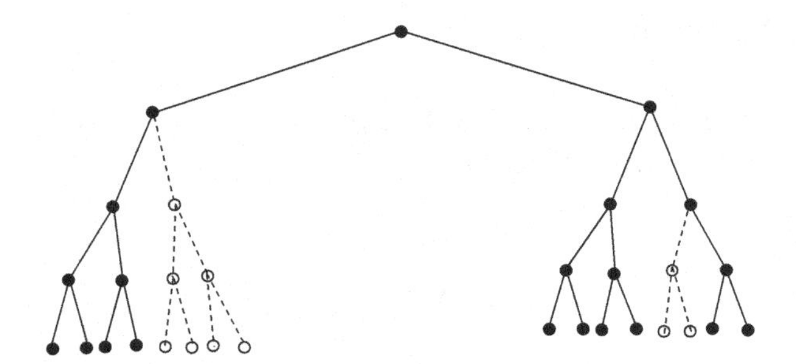

Figure 8.5. The first five levels of a possible pruned tree $\mathcal{T}([0,1])$.

Next we formulate and prove the atomic decomposition.

Theorem 8.27 *For every $f \in \mathcal{H}^1([0,1])$ there exists a sequence of scalars $\{c_j\}_{j=1}^\infty$ and atoms $\{a_j\}_{j=1}^\infty$ such that $f = \sum_j c_j a_j$ converges in $\mathcal{H}^1$ with $\sum_j |c_j| \le C\|f\|_{\mathcal{H}^1}$ for some absolute constant C.*

Proof Write $f = \sum_I b_I h_I$ and normalize in such a way that $\int_0^1 Sf = 1$. We shall now prune the binary tree by "cutting away" the "heavy" branches. To be specific, we let each node of the tree be weighted by b_I and define

$$\mathcal{B}([0,1]) := \big\{x \in [0,1] \,\big|\, (Sf)(x) > \lambda([0,1])\big\},$$

where $\lambda([0,1]) > 0$ is the minimal number in $\{2^k \,|\, k \ge 1\}$ such that $|\mathcal{B}([0,1])| \le \frac{1}{4}$. Next, let $\mathcal{F}([0,1]) := \big\{I_{1,j}\big\}_j$ be the maximal dyadic intervals in $\mathcal{B}([0,1])$. If $\mathcal{F}([0,1]) = \emptyset$ then we terminate the construction. These are the roots of binary trees $\mathcal{T}_{1,j}$. Removing all these trees from the full binary tree $\mathcal{D}$ yields a new tree $\mathcal{T}([0,1])$. Note that this is indeed a tree, i.e., a connected graph rooted at $[0,1]$ (it may be empty). Figure 8.5 shows the first five levels of a possible pruned tree $\mathcal{T}([0,1])$; only the solid lines and nodes are retained as part of $\mathcal{T}([0,1])$. By construction $f_{[0,1]} := \sum_{I\in\mathcal{T}([0,1])} b_I h_I$ satisfies $Sf_{[0,1]} \le \lambda([0,1])$, whence in particular

$$\big\|f_{[0,1]}\big\|_2^2 = \sum_{I\in\mathcal{T}([0,1])} b_I^2 \le \lambda^2([0,1]).$$

For each $I_{1,j}$ we now repeat the construction:

$$\mathcal{B}(I_{1,j}) := \{x \in I_{1,j} \,|\, (Sf)(x) > \lambda(I_{1,j})\}$$

satisfies $|\mathcal{B}(I_{1,j})| < \frac{1}{4}|I_{1,j}|$ for each j, with $\lambda(I_{1,j})$ again minimal and a power of 2. We again select the maximal dyadic intervals in $\mathcal{B}(I_{1,j})$ and denote them by $\mathcal{F}(I_{1,j})$. These are the roots of binary trees $\big\{\mathcal{T}_{2,k}(I_{1,j})\big\}_k$ that we remove from $\mathcal{T}_{1,j}$ as before to yield a tree $\mathcal{T}(I_{1,j})$. We use the latter to define a function

$$f_{I_{1,j}} := \sum_{I \in \mathcal{T}(I_{1,j})} b_I h_I$$

that satisfies $\| f_{I_{1,j}} \|_2^2 \leq \lambda^2(I_{1,j})|I_{1,j}|$. One now repeats this construction either until it terminates in finite time or indefinitely. This yields a family of pairwise disjoint subtrees $\{\mathcal{T}_\ell\}_\ell$ with roots $I_\ell \in \mathcal{D}$ and associated functions $f_\ell = \sum_{I \in \mathcal{T}_\ell} b_I h_I$ that satisfy $\| f_\ell \|_2 \leq 2^{n_\ell} |I_\ell|^{\frac{1}{2}}$ as well as

$$|I_\ell| \leq 4 \big| \{ I_\ell \mid Sf > 2^{n_\ell - 1} \} \big|;$$

the latter holds owing to the minimality of the n_ℓ. Finally, if $I_\ell \subsetneq I_{\ell'}$ for $\ell \neq \ell'$ then $n_\ell > n_{\ell'}$. Define $c_\ell := 2^{n_\ell} |I_\ell|$, $a_\ell := c_\ell^{-1} f_\ell$. Then $f = \sum_\ell c_\ell a_\ell$ and one has

$$\begin{aligned} \sum_\ell |c_\ell| &\leq \sum_\ell 2^{n_\ell} |I_\ell| \leq 4 \sum_\ell 2^{n_\ell} \big| \{ I_\ell \mid Sf > 2^{n_\ell - 1} \} \big| \\ &\leq 4 \sum_{k \in \mathbb{Z}^+} 2^k \big| \{ [0,1] \mid Sf > 2^{k-1} \} \big| \leq C \int_0^1 Sf = C, \end{aligned}$$

as desired. To pass to the second line we used the property that if $I_\ell \cap I_{\ell'} \neq \emptyset$ then $n_\ell \neq n_{\ell'}$. By construction the a_ℓ satisfy the properties of an atom, and we are done. □

The following exercise presents three immediate applications of the atomic decomposition.

Exercise 8.6

(a) Establish the continuous embedding

$$\mathcal{H}^1([0,1]) \hookrightarrow L^1([0,1])$$

with norm $\| f \|_{\mathcal{H}^1([0,1])} = \| Sf \|_1$ in $\mathcal{H}^1([0,1])$.

(b) Derive Theorem 8.22 from Theorem 8.27.

(c) Show that the Hilbert transform takes $\mathcal{H}^1([0,1])$ to $L^1([0,1])$.

Finally, we show that $\mathcal{H}^1$ norms BMO. In fact, it suffices to test a given $L^2([0,1])$ function against all atoms.

Lemma 8.28 *Let $f \in L^2([0,1])$ be such that $|\langle f, a \rangle| \leq 1$ for all $\mathcal{H}^1$ atoms a. Then $\| f \|_{\mathrm{BMO}} \leq 1$.*

Proof Pick any dyadic interval $I \subset [0, 1]$ and let $a := \lambda |I|^{-1}(f - f_I)\chi_I$, where $\lambda > 0$ is chosen such that $\|a\|_2 |I|^{1/2} = 1$, which is the same as

$$\lambda^2 \fint_I |f - f_I|^2 = 1. \tag{8.42}$$

In that case a is an atom, and we compute

$$\langle f, a \rangle = \lambda \fint_I |f - f_I|^2 \leq 1. \tag{8.43}$$

Comparison of (8.43) and (8.42) reveals that $\lambda \geq 1$, whence

$$\fint_I |f - f_I|^2 \leq 1$$

for all dyadic $I \subset [0, 1]$, as desired. □

Exercise 8.7 Show that C^∞ atoms norm BMO, i.e., we may assume that $a \in C^\infty$ in Lemma 8.28.

8.5. Oscillatory multipliers

To conclude this chapter we present a result of Lebedev and Olevskiĭ on Fourier multipliers of the form $e^{2\pi i \lambda \phi(\xi)}$, where $\phi : \mathbb{R}^d \to \mathbb{R}$ and $\lambda \geq 1$ (we will be particularly interested in $\lambda \to \infty$). The operators $T_{\lambda,\phi} f := \left(e^{2\pi i \lambda \phi(\xi)} \hat{f}(\xi)\right)^\vee$ have some simple properties:

- $\|T_{\lambda,\phi}\|_{2\to 2} = 1$ irrespective of the choice of λ or ϕ;
- if ϕ is an affine function, i.e., $\phi(\xi) = x_0 \cdot \xi + c$, then the same applies to all L^p with $1 \leq p \leq \infty$;
- if $\phi \in \mathcal{S}(\mathbb{R}^d)$ then $T_{\lambda,\phi}$ is bounded on $L^p(\mathbb{R}^d)$, for any $1 \leq p \leq \infty$, with norm $C_\phi(\lambda, p)$.

The second property follows from the fact that, for affine ϕ, T_λ acts as a translation composed with multiplication by a unimodular number. The third property follows from the observation that $e^{2\pi i \lambda \phi} - 1 \in \mathcal{S}(\mathbb{R}^d)$. We shall now show that these are basically best possible, in the following sense; the theorem is of course relevant only if ϕ is such that $T_{\lambda,\phi}$ is bounded on L^p.

Theorem 8.29 *If $\phi \in C^1(\mathbb{R}^d)$ is not an affine function then T_λ satisfies*

$$\|T_{\lambda,\phi}\|_{p\to p} \to \infty \quad \text{as } \lambda \to \infty$$

for any $1 \leq p \leq \infty$, $p \neq 2$.

Proof By duality and interpolation it suffices to treat the case $2 < p < \infty$. For simplicity, we will also assume $d = 1$. We remark that, by Problem 7.4,

$\|\phi'\|_1 < \infty$ is sufficient for $T_{\lambda,\phi}$ to be bounded on $L^p(\mathbb{R})$ for any $1 < p < \infty$, but this is irrelevant for the proof.

Assume that ϕ is not affine. The idea of the proof is to exhibit a function consisting of many disjoint bumps that $T_{\lambda,\phi}$ converts into a function with a large amount of "pileup" (the phenomenon whereby many functions add to give a large value). The underlying mechanism here is that on the graph of any nonaffine ϕ we can find infinitely many points with distinct slopes. If we think of ϕ as piecewise affine then, up to unimodular factors, $T_{\lambda,\phi}$ acts as a translation on functions whose Fourier supports are localized close to those points; as $\lambda \to \infty$ the amounts by which we translate become infinitely separated, leading to the desired conclusion.

To be specific, fix N to be a large positive integer and select $\{\xi_j\}_{j=1}^N \in \mathbb{R}$ such that $\phi'(\xi_j) =: k_j$ are pairwise distinct. Fix $\psi \in \mathcal{S}(\mathbb{R})$ with $\hat{\psi} \geq 0$, $\text{supp}(\hat{\psi}) \subset (-1, 1)$, and $\|\hat{\psi}\|_1 = 1$. Then for $\lambda \geq 1$ (in fact, for large λ), $\psi_\lambda(x) := \psi(\lambda^{-1}x)$ satisfies $\widehat{\psi_\lambda}(\xi) = \lambda\hat{\psi}(\lambda\xi)$, so that

$$\text{supp}\big(\widehat{\psi_\lambda}\big) \subset \big(-\lambda^{-1}, \lambda^{-1}\big).$$

Now define $g_j \in \mathcal{S}(\mathbb{R})$ by

$$\widehat{g_j}(\xi) = e^{-2\pi i\lambda(\phi(\xi_j)+k_j(\xi-\xi_j))}\widehat{\psi_\lambda}(\xi - \xi_j),$$

whence

$$g_j(x) = e^{-2\pi i\lambda\phi(\xi_j)}e^{2\pi i\xi_j x}\psi_\lambda(x - \lambda k_j).$$

To compute $T_{\lambda,\phi}g_j$ note that, for sufficiently large λ and all $1 \leq j \leq N$,

$$\big|\phi(\xi) - \phi(\xi_j) - k_j(\xi - \xi_j)\big| < \varepsilon_0\lambda^{-1} \quad \forall\, |\xi - \xi_j| < \lambda^{-1}.$$

Here $0 < \varepsilon_0 \ll 1$ is a fixed small parameter (of order, say, 10^{-3}). Hence, we may write

$$\widehat{T_{\lambda,\phi}g_j}(\xi) = e^{2\pi i\lambda E_j(\xi)}\widehat{\psi_\lambda}(\xi - \xi_j),$$

where $|E_j(\xi)| < \varepsilon_0$ for all ξ. Consequently, for all $|x| < \varepsilon_0\lambda$,

$$\begin{aligned}|(T_{\lambda,\phi}g_j)(x)| &= \left|\int e^{2\pi ix(\xi-\xi_j)}e^{2\pi i\lambda E_j(\xi)}\widehat{\psi_\lambda}(\xi - \xi_j)\,d\xi\right| \\ &\geq \cos(4\pi\varepsilon_0)\int \widehat{\psi_\lambda}(\xi - \xi_j)\,d\xi > \tfrac{1}{2}.\end{aligned}$$

Consequently, by Lemma 5.5,

$$\mathbb{E}\left\|\sum_{j=1}^{N} r_j T_{\lambda,\phi} g_j\right\|_p^p \geq \int_{[|x|<\varepsilon_0\lambda]} \mathbb{E}\left|\sum_{j=1}^{N} r_j T_{\lambda,\phi} g_j(x)\right|^p dx$$

$$\geq C^{-1}\int_{[|x|<\varepsilon_0\lambda]} \left(\sum_{j=1}^{N} |T_{\lambda,\phi} g_j(x)|^2\right)^{p/2} dx \geq C^{-1}\varepsilon_0\lambda N^{p/2}.$$

Thus, for a suitable choice of signs, $f := \sum_{j=1}^{N} \pm g_j$ satisfies on the one hand

$$\|T_{\lambda,\phi} f\|_p \geq C^{-1}\varepsilon_0^{1/p}\lambda^{1/p}N^{1/2}.$$

On the other hand, with functions v_j satisfying $|v_j(x)| = 1$, one has

$$\|f\|_p = \lambda^{1/p}\left\|\sum_{j=1}^{N} v_j(x)\psi(x-\lambda k_j)\right\|_p \leq C\lambda^{1/p}N^{1/p}.$$

The first equation here follows by a change of variables, whereas the upper bound holds for large λ (depending on N and the choice of ξ_j) since most of the L^p-mass of $\psi(\cdot - \lambda k_j)$ is localized in regions that move arbitrarily far from each other as $\lambda \to \infty$. Since N can be arbitrarily large this proves the theorem. □

Exercise 8.8 Does Theorem 8.29 hold if ϕ is Lipschitz rather than C^1?

Notes

The Littlewood–Paley theorem can be proved in a number of different ways; see for example Stein [109], which also contains a martingale-difference version of this theorem. For a good number applications of Littlewood–Paley theory (for example, to Besov spaces) and for connections with wavelet theory, see Frazier, Jawerth, and Weiss [43].

Concerning the question involving (8.10), it is a relatively easy consequence of the L^p-boundedness of the Hilbert transform that the answer is "yes" in one dimension. However, a very remarkable result of Charles Fefferman showed that the answer is "no" for dimensions $d \geq 2$. In fact, Fefferman [37] showed that the ball multiplier χ_B, where B is any ball in $\mathbb{R}^d$, is bounded on $L^p(\mathbb{R}^d)$ only for $p = 2$. This latter result is based on the existence of Kakeya sets; see for example Wolff's notes [128] for a construction of these sets. The positive answer for $d = 1$ is presented as Problem 8.4 below.

The "atoms" of Lemma 8.9 are exactly what are referred to as $\mathcal{H}^1$-*atoms*, where $\mathcal{H}^1(\mathbb{R}^d)$ is the real-variable Hardy space. Lemma 8.10 is an instance of the *atomic decomposition* that exists for any $f \in \mathcal{H}^1$; see Stein [111] and Koosis [71]. For an introduction to real-variable Hardy space theory with applications to PDEs, see Semmes' article [99].

The dyadic theory in Section 8.4 is very basic and well known; see for example Müller's book [85], which contains a wealth of deep material on $\mathcal{H}^1$. Slavin and Volberg [103] developed Fefferman's duality theorem in the context of Bellman functions. The final section is based on Lebedev and Olevskiĭ [73].

Problems

Problem 8.1 The conditions in Theorem 8.2 can be relaxed. Indeed, it suffices to assume the following:

$$\sup_{R>0} R^{2|\alpha|} \fint_{R<|\xi|<2R} |\partial_\xi^\alpha m(\xi)|^2 \, d\xi \le A^2 < \infty$$

for all $|\alpha| \le k$, where k is the smallest integer $> d/2$. *Hint:* Verify the Hörmander condition (ii) of Definition 7.1 directly rather than going through Lemma 7.2.

Problem 8.2 Let T be a bounded linear operator on $L^p(X, \mu)$, where (X, μ) is some measure space and $1 \le p < \infty$. Show the following vector-valued L^p estimate:

$$\big\| \{Tf_j\}_{\ell^2} \big\|_p \le C(p) \|T\|_{p\to p} \big\| \{f_j\}_{\ell^2} \big\|_p$$

for any sequence $\{f_j\}$ of measurable functions for which the right-hand side is finite. Such vector-valued extensions are very useful. A challenging question is how to extend this property to sub-linear operators such as the Hardy–Littlewood maximal function; see Stein [111, p. 51].

Problem 8.3 Let $1 \le p < 2$. By means of Khinchine's inequality (Lemma 5.5) show that for every $\varepsilon > 0$ there exists $f \in \mathcal{S}(\mathbb{R}^d)$ such that $\|\hat{f}\|_{p'} \le \varepsilon \|f\|_p$; cf. Lemma 4.10.

Problem 8.4 In this problem the dimension $d = 1$.

(a) Deduce (8.10) from Theorem 8.3 by means of the vector-valued inequality below. Let $\{I_j\}_{j\in\mathbb{Z}}$ be an arbitrary collection of intervals. Then, for any $1 < p < \infty$,

$$\begin{aligned} \big\| \{(\chi_{I_j} \hat{f}_j)^\vee\}_j \big\|_{L^p(\ell^2)} &:= \bigg\| \Big(\sum_j |(\chi_{I_j} \hat{f}_j)^\vee|^2 \Big)^{1/2} \bigg\|_p \\ &\le C_p \|\{f_j\}\|_{L^p(\ell^2)}, \end{aligned} \tag{8.44}$$

where $\{f_j\}_{j\in\mathbb{Z}}$ is an arbitrary collection of functions, in $\mathcal{S}(\mathbb{R})$, say.

(b) For any interval $I \subset \mathbb{R}$ write $S_I f = (\chi_I \hat{f})^\vee$ for any $f \in \mathcal{S}(\mathbb{R})$. Let (Z, ρ) be a measure space and consider maps $I(z)$ from Z to intervals in $\mathbb{R}$. Show that there exists a constant $C(p)$ for any $1 < p < \infty$ such that

$$\bigg\| \Big(\int_Z |S_{I(z)} f(z)|^2 \, \rho(dz) \Big)^{1/2} \bigg\|_{L^p(\mathbb{R})} \le C(p) \bigg\| \Big(\int_Z |f_z|^2 \, \rho(dz) \Big)^{1/2} \bigg\|_{L^p(\mathbb{R})}$$

for any $\{f_z\}_{z\in Z} \in L^p(\mathbb{R}; L^2(Z, \rho))$.

(c) Now deduce (8.44) from Theorem 3.17 using Khinchine's inequality, Lemma 5.5. *Hint:* Express the operator $f \mapsto (\chi_I \hat{f})^\vee$ by means of a Hilbert transform via the same procedure as that used in the proof of Theorem 3.20.

(d) By similar means prove the following Littlewood–Paley theorem for functions in $L^p(\mathbb{T})$. For any $f \in L^1(\mathbb{T})$, write

$$Sf = \left(\sum_{j=0}^{\infty} |P_j f|^2 \right)^{1/2},$$

where

$$(P_j f)(\theta) = \sum_{2^{j-1} \leq |n| < 2^j} \hat{f}(n) e(n\theta)$$

for $j \geq 1$ and $(P_0 f)(\theta) = \hat{f}(0)$. Show that, for any $1 < p < \infty$,

$$C_p^{-1} \|f\|_{L^p(\mathbb{T})} \leq \|Sf\|_{L^p(\mathbb{T})} \leq C_p \|f\|_{L^p(\mathbb{T})} \tag{8.45}$$

for all $f \in L^p(\mathbb{T})$.

(e) As a consequence of (8.10) with $d = 1$, show the following multiplier theorem. Let $m : \mathbb{R} \setminus \{0\} \to \mathbb{C}$ have the property that, for each $j \in \mathbb{Z}$,

$$m(\xi) = m_j = \text{ constant}$$

for all $2^{j-1} \leq |\xi| < 2^j$. Then, for any $1 < p < \infty$,

$$\|(m\hat{f})^\vee\|_{L^p(\mathbb{R})} \leq C_p \sup_{j \in \mathbb{Z}} |m_j| \|f\|_p$$

for all $f \in \mathcal{S}(\mathbb{R})$. Prove a similar theorem for $L^p(\mathbb{T})$ using (8.45).

Problem 8.5 Prove the following stronger version of Problem 7.4. In the present problem, by a dyadic interval we mean an interval of the form $\pm[2^k, 2^{k+1})$ where $k \in \mathbb{Z}$. Suppose that m is a bounded function on $\mathbb{R}$ which is locally of bounded variation. Furthermore, assume that, for some finite B, one has $\|m\|_\infty \leq B$ and

$$\int_I |dm| \leq B$$

for all dyadic intervals I. Show that m is bounded on $L^p(\mathbb{R})$ as a Fourier multiplier for any $1 < p < \infty$ with $\|(m\hat{f})^\vee\|_p \leq C(p) B \|f\|_p$. Give an example of m to which this problem, but not Problem 7.4, applies. For an extension to higher dimensions see Stein [108, p. 109].

Problem 8.6 Here we extend Corollary 8.11 to provide estimates for elliptic equations on a region $\Omega \subset \mathbb{R}^d$ with variable coefficients, i.e., such that

$$\sum_{i,j=1}^{n} a_{ij}(x) \frac{\partial^2 u}{\partial x_i \partial x_j}(x) = f(x) \quad \text{in } \Omega,$$

where $\sum_{i,j} a_{ij}\xi^i\xi^j \geq \lambda|\xi|^2$ and $a_{ij} \in C^\alpha(\Omega)$. By "freezing" x (i.e., fixing it at a particular point), prove the following a priori estimate from (8.25) for any $f \in C^\alpha(\Omega)$:

$$\sup_{1\leq i,j\leq d} \left[\frac{\partial^2 u}{\partial x_i \partial x_j}\right]_{C^\alpha(K)} \leq C(\alpha, d, K, \Omega)\big([f]_{C^\alpha(\Omega)} + \|f\|_{L^\infty(\Omega)}\big)$$

for any compact $K \subset \Omega$. *Hint:* See Gilbarg and Trudinger [48] for this estimate and much more.

Problem 8.7 Under the same assumptions as in Theorem 8.2, one has

$$[(m\hat{f})^\vee]_\alpha \leq C_\alpha B\,[f]_\alpha$$

for any $0 < \alpha < 1$ and $f \in C^\alpha(\mathbb{R}^d) \cap L^2(\mathbb{R}^d)$. *Hint:* This is a corollary of the proof of Theorem 8.2. Which properties of K does it establish, and what is needed in order for the proof of Theorem 8.8 to go through?

Problem 8.8 This problem introduces Sobolev and Besov spaces and studies some embeddings. We remark that the symbols with dots refer to homogeneous spaces, whereas those without dots refer to inhomogeneous spaces.

(a) Let $1 \leq p < \infty$ and $s \geq 0$. Define the Sobolev spaces $\dot{W}^{s,p}(\mathbb{R}^d)$ and $W^{s,p}(\mathbb{R}^d)$ as completions of $\mathcal{S}(\mathbb{R}^d)$ under the respective norms

$$\|(-\Delta)^{s/2} f\|_p \quad \text{and} \quad \|\langle\Delta\rangle^{s/2} f\|_p,$$

where $\langle a\rangle = \sqrt{1+a^2}$. The operators are interpreted here as Fourier multipliers. Show that for, any $1 < p < \infty$,

$$\|f\|_{\dot{W}^{s,p}} \simeq \left\|\Big(\sum_{j\in\mathbb{Z}} 2^{2js}|P_j f|^2\Big)^{1/2}\right\|_p,$$

with implicit constants that depend only on s, p, d. Formulate and prove an analogous statement for $W^{s,p}(\mathbb{R}^d)$. Does anything change for $s < 0$? Note that $s < 0$ arises naturally from the duality relation $(\dot{W}^{s,p})^* = \dot{W}^{-s,p'}$. For which s is this valid? What is the analogue for inhomogeneous spaces?

(b) State and prove the analogue of the Sobolev embedding estimate, Corollary 7.9, for $\dot{W}^{s,p}(\mathbb{R}^d)$.

(c) Define the Besov spaces $\dot{B}^s_{p,q}(\mathbb{R}^d)$ and $B^s_{p,q}(\mathbb{R}^d)$ by means of the respective norms (for q finite)

$$\Big(\sum_{j\in\mathbb{Z}} 2^{qjs}\|P_j f\|_p^q\Big)^{1/q} \quad \text{and} \quad \Big(\sum_{j\geq 0} 2^{qjs}\|P_j f\|_p^q\Big)^{1/q},$$

where in the second sum P_0 is interpreted as the projection onto all frequencies $|\xi| \leq C$. Using results from the present chapter, obtain embedding theorems between Besov and Sobolev spaces.

(d) Prove the embedding $\dot{B}^{\sigma}_{q,2}(\mathbb{R}^d) \hookrightarrow L^p(\mathbb{R}^d)$ for $2 \le q \le p < \infty$, $\sigma \ge 0$, and

$$\frac{1}{q} - \frac{1}{p} = \frac{\sigma}{d}.$$

Does the same hold for inhomogeneous spaces? What if anything can be changed in that case? By means of part (c), deduce embedding theorems for Sobolev spaces.

Problem 8.9 Prove that $[f]_\alpha \le C(\alpha, d)\|f\|_{\dot{W}^{1,p}(\mathbb{R}^d)}$ with $d < p < \infty$, $\alpha = 1 - d/p$ and $f \in \mathcal{S}(\mathbb{R}^d)$. This is called *Morrey's estimate*. For the case $p = d$, see Problem 10.7.

Problem 8.10 Show that $\dot{B}^{d/2}_{2,1}(\mathbb{R}^d) \hookrightarrow L^\infty(\mathbb{R}^d)$ but $\dot{B}^{d/2}_{2,q}(\mathbb{R}^d) \not\hookrightarrow L^\infty(\mathbb{R}^d)$ for any $1 < q \le \infty$. Obtain the embedding $\dot{B}^{d/2}_{2,\infty}(\mathbb{R}^d) \hookrightarrow \mathrm{BMO}(\mathbb{R}^d)$. Conclude that if $\|f\|_{\dot{B}^{d/2}_{2,q}(\mathbb{R}^d)} \le A$ for some $1 \le q < \infty$ then $w := e^f$ satisfies

$$⨍_Q w(x)\,dx ⨍_Q w^{-1}(x)\,dx \le C(A, d, q)$$

uniformly for all cubes $Q \subset \mathbb{R}^d$. This means precisely that w is an A_2-Mockenhaupt weight; see Stein [111, Chapter V] or Duoandikoetxea [31, Chapter 7].

9

Almost orthogonality

9.1. Cotlar's lemma

The proof of the L^2-boundedness of Calderón–Zygmund operators given in Chapter 7 was based on Fourier transforms. This is quite restrictive as it requires the operator to be translation invariant. We now present a device that allows one to avoid Fourier transforms in the context of L^2 theory, in many instances. This device, known as Cotlar's lemma (or the Cotlar–Stein lemma) has much wider applicability than to the L^2-boundedness of singular integrals. It has become an indispensable tool in harmonic analysis. This chapter presents a small sample of the many possible applications that Cotlar's lemma has found since its inception.

9.1.1. Motivation of almost orthogonality

Let us start from the basic observation that the operator norm of an infinite diagonal matrix viewed as an operator on $\ell^2(\mathbb{Z})$ is as large as its largest entry. To be specific, let

$$T\big(\{\xi_j\}_{j\in\mathbb{Z}}\big) = \{\lambda_j\xi_j\}_{j\in\mathbb{Z}} \quad \forall\{\xi_j\}_{j\in\mathbb{Z}} \in \ell^2(\mathbb{Z}),$$

where $\{\lambda_j\}_{j\in\mathbb{Z}}$ is a fixed sequence of complex numbers. Then

$$\|T\|_{\ell^2\to\ell^2} = \sup_j |\lambda_j|.$$

More generally, suppose that a Hilbert space $\mathcal{H}$ can be written as an infinite orthogonal sum:

$$\mathcal{H} = \bigoplus_j \mathcal{H}_j, \quad \mathcal{H} \ni f = \sum_j f_j, \quad f_j \in \mathcal{H}_j,$$

and suppose that the T_j are operators on $\mathcal{H}$ with $T_j\mathcal{H}_k = \{0\}$ if $k \neq j$. Furthermore, assume that the range of T_j is a subspace of $\mathcal{H}_j$. Then

$$\begin{aligned}\|Tf\|_{\mathcal{H}}^2 &= \sum_{j,k}\langle T_j f_j, T_k f_k\rangle = \sum_j \|T_j f_j\|_{\mathcal{H}_j}^2 \\ &\leq \sup_j \|T_j\|_{\mathcal{H}_j\to\mathcal{H}_j}^2 \sum_j \|f_j\|_{\mathcal{H}_j}^2 = M^2\|f\|_{\mathcal{H}}^2,\end{aligned}$$

where $M = \sup_j \|T_j\|_{\mathcal{H}_j\to\mathcal{H}_j}$.

Generalizing from this example, assume that T is a bounded operator on the Hilbert space $\mathcal{H}$ that admits the representation $T = \sum_j T_j$ such that we have $\mathrm{Ran}(T_j) \perp \mathrm{Ran}(T_k)$ as well as $\mathrm{Ran}(T_j^*) \perp \mathrm{Ran}(T_k^*)$ for $j \neq k$. In other words, we assume that $T_k^* T_j = 0$ and $T_k T_j^* = 0$ for $j \neq k$. Now recall that $\overline{\mathrm{Ran}(T_j^*)} = \ker(T_j)^\perp$ and denote by P_j the orthogonal projection onto that subspace. Then, for any $f \in \mathcal{H}$, one has

$$Tf = \sum_j T_j P_j f = \sum_j T_j f_j, \quad f_j = P_j f,$$

which implies that

$$\|Tf\|_{\mathcal{H}}^2 = \sum_j \|T_j f_j\|_{\mathcal{H}}^2 \leq \sup_j \|T_j\|^2 \sum_j \|f_j\|_{\mathcal{H}}^2 = M^2\|f\|_{\mathcal{H}}^2$$

with $M := \sup_j \|T_j\|$.

As a final step, we shall now relax the above orthogonality conditions $T_k^* T_j = 0$ and $T_k T_j^* = 0$ for $j \neq k$, since they turn out to be too restrictive for most applications. The idea is simply to replace this strong vanishing requirement by a condition that ensures sufficient decay in $|j-k|$. One can think of this as replacing diagonal matrices with matrices whose entries decay in a controllable fashion away from the diagonal.

9.1.2. The precise formulation

The following lemma, known as *Cotlar's lemma*, encapsulates the above idea.

Lemma 9.1 *Let $\{T_j\}_{j=1}^N$ be finitely many operators on some Hilbert space $\mathcal{H}$ such that, for some function $\gamma : \mathbb{Z} \to \mathbb{R}^+$, one has*

$$\|T_j^* T_k\| \leq \gamma^2(j-k), \quad \|T_j T_k^*\| \leq \gamma^2(j-k)$$

for any $1 \leq j, k \leq N$. Let

$$\sum_{\ell=-\infty}^{\infty} \gamma(\ell) =: A < \infty.$$

Then $\|\sum_{j=1}^N T_j\| \leq A$.

Proof For any positive integer n,

$$(T^*T)^n = \sum_{\substack{j_1,\dots,j_n=1\\k_1,\dots,k_n=1}}^{N} T_{j_1}^* T_{k_1} T_{j_2}^* T_{k_2} \cdots T_{j_n}^* T_{k_n}$$

We now take the operator norm of both sides of this identity and apply the triangle inequality to the right-hand side. Furthermore, we bound the norm of the products of operators by the products of the norms in two different ways and then take the square root of the product of the resulting estimates. To be specific, one has

$$\left\| T_{j_1}^* T_{k_1} T_{j_2}^* T_{k_2} \cdots T_{j_n}^* T_{k_n} \right\| \le \|T_{j_1}\| \|T_{k_n}\| \prod_{i=1}^{n-1} \left\| T_{k_i} T_{j_{i+1}}^* \right\|,$$

$$\left\| T_{j_1}^* T_{k_1} T_{j_2}^* T_{k_2} \cdots T_{j_n}^* T_{k_n} \right\| \le \prod_{i=1}^{n} \left\| T_{j_i}^* T_{k_i} \right\|.$$

Multiplying the previous two bounds and taking the square root yields

$$\|T_{j_1}^* T_{k_1} T_{j_2}^* T_{k_2} \cdots T_{j_n}^* T_{k_n}\| \le (\|T_{j_1}\| \|T_{k_n}\|)^{1/2} \prod_{i=1}^{n-1} \|T_{k_i} T_{j_{i+1}}^*\|^{1/2} \prod_{i=1}^{n} \|T_{j_i}^* T_{k_i}\|^{1/2}. \tag{9.1}$$

Therefore, with $\sup_{1\le j\le N} \|T_j\| =: B \le A$

$$\begin{aligned}
&\|(T^*T)^n\| \\
&\le \sum_{\substack{j_1,\dots,j_n=1\\k_1,\dots,k_n=1}}^{N} \|T_{j_1}\|^{1/2} \left\|T_{j_1}^* T_{k_1}\right\|^{1/2} \left\|T_{k_1} T_{j_2}^*\right\|^{1/2} \cdots \left\|T_{k_{n-1}} T_{j_n}^*\right\|^{1/2} \left\|T_{j_n}^* T_{k_n}\right\|^{1/2} \|T_{k_n}\|^{1/2} \\
&\le \sum_{\substack{j_1,\dots,j_n=1\\k_1,\dots,k_n=1}}^{N} \sqrt{B}\, \gamma(j_1-k_1)\gamma(k_1-j_2)\gamma(j_2-k_2)\cdots\gamma(k_{n-1}-j_n)\gamma(j_n-k_n)\sqrt{B} \\
&\le NBA^{2n-1}.
\end{aligned}$$

Since T^*T is self-adjoint, the spectral theorem implies that $\|(T^*T)^n\| = \|T^*T\|^n = \|T\|^{2n}$. Hence

$$\|T\| \le (NBA^{-1})^{1/2n} A.$$

Letting $n \to \infty$ yields the desired bound. □

Note that one needs to control *both* the sizes of $T_j^* T_k$ and $T_j T_k^*$ in the Cotlar lemma 9.1, as can be seen from the examples preceding the lemma.

9.1.3. Schur's lemma

In order to apply Lemma 9.1 one often invokes the following simple device.

Lemma 9.2 *Define an integral operator on a measure space $X \times Y$ with the positive product measure $\mu \otimes \nu$ via*

$$(Tf)(x) = \int_Y K(x, y) f(y)\, \nu(dy),$$

where K is a measurable kernel. One has the following bounds (the first three items constitute Schur's lemma*):*

(i) $\|T\|_{1\to 1} \le \sup_{y\in Y} \int_X |K(x, y)|\, \mu(dx) =: A$;
(ii) $\|T\|_{\infty\to\infty} \le \sup_{x\in X} \int_Y |K(x, y)|\, \nu(dy) =: B$;
(iii) $\|T\|_{p\to p} \le A^{1/p} B^{1/p'}$, *where* $1 \le p \le \infty$;
(iv) $\|T\|_{1\to\infty} \le \|K\|_{L^\infty(X\times Y)}$.

Proof Items (i) and (ii) are immediate from the definitions, and (iii) then follows by interpolation. Alternatively, one can use Hölder's inequality. Item (iv) is again evident from the definitions. □

Henceforth, we shall simply refer to this lemma as "Schur's test". We will now give an alternative proof of the L^2-boundedness of singular integrals for kernels as in Definition 7.1 that satisfy the stronger condition

$$|\nabla K(x)| \le B|x|^{-d-1};$$

cf. Lemma 7.2. The proof will be based on the same dyadic partition of unity that underlies the Littlewood–Paley decomposition; see Lemma 8.1.

9.1.4. Singular integrals on L^2

The point of the following corollary is the method of proof rather than the statement (which is weaker than the corresponding statement in Chapter 7).

Corollary 9.3 *Let K be as in Definition 7.1, with the additional assumption that $|\nabla K(x)| \le B|x|^{-d-1}$. Then*

$$\|T\|_{2\to 2} \le CB$$

with $C = C(d)$.

Proof We may take $B = 1$. Let ψ be a radial function as in Lemma 8.1 and set $K_j(x) = K(x)\psi(2^{-j}x)$. One now easily verifies that these kernels have the following properties. For all $j \in \mathbb{Z}$ one has

$$\int K_j(x)\, dx = 0,$$

$$\|\nabla K_j\|_\infty \le C\, 2^{-j} 2^{-jd}.$$

In addition, one has the estimates

$$\int |K_j(x)|\,dx < C,$$

$$\int |x|\,|K_j(x)|\,dx < C\,2^j,$$

with C some absolute constant. Define

$$(T_j f)(x) = \int_{\mathbb{R}^d} K_j(x-y) f(y)\,dy.$$

Observe that this integral is absolutely convergent for any $f \in L^1_{\mathrm{loc}}(\mathbb{R}^d)$. We shall now check the conditions in Lemma 9.1. Let $\widetilde{K}_j(x) := \overline{K}_j(-x)$. Then it is easy to see that

$$\left(T_j^* T_k f\right)(x) := \int_{\mathbb{R}^d} \left(\widetilde{K}_j * K_k\right)(y) f(x-y)\,dy.$$

and

$$\left(T_j T_k^* f\right)(x) = \int_{\mathbb{R}^d} \left(K_j * \widetilde{K}_k\right)(y) f(x-y)\,dy.$$

Hence, by Young's inequality,

$$\left\| T_j^* T_k \right\|_{2\to 2} \le \left\| \widetilde{K}_j * K_k \right\|_1$$

and

$$\left\| T_j T_k^* \right\|_{2\to 2} \le \left\| K_j * \widetilde{K}_k \right\|_1.$$

It suffices to consider the case $j \ge k$. Then, using the cancellation condition $\int K_k(y)\,dy = 0$, one obtains

$$\begin{aligned}
\left|\left(\widetilde{K}_j * K_k\right)(x)\right| &= \left| \int_{\mathbb{R}^d} \overline{K}_j(y-x) K_k(y)\,dy \right| \\
&= \left| \int_{\mathbb{R}^d} \left(\overline{K}_j(y-x) - \overline{K}_j(-x)\right) K_k(y)\,dy \right| \\
&\le \int_{\mathbb{R}^d} \|\nabla K_j\|_\infty |y|\,|K_k(y)|\,dy \\
&\le C\,2^{-j} 2^{-jd} 2^k.
\end{aligned}$$

Since

$$\operatorname{supp}\,(\widetilde{K}_j * K_k) \subset \operatorname{supp}(\widetilde{K}_j) + \operatorname{supp}(K_k) \subset B(0, C2^j)$$

we further conclude that

$$\left\| \widetilde{K}_j * K_k \right\|_1 \le C\,2^{k-j} = C\,2^{-|j-k|}.$$

Therefore, Lemma 9.1 applies with

$$\gamma^2(\ell) = C\,2^{-|\ell|},$$

whence $\|\sum_{j=-N}^{N} T_j\|_{2\to 2} \le C$ for all $N \ge 1$. For any $f \in \mathcal{S}(\mathbb{R}^d)$ one has $\sum_{j=-N}^{N} T_j f \to Tf$ pointwise, and thus from Fatou's lemma $\|Tf\|_2 \le C\|f\|_2$ with an absolute constant C. □

Exercise 9.1

(a) In the previous proof it suffices to consider the case $j > k = 0$. Provide the details of this reduction.

(b) Observe that Corollary 9.3 covers the Hilbert transform. In that case, draw the graph of $K(x) = x^{-1}$ and also $K_j(x) = x^{-1}\psi(2^{-j}x)$ and explain the previous argument by means of diagrams.

9.2. Calderón–Vaillancourt theorem

Another simple application of these almost orthogonality ideas is the Calderón–Vaillancourt theorem. This result concerns the L^2-boundedness of so-called pseudodifferential operators of the form

$$Tf(x) = \int_{\mathbb{R}^d} e^{ix\cdot\xi} a(x,\xi)\hat{f}(\xi)\,d\xi, \quad f \in \mathcal{S}(\mathbb{R}^d), \tag{9.2}$$

where $a \in C^\infty(\mathbb{R}^d \times \mathbb{R}^d)$ is such that

$$\sup_{x,\xi\in\mathbb{R}^d} \left(|\partial_x^\alpha a(x,\xi)| + |\partial_\xi^\alpha a(x,\xi)|\right) \le B \tag{9.3}$$

for all $|\alpha| \le 2d+1$.

Proposition 9.4 *Under the conditions* (9.3) *the operators in* (9.2) *are bounded on* $L^2(\mathbb{R}^d)$.

Proof We may assume that $B = 1$. Let χ be smooth and compactly supported such that $\{\chi(\cdot - k)\}_{k\in\mathbb{Z}^d}$ forms a partition of unity in $\mathbb{R}^d$:

$$\sum_{k\in\mathbb{Z}^d} \chi(\xi - k) = 1 \quad \forall \xi \in \mathbb{R}^d.$$

To construct such a χ, start from a compactly supported smooth $\eta \ge 0$ with the property that

$$\psi(x) := \sum_{k\in\mathbb{Z}^d} \eta(x-k) > 0 \quad \forall x \in \mathbb{R}^d.$$

Note that ψ is periodic with respect to $\mathbb{Z}^d$ and therefore uniformly lower bounded. Now define $\chi := \eta/\psi$.

Set

$$\chi_{k\ell}(x,\xi) := \chi(x-k)\chi(\xi-\ell), \quad a_{k,\ell}(x,\xi) := a(x,\xi)\chi_{k\ell}(x,\xi)$$

and define

$$(T_{k,\ell}f)(x) := \int_{\mathbb{R}^d} e^{ix\cdot\xi} a_{k,\ell}(x,\xi) f(\xi)\, d\xi$$

for $f \in \mathcal{S}(\mathbb{R}^d)$. Note that this involves f directly, not its Fourier transform, which is admissible by Plancherel's theorem. First, Lemma 9.2 implies that

$$\sup_{k,\ell} \|T_{k,\ell}\|_{2\to 2} \le C.$$

Furthermore, we claim that

$$\begin{aligned} \left\| T^*_{k',\ell'} T_{k,\ell} \right\|_{2\to 2} &\le C\langle k'-k\rangle^{-2d-1} \langle \ell'-\ell\rangle^{-2d-1}, \\ \left\| T_{k',\ell'} T^*_{k,\ell} \right\|_{2\to 2} &\le C\langle k'-k\rangle^{-2d-1} \langle \ell'-\ell\rangle^{-2d-1} \end{aligned} \tag{9.4}$$

for all $k, k', \ell, \ell' \in \mathbb{Z}^d$. Since the square roots of equations (9.4) are summable over $k, \ell \in \mathbb{Z}^d$, Lemma 9.1 implies that

$$\left\| \sum_{|k|<N} \sum_{|\ell|<N} T_{k,\ell} f \right\|_2 \le C\|f\|_2 \quad \forall f \in \mathcal{S}(\mathbb{R}^d)$$

for any positive integer N with some absolute constant C. Passing to the limit $N \to \infty$ concludes the proof. To prove (9.4) we start from

$$T^*_{k,\ell} g(\xi) = \int_{\mathbb{R}^d} e^{-ix\cdot\xi} \overline{a_{k\ell}(x,\xi)}\, g(x)\, dx,$$

which shows that $T^*_{k,\ell} T_{k',\ell'} \ne 0$ requires that $|k-k'| \le C$ and similarly, $T_{k,\ell} T^*_{k',\ell'} \ne 0$ requires that $|\ell - \ell'| \le C$. Figure 9.1 depicts the situation in phase space with three possible supports for different choices of k, ℓ. The squares denoted by B, C have empty intersections in terms of both their x-and ξ-projections, and therefore for the corresponding operators we have $T_B T_C^* = 0$ and $T_B^* T_C = 0$. The kernel of $T^*_{k,\ell} T_{k',\ell'}$ is

$$K_{k,\ell;k',\ell'}(\xi,\eta) := \int_{\mathbb{R}^d} e^{-ix\cdot(\xi-\eta)}\, \overline{a_{k,\ell}(x,\xi)}\, a_{k',\ell'}(x,\eta)\, dx. \tag{9.5}$$

To prove the decay in $|\ell - \ell'|$ we may assume that this difference exceeds some constant chosen such that $|\xi - \eta| \ge 1$ on the support $\overline{a_{k,\ell}(x,\xi)}\, a_{k',\ell'}(x,\eta)$. We now use the relation

$$i\frac{\xi-\eta}{|\xi-\eta|^2} \nabla_x e^{-ix\cdot(\xi-\eta)} = e^{-ix\cdot(\xi-\eta)}$$

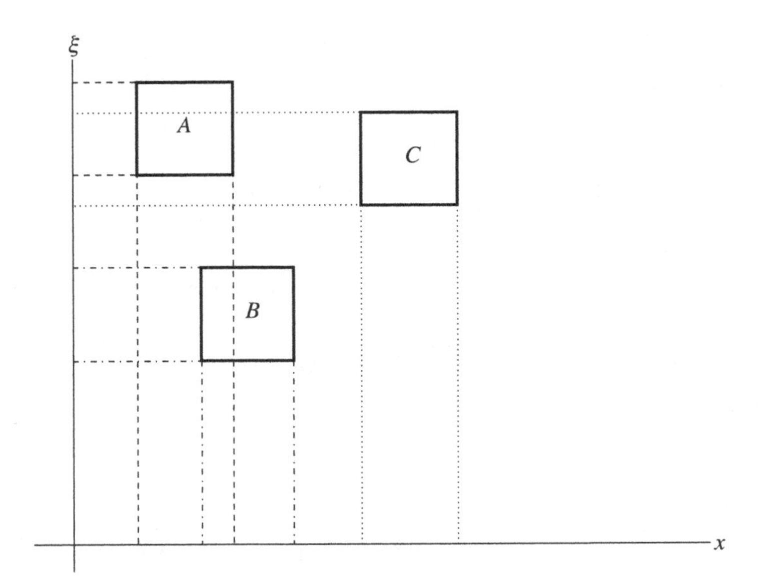

Figure 9.1. Supports of various $a_{k\ell}(x,\ \xi)$.

in order to integrate (9.5) by parts repeatedly, to be precise $2d+1$ times. This yields

$$|K_{k,\ell;k',\ell'}(\xi,\eta)| \le C|\ell-\ell'|^{-2d-1}.$$

Since the support of the kernel with respect to either variable is of size $\le C$, we conclude from Schur's test that

$$\left\| T^*_{k,\ell} T_{k',\ell'} \right\|_{2\to 2} \le C\langle \ell'-\ell\rangle^{-2d-1},$$

where we may assume that $|k-k'| \le C$. The same type of argument now implies the second relation in (9.4) and we are done, at least for compactly supported symbols $a(x,\xi)$ since then the sums over k,ℓ are finite. For the general case one applies Fatou's lemma and the pointwise convergence of $T_{k,\ell} f$ for a given $f \in \mathcal{S}(\mathbb{R}^d)$. □

9.3. Hardy's inequality

We now turn to *Hardy's inequality*, which combines ideas from this chapter with the simplest aspects of the uncertainty principle, to which Chapter 10 is devoted. The basic idea behind estimates such as (9.6) below is that the singular weight on the left-hand side can be traded for a derivative of the same order of magnitude. Note the scaling invariance of (9.6), which follows because both sides involve the same parameter s.

Theorem 9.5 *For any $0 \le s < d/2$ there is a constant $C(s,d)$ with the property that*

$$\big\| |x|^{-s} f \big\|_2 \le C(s,d)\, \|f\|_{\dot H^s(\mathbb{R}^d)} \tag{9.6}$$

for all $f \in \dot H^s(\mathbb{R}^d)$.

Proof Since $\mathcal{S}(\mathbb{R}^d)$ is dense in both L^2 and $\dot H^s(\mathbb{R}^d)$, it suffices to prove this estimate for $f \in \mathcal{S}(\mathbb{R}^d)$. It is clear that (9.6) requires that we work with two localizations simultaneously, namely localization in x and in its "dual" ξ, the Fourier variable. The natural way to localize the latter is through the Littlewood–Paley operators $P_k f = (\psi(2^{-k}\xi)\hat f)^\vee$ for any $k \in \mathbb{Z}$. Let us first consider the case of a single k, which amounts to showing that

$$\big\| |x|^{-s} P_k f \big\|_2 \le C(s,d)\, \|P_k f\|_{\dot H^s(\mathbb{R}^d)} \simeq 2^{sk} \|P_k f\|_2 \tag{9.7}$$

uniformly in k. By scaling it suffices to treat $k = 0$, which in turn reduces to verifying that the operator $(Tf)(x) = |x|^{-s} \tilde P_0 f$ is bounded on L^2; here $\tilde P_0 P_0 = P_0$ (see the proof of Theorem 8.3 for the notation). Schur's lemma fails when applied directly to T, but it succeeds for T^*T. Indeed, the latter operator has kernel

$$K(x,y) = \int_{\mathbb{R}^d} \phi(x-u)|u|^{-2s}\phi(u-y)\,du$$

for some $\phi \in \mathcal{S}(\mathbb{R}^d)$. The condition $s < d/2$ guarantees that this integral exists, and Schur's lemma implies that it yields an L^2-bounded operator.

Alternatively, we may use Bernstein's inequality; see Lemma 4.13 with $R = 1$. To be precise, one has

$$\begin{aligned} \big\| |x|^{-s} \tilde P_0 f \big\|_2 &\le \big\| |x|^{-s} \chi_{[|x|\le 1]} \tilde P_0 f \big\|_2 + \big\| |x|^{-s} \chi_{[|x|>1]} \tilde P_0 f \big\|_2 \\ &\le C \sum_{\ell \le 0} 2^{-s\ell} \|\chi_{[|x|\simeq 2^\ell]} \tilde P_0 f\|_2 + \|\tilde P_0 f\|_2 \\ &\le C \sum_{\ell \le 0} 2^{-s\ell} 2^{\ell d/2} \|\tilde P_0 f\|_\infty + \|f\|_2 \le C\|f\|_2, \end{aligned} \tag{9.8}$$

as desired. This argument reveals a simple and basic principle: the scale dual to the frequency scale $\{|\xi| \simeq 1\}$ is the physical space scale $\{|x| \simeq 1\}$. Indeed, as evidenced by (9.8), both smaller and larger scales in x contribute comparatively less to the L^2 bound. Chapter 10 elaborates several diverse aspects of this *uncertainty principle*.

At this point we face the problem of summing (9.7) over $k \in \mathbb{Z}$. In view of the definition of $\|f\|_{\dot H^s(\mathbb{R}^d)}$ the only way to do this would be by means of *square summation*; in other words we must use the (almost) orthogonality of the $\{P_k f\}_{k\in\mathbb{Z}}$. While this works perfectly for the right-hand side of (9.7), the left-hand side is less clear since we cannot expect the almost orthogonality of $\{|x|^{-s} P_k f\}_{k\in\mathbb{Z}}$.

In order to explore this issue it is natural to localize x. Thus, invoking the partition of unity in Lemma 8.1 we are asking whether

$$\|\chi_j f\|_2 \le C 2^{js} \|f\|_{\dot H^s(\mathbb{R}^d)} \quad \forall j \in \mathbb{Z}, \tag{9.9}$$

where $\chi_j(x) := \psi(2^{-j}x)$ in the terminology of Lemma 8.1. We may again rescale this to $j = 0$, which leads to

$$\|\psi f\|_2 \le C(s,d)\|f\|_{\dot H^s(\mathbb{R}^d)} \tag{9.10}$$

since $\chi_0 = \psi$. Note that the appearance of s on the right-hand side seems strange since it does not occur on the left-hand side. However, we do need to prove this for *all* $0 \le s < d/2$, which requires the Fourier decomposition $f = \sum_{k\in\mathbb{Z}} P_k f$. To begin, we note that

$$\|\psi P_k f\|_2 = \|\hat\psi * \widehat{P_k f}\|_2.$$

Owing to the rapid decay of $\hat\psi$ it is legitimate to think of it as being supported on a ball of some fixed radius, say 100. If k is positive, it then follows that the members of $\{\hat\psi * \widehat{P_k f}\}_{k\ge 0}$ retain their almost orthogonality, whereas for $k < 0$ the almost orthogonality is lost. This is illustrated by Figure 9.2, where the hatched center corresponds to the $P_k f$ that are smeared out by convolution with $\hat\psi$ whereas in the other shells almost orthogonality is retained.

The conclusion from these simple observations is that for $k < 0$ we cannot do better than the trivial triangle inequality, but for $k \ge 0$ we should definitely use Pythagoras (in other words, almost orthogonality). What will nevertheless save us from a divergence for $k < 0$ is the same Bernstein estimate as above, i.e.,

$$\begin{aligned}
\|\psi f\|_2^2 &\le C \sum_{k\in\mathbb{Z}^+} \|\psi P_k f\|_2^2 + \Big(\sum_{k\le 0} \|\psi P_k f\|_2\Big)^2 \\
&\le C \sum_{k\in\mathbb{Z}^+} 2^{2sk} \|P_k f\|_2^2 + \Big(\sum_{k\le 0} \|P_k f\|_\infty\Big)^2 \\
&\le C \|f\|_{\dot H^s(\mathbb{R}^d)}^2 + \Big(\sum_{k\le 0} 2^{kd/2} \|P_k f\|_2\Big)^2 \\
&\le C \|f\|_{\dot H^s(\mathbb{R}^d)}^2 + \Big(\sum_{k\le 0} 2^{k(d/2-s)} 2^{sk} \|P_k f\|_2\Big)^2 \\
&\le C \|f\|_{\dot H^s(\mathbb{R}^d)}^2,
\end{aligned}$$

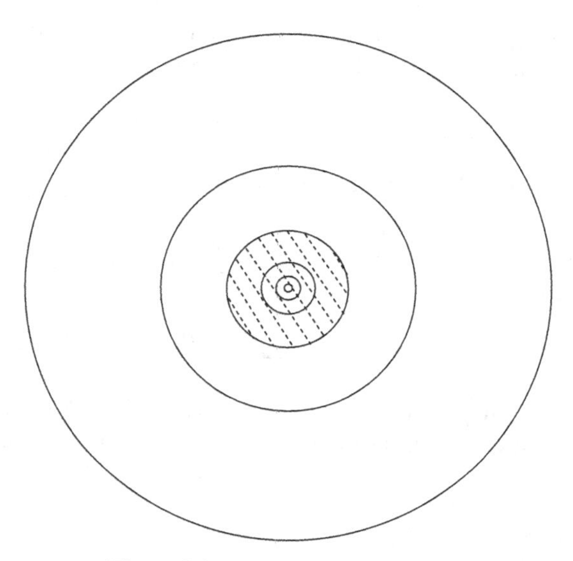

Figure 9.2. Littlewood–Paley shells.

where we have used $s < d/2$ and the Cauchy–Schwarz inequality to estimate the sum. To pass to the third line we used $s \geq 0$. The calculation for $\|\chi_j f\|_2$ is identical but has $k + j \geq 0$ and $k + j < 0$ as the two cases. Indeed, the convolution resulting from χ_j on the Fourier side moves ξ around randomly on a scale 2^{-j}. So we need to distinguish $k < -j$ from $k \geq -j$ as claimed.

We are now at the point where we can treat any fixed (dyadic) frequency summand $P_k f$ and also any x-localized shell $\{|x| \simeq 2^j\}$. However, we need to treat them *simultaneously*. It is important to realize, though, that we will not face the impossible task of summing over all k, j since, as we have just seen, the dominating configuration is $k + j = 0$ by the uncertainty principle.

The natural way to proceed is first to break up the function f dyadically in x and then to introduce a frequency partition:

$$\begin{aligned}\||x|^{-s} f\|_2^2 &\leq C \sum_{\ell} 2^{-2\ell s} \|\chi_\ell f\|_2^2 \\ &\leq C \sum_{\ell} 2^{-2\ell s} \Bigg(\sum_{k+\ell \leq 0} \|\chi_\ell P_k f\|_2 \Bigg)^2 + \sum_{\ell} 2^{-2\ell s} \|\chi_\ell P_{>-\ell} f\|_2^2,\end{aligned} \tag{9.11}$$

where $P_{>-\ell} := \sum_{j>-\ell} P_j$. For the sum involving both k and ℓ we invoke Bernstein's inequality in the same fashion as before to obtain the bound

$$\|\chi_\ell P_k f\|_2 \le C 2^{\ell d/2} \|P_k f\|_\infty \le C 2^{d(k+\ell)/2} \|P_k f\|_2. \tag{9.12}$$

Inserting this into the above-mentioned sum in (9.11) yields

$$\begin{aligned}\sum_\ell 2^{-2\ell s} \Big(\sum_{k+\ell\le 0} \|\chi_\ell P_k f\|_2 \Big)^2 &\le C \sum_\ell \Big(\sum_{k+\ell\le 0} 2^{(d/2-s)(\ell+k)} 2^{sk} \|P_k f\|_2 \Big)^2 \\ &\le C \sum_k 2^{2ks} \|P_k f\|_2^2 \le C \|f\|_{\dot H^s}^2,\end{aligned}$$

since $d/2 - s > 0$. To pass to the last line one may use Schur's test for sums, for example. The final sum in (9.11) uses that $s > 0$ (which suffices by interpolation with the case $s = 0$):

$$\begin{aligned}\sum_\ell 2^{-2\ell s} \|\chi_\ell P_{>-\ell} f\|_2^2 &\le \sum_\ell 2^{-2\ell s} \|P_{>-\ell} f\|_2^2 \\ &\le C \sum_\ell \sum_{k>-\ell} 2^{-2(\ell+k)s} 2^{2sk} \|P_k f\|_2^2 \\ &\le C \sum_k 2^{2sk} \|P_k f\|_2^2 \sum_{\ell>-k} 2^{-2(\ell+k)s} \le C \|f\|_{\dot H^s}^2,\end{aligned}$$

which is (9.6). □

An alternative, and useful, way to derive (9.12) is via Schur's lemma. In fact, we may write

$$(\chi_\ell P_k f)(x) = \int_{\mathbb{R}^d} \psi(2^{-\ell} x)\, 2^{kd} \check{\psi}(2^k(x-y)) f(y)\, dy$$

and then observe that

$$\begin{aligned}\sup_x \int_{\mathbb{R}^d} \big|\psi(2^{-\ell}x) 2^{kd} \check{\psi}(2^k(x-y))\big|\, dy &\le C, \\ \sup_y \int_{\mathbb{R}^d} \big|\psi(2^{-\ell}x) 2^{kd} \check{\psi}(2^k(x-y))\big|\, dx &\le C\, 2^{(k+\ell)d},\end{aligned}$$

which gives exactly (9.12) by Schur's test.

Theorem 9.5 is sharp in the following sense. Clearly, both sides scale in the same way, so it is impossible to use any other Sobolev space, for example. The condition $s \ge 0$ is needed as we cannot allow the weights on the left-hand side to grow. Furthermore, $s \ge d/2$ is not possible since in that case the estimate would fail for Schwartz functions that do not vanish at the origin (the norm on the left-hand side would then be infinite).

Several questions now pose themselves in a natural way. (i) Is there a substitute for $s > d/2$ in the previous theorem? (ii) Is there a version of Theorem 9.5 with L^p instead of L^2?

We shall address these questions in the problems at the end of this chapter.

9.4. The $T(1)$ theorem via Haar functions

We now turn to singular integrals on the line that are *not* given by convolution. To be specific, let $K : \mathbb{R}^2 \to \mathbb{R}$ be a measurable function which is locally bounded on $\mathbb{R}^2 \setminus \{x = y\}$ and which satisfies the pointwise bounds

$$
\begin{aligned}
|K(x,y) - K(x',y)| &\le \frac{|x-x'|^\delta}{|x-y|^{1+\delta}} \quad \forall |x-y| > 2|x-x'|, \\
|K(x,y) - K(x,y')| &\le \frac{|y-y'|^\delta}{|x-y|^{1+\delta}} \quad \forall |x-y| > 2|y-y'|,
\end{aligned}
\tag{9.13}
$$

where $0 < \delta \le 1$ is arbitrary but fixed.

Definition 9.6 By a *singular integral operator T with kernel K* we mean any linear operator $T : \mathcal{S}(\mathbb{R}) \to \mathcal{S}'(\mathbb{R})$ with the property that

$$\langle Tf, g\rangle = \int_{\mathbb{R}^2} K(x,y) f(y) g(x)\, dx dy \tag{9.14}$$

for all $f, g \in \mathcal{S}$ with disjoint supports. Here K satisfies (9.13).

Figure 9.3 illustrates (9.14), with $(f \otimes g)(x,y) = f(x)g(y)$. The essential feature here is that the rectangle, which equals $\mathrm{supp}(f \otimes g)$, does not intersect the diagonal.

Exercise 9.2 Show that for any $f \in \mathcal{S}(\mathbb{R}^d)$ with compact support

$$(Tf)(x) = \int_{\mathbb{R}} K(x,y) f(y)\, dy \qquad \text{for a.e. } x \in \mathbb{R}^d \setminus \mathrm{supp}(f).$$

Note that K is uniquely determined by T, but different T may correspond to the same kernel. Indeed, the zero operator and the identity operator as well as any differential operator T for which $Tf := \sum_{k=0}^{\infty} a_k f^{(k)}$ with scalars a_k of which only finitely many are nonzero, have $K \equiv 0$.

Because of this, the boundedness of T cannot be determined from K. Our goal is to reduce the boundedness problem on L^p to some basic criterion. The first step is the standard Calderón–Zygmund reduction to $p = 2$.

Proposition 9.7 *Let T be a singular integral operator. If T is bounded on $L^2(\mathbb{R})$ then it is bounded from L^1 to weak-L^1 and also on $L^p(\mathbb{R})$ for all $1 < p < \infty$.*

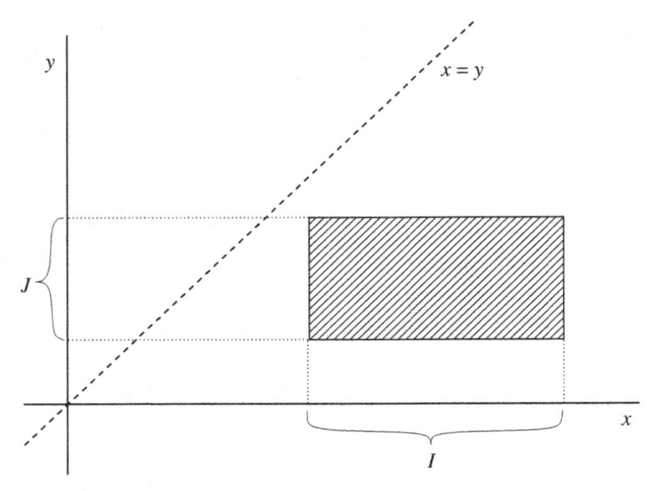

Figure 9.3. Support of $f \otimes g$, $I = \mathrm{supp}(f)$, $J = \mathrm{supp}(g)$.

Proof This is covered by Problem 7.7; here we will merely sketch the proof for the sake of completeness. First, by duality and the symmetry of the conditions (9.13), it suffices to prove the bound for L^1. The L^p case then follows by interpolation. Thus, let $f \in L^1(\mathbb{R})$ with $\|f\|_1 = 1$. We need to prove that

$$\left|\left\{x \in \mathbb{R} \,\middle|\, |(Tf)(x)| > \lambda\right\}\right| \le C\lambda^{-1} \quad \forall \lambda > 0.$$

Perform a Calderón–Zygmund decomposition: set $f = g + b$, $|g| \le 2\lambda$, $\|g\|_1 \le \|f\|_1$, and

$$b = \sum_{I \in \mathcal{B}} \chi_I(f - m_I(f)), \quad \sum_I |I| \le C\lambda^{-1}.$$

Then

$$\begin{aligned}\left|\left\{x \in \mathbb{R} \,\middle|\, |(Tf)(x)| > \lambda\right\}\right| &\le \left|\left\{x \in \mathbb{R} \,\middle|\, |(Tg)(x)| > \tfrac{1}{2}\lambda\right\}\right| \\ &\quad + \left|\left\{x \in \mathbb{R} \,\middle|\, |(Tb)(x)| > \tfrac{1}{2}\lambda\right\}\right|.\end{aligned}$$

For the first set on the right-hand side we use the L^2 bound, which gives the estimate

$$C\lambda^{-2}\|Tg\|_2^2 \le C\lambda^{-2}\|g\|_2^2 \le C\lambda^{-2}\|g\|_1\|g\|_\infty \le C\lambda^{-1}.$$

For the second set we can ignore the contribution of $\bigcup_{I \in \mathcal{B}} I^*$, where $I^* = 3I$ is a dilate. Indeed, that set gives $\le C\lambda^{-1}$ in measure. To conclude we now estimate, with x_I the center of I,

$$\begin{aligned}
&\left|\left\{x \in \mathbb{R} \setminus \cup_{I\in\mathcal{B}} I^* \,\middle|\, |(Tb)(x)| > \tfrac{1}{2}\lambda\right\}\right| \\
&\quad\le C\lambda^{-1} \sum_{I\in\mathcal{B}} \int_{\mathbb{R}\setminus I^*} \int_I |K(x,y) - K(x,y_I)|\, |f(y) - m_I(f)|\, dy dx \\
&\quad\le C\lambda^{-1} \sum_{I\in\mathcal{B}} \int_I |f| \le C\lambda^{-1},
\end{aligned}$$

by construction. Here we used

$$\sup_{y\in I} \int_{\mathbb{R}\setminus I^*} |K(x,y) - K(x,y_I)|\, dx \le C,$$

which follows from (9.13). Summing over I now yields the desired estimate. □

The main question is therefore how to establish L^2-boundedness. This is the content of the following $T(1)$ *theorem*. For simplicity, we restrict ourselves to the interval $[0, 1]$ although this can be easily extended to the whole line.

Theorem 9.8 *Let T be a singular integral operator such that $T, T^* : \mathcal{S}(\mathbb{R}) \to \mathcal{S}'(\mathbb{R}) \cap L^1_{\mathrm{loc}}(\mathbb{R})$. Assume that*

$$\max(\|Th_I\|_2, \|T^*h_I\|_2) \le \|h_I\|_2 = 1 \quad \forall I \in \mathcal{D} \tag{9.15}$$

and that $T(\chi_{[0,1]}) \in \mathrm{BMO}([0,1])$ and $T^*(\chi_{[0,1]}) \in \mathrm{BMO}([0,1])$. Then T is bounded on $L^2([0,1])$ with an operator norm bounded by some absolute constant depending only on δ, as in (9.13).

The assumptions in the theorem need to be interpreted in a suitable fashion, since they require the application of T to functions outside the allowed domain $\mathcal{S}(\mathbb{R})$. This is easily done by means of duality. Indeed, (9.15) can be interpreted in the following manner:

$$\max\left(|\langle h_I, T^*\varphi\rangle|, |\langle h_I, T\varphi\rangle|\right) \le \|\varphi\|_2 \quad \forall \varphi \in \mathcal{S}(\mathbb{R}).$$

Since we assumed that both T and T^* take φ to $L^1([0,1])$, the left-hand side is well defined. Furthermore, by Lemma 8.28 and Exercise 8.7, the BMO condition means that

$$|\langle \chi_{[0,1]}, T^*a\rangle| + |\langle \chi_{[0,1]}, Ta\rangle| \le C \tag{9.16}$$

for all *smooth* $\mathcal{H}^1([0,1])$ atoms a.

The name $T(1)$ refers to the fact that we have assumed that $T1 \in \mathrm{BMO}$; and $T^*1 \in \mathrm{BMO}$; the "1" lives on $[0, 1]$. The L^2 condition (9.15) is clearly necessary for L^2-boundedness. Note that it immediately excludes $Tf = f'$, although the latter satisfies the BMO condition. In fact, the BMO condition is also necessary.

Lemma 9.9 *If T is a singular integral operator that is bounded on $L^2([0,1])$ then $T(\chi_{[0,1]}) \in \mathrm{BMO}$ and $T^*(\chi_{[0,1]}) \in \mathrm{BMO}$.*

Proof By (9.16) it suffices to show that $\|Ta\|_1 + \|T^*a\|_1 \le C$ for all atoms. Thus, let $\mathrm{supp}(a) \subset I$. Then, on the one hand,

$$\|Ta\|_{L^1(I^*)} \le |I^*|^{1/2}\|Ta\|_2 \le C|I^*|^{1/2}\|a\|_2 \le C,$$

uniformly in a. On the other hand, by (9.13),

$$\begin{aligned}\int_{\mathbb{R}\setminus I^*} |(Ta)(x)|\,dx &\le \int_{\mathbb{R}\setminus I^*}\int_I |K(x,y) - K(x,y_I)|\ |a(y)|\,dydx \\ &\le C\int_{\mathbb{R}\setminus I^*}\int_I \frac{|I|^\delta}{\mathrm{dist}(x,I)^{1+\delta}}\,|a(y)|\,dydx \\ &\le C\int_{\mathbb{R}\setminus I^*} \frac{|I|^\delta}{\mathrm{dist}(x,I)^{1+\delta}}\,dx\,\|a\|_1 \le C\|a\|_2|I|^{1/2} \le C,\end{aligned}$$

and we are done. □

9.4.1. Carleson's lemma

In what follows, all intervals I, J, L will be understood to be dyadic in $[0,1]$. In other words, they are elements of $\mathcal{D} = \mathcal{D}_\infty$ in Definition 8.12.

Definition 9.10 A function $a : \mathcal{D} \to (0,\infty)$ is said to satisfy Carleson's condition if and only if, for all $I \in \mathcal{D}$,

$$\sum_{J\subset I} a(J) \le |I|.$$

The following result is a basic estimate on sums over all dyadic intervals due to Carleson.

Proposition 9.11 *Let $a, b : \mathcal{D} \to (0,\infty)$, where a satisfies Carleson's condition. Then*

$$\sum_J a(J)b(J) \le 4\int_0^1 \sup_{I\ni x} b(I)\,dx. \tag{9.17}$$

Proof The proof is based on exactly the same stopping-time argument as that in the proof of Theorem 8.22. Define the auxiliary function

$$A(I|x) := \sum_{x\in J\subset I} \frac{a(J)}{|J|} \quad \forall x \in \mathbb{R},\ I \in \mathcal{D}.$$

For simplicity we may assume that $a(J) = 0$ for all J with $|J| \le 2^{-N}$ for some arbitrary but fixed $N > 0$. Indeed, by means of monotone convergence we may then pass to the general case. Now let $I(x)$ be the maximal dyadic interval having $A(I(x)|x) \le 2$. Note that this is well defined, by the Carleson condition and the assumption of finite depth just made. We now make the following claim, which is based on the maximality of the intervals $I(x)$:

$$|\{x \in J \mid I(x) \supset J\}| \ge \tfrac{1}{2}|J| \quad \forall J. \tag{9.18}$$

Assuming this for now, we may then conclude as follows:

$$\begin{aligned}
\sum_J a(J)b(J) &\le 2\int_{-\infty}^{\infty} \sum_{x\in J\subset I(x)} |J|^{-1}a(J)b(J) \\
&\le 2\int_{-\infty}^{\infty} \sup_{x\in J} b(J) \sum_{x\in J\subset I(x)} |J|^{-1}a(J)\,dx \\
&= 2\int_{-\infty}^{\infty} \sup_{x\in J} b(J)A(I(x)\big|x)\,dx \\
&\le 4\int_{-\infty}^{\infty} \sup_{x\in J} b(J)\,dx,
\end{aligned} \tag{9.19}$$

as desired. To prove the claim, fix any J and define

$$E := \{x \in J \mid I(x) \subsetneq J\}.$$

Then, on the one hand,

$$\int_E A(J|x)\,dx \ge 2|E|$$

whereas, on the other hand,

$$\int_E A(J|x)\,dx \le \int_J A(J|x)\,dx = \sum_{L\subset J} a(L) \le |J|$$

by the Carleson condition, and we are done. □

9.4.2. Paraproducts

Given $b \in L^\infty([0,1])$ and $f \in L^2([0,1])$, we obviously have $\|bf\|_2 \leq \|b\|_\infty \|f\|_2$. No such estimate exists if $b \in \mathrm{BMO}$. The paraproducts derive their name from their resemblance to the product bf and their importance from their being bounded on L^2 if $b \in L^2$.

Definition 9.12 Given $b \in \mathrm{BMO}([0,1])$ and $f \in L^2$, we define the paraproduct of b and f as

$$\pi_b(f) = \sum_I \langle b, h_I \rangle m_I(f) h_I, \tag{9.20}$$

where $m_I(f) := \fint_I f$ is the average of f over I.

Note that $\pi_b(f)$ is a multiplier operator as defined in Section 8.4.2. Indeed, (9.20) agrees with (8.31) with b as argument and with weights $x_I := m_I(f)$. Further, note that $\pi_b(1) = b$ and $\pi_b^*(1) = 0$. The main result concerning these products is the following.

Proposition 9.13 *The operator π_b is bounded on $L^2([0,1])$ with norm $\leq C\|b\|_{\mathrm{BMO}}$. Furthermore, π_b is bounded on $L^p([0,1])$ for any $1 < p < \infty$.*

Proof One has

$$\|\pi_b(f)\|_2^2 = \sum_I \langle b, h_I \rangle^2 m_I(f)^2.$$

Now apply Proposition 9.11 with $a(J) := \langle b, h_J \rangle^2$ and $b(J) := m_J(f)^2$. The Carleson condition,

$$\sum_{J \subset I} a(J) \leq |I| \|b\|_{\mathrm{BMO}}^2,$$

holds and we obtain, with M the Hardy–Littlewood maximal operator,

$$\begin{aligned} \|\pi_b(f)\|_2^2 &\leq C\|b\|_{\mathrm{BMO}}^2 \int_0^1 \sup_{I \ni x} m_I(f)^2 \, dx \\ &\leq C\|b\|_{\mathrm{BMO}}^2 \int_0^1 (Mf)^2(x)\, dx \leq C\|b\|_{\mathrm{BMO}}^2 \|f\|_2^2, \end{aligned}$$

as claimed. The L^p case, as well as the L^1 to weak-L^1 bound, follow immediately from Corollary 8.17 since $\pi_b(h_I)$ is supported in I. □

9.4.3. Proof of Theorem 9.8

We begin by reducing the problem to the case where $T1 = 0$ and $T^*1 = 0$ (here the "1" lives on $[0,1]$). This is done by means of paraproducts, as follows. Let

$T1 = b_1$ and $T^*1 = b_2$. Then consider

$$S := T - \pi_{b_1} - \pi^*_{b_2},$$

which satisfies $S1 = 0$ and $S^*1 = 0$ by inspection. Further, by Proposition 9.13, T is L^2-bounded if and only if S is L^2-bounded; S also satisfies the L^2 condition (9.15). In order to complete the reduction, we need to know that π_{b_1} and π_{b_2} are singular integrals as in Definition 9.6. However, owing to the discontinuity of the Haar functions and of the indicator χ_I, this is not the case. The way around these conflicting requirements is to note that

$$\pi_{b_1} h_I = \pi^*_{b_1} h_I = \pi_{b_2} h_I = \pi^*_{b_2} h_I = 0 \quad \text{on } [0,1] \setminus I. \tag{9.21}$$

This will suffice, since we shall not use the full strength of (9.13) but, rather, its consequence

$$|(Th_I)(y)| + |(T^*h_I)(y)| \leq \frac{C|I|^{1/2+\delta}}{(|I| + \operatorname{dist}(y, I))^{1+\delta}} \quad \forall y \in [0,1] \setminus I^*. \tag{9.22}$$

Since the paraproducts satisfy this relation owing to (9.21), we may now indeed assume that $b_1 = b_2 = 0$. We now recall the following identity from Section 8.4: in $L^2([0,1])$,

$$\mathrm{Id} = \mathbb{E}_0 + \sum_{n=0}^{\infty} \Delta_n, \quad \Delta_n f := \sum_{I \in \mathcal{A}_n} \langle f, h_I \rangle h_I, \tag{9.23}$$

where "Id" is the representation of a function in the Haar basis. The sum in the left-hand equation converges in L^2. Further, $\Delta_n = \mathbb{E}_{n+1} - \mathbb{E}_n$; see (8.28). Our assumption that $T1 = T^*1 = 0$ means precisely that $T\mathbb{E}_0 = \mathbb{E}_0 = 0$. Thus,

$$\begin{aligned} T &= \left(\sum_{n=0}^{\infty} \Delta_n\right) T \left(\sum_{m=0}^{\infty} \Delta_m\right) \\ &= \sum_{n=0}^{\infty} (\Delta_n T \Delta_n + \mathbb{E}_n T \Delta_n + \Delta_n T \mathbb{E}_n), \end{aligned} \tag{9.24}$$

where we have used $\mathbb{E}_n = \mathbb{E}_0 + \sum_{m<n} \Delta_m$ and the partition

$$\{m \geq 0,\ n \geq 0\} = \{m = n \geq 0\} \cup \{m > n \geq 0\} \cup \{n > m \geq 0\}$$

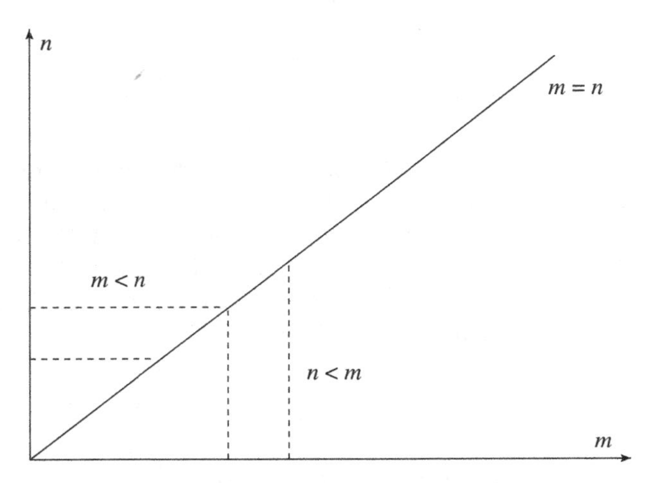

Figure 9.4. The decomposition of T.

illustrated by Figure 9.4. We shall now establish separately that

$$\left\| \sum_{n=0}^{\infty} \Delta_n T \Delta_n \right\|_{2\to 2} \le C, \quad \left\| \sum_{n=0}^{\infty} \mathbb{E}_n T \Delta_n \right\|_{2\to 2} \le C, \quad \left\| \sum_{n=0}^{\infty} \Delta_n T \mathbb{E}_n \right\|_{2\to 2} \le C, \tag{9.25}$$

which will conclude the proof, in view of (9.24). By symmetry it suffices to prove the first two bounds in (9.25). For the first, we note the orthogonality property

$$(\Delta_n T \Delta_n)(\Delta_m T \Delta_m)^* = 0 \quad \forall n \neq m,$$

whence

$$\left\| \sum_{n=0}^{\infty} \Delta_n T \Delta_n \right\|_{2\to 2} \le \sup_{n\ge 0} \|\Delta_n T \Delta_n\|_{2\to 2}.$$

In the basis $\{h_I\}_{I\in\mathcal{A}_n}$, the operator $\Delta_n T \Delta_n$ is given by the matrix with entries $\langle h_I, T h_J\rangle$. In view of (9.22) and (9.15) they satisfy the following bound, with absolute constant C,

$$|\langle h_I, T h_J\rangle| \le C(1+k)^{-\delta},$$

where $k = 0$ if $I = J$ or if they are nearest neighbors and where k is the smallest integer exceeding $|I|^{-1}\mathrm{dist}(I, J)$ otherwise. Schur's lemma therefore implies that $\Delta_n T \Delta_n$ is L^2-bounded uniformly in n as desired.

Next we show that $\|\mathbb{E}_n T \Delta_n\|_{2\to 2}$ is uniformly bounded. The matrix in this case is $\langle Th_I, |J|^{-1/2}\chi_J\rangle$ with $I, J \in \mathcal{D}_n$. We obtain the estimate

$$\begin{aligned}
|\langle Th_I, |J|^{-1/2}\chi_J\rangle| &\le |\langle Th_I, |J|^{-1/2}\chi_{I^*\cap J}\rangle| + |\langle Th_I, |J|^{-1/2}\chi_{J\setminus I^*}\rangle| \\
&\le C\chi_{[I^*\cap J\neq\emptyset]} + \int_{J\setminus I^*} \frac{C|I|^{1/2+\delta}|J|^{-1/2}}{\operatorname{dist}(x,I)^{1+\delta}}\,dx \\
&\le C\chi_{[I^*\cap J\neq\emptyset]} + C\frac{|I|^{1+\delta}}{\operatorname{dist}(I,J)^{1+\delta}}\chi_{[I^*\cap J=\emptyset]}.
\end{aligned}$$

Schur's lemma now implies that

$$\sup_{n\ge 0} \|\mathbb{E}_n T\Delta_n\|_{2\to 2} \le C.$$

In contrast with the case for the first bound in (9.25), this does not suffice, and we need to invoke Cotlar's lemma. Since

$$(\mathbb{E}_n T\Delta_n)(\mathbb{E}_m T\Delta_m)^* = 0 \quad \forall n \neq m,$$

it suffices to bound

$$S_{nm} := (\mathbb{E}_n T\Delta_n)^*(\mathbb{E}_m T\Delta_m) = (\Delta_n T^*)(\mathbb{E}_{m\wedge n} T\Delta_m).$$

By symmetry, we may assume that $m < n$, whence $S_{nm} = (\Delta_n T^*)(\mathbb{E}_m T\Delta_m)$. The matrix coefficients of S_{nm} are of the form

$$\langle \mathbb{E}_n Th_I, \mathbb{E}_m Th_J\rangle, \quad I \in \mathcal{A}_n, \quad J \in \mathcal{A}_m.$$

We shall apply Schur's lemma to the absolute values of the matrix entries $M_{IJ} := |\langle \mathbb{E}_m Th_I, \mathbb{E}_n Th_J\rangle|$, which we now estimate. We rewrite (9.22) in the form

$$|\mathbb{E}_n Th_I| \le |I|^{-1/2}\omega_I, \qquad \omega_I(x) := C\left(\tfrac{|I|}{|I|+\operatorname{dist}(x,I)}\right)^{1+\delta} \tag{9.26}$$
$$|\mathbb{E}_m Th_J| \le |J|^{-1/2}\omega_J.$$

Then, for all I, J as above,

$$M_{IJ} \le (|I||J|)^{-1/2}\int_{-\infty}^{\infty} \omega_I(x)\omega_J(x)\,dx.$$

By means of the calculus inequality (see Exercise 9.3)

$$\int_{-\infty}^{\infty} \omega_I(x)\omega_J(x)\,dx \le C\min(|I|,|J|)\left(\frac{|I|+|J|}{|I|+|J|+\operatorname{dist}(I,J)}\right)^{1+\delta}, \tag{9.27}$$

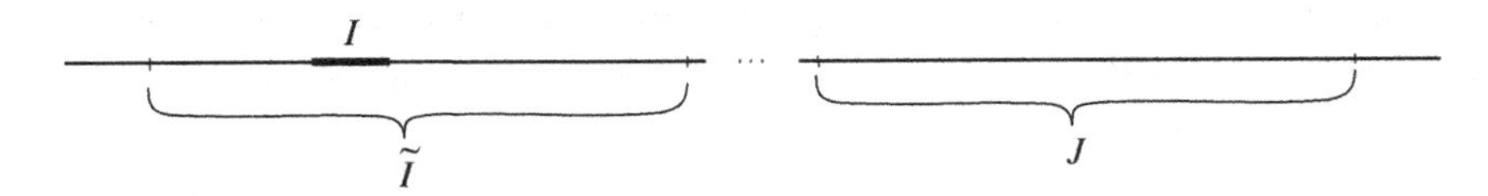

Figure 9.5. Intervals I, $\tilde{I}$ and J.

which is valid for any two intervals $I, J \subset \mathbb{R}$ with some absolute constant C, we conclude that

$$M_{IJ} \leq C\,2^{(m-n)/2}\left(\frac{|J|}{|J| + \operatorname{dist}(I, J)}\right)^{1+\delta}, \tag{9.28}$$

which implies that

$$A(m,n) := \sup_{I \in \mathcal{A}_n} \sum_{J \in \mathcal{A}_m} M_{IJ} \leq C\,2^{(m-n)/2}. \tag{9.29}$$

However, the bound (9.28) *cannot be summed* over all $I \subset J$, $I \in \mathcal{A}_n$. Indeed, since there are 2^{n-m} such intervals for any given J, this summation would yield, for any $J \in \mathcal{A}_m$,

$$\sum_{I \subset J} M_{IJ} \leq C\,2^{(n-m)/2},$$

exactly canceling the gain provided by (9.29). However, something *can be salvaged* from this estimate if one sums over only a small subclass of the short intervals I. We define such a small subclass by writing

$$\mathcal{A}_n = \mathcal{A}_{nm} \cup \mathcal{B}_{nm}, \quad \mathcal{A}_{nm} := \left\{ I \in \mathcal{A}_n \,\middle|\, \operatorname{dist}(I, \partial J) \leq \mu |I|, \ \forall J \in \mathcal{A}_m \right\},$$

where $1 \leq \mu \leq 2^{n-m}$ is to be determined and $\mathcal{A}_{nm} \cap \mathcal{B}_{nm} = \emptyset$. Then (9.28) yields

$$B_1(m,n) := \sup_{J \in \mathcal{A}_m} \sum_{I \in \mathcal{A}_{nm}} M_{IJ} \leq C\,2^{(m-n)/2}\mu. \tag{9.30}$$

The sum over $I \in \mathcal{B}_{n,m}$ therefore requires a different estimate on M_{IJ}. The only way to obtain such an estimate is to invoke cancellation in $\langle \mathbb{E}_n T h_I, \mathbb{E}_m T h_J \rangle$. To be specific, we shall use

$$\int_0^1 \mathbb{E}_n T h_I = \int_0^1 T h_I = \langle 1, T h_I \rangle = \langle T^* 1, h_I \rangle = 0.$$

Indeed, let $\tilde{I} \in \mathcal{A}_m$ be the unique ancestor of $I \in \mathcal{A}_n$, and define

$$(T h_J)(\tilde{I}) := \fint_{\tilde{I}} \mathbb{E}_m T h_J = \fint_{\tilde{I}} T h_J.$$

Then, for any $J \in \mathcal{A}_m$ and $I \in \mathcal{B}_{nm}$, the aforementioned cancellation together with the estimates (9.26) imply that

$$\begin{aligned} M_{IJ} &= \left| \int_{\tilde{I}} (\mathbb{E}_n T h_I)(T h_J)(\tilde{I}) + \int_{\tilde{I}^c} (\mathbb{E}_n T h_I)(\mathbb{E}_m T h_J) \right| \\ &= \left| - \int_{\tilde{I}^c} (\mathbb{E}_n T h_I)(T h_J)(\tilde{I}) + \int_{\tilde{I}^c} (\mathbb{E}_n T h_I)(\mathbb{E}_m T h_J) \right| \\ &\le C \int_{\tilde{I}^c} |I|^{-1/2} \omega_I \fint_{\tilde{I}} |J|^{-1/2} \omega_J + \int_{\tilde{I}^c} |I|^{-1/2} \omega_I \, |J|^{-1/2} \omega_J \\ &\le C \left(\frac{|I|}{|J|} \right)^{1/2} \mu^{-\delta} \left(\frac{|J|}{|J| + \operatorname{dist}(\tilde{I}, J)} \right)^{1+\delta}, \end{aligned} \tag{9.31}$$

where the final step again amounts to the verification of a calculus inequality; see Exercise 9.3. Hence,

$$B_2(m,n) := \sup_{J \in \mathcal{A}_m} \sum_{I \in \mathcal{B}_{nm}} M_{IJ} \le C\, 2^{(n-m)/2} \mu^{-\delta}. \tag{9.32}$$

Combining (9.29), (9.30), and (9.32) implies that

$$A(m,n)(B_1(m,n) + B_2(m,n)) \le C(2^{m-n}\mu + \mu^{-\delta}) \le C 2^{(n-m)\delta/(1+\delta)}$$

by optimizing over μ. This suffices for Cotlar's lemma to apply, and we are done.

Exercise 9.3 Verify the calculus facts consisting of (9.27) and the final step of (9.31); for the latter it is essential that $I \in \mathcal{B}_{nm}$. *Hint:* For (9.27) reduce to $I = [0, 1]$ and $|J| \le 1$ by scaling and then consider different cases depending on the size of $\operatorname{dist}(0, I)$.

9.4.4. $T(1)$ on the line

It is clear that we may choose any interval in place of $[0, 1]$. Not only does the proof of Theorem 9.8 apply but one may also pass to other intervals by translation and rescaling. In this way we obtain the following corollary. For the sake of simplicity we formulate it on a dyadic interval I_0, say of the form $I_0 = [0, 2^n)$ for some integer $n \in \mathbb{Z}$, since the Haar functions h_I for $I \subset I_0$ dyadic then agree with our previous definition.

Corollary 9.14 *Let T be a singular integral operator and assume that $T\chi_{I_0} \in \mathrm{BMO}(I_0)$ and $T^*\chi_{I_0} \in \mathrm{BMO}(I_0)$, where I_0 is some dyadic interval. Furthermore, assume that $\|Th_I\|_2 + \|T^*h_I\|_2 \le 1$ for all Haar functions h_I with $I \subset I_0$. Then T is bounded on $L^2(I_0)$ with an operator norm that depends only on the* BMO *norms involved, not on the size of the interval I_0.*

More care is required in the case of the whole line $\mathbb{R}$. However, again essentially the same proof applies. To begin, we now consider *all* dyadic intervals on the line, starting from $\{[n, n+1) \mid n \in \mathbb{Z}\}$ and those obtained from this by the bisection procedure. On a larger scale, we combine nearest neighbors by agreeing on the convention that $[0, 1)$ has $[0, 2)$ as parent, which in turn has $[0, 4)$ as parent, etc. The Haar functions are now defined exactly as before and, for any $n \in \mathbb{Z}$, the $\mathcal{A}_n$ are all dyadic intervals of size 2^{-n}.

Given any singular integral operator T on the line, as in Definition 9.6, we say that $T1 \in \mathrm{BMO}(\mathbb{R})$ if and only if there exists a constant M such that

$$|\langle 1, T^*a\rangle| \leq M \quad \text{for all smooth dyadic } \mathcal{H}^1(\mathbb{R}) \text{ atoms } a.$$

We also impose the condition $T^*a \in L^1(\mathbb{R})$ for all smooth atoms a, where the latter are defined exactly as in Definition 8.25 but for any dyadic interval $I \subset \mathbb{R}$. Similarly, one defines $T^*1 \in \mathrm{BMO}(\mathbb{R})$.

Then the full-line version of Theorem 9.8 reads as follows.

Theorem 9.15 *Let T be a singular integral operator and assume that $T1 \in \mathrm{BMO}(\mathbb{R})$ and $T^*1 \in \mathrm{BMO}(\mathbb{R})$. Furthermore, assume that $\|Th_I\|_2 + \|T^*h_I\|_2 \leq 1$ for all Haar functions h_I on the line. Then T is bounded on $L^2(\mathbb{R})$ with an operator norm that depends only on the* BMO *norms involved.*

As far as the formalism is concerned, we remark that (9.23) is replaced by the identity, valid for all $f \in L^2(\mathbb{R})$,

$$f = \sum_{n\in\mathbb{Z}} \Delta_n f,$$

where

$$\Delta_n f = \sum_{I\in\mathcal{A}_n} \langle f, h_I\rangle h_I,$$

where the sums converge in the sense of $L^2(\mathbb{R})$.

Exercise 9.4 Adapt the proof of Theorem 9.8, to be found in subsection 9.4.3, to Theorem 9.15.

In general dimensions the proof of Theorem 9.15 does not apply, since it relies on the Haar system. Nevertheless, the standard proof for the Euclidean case exhibits many similarities to the Haar function argument in subsection 9.4.3. Indeed, the Haar expansion is replaced by a continuous Littlewood–Paley decomposition and paraproducts take a similar form to that before. The Carleson condition (8.35) is replaced by the notion of a Carleson measure. While we will not give a detailed proof of the $T(1)$ theorem in dimensions 2 and higher, we now introduce the main notions involved in it.

9.5. Carleson measures, BMO, and $T(1)$

This section discusses a connection between BMO (nondyadic, as defined in Chapter 7) and Littlewood–Paley theory. This will give us the opportunity to introduce some notions that historically have played an important role in the development of the theory, such as the *Carleson measure* and also the *Calderón reproducing formula*, which can be seen as a continuum version of the discrete Littlewood–Paley decomposition.

9.5.1. Carleson measures

Definition 9.16 Let μ be a (complex) Borel measure on the upper half-space $\mathbb{R}^{d+1}_+$. Denote by I a general dyadic cube in $\mathbb{R}^d$ and by $Q(I)$ the cube in $\mathbb{R}^{d+1}_+ = \mathbb{R}^d \times (0, \infty)$ with base I and of height $\ell(I)$, where ℓ is the side length of I. In other words, $Q(I) := \{(x,t) \mid x \in I,\ 0 < t < \ell(I)\}$. One says that μ is a Carleson measure if

$$\|\mu\|_C := \sup_I \frac{|\mu|(Q(I))}{|I|} < \infty.$$

Equivalently, we may formulate this condition for all Euclidean balls B.

The main result for Carleson measures links them to the nontangential maximal function F^*, which is defined as follows. Given a function $F(x,t)$ defined on the upper half-space $\mathbb{R}^{d+1}_+$, we set

$$F^*(x) := \sup_{|x-y|<t} |F(y,t)|.$$

Further, we define the *tent* $T(B)$ over any ball $B \subset \mathbb{R}^d$ as

$$T(B) := \left\{(y,t) \in \mathbb{R}^{d+1}_+ \big| |x-y| < r-t\right\}, \quad B = B(x,r).$$

We immediately see that the Carleson condition on a measure μ is equivalent to the following estimate:

$$\mu(T(B)) \leq C|B| \quad \forall B \subset \mathbb{R}^d,$$

where B is a ball; see Figure 9.6.

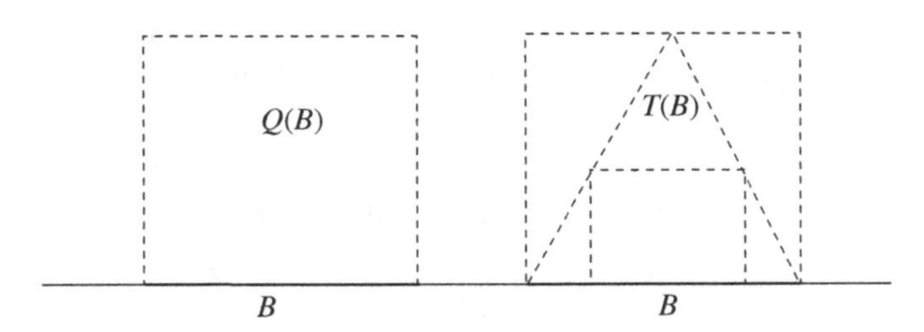

Figure 9.6. Carleson cube $Q(B)$ and tent $T(B)$.

Lemma 9.17 *For any measurable function F defined on $\mathbb{R}^{d+1}_+$, one has, for every $1 \le p < \infty$,*

$$\int_{\mathbb{R}^{d+1}_+} |F(x,t)|^p \mu(d(x,t)) \le C\|\mu\|_C \int_{\mathbb{R}^d} |F^*(x)|^p \, dx \tag{9.33}$$

provided the right-hand side is finite. The constant C is absolute.

Proof It suffices to consider $F \ge 0$ of compact support and bounded. The general case then follows on cutting F off and invoking monotone convergence. Suppose that $F(x,t) > \lambda$ for some $x \in \mathbb{R}^d$ and $t > 0$. Then $F^*(y) > \lambda$ for all $|y-x| < t$. This implies that, for any compact set $K \subset \{F(x,t) > \lambda\}$, one has

$$K \subset \bigcup_{B \subset E} T(B), \qquad E := \{F^* > \lambda\},$$

where B is a ball. By the compactness of K a finite cover suffices. By Wiener's covering theorem we can pass to a finite subset of pairwise disjoint balls $B_j \subset E$ such that

$$K \subset \bigcup_j T(3B_j).$$

In summary,

$$\mu(\{F > \lambda\}) \le \sum_j \mu(T(3B_j)) \le C\|\mu\|_C \sum_j |B_j| \le C\|\mu\|_C \left|\{F^* > \lambda\}\right|.$$

Integrating this estimate over $\lambda^{p-1}\, d\lambda$ yields (9.33). □

As an immediate consequence of Lemma 9.17 and the boundedness of the maximal operator we arrive at the following statement, which will replace Proposition 9.11 as the main tool that leads to the L^p-boundedness of paraproducts.

Corollary 9.18 *Fix $\phi \in \mathcal{S}(\mathbb{R}^d)$ and define $\phi_t(x) := t^{-d}\phi(x/t)$ for all $t > 0$. Then, for any Carleson measure μ and $1 < p < \infty$, one has*

$$\int_0^\infty \int_{\mathbb{R}^d} |\phi_t * f|^p(x)\,\mu(d(x,t)) \le C(d,p,\phi)\|\mu\|_{\mathcal{C}} \int_{\mathbb{R}^d} |f|^p(x)\,dx$$

for all $f \in \mathcal{S}(\mathbb{R}^d)$.

9.5.2. Bounded mean oscillation space and Carleson measures

The following result is a continuum version of the decomposition of a BMO function relative to the Haar system. Indeed, the functions b_I below play the role of the Haar functions, and the discrete Carleson condition (8.35) is replaced by the Carleson measure property (v) in Proposition 9.19 below.

Proposition 9.19 *Let $f \in \mathrm{BMO}(\mathbb{R}^d)$ with $\|f\|_{\mathrm{BMO}} \le 1$ have compact support and mean zero. There exist functions $\{b_I\}_I$, indexed by the dyadic cubes in $\mathbb{R}^d$, and scalars λ_I such that*

(i) $f = \sum_I \lambda_I b_I$,
(ii) $\mathrm{supp}(b_I) \subset 3I$,
(iii) $\int_{\mathbb{R}^d} b_I(x)\,dx = 0$,
(iv) $\|\nabla b_I\|_\infty \le \ell(I)^{-1}$,
(v) $\sum_{I \subset J} |\lambda_I|^2 |I| \le |J|$ *for all dyadic cubes J.*

The final property implies that $\sum_I |\lambda_I|^2 |I| \delta_{(x_I, \ell(I))}$ is a Carleson measure, where x_I is the center of I.

Proof Let $\varphi \in \mathcal{S}(\mathbb{R}^d)$ be real-valued, radial, and such that $\mathrm{supp}(\varphi) \subset B(0,1)$ with

$$\int_{\mathbb{R}^d} \hat{\varphi}(t\xi)^2 \,\frac{dt}{t} = 1 \quad \forall \xi \ne 0. \tag{9.34}$$

Note that the left-hand side is a radial function and, by a scaling argument, does not even depend on $|\xi|$. In other words, it does not depend on the choice of $\xi \ne 0$. The only requirement is that $\hat{\varphi}(0) = 0$ or, equivalently, $\int \varphi = 0$. Moreover, any $\varphi \not\equiv 0$ that satisfies this condition can be normalized in such a way as to fulfill (9.34).

Clearly, (9.34) is a continuum analogue of the discrete partition of unity in Lemma 8.1. The Calderón reproducing formula now becomes the following statement, which is implied by (9.34) and by $\int f = 0$:

$$\begin{aligned} f(x) &= \int_0^\infty (\varphi_t * \varphi_t * f)(x)\,\frac{dt}{t} \\ &= \sum_I \int_{T(I)} \varphi_t(x-y)(\varphi_t * f)(y)\,\frac{dt}{t} dy = \sum_I \tilde{b}_I(x), \end{aligned} \tag{9.35}$$

where $T(I) := \{(x,t)|x \in I,\ t \in (\ell(I)/2, \ell(I))\}$ and $\varphi_t(x) = t^{-d}\varphi(x/t)$. By construction, $\int \tilde{b}_I = 0$ and $\mathrm{supp}(\tilde{b}_I) \subset 3I$. Moreover, if we define, with a suitable constant C,

$$\lambda_I := C\left(|I|^{-1}\int_{T(I)} |\varphi_t * f(y)|^2 t^{-1}\,dtdy\right)^{1/2}$$

then one can check by means of the Cauchy–Schwarz inequality that

$$|\nabla\tilde{b}_I(x)| = \left|\int_{T(I)} \nabla\varphi_t(x-y)(\varphi_t * f)(y)\,\frac{dt}{t}dy\right| \le \lambda_I\,\ell(I)^{-1}.$$

Define $b_I := \lambda_I^{-1}\tilde{b}_I$. We have verified all properties up to the Carleson measure. To obtain this final claim we first show that, for any $f \in \mathrm{BMO}(\mathbb{R}^d)$ and any (dyadic) cube $I \subset \mathbb{R}^d$,

$$\int_{Q(I)} |(\varphi_t * f)(y)|^2\,t^{-1}\,dtdy \le C\|f\|_{\mathrm{BMO}}^2 |I|. \tag{9.36}$$

By the scaling and translation symmetries we may assume that I is the cube $[-1,1]^d$. Define $I^* := [-2,2]^d$ and split f as follows:

$$f = f_{I^*} + (f - f_{I^*})\chi_{I^*} + (f - f_{I^*})\chi_{\mathbb{R}^d\setminus I^*} =: f_1 + f_2 + f_3.$$

The constant f_1 does not contribute to (9.36), and f_2 is estimated by

$$\int_{\mathbb{R}^{d+1}_+} |(\varphi_t * f_2)(y)|^2\,t^{-1}\,dtdy = \|f_2\|_2^2 \le C\|f\|_{\mathrm{BMO}}^2,$$

as can be seen by passing to Fourier transforms and invoking the bound

$$|\hat{\varphi}(\xi)| \le C_N \min(|\xi|, |\xi|^{-N}) \qquad \forall \xi \in \mathbb{R}^d.$$

Finally,

$$\begin{aligned}
&\int_{Q(I)} |(\varphi_t * f_3)(x)|^2\,t^{-1}\,dtdx \\
&\le \int_0^2 \frac{dt}{t}\int_{(-1,1)^d} dx\,\left|\int_{\mathbb{R}^d\setminus(-2,2)^d} (f(y) - f_{I^*})\,t^{-d}\varphi\left(\frac{x-y}{t}\right)dy\right|^2 \\
&\le C\int_0^2 \frac{dt}{t}\int_{(-1,1)^d} dx\int_{\mathbb{R}^d\setminus(-2,2)^d} |f(y) - f_{I^*}|^2\,t^{-d}\left|\varphi\left(\frac{x-y}{t}\right)\right| dy \\
&\le C\int_{\mathbb{R}^d} \frac{|f(y) - f_{I^*}|^2}{1+|y|^{d+1}}\,dy \le C\|f\|_{\mathrm{BMO}}^2,
\end{aligned}$$

as desired. For the final step see Problem 7.5. Hence (9.36) holds and therefore

$$\sum_{J\subset I} |\lambda_J|^2 |J| = \sum_{J\subset I} \int_{T(J)} |(\varphi_t * f)(x)|^2 \, t^{-1} \, dt dx$$
$$\le C \int_{Q(I)} |(\varphi_t * f)(x)|^2 \, t^{-1} \, dt dx \le C \|f\|_{\mathrm{BMO}}^2 |I|;$$

thus, we are done. □

Exercise 9.5 Give a detailed proof of (9.35).

Another closely related statement, which is proved in essentially the same way as Proposition 9.19(v), reads as follows.

Proposition 9.20 *For any $\varphi \in \mathcal{S}(\mathbb{R}^d)$ with $\int \varphi = 0$ one has the following property: for any $b \in \mathrm{BMO}(\mathbb{R}^d)$, the measure*

$$\mu(dt, dx) := |\varphi_t * b|^2(x) \, dx dt / t$$

on $\mathbb{R}^{d+1}_+$ is a Carleson measure with $\|\mu\|_C \le C(d, \varphi) \|b\|^2_{\mathrm{BMO}}$.

In conjunction with Corollary 9.18 we arrive at the following conclusion. For any $b \in \mathrm{BMO}(\mathbb{R}^d)$, $f \in L^p(\mathbb{R}^d)$, one has

$$\int_0^\infty \int_{\mathbb{R}^d} |\phi_t * f|^p(x) |\varphi_t * b|^2(x) \frac{dx dt}{t} \le C \|b\|^2_{\mathrm{BMO}} \|f\|_p^p, \qquad (9.37)$$

where the constant $C = C(p, d, \phi, \varphi)$, $1 < p < \infty$. Here $\varphi, \phi \in \mathcal{S}(\mathbb{R}^d)$ and $\int \varphi = 0$.

9.5.3. Paraproducts and $T(1)$

In analogy with the paraproducts above we define, at least formally,

$$\pi_b(f) := \int_0^\infty Q_t(P_t f \cdot Q_t b) \, \frac{dt}{t}$$

where $P_t f = \phi_t * f$ and $Q_t b := \varphi_t * b$, with $\phi, \varphi \in \mathcal{S}$ as in (9.37). Then (9.37) shows that $\|\pi_b(f)\|_2 \le C \|b\|_{\mathrm{BMO}} \|f\|_2$ in analogy with Proposition 9.13 above. Alternatively, we may define

$$\pi_b(f) = \sum_{j\in\mathbb{Z}} \widetilde{P}_j (P_j b \cdot P_{\le j-C} f),$$

where P_j is the Littlewood–Paley projector, as usual, $P_{\le j} := \sum_{k\le j} P_k$, and $\widetilde{P}_j \circ P_j = P_j$. This, too, satisfies the desired L^2 bound.

As in Theorem 9.15, we require $T(1), T^*(1) \in \mathrm{BMO}(\mathbb{R}^d)$, which can be defined via $H^1(\mathbb{R}^d)$ atoms as in the case of the line. Indeed, we require that

$Ta \in L^1(\mathbb{R}^d)$ and $T^*a \in L^1(\mathbb{R}^d)$ for all such atoms and that

$$|\langle 1, Ta\rangle| + |\langle 1, T^*a\rangle| \le C \quad \text{for all atoms } a \text{ in } H^1,$$

where C is some fixed constant. In addition, we impose the following *weak L^2-boundedness condition* (WBC), which replaces the Haar function condition $\|Th_I\|_2 + \|T^*h_I\|_2 \le 1$ in Theorem 9.8. To formulate the WBC, we let B_N be the class of $C^\infty(\mathbb{R}^d)$ functions f supported in the unit ball that satisfy $\|f\|_{C^N} \le 1$ for some fixed integer N. The WBC requires that there exist N, C such that, for all $\phi, \psi \in B_N$ and all $x \in \mathbb{R}^d$, $t > 0$, one has

$$|\langle T\phi^{x,t}, \psi^{x,t}\rangle| \le Ct^d,$$

where $\phi^{x,t}(y) := \phi((x-y)/t)$, $\psi^{x,t}(y) := \psi((x-y)/t)$. This condition is also necessary for the L^2-boundedness.

The $T(1)$ theorem in $\mathbb{R}^d$ is given by the following statement. Under a singular integral we mean the same object as in $d = 1$ but with (9.13) replaced by

$$|K(x,y)| \le |x-y|^{-d} \quad \forall x \ne y,$$

$$|K(x,y) - K(x',y)| \le \frac{|x-x'|^\delta}{|x-y|^{d+\delta}} \quad \forall\, |x-y| > 2|x-x'|,$$

$$|K(x,y) - K(x,y')| \le \frac{|y-y'|^\delta}{|x-y|^{d+\delta}} \quad \forall\, |x-y| > 2|y-y'|.$$

As before, $0 < \delta \le 1$ is arbitrary but fixed.

Theorem 9.21 *Let T be a singular integral operator that obeys the WBC as well as $T1$, $T^*1 \in$ BMO. Then T is a bounded operator on $L^2(\mathbb{R}^d)$.*

The proof proceeds along lines similar to those in the proof of Theorem 9.8. Indeed, by means of $S = T - \pi_{b_1} - \pi^*_{b_2}$ with $b_1 = T1$, $b_2 = T^*1$, we can reduce matters to $b_1 = b_2 = 0$. The new twist here, as compared with the Haar function proof, is that the paraproducts π_{b_1}, π_{b_2} are themselves singular integral operators. The proof then proceeds as in that presented in subsection 9.4.3 for Theorem 9.8. The WBC arises in the representation of functions according to a Littlewood–Paley decomposition; see for example (9.35).

Notes

For more on the method of almost orthogonality, in particular the use of Cotlar's lemma, see Stein [111]. The original Cotlar's reference is [26], where Cotlar's lemma was stated for commuting operators. The version as it appears here is due to Knapp and Stein [68]. Problem 9.7 states a continuum version of Cotlar's lemma, which is from Calderón [14].

Hardy's inequalities are basic tools that have been generalized in many different directions; see for example Kufner and Opic [89]. The problems at the end of this chapter develop this topic further. Proposition 9.19 is from Uchiyama [124].

Paraproducts first appeared, at least implicity, in the work of Calderón (see [10] in Vol. II). See also the paper of Coifman and Meyer ([24] in Vol. II), in which certain bilinear singular integrals related to paraproducts were studied. Paraproducts appeared also in the work of Pommerenke [91] and were developed systematically in a nonlinear PDE context by Bony [9] under the name paradifferential calculus. This theory has reached a high level of sophistication, as evidenced for example by Taylor's books [121, 122].

The $T(1)$ theorem is due to David and Journé [27]. Meyer [82] gave a proof based on the Franklin wavelet, and his argument is similar to that leading to Theorem 9.8. The adaptation of Meyer's wavelet proof to the (discontinuous) Haar system is essentially the proof given by Katz in his lecture notes [64]. See also Coifman, Jones, and Semmes [23] for the use of the Haar system in connection with an elementary proof of the boundedness of the Cauchy integral on Lipschitz curves, which has similar features. The use of Haar functions arguably does not simplify the traditional proof of $T(1)$ based on Littlewood–Paley decompositions. However, we do feel that the Haar function version requires very little in terms of preparation and leads to both a clean statement and a transparent proof without any loss in the strength of the result.

The books by Stein [111] and Duoandikoetxea [31] contain detailed presentations of the $T(1)$ theorem in an $\mathbb{R}^d$ setting, and omitted details can be found there. The $T(1)$ theorem has a variant, which is essential for applications, called the $T(b)$ theorem. See Christ's book [21] for more on this topic, as well as many applications to such problems as the Cauchy integral on Lipschitz curves and Calderón's commutators. We shall study the latter two areas in great detail in the second volume of this book, but without making use of the $T(1)$ or $T(b)$ theorems. Generally speaking, the type of dyadic harmonic analysis developed in Sections 8.4 and 9.4 will play a prominent role in the second volume. For a recent, as well as unifying, point of view on $T(b)$, Carleson measures, etc., see Auscher, Hofmann, Muscalu *et al.* [4].

Problems

Problem 9.1 Let T be of the form (9.2), where $a \in C^\infty(\mathbb{R}^d \times \mathbb{R}^d)$ is such that

$$\sup_{x\in\mathbb{R}^d} \left(\left|\partial_x^\alpha a(x,\xi)\right| + \left|\partial_\xi^\alpha a(x,\xi)\right|\right) \le B\langle\xi\rangle^s$$

for all $|\alpha| \le 2d+1$ and all $\xi \in \mathbb{R}^d$. Here $s \ge 0$ is arbitrary but fixed. Show that T is bounded from $H^s(\mathbb{R}^d)$ to $L^2(\mathbb{R}^d)$.

Problem 9.2 Suppose $K(x, y)$ is defined on $(0,\infty)^2$ and is homogeneous of degree -1, i.e., $K(\lambda x, \lambda y) = \lambda^{-1}K(x, y)$ for all $\lambda > 0$. Further, assume that, for an arbitrary but fixed $1 \le p \le \infty$, one has

$$\int_0^\infty |K(1, y)|y^{-1/p}\,dy = A.$$

Show that $(Tf)(x) := \int_0^\infty K(x, y) f(y) dy$ satisfies $\|Tf\|_p \le A\|f\|_p$. Apply this to $K(x, y) = 1/(x + y)$. For which value of p is

$$\int_0^\infty \frac{f(y)}{x + y} dy =: (Tf)(x)$$

an L^p-bounded operator?

Problem 9.3 Prove Hardy's original inequalities, valid for every measurable $f \ge 0$:

$$\int_0^\infty \left(\int_0^x f(t) dt\right)^p x^{-r-1} dx \le \left(\frac{p}{r}\right)^p \int_0^\infty (xf(x))^p x^{-r-1} dx,$$

$$\int_0^\infty \left(\int_x^\infty f(t) dt\right)^p x^{r-1} dx \le \left(\frac{p}{r}\right)^p \int_0^\infty (xf(x))^p x^{r-1} dx$$

for every $1 \le p < \infty$, $r > 0$. Equality holds if and only if $f = 0$ almost everywhere. *Hint:* Use Problem 9.2.

Problem 9.4 Show the following Hardy's inequality in $\mathbb{R}^d$:

$$\int_{\mathbb{R}^d} |\nabla u(x)|^p dx \ge \left(\frac{|d - p|}{p}\right)^p \int_{\mathbb{R}^d} \frac{|u(x)|^p}{|x|^p} dx \tag{9.38}$$

for all $u \in \mathcal{S}(\mathbb{R}^d)$, if $1 \le p < d$, and for all such u that also vanish on a neighborhood of the origin if $\infty > p > d$. The constants here are sharp. *Hint:* Use Problem 9.3.

Problem 9.5 Show that if $1 \le p < d$ and

$$\frac{1}{p} - \frac{1}{p^*} = \frac{1}{d}$$

then

$$\int_{\mathbb{R}^d} \frac{|u(x)|^p}{|x|^p} dx = C(p, d)\|u\|_{p^*}^p,$$

provided that $u \in \mathcal{S}(\mathbb{R}^d)$ is radial and decreasing. Conclude that for such functions the Sobolev embedding inequality $\|u\|_{p^*} \le C(p, d)\|\nabla u\|_p$ holds. By means of the device of *nonincreasing rearrangement* this special case implies the general case. In other words, Sobolev embedding holds without the radial-decrease assumption. The point here is that the rearrangement does not change $\|u\|_{p^*}$ and at most decreases $\|\nabla u\|_p$; see Lieb and Loss [77]. This is useful for finding the optimal constants; see Frank and Seiringer [42].

Problem 9.6 Prove the following more general form of Cotlar's lemma. Let (X, μ) be a measure space, and suppose that $\{T_x\}_{x \in X}$ is a measurable family of operators between Hilbert spaces H_1 and H_2. This means that $\langle T_x u, v\rangle$ is measurable for any $u \in H_1$ and $v \in H_2$. Assume further that

- $\sup_{x \in X} \|T_x\| < \infty$,
- $\|T_x^* T_y\| \le h(x, y)^2$ and $\|T_x T_y^*\| \le h(x, y)^2$, where $\int_X h(x, y) \mu(dx) \le A$ for all $y \in X$.

Then $T := \int_X T_x \, \mu(dx)$ is well defined and $\|T\| \le A$. That the integral over T_x is well defined has the following meaning: $\langle T_x u, v\rangle \in L^1(X, \mu)$ for all $u \in H_1, b \in H_2$, and $B(u, v) := \int_X \langle T_x u, v\rangle \, \mu(dx)$ is bounded, i.e., $|B(u, v)| \le C\|u\|\|v\|$. By the Riesz representation theorem there exists a unique bounded operator T with $B(u, v) = \langle Tu, v\rangle$, which we equate with the integral of T_x.

Problem 9.7 Using the ideas of Chapter 8 as well as the proof strategy of Theorem 9.5, try to prove the following Hardy's inequality on L^p, for $1 < p < d/s$:

$$\||x|^{-s} f\|_p \le C(s, d, p)\|f\|_{\dot{W}^{s,p}(\mathbb{R}^d)}.$$

For which values of p does this approach succeed? Now try to prove the inequality by "factoring" through L^q. This refers to a stronger bound, where the right-hand side is replaced by L^q, q being the Sobolev embedding exponent, cf. Problem 8.8. Note that it suffices to establish a weak-type (q, p) estimate.

Problem 9.8 Prove the following fractional Leibniz rule: for any $s \ge 0$ one has

$$\|fg\|_{\dot{H}^s(\mathbb{R}^d)} \le C(s, d) \left(\|f\|_{\dot{H}^s(\mathbb{R}^d)}\|g\|_\infty + \|f\|_\infty\|g\|_{\dot{H}^s(\mathbb{R}^d)}\right).$$

Now do the same with H^s instead of $\dot{H}^s$. Conclude that H^s for $s > d/2$ is an algebra. Does this hold for $s = d/2$? *Hint:* Write $fg = F_1 + F_2 + F_3$, where F_1, F_2 are paraproducts and F_3 is the "diagonal" portion of fg, which contains the contributions to fg for which f and g have comparable frequencies. See also the second volume, where this type of estimate is treated systematically and in much greater generality.

10

The uncertainty principle

10.1. Bernstein's bound and Heisenberg's uncertainty principle

10.1.1. Motivation

This chapter deals with various manifestations of the heuristic principle that *it is not possible for both f and its Fourier transform $\hat{f}$ the be localized on small sets*. We have already encountered a rigorous, albeit qualitative and rather weak, version of this principle: it is impossible for both $f \in L^2(\mathbb{R}^d)$ and $\hat{f}$ to be compactly supported. Here we are more interested in quantitative versions and, moreover, the tightness of these quantitative versions. What is meant by that can be seen from the Fourier transformation (4.1): let χ be a smooth bump function and A be an invertible affine transformation that takes the unit ball onto an ellipsoid

$$\mathcal{E} := \left\{ x \in \mathbb{R}^d \;\middle|\; \sum_{j=1}^{d} (x_j - y_j)^2 r_j^{-2} \le 1 \right\}$$

where the y_j are arbitrary constants and $r_j > 0$. In other words, assuming $y_j = 0$ for simplicity,

$$A = \operatorname{diag}(r_1, \ldots, r_d),$$

in standard Euclidean coordinates. Then $\chi_A := \chi \circ A^{-1}$ can be thought of as a smoothed out indicator function of $\mathcal{E}$. By (4.1), $\widehat{\chi_A}$ then essentially "lives" on the dual ellipsoid

$$\mathcal{E}^* := \left\{ \xi \in \mathbb{R}^d \;\middle|\; \sum_{j=1}^{d} \xi_j^2 r_j^2 \le 1 \right\}.$$

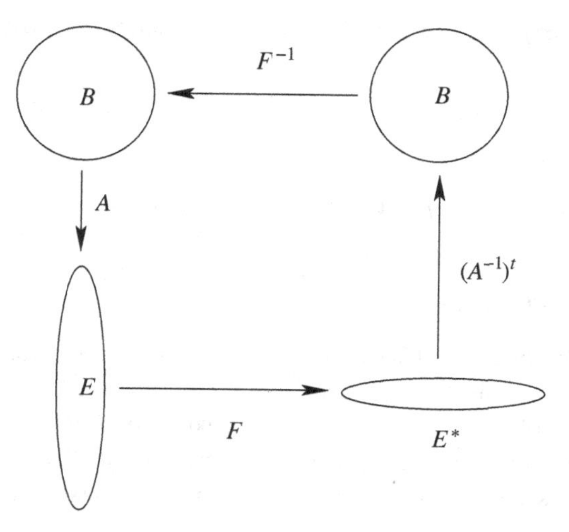

Figure 10.1. An ellipse E with its dual E^*.

Moreover, while χ_A is L^∞-normalized, $\widehat{\chi_A}$ is L^1-normalized. To "essentially live on" $\mathcal{E}^*$ means that

$$|\widehat{\chi_A}(\xi)| \le C_N |\mathcal{E}| \big(1 + |\xi|^2_{\mathcal{E}^*}\big)^{-N},$$

where

$$|\xi|^2_{\mathcal{E}^*} := \sum_{j=1}^{d} \xi_j^2 r_j^2 \quad \forall \xi$$

is the Euclidean norm relative to $\mathcal{E}^*$. The geometry here is depicted by Figure 10.1; the horizontal arrows correspond to Fourier transforms and B denotes the Euclidean balls.

What we can see from this is another, still heuristic but more quantitative, form of the uncertainty principle, namely the following: *if $f \in L^2(\mathbb{R}^d)$ is supported in a ball of size R then it is not possible for $\hat{f}$ to be "concentrated" on a scale much less than R^{-1}.*

10.1.2. Bernstein's bound

We begin a rigorous discussion with another form of Bernstein's estimate. In what follows $B(x_0, r)$ denotes the Euclidean ball in $\mathbb{R}^d$ that is of radius r and centered at x_0.

Lemma 10.1 *Suppose that $f \in L^2(\mathbb{R}^d)$ satisfies $\operatorname{supp}(f) \subset B(0,r)$. Then*

$$\|\partial^\alpha \hat{f}\|_2 \le (2\pi r)^{|\alpha|} \|f\|_2 \tag{10.1}$$

for all multi-indices α.

Proof This follows from $\hat{f} \in C^\infty(\mathbb{R}^d)$, the relation

$$\partial^\alpha \hat{f}(\xi) = (-2\pi i)^{|\alpha|} \widehat{x^\alpha f}(\xi),$$

and Plancherel's theorem. □

Heuristically speaking, this can be seen as a version of the nonconcentration statement from the end of the last subsection. Indeed, if $\hat{f}$ were to be concentrated on a scale of size $\ll r^{-1}$ then we would expect the left-hand side of (10.1) to be much larger than $r^{-|\alpha|}$. Later in this chapter we shall see a better reason why Bernstein's estimate should be viewed in the context of the uncertainty principle.

10.1.3. Heisenberg's inequality

Next we prove Heisenberg's uncertainty principle.

Proposition 10.2 *For any $f \in \mathcal{S}(\mathbb{R})$ one has*

$$\|f\|_2^2 \le 4\pi \, \|(x - x_0) f\|_2 \, \|(\xi - \xi_0)\hat{f}\|_2 \tag{10.2}$$

for all $x_0, \xi_0 \in \mathbb{R}$. This inequality is sharp, the extremizers being precisely given by the modulated Gaussians

$$f(x) = C e^{2\pi i \xi_0 x} e^{-\pi\delta(x-x_0)^2},$$

where C and $\delta > 0$ are arbitrary.

Proof Without loss of generality we set $x_0 = \xi_0 = 0$. Define

$$D = \frac{1}{2\pi i}\frac{d}{dx},$$

and let $(Xf)(x) = xf(x)$. Then we have the commutator

$$[D, X] = DX - XD = \frac{1}{2\pi i},$$

whence, for any $f \in \mathcal{S}(\mathbb{R})$,

$$\begin{aligned} \|f\|_2^2 &= 2\pi i \langle [D,X] f, f\rangle = 2\pi i(\langle Xf, Df\rangle - \langle Df, Xf\rangle) \\ &= 4\pi \operatorname{Im}\langle Df, Xf\rangle \le 4\pi \|Df\|_2 \, \|Xf\|_2, \end{aligned} \tag{10.3}$$

as claimed.

For the sharpness, note that if equality is achieved in (10.3) then from the condition for equality in the Cauchy–Schwarz relation one has $Df = \lambda Xf$ for some $\lambda \in \mathbb{C}$. Then $\langle Df, Xf\rangle = \lambda\|Xf\|_2^2$, which means that $\lambda = i\delta$ with some $\delta > 0$. Finally, this implies that

$$f'(x) = -2\pi\delta f(x), \quad f(x) = Ce^{-\pi\delta x^2}.$$

The general case follows by translation in x and translation in ξ (the latter being the same as multiplication by $e^{ix\xi_0}$). □

There is an analogous statement in higher dimensions, which we leave to the reader. We can see Proposition 10.2 as, once again, a manifestation of the principle that if f is concentrated on a scale $R > 0$ then $\hat{f}$ cannot be concentrated on a scale $\ll R^{-1}$. Hardy's inequality from Chapter 9 can be seen as another instance of this principle. Indeed, in $\mathbb{R}^3$ it implies that

$$\||x - x_0|^{-1} f\|_2 \le C\|\nabla f\|_2 = C\|\xi\hat{f}\|_2 \quad \forall f \in \mathcal{S}(\mathbb{R}^3) \tag{10.4}$$

for any $x_0 \in \mathbb{R}^3$. Thus, if $\hat{f}$ is supported on $B(0, R)$ then

$$\sup_{x_0\in\mathbb{R}^3} \|f\|_{L^2(B(x_0,\rho))} \le C\rho R\|f\|_2 \qquad 0 < \rho < R^{-1}.$$

Thus, if $f \neq 0$ then we cannot have $\rho R \ll 1$. Again, we leave it to the reader to derive similar statements in all dimensions. In fact, the proof of Hardy's inequality which we gave in Section 9.3 clearly shows the dominant role of the dual scales R in x and R^{-1} in ξ.

Let us reformulate (10.4) with $x_0 = 0$, as follows:

$$\langle |x|^{-2} f, f\rangle \le C\langle -\Delta f, f\rangle \quad \forall f \in \mathcal{S}(\mathbb{R}^3).$$

This can be rewritten as $-\Delta \ge c/|x|^2$ for some $c > 0$ as an operator expression (in other words, $-\Delta - c/|x|^2$ is a nonnegative symmetric operator).

It is interesting to note that we can now give a proof of this special case of Hardy's inequality using commutator arguments similar to those in the proof of Proposition 10.2. In addition, this will yield the optimal value of the constant.

Proposition 10.3 *One has*

$$-\Delta \ge \frac{(d-2)^2}{4|x|^2}$$

in $\mathbb{R}^d$ with $d \ge 3$, which means that, for any $f \in \mathcal{S}(\mathbb{R}^d)$,

$$\langle -\Delta f, f\rangle \ge \frac{(d-2)^2}{4}\langle |x|^{-2} f, f\rangle \iff \||x|^{-1} f\|_2 \le \frac{2}{d-2}\|\nabla f\|_2.$$

Proof Let $p = -i\nabla$. The relevant commutator here is, with $r = |x|$,

$$i[r^{-1}pr^{-1}, X] = dr^{-2}.$$

This implies that

$$d\|r^{-1}f\|_2^2 = 2\,\mathrm{Im}\,\langle r^{-1}pr^{-1}f, Xf\rangle \quad \forall f \in \mathcal{S}(\mathbb{R}^d).$$

Since

$$p|x|^{-1} = |x|^{-1}p + i\frac{x}{|x|^3},$$

one obtains

$$(d-2)\|r^{-1}f\|_2^2 = 2\,\mathrm{Im}\,\langle pf, Xr^{-2}f\rangle \quad \forall f \in \mathcal{S}(\mathbb{R}^d).$$

Application of the Cauchy–Schwarz inequality concludes the proof. □

The bound in the previous proposition is optimal but is not attained in the largest admissible class, i.e., $\dot{H}^1(\mathbb{R}^d)$. We shall not go further into this issue here.

10.2. The Amrein–Berthier theorem

Next, we investigate the following question: *if $E, F \subset \mathbb{R}^d$ are of finite measure, can there be a nonzero $f \in L^2(\mathbb{R}^d)$ with* $\mathrm{supp}(f) \subset E$ *and* $\mathrm{supp}(\hat{f}) \subset F$*?*

Note that if there were such an f then $Tf := \chi_E(\chi_F\hat{f})^\vee$ would satisfy $Tf = f$. In particular, this would imply that $\|T\|_{2\to 2} \geq 1$. We might expect that T would have a small norm if E and F are small, and this would lead to a contradiction and negative answer to our question. Now note that

$$(Tf)(x) = \int_{\mathbb{R}^d} \chi_E(x)\check{\chi}_F(x-y)f(y)\,dy.$$

In other words, the kernel of T is $K(x,y) = \chi_E(x)\check{\chi}_F(x-y)$ with Hilbert–Schmidt norm

$$\int_{\mathbb{R}^{2d}} |K(x,y)|^2\,dxdy = |E|\,|F| =: \sigma^2 < \infty.$$

We remark that σ is a scaling-invariant quantity. Hence, T is a compact operator and its L^2 operator norm satisfies $\|T\| \leq \min(\sigma, 1)$. So, if $\sigma < 1$ then we see that we cannot have $f \neq 0$ as in the question. Interestingly, the answer to the question is "no" in all cases and one has the following quantitative expression of this property.

Theorem 10.4 *Let E and F be sets of finite measure in $\mathbb{R}^d$. Then*

$$\|f\|_{L^2(\mathbb{R}^d)} \leq C(\|f\|_{L^2(E^c)} + \|\hat{f}\|_{L^2(F^c)}) \tag{10.5}$$

for some constant $C(E, F, d)$.

The reader will easily be able to prove the estimate in the case $\sigma < 1$. For the general case, we formulate the following equivalences with the goal of showing that $\|T\| < 1$ for any E, F as in the theorem. In the following lemma, "supp" refers to the essential support.

Lemma 10.5 *Let E, F be measurable subsets of $\mathbb{R}^d$. Then the following are equivalent (C_1 and C_2 denote constants):*

(i) $\|f\|_{L^2(\mathbb{R}^d)} \le C_1(\|f\|_{L^2(E^c)} + \|\hat{f}\|_{L^2(F^c)})$ *for all* $f \in L^2(\mathbb{R}^d)$;
(ii) *there exists* $\varepsilon > 0$ *such that* $\|f\|^2_{L^2(E)} + \|\hat{f}\|^2_{L^2(F)} \le (2-\varepsilon)\|f\|_2^2$ *for all* $f \in L^2(\mathbb{R}^d)$;
(iii) *if* $\mathrm{supp}(\hat{f}) \subset F$ *then* $\|f\|_2 \le C_2\|f\|_{L^2(E^c)}$;
(iv) *if* $\mathrm{supp}(f) \subset E$ *then* $\|\hat{f}\|_2 \le C_2\|\hat{f}\|_{L^2(F^c)}$;
(v) *there exists* $0 < \rho < 1$ *such that* $\|\chi_E(\chi_F\hat{f})^\vee\|_2 \le \rho\|f\|_2$ *for all* $f \in L^2(\mathbb{R}^d)$.

Proof (i) $\Longrightarrow$ (ii): one has

$$\begin{aligned}\|f\|^2_{L^2(E)} + \|\hat{f}\|^2_{L^2(F)} &= 2\|f\|_2^2 - \|f\|^2_{L^2(E^c)} - \|\hat{f}\|^2_{L^2(F^c)} \\ &\le (2-(2C_1^2)^{-1})\|f\|_2^2\end{aligned}$$

(ii) $\Longrightarrow$ (iii): for f as in (iii), $\|f\|^2_{L^2(E)} \le (1-\varepsilon)\|f\|_2^2$ or $\|f\|_2^2 \le \varepsilon^{-1}\|f\|^2_{L^2(E^c)}$.

(iii) $\Longrightarrow$ (i): write $f = P_F f + P_{F^c} f$ where $P_F f = (\chi_F\hat{f})^\vee$ and set $Q_{E^c} f = \chi_{E^c} f$. Then

$$\begin{aligned}\|f\|_2 &\le \|P_F f\|_2 + \|P_{F^c} f\|_2 \le C_2\|Q_{E^c}P_F f\|_2 + \|P_{F^c} f\|_2 \\ &\le C_2\|Q_{E^c} f\|_2 + C_2\|Q_{E^c}P_{F^c} f\|_2 + \|P_{F^c} f\|_2 \\ &\le C_2\|Q_{E^c} f\|_2 + (C_2+1)\|P_{F^c} f\|_2.\end{aligned}$$

Interchanging f and $\hat{f}$ one sees that property (iv) is equivalent to the first three.

(iii) $\Longrightarrow$ (v): with the projection P, Q as above, one has

$$\begin{aligned}\|P_F f\|_2^2 &= \|Q_E P_F f\|_2^2 + \|Q_{E^c} P_F f\|_2^2 \\ &\ge \|Q_E P_F f\|_2^2 + C_2^{-2}\|P_F f\|_2^2,\end{aligned}$$

which implies the desired bound with $\rho = (1 - C_2^{-2})^{1/2}$.

(v) $\Longrightarrow$ (iii): this is clear. □

Proof of Theorem 10.4 We have already observed that $T = Q_E P_F$ is compact with Hilbert–Schmidt norm $\|T\|_{\mathrm{HS}} \le \sigma$. In particular,

$$\dim\{f \in L^2(\mathbb{R}^d) \mid Tf = \lambda f\} \le \lambda^{-2}\sigma^2. \tag{10.6}$$

To prove this bound, let $\{f_j\}_{j=1}^m$ be an orthonormal sequence in the eigenspace on the left. Then, with K also denoting the kernel of K, one has

$$m\lambda^2 = \sum_j \left| \int_{\mathbb{R}^{2d}} K(x, y) f_j(x) \overline{f_j}(y)\, dx dy \right|^2 \leq \|K\|^2_{L^2(\mathbb{R}^{2d})}, \tag{10.7}$$

by Bessel's inequality. Furthermore, as a product of projections, the operator norm $\|T\| \leq 1$. If $\|T\| = 1$ then it follows from compactness of T that there exists $f \in L^2(\mathbb{R}^d)$ with $f \neq 0$ and $\|Tf\|_2 = \|f\|_2$.

Thus, there exists f with $\mathrm{supp}(f) \subset E$ and $\mathrm{supp}(\hat{f}) \subset F$ (as essential supports). We shall now obtain a contradiction to (10.6) for $\lambda = 1$. Inductively, define

$$S_0 := \mathrm{supp}(f), \quad S_1 := S_0 \cup (S_0 - x_0),$$
$$S_{k+1} := S_k \cup (S_0 - x_k), \quad k \geq 1,$$

where the translations x_k are chosen such that

$$|S_k| < |S_{k+1}| < |S_k| + 2^{-k}.$$

It follows that the collection $f_k := f(\cdot + x_k)$ with $k \geq 0$ is linearly independent, with

$$\mathrm{supp}(f_k) \subset S_\infty := \bigcup_{\ell=0}^{\infty} S_\ell \quad \forall k \geq 0$$

where $|S_\infty| < \infty$, while $\mathrm{supp}(\hat{f}_k) \subset F$ is maintained. This gives the desired contradiction. □

10.3. The Logvinenko–Sereda theorem

Next we formulate some results that provide further evidence of the nonconcentration property of functions with Fourier support in $B(0, 1)$.

10.3.1. A simple version

Proposition 10.6 *Let $\alpha > 0$ and suppose that $S \subset \mathbb{R}^d$ satisfies*

$$|S \cap B| < \alpha |B| \quad \textit{for all balls } B \textit{ of radius } 1.$$

If $f \in L^2(\mathbb{R}^d)$ satisfies $\mathrm{supp}(\hat{f}) \subset B(0, 1)$ *then*

$$\|f\|_{L^2(S)} \leq \delta(\alpha) \|f\|_2, \tag{10.8}$$

where $\delta(\alpha) \to 0$ as $\alpha \to 0$.

Proof Let ψ be a smooth compactly supported function with $\psi = 1$ on $B(0, 1)$. Define $Tg := \chi_S(\psi \hat{g})^\vee$. The kernel of this operator is

$$K(x, y) = \chi_S(x)\check{\psi}(x - y)$$

and satisfies the bound, for any $N \geq 1$,

$$|K(x, y)| \leq C_N \chi_S(x)(1 + |x - y|)^{-N}.$$

We now apply Schur's lemma to this kernel. To this end let $\varepsilon > 0$ be arbitrary but fixed. First, $\|K(x, \cdot)\|_{L^1_y} \leq C$. Second, take $N > d$ and choose R sufficiently large that

$$\sup_y \int_{[|x-y|>R]} C_N(1 + |x - y|)^{-N}\, dx \leq \varepsilon.$$

Then we see by covering $B(y, R)$ with finitely many balls of radius 1 that

$$\int_{[|x-y|<R]} C_N \chi_S(x)(1 + |x - y|)^{-N}\, dx \leq \varepsilon,$$

provided that $\alpha > 0$ is small enough. Therefore, by Schur's lemma 9.2 we conclude that $\|T\| \leq C\sqrt{\varepsilon}$. Hence if f is as in the assumption then $\chi_S f = Tf$, whence

$$\|f\|_{L^2(S)} \leq C\sqrt{\varepsilon}\|f\|_2$$

as claimed. □

10.3.2. A refined version

We now establish a somewhat more refined version, the *Logvinenko–Sereda theorem*, of the previous result.

Theorem 10.7 *Suppose that a measurable set $E \subset \mathbb{R}^d$ satisfies the following "thickness" condition: there exists $\gamma \in (0, 1)$ such that*

$$|E \cap B| > \gamma|B| \qquad \textit{for all balls } B \textit{ of radius } R^{-1},$$

where $R > 0$ is arbitrary but fixed. Assume that $\operatorname{supp}(\hat{f}) \subset B(0, R)$. *Then*

$$\|f\|_{L^2(\mathbb{R}^d)} \leq C\|f\|_{L^2(E)}, \tag{10.9}$$

where the constant C depends only on d and γ.

One can prove this result for γ near 1, using the same argument as in Proposition 10.6 (or directly from that proposition).

Exercise 10.1 Prove Theorem 10.7 if γ is close to 1.

We shall base the proof of Theorem 10.7 on an argument involving analytic functions of several complex variables. This is natural in view of the Paley–Wiener theorem, cf. Problem 4.5, which we shall state below. However, we shall not obtain explicit constants $C(d,\gamma)$ from this argument.

10.3.3. Some facts from the theory of several complex variables

In order to keep our discussion self-contained, we briefly review the few facts about analytic functions of several complex variables that will be needed. Let $\Omega \subset \mathbb{C}^d$ be open and connected (a "domain"), and suppose that $G \in C(\Omega)$ is complex-valued. We say that G is analytic in Ω if and only if

$$z_j \mapsto G(z_1, z_2, \ldots, z_{j-1}, z_j, z_{j+1}, \ldots, z_d)$$

is analytic on the corresponding sections of Ω, the z_k for $k \neq j$ being kept fixed. We denote this class by $\mathcal{A}(\Omega)$. The Cauchy integral formula applies via integration along contours in the individual variables. Examples of such functions are the power series

$$f(z) = \sum_\alpha a_\alpha (z - z_0)^\alpha, \quad z \in \Delta := \{z \mid \max_{1\le j\le d} |z_j - z_j^0| < M\},$$

provided that $|a_\alpha| \le M^{-|\alpha|}$ for all multi-indices α and a constant $M > 0$. Indeed, it is easy to see that the power series converges uniformly in Δ and thus the sum $f(z)$ is analytic in Δ. In particular, if

$$|a_\alpha| \le M \frac{B^{|\alpha|}}{\alpha!} \quad \forall \alpha$$

then the power series represents an entire function in $\mathbb{C}^d$.

Conversely, if $f \in \mathcal{A}(\Omega)$ then

$$f(z) = \sum_\alpha \frac{\partial_z^\alpha f(z_0)}{\alpha!} (z - z_0)^\alpha, \quad |z - z_0| < r$$

for some $r > 0$. Therefore the *uniqueness theorem* also holds in this setting: if $\partial_z^\alpha f(z_0) = 0$ for all α then $f \equiv 0$ in Ω. Moreover, if $\Omega \cap \mathbb{R}^d$ contains a set of positive measure on which f vanishes then $f \equiv 0$ in Ω. Owing to the Cauchy estimates, one has Montel's theorem on *normal families:* if $\mathcal{F} \subset \mathcal{A}(\Omega)$ is a locally bounded family then $\mathcal{F}$ is precompact, i.e., every sequence in $\mathcal{F}$ has a subsequence that converges locally uniformly on Ω to an element of $\mathcal{A}(\Omega)$. Finally, we recall the (easy half of the) Paley–Wiener theorem. Suppose that $f \in L^2(\mathbb{R}^d)$ satisfies $\mathrm{supp}(\hat{f}) \subset B(0,R)$. Then f extends to an entire function in $\mathbb{C}^d$ with exponential growth:

$$|f(z)| \le C\, R^{d/2}\, e^{2\pi|R||z|} \|f\|_{L^2(\mathbb{R}^d)}.$$

This can be proved by placing absolute values inside the integral defining f in terms of $\hat{f}$ and applying the Cauchy–Schwarz inequality.

10.3.4. The proof of the Logvinenko–Sereda theorem

Proof of Theorem 10.7 We may assume that $R = 1$. Let $A \geq 2$ be a constant that we shall fix later. Partition $\mathbb{R}^d$ into congruent cubes of side length 2, which we shall denote by Q. We say that one such Q is "good" if

$$\|\partial^\alpha f\|_{L^2(Q)} \leq A^{|\alpha|}\|f\|_{L^2(Q)} \quad \forall\alpha.$$

We now claim that

$$\|f\|_{L^2(\bigcup_{\text{bad}} Q)} \leq CA^{-1}\|f\|_2, \tag{10.10}$$

where the cubes Q that are not good are designated "bad". To prove (10.10), consider the sum

$$\sum_{\alpha\neq 0}\sum_{Q\text{ bad}} A^{-2|\alpha|}\|D^\alpha f\|^2_{L^2(Q)}. \tag{10.11}$$

Since here the cubes Q are bad, there is a lower bound $\|f\|^2_{L^2(\bigcup_{\text{bad}} Q)}$. However, by Lemma 10.1, one has an upper bound of the form

$$\begin{aligned}
&\sum_{\alpha\neq 0} A^{-2|\alpha|}(2\pi)^{|\alpha|}\|f\|_2^2 \\
&\quad \leq \sum_{k=1}^{\infty}(k+1)^d(2\pi A^{-2})^k\|f\|_2^2 \leq CA^{-2}\|f\|_2^2,
\end{aligned}$$

which proves the claim.

Next, we claim that if Q is a good cube then there exists $x_0 \in Q$ such that

$$|\partial^\alpha f(x_0)| \leq A^{2|\alpha|+1}\|f\|_{L^2(Q)}, \quad \forall\alpha. \tag{10.12}$$

Indeed, if (10.12) does not hold then

$$\begin{aligned}
A^2|Q|\|f\|^2_{L^2(Q)} &\leq \int_Q \sum_\alpha A^{-3|\alpha|}A^{4|\alpha|+2}\|f\|^2_{L^2(Q)}\,dx \\
&\leq \int_Q \sum_\alpha A^{-3|\alpha|}|\partial^\alpha f(x)|^2\,dx \\
&\leq \sum_\alpha A^{-3|\alpha|}A^{2|\alpha|}\|f\|^2_{L^2(Q)} \leq \sum_{k=0}^{\infty}(k+1)^d 2^{-k}\|f\|^2_{L^2(Q)},
\end{aligned}$$

which is a contradiction for large $A \geq 2$.

Finally, we claim that there exists $\eta > 0$ that depends only on d, γ, A such that

$$\|f\|_{L^2(E\cap Q)} \geq \eta\|f\|_{L^2(Q)} \qquad \text{for all good } Q. \tag{10.13}$$

If not, there exist $\{f_n\}_{n=1}^{\infty}$ in $L^2(\mathbb{R}^d)$ with $\text{supp}(\hat{f}_n) \subset B(0, 1)$, measurable sets $E_n \subset \mathbb{R}^d$ with

$$|E_n \cap B(x, 1)| \geq \gamma |B(x, 1)| \quad \forall x \in \mathbb{R}^d,$$

and good cubes Q_n such that

$$\|f_n\|_{L^2(Q_n)} = 1, \quad \eta_n := \|f_n\|_{L^2(E_n\cap Q_n)} \to 0$$

as $n \to \infty$. After translation, $Q_n = [-1, 1]^d =: Q_0$. Denote the entire extension of f_n given by the Paley–Wiener theorem by F_n. We expand F_n around a "good" x_n, as in (10.12), to conclude that for any $|z| \leq R$ one has

$$|F_n(z)| \leq \sum_{\alpha} \frac{A^{2|\alpha|+1}}{\alpha!} |(z - x_n)^{\alpha}|$$

$$\leq \left(\sum_{\ell=0}^{\infty} \frac{A^{2\ell+1}}{\ell!}(2 + |z|)^{\ell}\right)^d \leq C(A, d, R).$$

So on the one hand $\{F_n\}_{n=1}^{\infty}$ is a normal family in $\mathbb{C}^d$ and we may extract a locally uniform limit $F_n \to F_\infty$ (if necessary we may pass to a subsequence), whence $\|F_\infty\|_{L^2(Q_0)} = 1$. On the other hand,

$$\left|\left\{x \in Q_0 \cap E_n \;\middle|\; |f_n(x)| \geq \lambda_n\right\}\right| \leq \lambda_n^{-2}\|f_n\|^2_{L^2(E_n\cap Q_0)} = \lambda_n^{-2}\eta_n^2$$

$$\leq \frac{\gamma}{2}|Q_0|,$$

where

$$\eta_n = \lambda_n \sqrt{\frac{\gamma}{2}|Q_0|}.$$

In particular $\lambda_n \to 0$. By the thickness condition,

$$X_n := \left\{x \in Q_0 \cap E_n \;\middle|\; |f_n(x)| \leq \lambda_n\right\}$$

satisfies

$$|X_n| > \frac{\gamma}{2}|B(0, 1)|.$$

But then

$$X_\infty := \limsup_{n\to\infty} X_n$$

satisfies

$$|X_\infty| \geq \frac{\gamma}{2}|B(0,1)| > 0.$$

By construction $\lambda_n \to 0$, whence $F_\infty = 0$ on X_∞. By the uniqueness theorem for analytic functions of several variables, $F_\infty \equiv 0$ on Q_0.

This contradiction establishes our final claim. We can now finish the proof of the theorem. Indeed, one has from (10.13) and (10.10)

$$\begin{aligned}
\|f\|^2_{L^2(E\cap\bigcup_{Q \text{ good}} Q)} &= \sum_{Q \text{ good}} \|f\|^2_{L^2(E\cap Q)} \\
&\geq \sum_{Q \text{ good}} \eta^2 \|f\|^2_{L^2(Q)} = \eta^2 \|f\|^2_{L^2(\bigcup_{\text{good}} Q)} \\
&\geq \eta^2\big(\|f\|_2^2 - C^2 A^{-2}\|f\|_2^2\big) \\
&\geq \frac{\eta^2}{2}\|f\|_2^2,
\end{aligned}$$

if A is chosen large enough. This implies the desired estimate. □

We will now strengthen Theorem 10.7 in two ways. First, we do this by relaxing the thickness condition.

Exercise 10.2 Prove that Theorem 10.7 with $R = 1$ can also be proved under the following weaker hypothesis: there exist $\gamma > 0$ and $r > 0$ such that

$$|E \cap B(x,r)| > \gamma \quad \forall x \in \mathbb{R}^d.$$

The constant in (10.9) then of course depends also on r.

Second, we note that the use of L^2 in Theorem 10.7 is not essential. This deduction rests on the fact that Bernstein's inequality holds uniformly in $1 \leq p \leq \infty$.

Exercise 10.3 Generalize Theorem 10.7 to $L^p(\mathbb{R}^d)$ with $1 \leq p \leq \infty$.

A more difficult question concerns the nature of the constant in Theorem 10.7. Our indirect argument does not yield an effective constant but, using basic lower bounds on the modulus of analytic functions on a disk, this can indeed be done. This issue is explored further in the problems at the end of this chapter.

10.3.5. Negligible sets, and an equivalent version

We now formulate a second version of the Logvinenko–Sereda theorem. We begin with a definition.

Definition 10.8 A set $F \subset \mathbb{R}^d$ is called $B(0,1)$-negligible if and only if there exists $\varepsilon > 0$ such that for all $f \in L^2(\mathbb{R}^d)$ one has either

$$\|\hat{f}\|_{L^2(\mathbb{R}^d \setminus B(0,1))} \geq \varepsilon \|f\|_2 \quad \text{or} \quad \|f\|_{L^2(\mathbb{R}^d \setminus F)} \geq \varepsilon \|f\|_2. \tag{10.14}$$

We can characterize negligible sets by means of Theorem 10.7.

Theorem 10.9 *A set F is $B(0,1)$-negligible if and only if there exist $r > 0$ and $\beta \in (0,1)$ such that, for all $x \in \mathbb{R}^d$,*

$$|F \cap B(x,r)| < \beta |B(x,r)|. \tag{10.15}$$

Proof First we note that the condition (10.15) is equivalent to the following: there exist $r > 0$ and $\gamma > 0$ such that

$$|F^c \cap B(x,r)| > \gamma \quad \forall x \in \mathbb{R}^d. \tag{10.16}$$

If (10.16) fails then there exists $x_j \in \mathbb{R}^d$ and $r_j \to \infty$ with

$$|F^c \cap B(x_j, r_j)| \to 0, \quad j \to \infty.$$

Now take φ to be Schwartz with $\mathrm{supp}(\hat{\varphi}) \subset B(0,1)$ and set $\widehat{\varphi_j}(\xi) := e^{2\pi i x_j \cdot \xi} \hat{\varphi}$. Then $\widehat{\varphi_j}$ is still supported in $B(0,1)$ and, with $B_j := B(x_j, r_j)$, we have

$$\begin{aligned} \|\varphi_j\|_{L^2(\mathbb{R}^d \setminus F)} &\leq \|\varphi_j\|_{L^2(B_j \setminus F)} + \|\varphi_j\|_{L^2(\mathbb{R}^d \setminus B_j)} \\ &\leq \|\varphi\|_\infty \sqrt{|B_j \setminus F|} + \left(\int_{[|x - x_j| > r_j]} C|x - x_j|^{-d-1}\, dx \right)^{1/2}, \end{aligned} \tag{10.17}$$

which vanishes in the limit $j \to \infty$. But this contradicts (10.14), which finishes the proof of the necessity.

For the sufficiency, let $E := F^c$. Applying Theorem 10.7, see also Exercise 10.2, we infer that, for any $f \in L^2(\mathbb{R}^d)$ with $\mathrm{supp}(\hat{f}) \subset B(0,1)$, the following estimate holds:

$$\|f\|_2 \leq C \|f\|_{L^2(E)}$$

with $C = C(\gamma, r, d)$. Now set $f = f_1 + f_2$ with $f_1 = (\chi_{B(0,1)} \hat{f})^\vee$. If $\|f_2\|_2 \leq \varepsilon \|f\|_2$ then

$$\begin{aligned} \|f\|_{L^2(E)} &\geq \|f_1\|_{L^2(E)} - \|f_2\|_2 \\ &\geq C^{-1} \|f\|_2 - \varepsilon \|f\|_2 > \varepsilon \|f\|_2, \end{aligned}$$

provided that $\varepsilon > 0$ is sufficiently small, and we are done. □

10.4. Solvability of constant-coefficient linear PDEs

We shall now apply these uncertainty principle ideas to the local solvability of partial differential equations with constant coefficients. To be specific, suppose that we are given a polynomial

$$p(\xi) = \sum_{\alpha} a_\alpha \, \xi^\alpha .$$

Then with

$$D := \frac{1}{2\pi i} \partial_x$$

one has

$$p(D) := \sum_{\alpha} a_\alpha (2\pi i)^{-|\alpha|} \partial^\alpha .$$

This definition is chosen so that, for all $f \in \mathcal{S}(\mathbb{R}^d)$,

$$\widehat{p(D)f}(\xi) = p(\xi)\hat{f}(\xi) \quad \forall \xi \in \mathbb{R}^d .$$

The following theorem is known as the Malgrange–Ehrenpreis theorem.

10.4.1. The Malgrange–Ehrenpreis theorem

Theorem 10.10 *Let Ω be a bounded domain in $\mathbb{R}^d$ and let $p \neq 0$ be a polynomial. Then, for all $g \in L^2(\Omega)$, there exists $f \in L^2(\Omega)$ such that $p(D)f = g$ in the distributional sense.*

The "distributional sense" means that, for all $\varphi \in C^\infty(\Omega)$ with compact support $\mathrm{supp}(\varphi) \subset \Omega$, one has

$$(g, \varphi) = (f, \overline{p}(D)\varphi),$$

where $\overline{p}(\xi) := \sum_\alpha \overline{a_\alpha}\, \xi^\alpha$. The pairing $(\cdot, \cdot)$ is an L^2 pairing.

We emphasize that we are not making any ellipticity assumptions in the theorem. Even so, one can show that f can be taken to belong to $C^\infty(\Omega)$; see Problem 10.6 below. This is very different from the assertion that every solution f is C^∞, which is known as *hypoellipticity*. This is the subject of not only a large area of research in itself but also is false in the generality of the Malgrange–Ehrenpreis theorem.

The proof of Theorem 10.10 follows the common principle in functional analysis that *the uniqueness of solutions to the adjoint equation implies the existence of solutions to the equation itself.* Recall that this is the essence of Fredholm theory, which established this fact for compact perturbations of the identity operator. Of course, we cannot expect uniqueness in our case unless we choose boundary conditions, so we introduce a strong vanishing condition at the boundary.

Henceforth we take Ω to be a ball, which we may do since Ω is bounded. By dilation and translation symmetries the ball may furthermore be taken to be $B(0, 1)$.

Proposition 10.11 *One has the bound*

$$\|\overline{p}(D)\varphi\|_2 \geq C^{-1}\|\varphi\|_2,$$

for all $\varphi \in C^\infty(B(0, 1))$ that are compactly supported in $B(0, 1)$. The constant C depends on p and the dimension.

Proof of Theorem 10.10 We will now show how to deduce the theorem from the proposition. Let

$$X := \left\{\overline{p}(D)\varphi \;\middle|\; \varphi \in C^\infty_{\mathrm{comp}}(B(0, 1))\right\}$$

be viewed as a linear subspace of $L^2(B(0, 1))$. Define a linear functional ℓ on X by

$$\ell(\overline{p}(D)\varphi) = (g, \varphi),$$

where the right-hand side is an L^2-pairing. Proposition 10.11 implies that ℓ is well defined and, in fact, one has the bound

$$|\ell(\overline{p}(D)\varphi)| \leq \|g\|_2 \,\|\varphi\|_2 \leq C\|g\|_2 \,\|\overline{p}(D)\varphi\|_2.$$

By the Hahn–Banach theorem we may extend ℓ as a bounded linear functional to $L^2(B(0, 1))$ with the same norm. So, by the Riesz representation of such functionals, there exists $f \in L^2(B(0, 1))$ with the property that

$$\ell(\overline{p}(D)\varphi) = (f, \overline{p}(D)\varphi) = (g, \varphi)$$

for all φ as above, and this proves the theorem. □

The proof of Proposition 10.11 will be based on Theorem 10.7 and requires some preparatory lemmas on polynomials. The logic here is very simple: writing

$$\|\bar{p}(D)\varphi\|_2 = \|\bar{p}(\xi)\hat{\varphi}\|_2,$$

we see that the obstacle consists of functions $\hat{\varphi}$ that are highly localized around the sets where the polynomial p vanishes. However, φ is supported on the unit ball whence $\hat{\varphi}$ cannot be highly localized on small sets, by the uncertainty principle. So we need to verify that the set where $|p|$ is very small is negligible in the sense of the previous section. This is what we will now do.

10.4.2. A bound on polynomials

The following lemmas on level sets of polynomials are formulated so as to be invariant under translation. This is crucial for applications of the notion of negligible sets.

Lemma 10.12 *Let p be a nonzero polynomial in $\mathbb{R}^d$. There exists $\beta \in (0,1)$ that depends only on the degree of p and the dimension such that*

$$\left|\left\{x \in B \,\middle|\, 2|p(x)| \le \max_B |p|\right\}\right| < \beta |B|$$

for all balls B of radius 1.

Proof Denote the degree of p by N. The proof exploits the fact that any two norms on a finite-dimensional space are comparable (in our case the space of polynomials of degree at most N). Fix a unit ball B. Then there exists a constant $C(N)$ such that

$$\max_{x\in B}(|p(x)| + |\nabla p(x)|) \le C(N) \max_{x \in B} |p(x)|,$$

whence

$$\max_{x\in B} |\nabla p(x)| \le C(N) \max_{x\in B} |p(x)|.$$

By the mean-value theorem, we have

$$|p(x) - p(y)| \le \max_B |\nabla p|\, |x-y| \le C(N)|x-y| \max_B |p|,$$

from which we infer that

$$|p(y)| \ge |p(x_{\max})| - C(N) \max_B |p|\, |x_{\max} - y| \ge \tfrac{1}{2}|p(x_{\max})|$$

for any $y \in B(x_{\max}, (2C(N))^{-1}) \cap B$. In particular,

$$\left|\left\{y \in B \,\middle|\, 2|p(y)| \ge \max_B |p|\right\}\right| \ge c(N,d) > 0.$$

Finally, we may translate B without affecting any of the constants, and this concludes the proof. □

The following lemma is formulated to fit Theorem 10.9.

Lemma 10.13 *Let p be any nonzero polynomial in $\mathbb{R}^d$. Then there exists $\varepsilon_0 > 0$ depending on p such that*

$$\left|\left\{x \in B \,\middle|\, |p(x)| \le \varepsilon_0\right\}\right| \le \beta |B|$$

for all unit balls B. Here $\beta \in (0,1)$ is from Lemma 10.12.

Proof By Lemma 10.12 it suffices to show that

$$\inf_B \max_{x\in B} |p(x)| \ge \varepsilon_0,$$

where the infimum is taken over all unit balls. First, with N the degree of p,

$$\max_{B(0,1)} |p(x)| \ge C(N)^{-1} \sum_\alpha |a_\alpha|,$$

where a_α are the coefficients of p. Furthermore, one has

$$\max_B |p(x)| \geq C(N)^{-1} \sum_{|\alpha|=N} |a_\alpha| =: \varepsilon_0$$

uniformly in B. This follows from the previous inequality and the property that the highest-order coefficients are invariant under translation. This concludes the proof. □

It is now an easy matter to finish the proof of local solvability, more precisely that of the lower L^2 bound of Proposition 10.11.

Proof of Proposition 10.11 Lemma 10.13 guarantees that

$$F := \left\{x \in \mathbb{R}^d \,\middle|\, |p(x)| \leq \varepsilon_0\right\}$$

is $B(0,1)$-negligible. Thus, by Theorem 10.9,

$$\begin{aligned}\|\bar{p}(D)\varphi\|_2 &= \|\bar{p}(\xi)\hat{\varphi}\|_2 \geq \|\bar{p}\hat{\varphi}\|_{L^2(F^c)} \\ &\geq \varepsilon_0\|\hat{\varphi}\|_{L^2(F^c)} \geq \varepsilon_1\|\hat{\varphi}\|_{L^2(\mathbb{R}^d)} = \varepsilon_1\|\varphi\|_2,\end{aligned}$$

with some constant ε_1 that depends on p and d. □

There is no analogue of Theorem 10.10 for partial differential operators with nonconstant smooth coefficients. This is known as *Lewy's* example; see for example Folland [41]. Note the contrast with the Cauchy–Kowalewski theorem, which guarantees the solvability of the Cauchy problem for PDEs in the purely analytic category.

Notes

As already seen in the proof of Hardy's inequality in Chapter 9, uncertainty principle ideas are ubiquitous in harmonic analysis. More precisely, they are essential to any considerations involving phase space and in consequence arguments based on working with f and $\hat{f}$ simultaneously. The second volume is largely concerned with precisely this type of analysis.

The simple observations at the beginning of this chapter involving ellipses $\mathcal{E}$ and their duals $\mathcal{E}^*$ are relevant also to general normed finite-dimensional linear spaces, which we may take to be $\mathbb{R}^n$. There the unit ball takes the form $B = \{\|x\| \leq 1\}$ for some norm $\|\cdot\|$ that is a symmetric convex set and, conversely, any such set defines a norm of which it is the unit ball. The dual norm defines the dual unit ball B^*. Then there exists an ellipsoid $\mathcal{E} \subset B$ such that $B \subset \sqrt{n}\,\mathcal{E}$. One calls $\mathcal{E}$ the *John* ellipsoid, since it was discovered by Fritz John. See Hörmander [56, Lemma 1.4.3]. By means of the John ellipsoid we can apply the same type of considerations to smooth functions living on general symmetric convex sets as those discussed at the beginning of this chapter. A standard reference for the material in this chapter is the book of Havin and Jöricke [54]. Another basic resource in this area is Fefferman's classic survey [38]. The survey of Bonami and Demange [8] contains a more recent as well as different perspective of the uncertainty principle. An explicit constant in Theorem 10.4, of the form $e^{C|E||F|}$ in dimension 1, was obtained by Nazarov [87]; a result for higher dimensions was obtained by Jaming [60].

Kovrijkine's paper [72] gives effective, and in some cases optimal, constants in the Logvinenko–Sereda theorem. These results were achieved by the use of a quantitative lower bound on the size of analytic functions on a disk. In Problem 10.8 below we present one such standard device (it leads to the same bounds as those in [72]).

For the solvability of linear constant-coefficient PDEs, see Jerison [62].

Problems

Problem 10.1 Does there exist a nonzero function $f \in L^2(\mathbb{R}^d)$ such that we have both $f = 0$ and $\hat{f} = 0$ on nonempty open sets?

Problem 10.2 Does there exist a nonzero $f \in L^2(\mathbb{R})$ with $f = 0$ on $[-1, 1]$ and $\hat{f} = 0$ on a half-line? Can we have $f = 0$ on a set of positive measure and $\hat{f} = 0$ on a half-line? *Hint:* Consider the circle case, and relate functions of this type to a suitable nonvanishing theorem in Chapter 3.

Problem 10.3 Suppose that $E, F \subset \mathbb{R}^d$ have finite measure. Show that for any $g_1, g_2 \in L^2(\mathbb{R}^d)$ there exists $f \in L^2(\mathbb{R}^d)$ with

$$f = g_1 \text{ on } E, \qquad \hat{f} = g_2 \text{ on } F.$$

This can be seen as a statement that $f|_E$ and $\hat{f}|_F$ are independent.

Problem 10.4 Prove that, for $E, F \subset \mathbb{R}^d$ of finite measure, one has

$$\dim\{f \in L^2(\mathbb{R}^d) \mid f = 0 \text{ on } E, \hat{f} = 0 \text{ on } F\} = \infty.$$

Problem 10.5 There are several ways of proving Proposition 10.6. As an alternative to the Schur's lemma proof we gave, one can use Sobolev embedding on cubes together with Bernstein's theorem. In this problem we suggest yet another route.

(a) Prove Poincaré's inequality: for any $f \in H^1(\mathbb{R}^d)$ one has, for every $R > 0$,

$$\int_{B(0,R)} |f(x) - \langle f \rangle_R|^2 \, dx \le CR^2 \int_{B(0,R)} |\nabla f(x)|^2 \, dx$$

with an absolute constant. Here $\langle f \rangle_R$ is the average of f on $B(0, R)$.

(b) Combine this inequality with Bernstein's estimate as in Lemma 10.1 to give an independent proof of Proposition 10.6.

Problem 10.6 Show that f in Theorem 10.10 may be taken to belong to $C^\infty(\Omega)$. *Hint:* Use Sobolev spaces in the existence part, in addition to L^2.

Problem 10.7 Show Poincaré's inequality for general powers $1 \le p < \infty$, i.e., verify that, for any $f \in \mathcal{S}(\mathbb{R}^d)$, one has

$$\int_{B(0,R)} |f(x) - \langle f \rangle_R|^p \, dx \le C(p, d) R^p \int_{B(0,R)} |\nabla f(x)|^p \, dx.$$

Deduce from this that $\dot{W}^{1,d}(\mathbb{R}^d) \hookrightarrow \mathrm{BMO}(\mathbb{R}^d)$, where $\|f\|_{\dot{W}^{1,d}(\mathbb{R}^d)} = \|\nabla f\|_{L^d(\mathbb{R}^d)}$.

Problem 10.8 Prove the following bound on analytic functions of one variable. Assume that f is analytic in the disk $\{|z| \le 10\}$ and satisfies $|f(0)| = 1$ as well as $M \ge \sup_{|z|=10} |f(z)|$. Show that there exists an absolute constant C such that, for any $0 < \eta < 1$, one has

$$\log|f(z)| > -H(\eta)\log M, \quad H(\eta) := \log\left(\frac{C}{\eta}\right),$$

for all $z \in \{|z| \le 1\} \setminus \bigcup_j D_j$, where D_j are disks of radius r_j such that $\sum_j r_j \le \eta$. *Hint:* Apply the Riesz representation to $\log|f(z)|$, cf. Problem 3.2, and express both the sizes of μ and the harmonic function on $D(0, 2)$ in terms of $\log M$. Then use Cartan's estimate, see Problem 3.4, to bound the logarithmic potential from below and, for the harmonic function, use Harnack's inequality. See also Levin [75, p. 79].

Problem 10.9 Using Problem 10.8, give a direct proof of Theorem 10.7 with an effective constant. To be more specific, show the following. Let $E \subset \mathbb{R}$ be "thick", in the sense that $|E \cap I| \ge \gamma|I|$ for all intervals I of a fixed length $a > 0$. Here $\gamma > 0$ is fixed. Assume that $f \in L^p(\mathbb{R})$ with $1 \le p \le \infty$ satisfies $\mathrm{supp}(\hat{f}) \subset J$, where J is an interval of length b. Then, show that

$$\|f\|_{L^p(E)} \ge \left(\frac{\gamma}{C}\right)^{C(ab+1)} \|f\|_{L^p(\mathbb{R})},$$

where C is some absolute constant. Also show that this bound is optimal up to the choice of C. *Hint:* This is obtained by the same proof strategy as before, but the estimate in Problem 10.8 replaces the use of normal families. See Kovrijkine's paper [72] for details, as well as for an adaptation to higher dimensions and functions that are Fourier supported in more complicated sets (finite unions of intervals or parallelepipeds).

11

Fourier restriction and applications

11.1. The Tomas–Stein theorem

11.1.1. The restriction question

This chapter is based on the following question and its ramifications: is it possible to restrict the Fourier transform $\hat{f}$ of a function $f \in L^p(\mathbb{R}^d)$ with $1 \le p \le 2$ to the sphere S^{d-1} as a function in $L^q(S^{d-1})$ for some $1 \le q \le \infty$? By the uniform boundedness principle this is the same as asking about the validity of the inequality

$$\|\hat{f} \restriction S^{d-1}\|_{L^q(S^{d-1})} \le C\|f\|_{L^p(\mathbb{R}^d)} \tag{11.1}$$

(here the half-arrow indicates restriction to the set on its right), for all $f \in \mathcal{S}(\mathbb{R}^d)$, with a constant $C = C(d, p, q)$? As an example take $p = 1$, $q = \infty$, and $C = 1$. However, $p = 2$ is impossible as $\hat{f}$ possesses no more regularity than a general L^2-function, by Plancherel's theorem. During the 1960s Stein asked whether it is possible to find $1 < p < 2$ such that, for some finite q, one has the estimate (11.1).

We have encountered an estimate of this type in Chapter 4, under the name of a *trace estimate*. In fact, by Lemma 4.12 one has

$$\|\hat{f} \restriction S^{d-1}\|_2 \le C\|\langle x\rangle^\sigma f\|_2,$$

as long as $\sigma > \frac{1}{2}$ (some work is required to pass from the planar case to the curved case). However, (11.1) is translation invariant and therefore much more useful for applications to nonlinear partial differential equations. In addition, as we shall see, the answers to (11.1) crucially involve *curvature*, whereas the trace lemmas do not distinguish between flat and curved surfaces.

Exercise 11.1 Suppose that S is a bounded subset of a hyperplane in $\mathbb{R}^d$. Prove that if $\|\hat{f} \restriction S\|_1 \le C\|f\|_{L^p(\mathbb{R}^d)}$ for all $f \in \mathcal{S}(\mathbb{R}^d)$ then necessarily

$p = 1$, in other words, there cannot be a nontrivial restriction theorem for flat (affine) surfaces.

The *Tomas–Stein theorem*, Theorem 11.1, settles the important case $q = 2$ of Stein's question. For simplicity we state it for the sphere but it applies equally well to a bounded subset of any smooth hypersurface in $\mathbb{R}^d$ with nonvanishing Gaussian curvature.

Theorem 11.1 *For every dimension $d \geq 2$ there is a constant $C(d)$ such that, for all $f \in L^p(\mathbb{R}^d)$,*

$$\|\hat{f} \restriction S^{d-1}\|_{L^2(S^{d-1})} \leq C(d)\|f\|_{L^p(\mathbb{R}^d)}, \tag{11.2}$$

with

$$p \leq p_d := \frac{2d+2}{d+3}.$$

Moreover, this bound fails for $p > p_d$.

The left-hand side in (11.2) is

$$\left(\int_{S^{d-1}} |\hat{f}(w)|^2 \, \sigma(dw)\right)^{1/2}$$

where σ is the surface measure on S^{d-1}.

We shall prove this theorem below, after making some initial remarks. First, there is nothing special about the sphere. In fact, if S_0 is a compact subset of a hypersurface S with nonvanishing Gaussian curvature then

$$\|\hat{f} \restriction S_0\|_{L^2(S_0)} \leq C(d, S_0)\|f\|_{L^{p_d}(\mathbb{R}^d)} \tag{11.3}$$

for any $f \in \mathcal{S}(\mathbb{R}^d)$. For example, take the truncated paraboloid

$$S_0 := \left\{(\xi', |\xi'|^2) \,\middle|\, \xi' \in \mathbb{R}^{d-1}, |\xi'| \leq 1\right\},$$

which is the characteristic variety of the Schrödinger equation. However, (11.3) fails for

$$S_0 := \left\{(\xi', |\xi'|) \,\middle|\, \xi' \in \mathbb{R}^{d-1}, 1 \leq |\xi'| \leq 2\right\},$$

since this section of the cone has exactly one vanishing principal curvature, namely the one along a generator of the cone. This latter example is relevant to the wave equation, as the characteristic surface of that equation is a cone. We shall need to find an estimate corresponding to (11.3) for the wave equation case that takes the vanishing of one principal curvature into account. A much simpler remark concerns the range $1 \leq p \leq p_d$ in Theorem 11.1. For $p = 1$, one has

$$\|\hat{f} \restriction S^{d-1}\|_{L^2(S^{d-1})} \leq \|\hat{f} \restriction S^{d-1}\|_\infty |S^{d-1}|^{1/2} \leq \|f\|_{L^1(R^d)}|S^{d-1}|^{1/2}. \tag{11.4}$$

Hence, it suffices to prove Theorem 11.1 for $p = p_d$ since the range $1 \leq p < p_d$ follows by interpolation with (11.4). The Tomas–Stein theorem is more accessible than the general restriction problem with $q > 2$; in fact, we shall now observe that the restriction to $L^2(S^{d-1})$ on the left-hand side of (11.2) allows one to use duality in the proof.

11.1.2. Duality and equivalent formulation

To use duality, we need to identify the adjoint of the restriction operator

$$R : f \mapsto \hat{f} \restriction S^{d-1}.$$

Lemma 11.2 *For any finite measure μ in R^d and any $f, g \in \mathcal{S}(\mathbb{R}^d)$ one has the identity*

$$\int_{R^d} \hat{f}(\xi)\overline{\hat{g}}(\xi)\,\mu(d\xi) = \int_{R^d} f(x)\,(\overline{g} * \hat{\mu})(x)\,dx.$$

Proof We use the following elementary identity for tempered distributions. If μ is a finite measure and $\phi \in \mathcal{S}$ then

$$\widehat{\phi\mu} = \hat{\phi} * \hat{\mu}.$$

Therefore (and occasionally using $\mathcal{F}(f)$ synonymously with $\hat{f}$ for notational reasons),

$$\begin{aligned}\int_{R^d} \hat{f}(\xi)\overline{\hat{g}}(\xi)\,\mu(d\xi) &= \int_{R^d} f(x)\mathcal{F}\left(\overline{\hat{g}}\,\mu\right)(x)\,dx \\ &= \int_{R^d} f(x)\left(\widehat{\overline{\hat{g}}} * \hat{\mu}\right)(x)\,dx = \int_{R^d} f(x)\,(\overline{g} * \hat{\mu})(x)\,dx,\end{aligned}$$

since $\widehat{\overline{\hat{g}}} = \widehat{\overline{\check{g}}} = \overline{g}$. □

Lemma 11.3 *Let μ be a finite measure on $\mathbb{R}^d$, and $d \geq 2$. Then the following are equivalent:*

(i) $\|\widehat{f\mu}\|_{L^q(\mathbb{R}^d)} \leq C\|f\|_{L^2(\mu)}$ *for all* $f \in \mathcal{S}(\mathbb{R}^d)$;
(ii) $\|\hat{g}\|_{L^2(\mu)} \leq C\|g\|_{L^{q'}(\mathbb{R}^d)}$ *for all* $g \in \mathcal{S}(\mathbb{R}^d)$;
(iii) $\|\hat{\mu} * f\|_{L^q(\mathbb{R}^d)} \leq C^2\|f\|_{L^{q'}(\mathbb{R}^d)}$ *for all* $f \in \mathcal{S}(\mathbb{R}^d)$.

Proof By Lemma 11.2, for any $g \in \mathcal{S}(\mathbb{R}^d)$ we have

$$\begin{aligned}\|\hat{g}\|_{L^2(\mu)} &= \sup_{f\in\mathcal{S},\,\|f\|_{L^2(\mu)}=1} \left|\int_{\mathbb{R}^d} \hat{g}(\xi)f(\xi)\,\mu(d\xi)\right| \\ &= \sup_{f\in\mathcal{S},\,\|f\|_{L^2(\mu)}=1} \left|\int_{\mathbb{R}^d} g(x)\widehat{f\mu}(x)\,dx\right|. \qquad (11.5)\end{aligned}$$

Hence, if (i) holds then the right-hand side of (11.5) is no larger than $\|g\|_{L^{q'}(\mathbb{R}^d)}$ and (ii) follows. Conversely, if (ii) holds then the entire expression in (11.5) is no larger than $C\|g\|_{L^{q'}(\mathbb{R}^d)}$, which implies (i). Thus (i) and (ii) are equivalent with the same choice of C. Clearly, applying first (ii) and then (i) with

$f = \hat{g}$ yields

$$\|\mathcal{F}(\hat{g}\mu)\|_{L^q(\mathbb{R}^d)} \le C\|g\|_{L^{q'}(\mathbb{R}^d)}$$

for all $g \in \mathcal{S}(\mathbb{R}^d)$. Since $\mathcal{F}(\hat{g}\mu) = g(-\,\cdot) * \hat{\mu}$, part (iii) follows. Finally, we note the relation

$$\int g(x)(\hat{\mu} * f)(x)\,dx = \int g(x)\mathcal{F}(\mu\check{f})(x)\,dx = \int \hat{g}(\xi)\check{f}(\xi)\,\mu(d\xi)$$

for any $f, g \in \mathcal{S}(\mathbb{R}^d)$. Hence, if (iii) holds then

$$\left|\int_{\mathbb{R}^d} \hat{g}(\xi)\check{f}(\xi)\,\mu(d\xi)\right| \le C^2\|g\|_{L^{q'}(\mathbb{R}^d)}\|f\|_{L^{q'}(\mathbb{R}^d)}.$$

Now set $f(x) = g(-x)$. Then

$$\int_{\mathbb{R}^d} |\hat{g}(\xi)|^2\,\mu(d\xi) \le C^2\|g\|^2_{L^{q'}(\mathbb{R}^d)},$$

which is (ii). □

Setting $\mu = \sigma = \sigma_{S^{d-1}}$, where the last quantity is the surface measure of the unit sphere in $\mathbb{R}^d$, we now arrive at this conclusion.

Corollary 11.4 *The following assertions are equivalent.*

(i) *The Tomas–Stein theorem in the "restriction form"*

$$\|\hat{f} \restriction S^{d-1}\|_{L^2(\sigma)} \le C\|f\|_{L^{q'}(\mathbb{R}^d)},$$

for $q' = (2d+2)/(d+3)$ and all $f \in \mathcal{S}(\mathbb{R}^d)$.

(ii) *The "extension form" of the Tomas–Stein theorem*

$$\|\widehat{g\sigma_{S^{d-1}}}\|_{L^q(\mathbb{R}^d)} \le C\|g\|_{L^2(\sigma)},$$

for $q = (2d+2)/(d-1)$ and all $g \in \mathcal{S}(\mathbb{R}^d)$.

(iii) *The composition of (i) and (ii): for all $f \in \mathcal{S}(\mathbb{R}^d)$,*

$$\|f * \widehat{\sigma_{S^{d-1}}}\|_{L^q(\mathbb{R}^d)} \le C^2\|f\|_{L^{q'}(\mathbb{R}^d)},$$

with $q = (2d+2)/(d-1)$.

Proof Set $\mu = \sigma_{S^{d-1}} = \sigma$ in Lemma 5.4. □

Exercise 11.2 Show that parts (i) and (ii) of Corollary 11.4 are true in general whereas (iii) requires $L^2(\sigma)$. More precisely, show the following. The restriction estimate

$$\|\hat{f} \restriction S^{d-1}\|_{L^p(\sigma)} \le C\|f\|_{L^{q'}(\mathbb{R}^d)} \quad \forall f \in \mathcal{S}$$

is equivalent to the extension estimate

$$\|\widehat{g\sigma_{S^{d-1}}}\|_{L^q(\mathbb{R}^d)} \le C\|g\|_{L^{p'}(\sigma)} \quad \forall g \in \mathcal{S}.$$

11.1.3. Optimality, Knapp example

We shall now show, using part (ii) of Corollary 11.4 that the Tomas–Stein theorem is optimal. This is the well-known Knapp example.

Lemma 11.5 *The exponent* $p_d = (2d+2)/(d+3)$ *in Theorem 11.1 is optimal.*

Proof This is equivalent to saying that the exponent $q = \frac{2d+2}{d-1}$ in part (ii) of Corollary 11.4 is optimal. Fix a small $\delta > 0$ and let $g \in \mathcal{S}$ be such that $g = 1$ on $B(e_d; \sqrt{\delta})$, $g \geq 0$, and $\mathrm{supp}(g) \subset B(e_d; 2\sqrt{\delta})$ where $e_d = (0, \dots, 0, 1)$ is a unit vector.

Then

$$\begin{aligned} |\widehat{g\sigma}(\xi)| &= \left| \int \exp\Big(-2\pi i\big(x' \cdot \xi' + \xi_d\big(\sqrt{1-|x'|^2} - 1\big)\big)\Big) \frac{g(x', \sqrt{1-|x'|^2})}{\sqrt{1-|x'|^2}}\, dx' \right| \\ &\geq \left| \int \cos\Big(2\pi\big(x' \cdot \xi' + \xi_d\big(\sqrt{1-|x'|^2} - 1\big)\big)\Big) \frac{g(x', \sqrt{1-|x'|^2})}{\sqrt{1-|x'|^2}}\, dx' \right| \\ &\geq \cos\frac{\pi}{4} \int g\, d\sigma \geq C^{-1}\delta^{(d-1)/2} \end{aligned} \tag{11.6}$$

provided that

$$|\xi'| \leq \frac{(\sqrt{\delta})^{-1}}{100}\quad |\xi_d| \leq \frac{\delta^{-1}}{100}.$$

Indeed, under these assumptions, and for $\delta > 0$ small,

$$\big|x' \cdot \xi' + \xi_d\big(\sqrt{1-|x'|^2} - 1\big)\big| \leq \sqrt{\delta} \cdot \frac{(\sqrt{\delta})^{-1}}{100} + \frac{\delta^{-1}}{100}\left(\sqrt{\delta}\right)^2 \leq \frac{1}{50},$$

so that the argument of the cosine in (11.6) is smaller than $2\pi/50 \leq \pi/4$ in absolute value, as claimed. This is illustrated in Figure 11.1, where the horizontal rectangle depicts a cross-section through the support of g. The "support" of the Fourier transform of $g\sigma_{S^{d-1}}$ is given by the dual rectangle, of dimensions $\delta^{-1/2} \times \delta^{-1}$ (the latter being the vertical direction). Hence,

$$\begin{aligned} \|\widehat{g\sigma_{S^{d-1}}}\|_{L^q(\mathbb{R}^d)} &\geq C^{-1}\delta^{(d-1)/2}\left(\delta^{-(d-1)/2}\delta^{-1}\right)^{1/q} \\ &= C^{-1}\delta^{(d-1)/2}\delta^{-(d+1)/2q} \end{aligned}$$

whereas $\|g\|_{L^2(\sigma)} \leq C\delta^{(d-1)/4}$. It is therefore necessary that

$$\frac{d-1}{4} \leq \frac{d-1}{2} - \frac{d+1}{2q}$$

or $q \geq (2d+2)/(d-1)$, as claimed. □

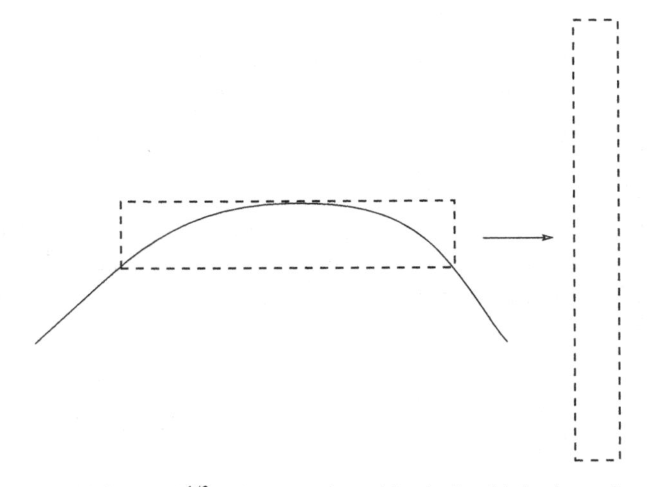

Figure 11.1. The $\delta^{1/2} \times \delta$ rectangle and its dual, with horizontal and vertical alignments, respectively.

11.1.4. Decay of $\widehat{\sigma_{S^{d-1}}}$

For the proof of Theorem 11.1 we need the following decay estimate for the Fourier transform of the surface measure $\sigma_{S^{d-1}}$, see Corollary 4.17:

$$\left|\widehat{\sigma_{S^{d-1}}}(\xi)\right| \leq C(1 + |\xi|)^{-(d-1)/2}. \tag{11.7}$$

It is easy to see that (11.7) imposes a restriction, on the possible exponents for an extension theorem, of the form

$$\|\widehat{f\sigma_{S^{d-1}}}\|_{L^q(\mathbb{R}^d)} \leq C\|f\|_{L^p(S^{d-1})}. \tag{11.8}$$

Indeed, setting $f = 1$ implies that on the one hand one needs

$$q > \frac{2d}{d-1},$$

by (11.7). On the other hand, one has the following result.

Exercise 11.3 Check by means of Knapp's example from Lemma 11.5 that (11.8) can only hold for

$$q \geq \frac{d+1}{d-1}p'. \tag{11.9}$$

The *restriction conjecture*, still unproved (in dimensions $d \geq 3$), states that (11.8) holds under these conditions, i.e., provided that

$$\infty \geq q > \frac{2d}{d-1} \quad \text{and} \quad q \geq \frac{d+1}{d-1} p'. \tag{11.10}$$

Observe that the Tomas–Stein theorem with $q = (2d+2)/(d-1)$ and $p = 2$ is a partial result in this direction. In dimension $d = 2$ the conjecture is true and can be proved by elementary means; see Chapter 12.

11.1.5. Proof of the Tomas–Stein theorem, non-endpoint version

Proof of the Tomas–Stein theorem for $p < (2d+2)(d+3)$ Let

$$\sum_{j \in \mathbb{Z}} \psi(2^{-j} x) = 1,$$

for all $x \neq 0$, be the usual Littlewood–Paley partition of unity. By Lemma 5.4 and Corollary 11.4 it is necessary and sufficient to prove that

$$\| f * \widehat{\sigma_{S^{d-1}}} \|_{L^{p'}(\mathbb{R}^d)} \leq C \| f \|_{L^p(\mathbb{R}^d)}$$

for all $f \in \mathcal{S}(\mathbb{R}^d)$. First, let

$$\varphi(x) = 1 - \sum_{j \geq 0} \psi(2^{-j} x).$$

Clearly $\varphi \in C_0^\infty(\mathbb{R}^d)$, and

$$1 = \varphi(x) + \sum_{j \geq 0} \psi(2^{-j} x) \quad \text{for all} \quad x \in \mathbb{R}^d.$$

Now observe that $\varphi \, \widehat{\sigma_{S^{d-1}}} \in C_0^\infty(\mathbb{R}^d)$, so that by Young's inequality

$$\| f * \varphi \, \widehat{\sigma_{S^{d-1}}} \|_{L^{p'}(\mathbb{R}^d)} \leq C \| f \|_{L^p(\mathbb{R}^d)} \tag{11.11}$$

with $C = \| \varphi \, \widehat{\sigma_{S^{d-1}}} \|_{L^r}$, where $1 + 1/p' = 1/r + 1/p$, i.e., $2/p' = 1/r$. It therefore remains to bound the operator given by convolution with

$$K_j := \psi(2^{-j} x) \, \widehat{\sigma_{S^{d-1}}}(x).$$

To be precise, we need to prove an estimate of the form

$$\| f * K_j \|_{L^{p'}(\mathbb{R}^d)} \leq C 2^{-j\varepsilon} \| f \|_{L^p(\mathbb{R}^d)} \quad \forall f \in \mathcal{S}(\mathbb{R}^d) \tag{11.12}$$

for all $j \geq 0$ and some small $\varepsilon > 0$. It is clear that the desired bound follows on summing (11.11) and (11.12) over $j \geq 0$. To prove (11.12), we interpolate

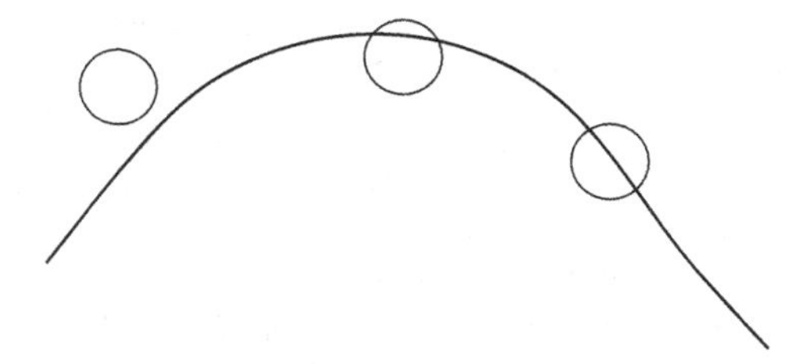

Figure 11.2. Various small balls relative to the surface S.

an $L^2 \to L^2$ with an $L^1 \to L^\infty$ bound as follows:

$$\begin{aligned}\|f * K_j\|_{L^2} &= \|\hat{f}\|_{L^2}\|\hat{K}_j\|_{L^\infty} \\ &= \|f\|_{L^2}\|2^{dj}\hat{\psi}(2^j\cdot) * \sigma_{S^{d-1}}\|_{L^\infty} \\ &\le C\|f\|_{L^2}2^{dj}\cdot 2^{-j(d-1)} = C2^j\|f\|_{L^2}.\end{aligned} \tag{11.13}$$

To pass to the estimate (11.13) one uses that on the one hand

$$\sup_x \sigma_{S^{d-1}}(B(x,r)) \le Cr^{d-1}, \tag{11.14}$$

as well as the fact that $\hat{\psi}$ has rapidly decaying tails; the latter implies that the estimate is the same as for compactly supported $\hat{\psi}$. Figure 11.2 illustrates the estimate (11.14). On the other hand,

$$\|f * K_j\|_\infty \le \|K_j\|_\infty\|f\|_1 \le C2^{-j(d-1)/2}\|f\|_1, \tag{11.15}$$

since the size of K_j is controlled by (11.7). Interpolating (11.13) with (11.15) yields

$$\|f * K_j\|_{p'} \le C2^{-j\theta(d-1)/2}2^{j(1-\theta)}\|f\|_p,$$

where

$$\frac{1}{p'} = \frac{\theta}{\infty} + \frac{1-\theta}{2} = \frac{1-\theta}{2}.$$

We thus obtain (11.12), provided that

$$\begin{aligned}0 &< \frac{d-1}{2}\theta - (1-\theta) = \frac{d+1}{2}\theta - 1 \\ &= (d+1)\left(\frac{1}{2} - \frac{1}{p'}\right) - 1 = \frac{d-1}{2} - \frac{d+1}{p'}.\end{aligned}$$

This is the same as $p' > (2d+2)/(d-1)$ or $p < (2d+2)/(d+3)$, as claimed. □

Exercise 11.4 Provide the details concerning the rapidly decaying tails in the previous proof.

11.2. The endpoint

The above proof strategy for the Tomas–Stein theorem does not achieve the sharp exponent $p = (2d + 2)/(d + 3)$ because (11.12) leads to a divergent series when $\varepsilon = 0$. In order to achieve this endpoint exponent, one therefore has to avoid interpolating the operator bounds on each dyadic piece separately. In this section we present two different methods leading to the endpoint. The first method is based on Stein's complex interpolation theorem whereas the second rests on a splitting of the coordinate directions and fractional integration in one of these directions. The second method has proven to be more robust and powerful, as will be seen in the next section, devoted to the Strichartz estimates. Nevertheless, the complex interpolation method as a tool appears frequently in harmonic analysis. Since the Tomas–Stein endpoint serves to illustrate this method reasonably well, we begin by following this more laborious route.

11.2.1. Complex interpolation

One general and powerful idea directed at resolving the aforementioned divergence is basically *first* to sum and *then* to interpolate, rather than vice versa. Of course, one needs to explain what it means to sum first: recall that the proof of the Riesz–Thorin interpolation theorem is based on the three-lines theorem from complex analysis. The key idea in the present context is to sum the dyadic pieces $T_j : f \mapsto f * K_j$ *together with complex weights* $w_j(z)$ in such a way that

$$T_z := \sum_{j\geq 0} w_j(z)T_j$$

converges on the strip $0 \leq \operatorname{Re} z \leq 1$ to an analytic operator-valued function with the property that

$$\begin{aligned} T_z : L^1 &\to L^\infty \quad \text{for } \operatorname{Re} z = 1, \\ T_z : L^2 &\to L^2 \quad \text{for } \operatorname{Re} z = 0. \end{aligned}$$

It then follows that $T_\theta = L^p(\mathbb{R}^d) \to L^{p'}(\mathbb{R}^d)$ for $1/p' = (1-\theta)/2$. Careful judgement is then required in choosing the weights $w_j(z)$ so that T_θ at a specific θ is a prescribed operator, in our case convolution by $\widehat{\sigma_{S^{d-1}}}$.

While it is conceptually correct and helpful to describe complex interpolation as a method of summing divergent series, it is rarely implemented in this fashion. Rather, one tries from the start to embed the operator under consideration into an analytic family, without breaking it up into dyadic pieces. There are standard ways of doing this that involve a little complex analysis and distribution theory (at the level of integration by parts and analytic continuation). We shall present such an approach now.

First proof of the endpoint for the Tomas–Stein theorem We consider a surface of nonzero curvature which can be written locally as a graph: $\xi_d = h(\xi')$, $\xi' \in \mathbb{R}^{d-1}$. Define

$$M_z(\xi) = \frac{1}{\Gamma(z)}\big(\xi_d - h(\xi')\big)_+^{z-1}\chi_1(\xi')\chi_2\big(\xi_d - h(\xi')\big), \tag{11.16}$$

where $\chi_1 \in C_0^\infty(\mathbb{R}^{d-1})$, $\chi_2 \in C_0^\infty(\mathbb{R})$ are smooth-cutoff functions, Γ is the gamma function, and $\operatorname{Re} z > 0$. Moreover, $(\cdot)_+$ refers to the positive part of a function. We will show that

$$T_z f := (M_z \hat{f})^\vee \tag{11.17}$$

can be defined by means of analytic continuation to $\operatorname{Re} z \le 0$. The main estimates are now

$$\|T_z\|_{2\to 2} \le B(z) \quad \text{for } \operatorname{Re} z = 1, \tag{11.18}$$

$$\|T_z\|_{1\to\infty} \le A(z) \quad \text{for } \operatorname{Re} z = -\frac{d-1}{2} \tag{11.19}$$

where $A(z)$, $B(z)$ grow faster than $e^{C|z|^2}$ as $|\operatorname{Im} z| \to \infty$. We will see that the singularity of $\big(\xi_d - h(\xi')\big)_+^{z-1}$ at $z = 0$ cancels out the simple zero of $\Gamma(z)^{-1}$ at $z = 0$, to produce

$$M_0(\xi) = \chi_1(\xi')\delta_0(\xi_d - h(\xi'))\, d\xi'; \tag{11.20}$$

see (11.23) below; this means that $M_0(\xi)$ is proportional to the surface measure on the graph. It then follows from Stein's complex interpolation theorem that

$$f \mapsto \widehat{M_0} * f$$

is bounded from $L^p \to L^{p'}$, where

$$\frac{1}{p'} = \frac{\theta}{\infty} + \frac{1-\theta}{2}, \qquad 0 = -\theta\frac{d-1}{2} + 1 - \theta;$$

that implies that

$$\frac{1}{p'} = \frac{d-1}{2d+2},$$

as desired. It remains to check (11.18)–(11.20). To do so, recall first that $\Gamma(z)^{-1}$ is an entire function with simple zeros at $z = 0, -1, -2, \ldots$ It has the product representation

$$\frac{1}{\Gamma(z)} = ze^{\gamma z} \prod_{\nu=1}^{\infty} \left(1 + \frac{z}{\nu}\right) e^{-z/\nu}, \qquad z = x + iy,$$

which converges everywhere in $\mathbb{C}$. Thus,

$$\begin{aligned}\left|\frac{1}{\Gamma(z)}\right|^2 &\le |z|^2 e^{2\gamma x} \prod_{\nu=1}^{\infty} \left(\left(1 + \frac{x}{\nu}\right)^2 + \frac{y^2}{\nu^2}\right) e^{-2x/\nu} \\ &\le |z|^2 e^{2\gamma x} \prod_{\nu=1}^{\infty} \left(e^{2x/\nu + |z|^2/\nu^2} e^{-2x/\nu}\right) = |z|^2 e^{2\gamma x} e^{|z|^2\pi^2/6}. \qquad (11.21)\end{aligned}$$

In particular, if $\operatorname{Re} z = 1$ then, for all $\xi \in \mathbb{R}^d$,

$$|M_z(\xi)| \le (1 + y^2) e^{2\gamma} e^{(1+y^2)\pi^2/6} \chi_1(\xi') \chi_2\big(\xi_d - h(\xi')\big) \le Ce^{cy^2}.$$

Thus (11.18) holds with the stated bound on $B(z)$. We remark that the bound we have just obtained is far from optimal. The true growth, given by the Stirling formula, is of the form $|y|^{1/2} e^{\pi |y|/2}$ as $|y| \to \infty$, but this makes no difference in our particular case.

Now let $\varphi \in \mathcal{S}(\mathbb{R}^d)$. Thus, for $\operatorname{Re} z > 0$,

$$\begin{aligned}&\int_{\mathbb{R}^d} M_z(\xi) \varphi(\xi) \, d\xi \\ &= \frac{1}{\Gamma(z)} \int_{\mathbb{R}^{d-1}} \int_0^{\infty} \chi_2(t) \varphi(\xi', t + h(\xi')) t^{z-1} \, dt \, \chi_1(\xi') \, d\xi' \\ &= -\frac{1}{z\Gamma(z)} \int_{\mathbb{R}^{d-1}} \int_0^{\infty} \frac{d}{dt} \big(\chi_2(t) \varphi(\xi', t + h(\xi'))\big) \, t^z \, dt \, \chi_1(\xi') \, d\xi'. \end{aligned} \qquad (11.22)$$

Observe that the right-hand side is well defined for $\operatorname{Re} z > -1$. Furthermore, at $z = 0$, using $z\Gamma(z)|_{z=0} = 1$, we have

$$\int_{\mathbb{R}^d} M_0(\xi) \varphi(\xi) \, d\xi = \int_{\mathbb{R}^{d-1}} \chi_2(0) \varphi(\xi', h(\xi')) \chi_1(\xi') \, d\xi',$$

which shows that the analytic continuation of M_z to $z = 0$ is equal to (setting $\chi_2(0) = 1$)

$$M_0(\xi) = \chi_1(\xi') \, d\xi' \delta_0(\xi_d - h(\xi')). \qquad (11.23)$$

Clearly, M_0 is proportional to the surface measure on an area S of the surface, where

$$S = \left\{(\xi', h(\xi')) \mid \xi' \in \mathbb{R}^{d-1}\right\}.$$

This is exactly what we want, since we need to bound $\widehat{\sigma_S} * f$.

Observe that (11.22) defines analytic continuation to $\operatorname{Re} z > -1$. Integrating by parts again extends this to $\operatorname{Re} z > -2$, and so forth. Indeed, the right-hand side of

$$\int_{\mathbb{R}^d} M_z(\xi)\varphi(\xi)\,d\xi = \frac{(-1)^k}{z(z+1)\cdots(z+k-1)\Gamma(z)} \int_{\mathbb{R}^{d-1}} \chi_1(\xi') \times \int_0^\infty t^{z+k-1} \frac{d^k}{dt^k}\big(\chi_2(t)\varphi(\xi', \xi_d + k(\xi'))\big)\,dt\,d\xi'$$

is well defined for all $\operatorname{Re} z > -k$.

Next, we prove (11.19) by means of an estimate on $\|\widehat{M_z}\|_\infty$. This requires the following preliminary calculation: let N be a positive integer such that $N > \operatorname{Re} z + 1 > 0$. Then we claim that

$$\left|\int_0^\infty e^{-2\pi i t\tau} t^z \chi_2(t)\,dt\right| \le \frac{C_N(1+|z|)^N}{1+\operatorname{Re} z}(1+|\tau|)^{-\operatorname{Re} z-1}. \tag{11.24}$$

To prove (11.24) we will distinguish large and small $t\tau$. Let $\psi \in C_0^\infty(\mathbb{R})$ be such that $\psi(t) = 1$ for $|t| \le 1$ and $\psi(t) = 0$ for $|t| \ge 2$. Then, since $0 \le \chi_2 \le 1$, we have

$$\begin{aligned}\left|\int_0^\infty e^{-2\pi i t\tau} t^z \psi(t\tau)\chi_2(t)\,dt\right| &\le \int_0^\infty t^{\operatorname{Re} z}\psi(t|\tau|)\,dt \\ &\le |\tau|^{-\operatorname{Re} z-1}\int_0^2 t^{\operatorname{Re} z}\,dt \le \frac{C}{\operatorname{Re} z+1}|\tau|^{-\operatorname{Re} z-1}.\end{aligned} \tag{11.25}$$

If $|\tau| \le 1$ then (11.25) is no larger than

$$\int_0^\infty t^{\operatorname{Re} z}\chi_2(t)\,dt \le \frac{C}{\operatorname{Re} z+1}.$$

Hence

$$(11.25) \le \frac{C}{1+\operatorname{Re} z}(1+|\tau|)^{-\operatorname{Re} z-1} \tag{11.26}$$

in all cases. To treat the case of large $t\tau$, for which τ is large, we exploit cancellation in the phase. More precisely,

$$
\begin{aligned}
&\left|\int_0^\infty e^{-2\pi i t\tau} t^z(1-\psi(t\tau))\chi_2(t)\,dt\right| \\
&\le \left(\frac{1}{2\pi|\tau|}\right)^N \int_0^\infty \left|\frac{d^N}{d\tau^N}\left(t^z(1-\psi(t\tau))\chi_2(t)\right)\right| dt \\
&\le C_N\left(\frac{1}{2\pi|\tau|}\right)^N \int_0^\infty dt\big(|z(z-1)\cdots(z-N+1)|t^{\operatorname{Re}z-N}(1-\psi(t\tau))\chi_2(t) \\
&\qquad + t^{\operatorname{Re}z}\big|\psi^{(N)}(t\tau)\big|\tau^N\chi_2(t) + t^{\operatorname{Re}z}(1-\psi(t\tau))\big|\chi_2^{(N)}(t)\big|\big) \\
&\le C_N\left(\frac{1}{2\pi|\tau|}\right)^N |z(z-1)\cdots(z-N+1)| \int_{1/\tau}^\infty t^{\operatorname{Re}z-N}\,dt \\
&\quad + \frac{C}{(2\pi|\tau|)^N}|\tau|^N|\tau|^{-\operatorname{Re}z-1} + C|\tau|^{-N}\int_0^1 t^{\operatorname{Re}z}\big|\chi_2^{(N)}(t)\big|\,dt \\
&\le C_N\big(|z(z-1)\cdots(z-N+1)|+1\big)|\tau|^{-\operatorname{Re}z-1} + C_N|\tau|^{-N}.
\end{aligned}
\tag{11.27}
$$

Observe that the indefinite integrals here converge because $\operatorname{Re}z - N < -1$. Moreover, the second term in (11.27) is $\le |\tau|^{\operatorname{Re}z-1}$ by the same condition (recall that we are taking τ to be large). Hence (11.24) follows from (11.26) and (11.27). We now compute $\widehat{M_z}(x)$. Let k be a positive integer with $\operatorname{Re}z > -k$. Then $\widehat{M_z}(x)$ equals

$$
\begin{aligned}
&\int e^{-2\pi i x\cdot\xi}\frac{1}{\Gamma(z)}(\xi_d - h(\xi'))_+^{z-1}\chi_1(\xi')\chi_2(\xi_d-h(\xi'))\,d\xi'd\xi_d \\
&= \frac{1}{\Gamma(z)}\int_0^\infty e^{-2\pi i x_d t}\,t^{z-1}\,\chi_2(t)\,dt \int_{\mathbb{R}^{d-1}} e^{-2\pi i(x'\cdot\xi'+x_d h(\xi'))}\,\chi_1(\xi')\,d\xi' \\
&= \frac{(-1)^k}{\Gamma(z)z(z-1)\cdots(z-k+1)}\int_0^\infty \left(e^{-2\pi i x_d t}\chi_2(t)\right)^{(k)} t^{z+k-1}\,dt \\
&\quad \times \int_{\mathbb{R}^{d-1}} e^{-2\pi i(x'\cdot\xi'+x_d h(\xi'))}\chi_1(\xi')\,d\xi',
\end{aligned}
\tag{11.28}
$$

where the final expression is well defined for $\operatorname{Re}z > -k$. Now suppose that $\operatorname{Re}z = -(d-1)/2$ and pick $k \in \mathbb{Z}^+$ such that $1-k \le \operatorname{Re}z < -k+2$, i.e., $(d+1)/2 \le k < (d+1)/2+1$. Apply (11.24) with $z+k-1$ instead of z

and with $N = 2$. Then the first integral in the last line of (11.28) is bounded by

$$\left| \int_0^\infty \left(e^{-2\pi i x_d t} \chi_2(t)\right)^{(k)} t^{z+k-1}\, dt \right|$$
$$\le \frac{C(1+|z|)^2}{\operatorname{Re} z + k} (1+|x_d|)^{-\operatorname{Re} z - k} C_k (1+|x_d|)^k$$
$$\le C_k\, (1+|z|)^2 (1+|x_d|)^{-\operatorname{Re} z}. \tag{11.29}$$

The second integral, however, is controlled by the stationary phase estimate, cf. (11.7),

$$\left| \int_{\mathbb{R}^{d-1}} e^{-2\pi i (x'\cdot\xi' + x_d h(\xi'))} \chi_1(\xi')\, d\xi' \right| \le C(1+|x|)^{-(d-1)/2}. \tag{11.30}$$

Observe that the growth in $|x_d|$ for $\operatorname{Re} z = -(d-1)/2$ is exactly balanced by the decay in (11.30). One concludes that, for $\operatorname{Re} z = -(d-1)/2$ and with k as in (11.29),

$$\|\widehat{M_z}\|_\infty \le C_d \left| \frac{1}{\Gamma(z) z(z-1)\cdots(z-k+1)} \right| (1+|z|)^2;$$

see (11.28)–(11.30). Thus (11.19) follows from the growth estimate (11.21) or Stirling's formula, and we are done. □

11.2.2. The fractional integration method

We now present an easier method of proof for the Tomas–Stein theorem, based on fractional integration in a direction transverse to the hypersurface. It will be applied again in Section 11.3.

Let $S \subset \mathbb{R}^d$ be a smooth hypersurface of nonvanishing Gaussian curvature. For the meaning of the latter condition, we refer the reader to subsection 4.2.3. Choose any ψ of compact support and let $\mu := \psi \sigma_S$, where σ_S is the Riemannian surface measure on S. Then as before we need to verify that

$$\|f * \hat{\mu}\|_{L^{p'}(\mathbb{R}^d)} \le C(d) \|f\|_{L^p(\mathbb{R}^d)}, \quad f \in \mathcal{S}(\mathbb{R}^d),$$

for $p = p_d = (2d+2)/(d+3)$. Assuming, as we may, that the support of ψ is sufficiently small, we change coordinates and write S as the graph of some smooth function $\Phi(x')$, defined near $x' = 0 \in \mathbb{R}^{d-1}$, such that $\Phi(0) = 0$, $\nabla\Phi(0) = 0$, and $\Phi(x')$ has a nonvanishing Hessian at $x' = 0$. We now split the variables in $\mathbb{R}^d$, setting $x = (x', t)$ with $t \in \mathbb{R}$. Then

$$(f * \hat{\mu})(x) = \int_{\mathbb{R}^d} K(x'-y', t-s) f(y', s)\, dy' ds,$$

where $K = \hat{\mu}$. As shown in subsection 4.2.3, one has the pointwise bound

$$|K(x' - y', t - s)| \le C\langle t - s\rangle^{-(d-1)/2} \quad \forall x', y' \in \mathbb{R}^{d-1}.$$

Therefore, if we write

$$(U(t)g)(x') := \int_{\mathbb{R}^{d-1}} K(x' - y', t)g(y')\,dy', \quad g \in \mathcal{S}(\mathbb{R}^{d-1}),$$

then on the one hand

$$\|U(t)g\|_\infty \le C\langle t\rangle^{-(d-1)/2}\|g\|_1 \tag{11.31}$$

and on the other hand

$$\|U(t)g\|_2 \le C\|g\|_2, \tag{11.32}$$

for all $t \in \mathbb{R}$. The latter estimate can be obtained from Plancherel's theorem, since

$$\hat{K}(\cdot, t) \in L^\infty(\mathbb{R}^{d-1}) \quad \text{uniformly in } t \in \mathbb{R},$$

where the Fourier transform is taken only in the x'-coordinate. Interpolating between (11.31) and (11.32) yields

$$\|U(t)g\|_{p'} \le C\langle t\rangle^{-\alpha(p)}\|g\|_p$$

for any $1 \le p \le 2$, where

$$\alpha(p) = \frac{d-1}{2}\left(\frac{1}{p} - \frac{1}{p'}\right).$$

Applying fractional integration in the t-variable yields

$$\begin{aligned}\|f * \hat{\mu}\|_{L^{p'}(\mathbb{R}^d)} &\le \left\| \int_{-\infty}^{\infty} \big\|U(t-s)f(s)\big\|_{L^{p'}(\mathbb{R}^{d-1})}\,ds \right\|_{L^{p'}(\mathbb{R})} \\ &\le C\left\| \int_{-\infty}^{\infty} \langle t-s\rangle^{-\alpha(p)}\|f(s)\|_{L^p(\mathbb{R}^{d-1})}\,ds \right\|_{L^{p'}(\mathbb{R})} \le C\|f\|_{L^p(\mathbb{R}^d)}.\end{aligned}$$

The final step involves fractional integration and therefore requires that $1 + 1/p' = \alpha(p) + 1/p$ as well as $0 < \alpha(p) < 1$. However, this means that we have precisely $p = p_d$ and this implies that $\alpha(p) = (d-1)/(d+1)$ which is admissible for all $d \ge 2$. This concludes the second, simpler and more robust, proof of the Tomas–Stein theorem.

11.3. Restriction and PDE; Strichartz estimates

11.3.1. Schrödinger evolution and Fourier restriction

The endpoint of the Tomas–Stein theorem is important because of *its scaling invariance*. To see this, let $u(x, t)$ be a smooth solution of the Schrödinger equation,

$$\begin{cases} \frac{1}{i}\partial_t u + \frac{1}{2\pi}\Delta_{\mathbb{R}^d} u = h, \\ \qquad\qquad u|_{t=0} = f, \end{cases} \tag{11.33}$$

where $f \in \mathcal{S}(\mathbb{R}^d)$, say. The constant $1/(2\pi)$ is chosen for cosmetic reasons having to do with the normalization of the Fourier transform. It could be replaced by any other positive constant by rescaling; one can even change the sign of the constant by passing to complex conjugates. We first treat the case $h = 0$. Then

$$u(t,x) = \int_{\mathbb{R}^d} e^{2\pi i(x\cdot\xi + t|\xi|^2)} \hat{f}(\xi)\, d\xi = (\hat{f}\mu)^\vee(x,t), \tag{11.34}$$

where μ is the measure in $\mathbb{R}^{d+1}$ defined by the integral

$$\int_{\mathbb{R}^{d+1}} h(\xi,\tau)\,\mu(d(\xi,\tau)) = \int_{\mathbb{R}^d} h(\xi,|\xi|^2)\,d\xi,$$

for all $h \in C^0(\mathbb{R}^{d+1})$. Now let $\varphi \in C_0^\infty(\mathbb{R}^{d+1})$ with $\varphi(\xi,\tau) = 1$ if $|\xi| + |\tau| \le 1$. Then the Tomas–Stein endpoint applies and one concludes that

$$\|(\hat{f}\varphi\mu)^\vee\|_{L^q(\mathbb{R}^{d+1})} \le C\|\hat{f}\|_{L^2(\varphi\mu)},$$

where $q = (2d+4)/d = 2 + 4/d$. In other words, if $\mathrm{supp}(\hat{f}) \subset B(0,1)$ then

$$\|(\hat{f}\mu)^\vee\|_{L^{2+4/d}(\mathbb{R}^{d+1})} \le C\|\hat{f}\|_{L^2(\mathbb{R}^d)} = C\|f\|_{L^2(\mathbb{R}^d)}. \tag{11.35}$$

To remove the condition $\mathrm{supp}(\hat{f}) \subset B(0,1)$ we need to rescale. Indeed, the rescaled functions

$$\begin{aligned} f_\lambda(x) &= f(x/\lambda), \\ u_\lambda(x,t) &= u(x/\lambda, t/\lambda^2) \end{aligned} \tag{11.36}$$

satisfy the equation

$$\begin{cases} \frac{1}{i}\partial_t u_\lambda + \frac{1}{2\pi}\Delta u_\lambda = 0, \\ \qquad\qquad u_\lambda|_{t=0} = f_\lambda. \end{cases}$$

If $\mathrm{supp}(\hat{f})$ is compact then $\mathrm{supp}(\hat{f_\lambda}) \subset B(0,1)$ if λ is large. Hence, in view of (11.34) and (11.35),

$$\|u_\lambda\|_{L^q(\mathbb{R}^{d+1})} \le C\|f_\lambda\|_{L^2(\mathbb{R}^d)}. \tag{11.37}$$

However,

$$\|f_\lambda\|_{L^2(\mathbb{R}^d)} = \lambda^{d/2}\|f\|_{L^2(\mathbb{R}^d)}$$

and

$$\|u_\lambda\|_{L^q(\mathbb{R}^{d+1})} = \lambda^{(d+2)/q}\|u\|_{L^q} = \lambda^{d/2}\|u\|_{L^q}.$$

Thus, (11.37) is the same as

$$\|u\|_{L^q(\mathbb{R}^{d+1})} \le C\|f\|_{L^2(\mathbb{R}^d)} \tag{11.38}$$

for all $f \in \mathcal{S}$ with $\mathrm{supp}(\hat{f})$ compact. These functions are dense in $L^2(\mathbb{R}^d)$. Therefore, (11.38), which is the original *Strichartz bound* for the Schrödinger equation in d spatial dimensions, follows for all $f \in L^2(\mathbb{R}^d)$. Note the main difference between the Strichartz estimate (11.37) and the Tomas–Stein endpoint estimate: while the latter is confined to *compact* surfaces, the former applies to functions that live on the noncompact characteristic version of the Schrödinger equation, i.e., the paraboloid $\{(\xi, |\xi|^2) \mid \xi \in \mathbb{R}^d\}$. Therefore the Strichartz estimate necessarily obeys the scaling symmetry of that version, and this is all we need to pass from compact surfaces to this *specific* noncompact surface.

11.3.2. General Strichartz estimates

The approach that we followed to derive the estimate (11.38) for solutions of (11.33) with $h = 0$ is the original one used by Strichartz and is instructive owing to its reliance on the restriction property of the Fourier transform. It is, however, technically advantageous for many reasons to use a shorter and somewhat different argument, which was introduced and developed by Ginibre and Velo in a series of papers. In essence this method is exactly that presented in subsection 11.2.2. We will use this argument to derive the following theorem. We call a pair (p, q) *Strichartz admissible* if and only if

$$2/p + d/q = d/2 \tag{11.39}$$

and if $2 \le p \le \infty$ with $(p, q) \ne (2, \infty)$.

Theorem 11.6 *Let h be a space–time Schwartz function in $d + 1$ dimensions and f a spatial Schwartz function. Let $u(x, t)$ solve* (11.33). *Then*

$$\|u\|_{L_t^p L_x^q(\mathbb{R}^{d+1})} \le C\big(\|f\|_{L^2(\mathbb{R}^d)} + \|h\|_{L_t^{a'} L_x^{b'}(\mathbb{R}^{d+1})}\big), \tag{11.40}$$

where (p, q) and (a, b) are Strichartz admissible with $a > 2$ and $p > 2$. Finally, these estimates localize in time: if on the left-hand side u is restricted to some time interval $I \ni 0$ then on the right-hand side h can be restricted to I.

Some remarks are in order: first, the estimate (11.40) is scaling invariant under the law (11.36) if and only if (p, q) and (a, b) obey the relation (11.39). Hence the latter is *necessary* for the Strichartz estimates (11.40) to hold. Note that the pair $p = q = 2 + 4/d$, which we derived above for $h = 0$, is of this type. However, the restriction $p > 2$ is technical and the important endpoint estimate for $p = 2$ and $d \geq 3$ was proved by Keel and Tao. Finally, under strong conditions on h and f it is easy to solve (11.33) by means of an explicit expression,

$$u(t) = e^{-i\Delta t/2\pi} f + \int_0^t e^{-i\Delta(t-s)/2\pi} h(s)\, ds, \tag{11.41}$$

with the understanding that the operator $e^{-i\Delta t/2\pi}$ is to be interpreted as the Fourier multiplier given by

$$\big(e^{-i\Delta t/2\pi} f\big)(x) := \int e^{2\pi i(x\cdot\xi + |\xi|^2 t)} \hat{f}(\xi)\, d\xi,$$

which is well defined for $f \in \mathcal{S}(\mathbb{R}^d)$, say.

Proof of Theorem 11.6 We again start with $h = 0$ and let $U(t)$ denote the propagator, i.e., $U(t)f$ is the solution of (11.33) given by (11.41): $U(t) = e^{-i\Delta t/2\pi}$. By (4.15) we have the bound

$$\|U(t)f\|_{L^q(\mathbb{R}^d)} \leq C|t|^{(-d/2)(1/q'-1/q)} \|f\|_{L^{q'}(\mathbb{R}^d)} \tag{11.42}$$

for $1 \leq q' \leq 2$. Indeed, for $q' = 1$ this bound follows when absolute values are placed inside (4.15), whereas $q' = 2$ gives Plancherel's theorem. The intermediate-q' values then follow by interpolation. Our goal for now is to prove that, for any (p, q) as in the theorem,

$$\|Uf\|_{L_t^p L_x^q} \leq C\|f\|_2,$$

where $(Uf)(x, t) = (U(t)f)(x)$, with a slight abuse of the notation. In analogy with the duality equivalence lemma 11.3, one now has that each of the following estimates implies the other two:

$$\|Uf\|_{L_t^p L_x^q} \leq C\|f\|_{L_x^2},$$
$$\|U^*F\|_{L_x^2} \leq C\|F\|_{L_t^{p'} L_x^{q'}},$$
$$\|U \circ U^*F\|_{L_t^p L_x^q} \leq C^2 \|F\|_{L_t^{p'} L_x^{q'}}.$$

As in the case of the Tomas–Stein theorem, the key is to prove the third property. Note that one has

$$U^*F = \int_{-\infty}^{\infty} U(-s)F(s)\, ds$$

whence, by the group property of $U(t)$,

$$(U \circ U^* F)(t) = \int_{-\infty}^{\infty} U(t-s)F(s)\,ds.$$

By (11.42),

$$\|(U \circ U^* F)(t)\|_{L_x^q} \le C \int_{-\infty}^{\infty} |t-s|^{(-d/2)(1/q'-1/q)} \|F(s)\|_{L_x^{q'}}\,ds.$$

Then, by fractional integration over time, with the implication

$$1 + \frac{1}{p} = \frac{1}{p'} + \frac{d}{2}\left(\frac{1}{q'} - \frac{1}{q}\right) \iff (11.39)$$

provided that

$$0 < \frac{d}{2}\left(\frac{1}{q'} - \frac{1}{q}\right) < 1 \iff p > 2, \tag{11.43}$$

one has

$$\|U \circ U^* F\|_{L_t^p L_x^q} \le C \|F\|_{L_t^{p'} L_x^{q'}},$$

the desired estimate for the case $h = 0$. Now suppose that $f = 0$, and write the solution of (11.33) as (writing $h = F$)

$$u(t) = \int_0^t U(t-s)F(s)\,ds = \int_{-\infty}^{\infty} \chi_{[0<s<t]}\, U(t-s)F(s)\,ds.$$

We now claim two estimates on u as a space–time function:

$$\|u\|_{L_t^p L_x^q} \le C \|F\|_{L_t^{p'} L_x^{q'}},$$

$$\|u\|_{L_t^p L_x^q} \le C \|F\|_{L_t^1 L_x^2}.$$

The first one can be proved by the same argument as before whereas the second one is proved as follows:

$$\begin{aligned} \|u(t)\|_{L_x^q} &\le \int_{-\infty}^{\infty} \chi_{[0<s<t]} \|U(t-s)F(s)\|_{L_x^q}\,ds \\ &\le \int_{-\infty}^{\infty} \|U(t-s)F(s)\|_{L_x^q}\,ds, \end{aligned}$$

whence

$$\begin{aligned} \|u\|_{L_t^p L_x^q} &\le C \int_{-\infty}^{\infty} \|U(t-s)F(s)\|_{L_t^p L_x^q}\,ds \\ &\le C \int_{-\infty}^{\infty} \|F(s)\|_{L_x^2}\,ds = \|F\|_{L_s^1 L_x^2}. \end{aligned}$$

By duality and interpolation one now obtains

$$\|u\|_{L_t^p L_x^q} \le C\|F\|_{L_s^{a'} L_x^{b'}}$$

for any admissible pair (a, b) and (p, q) with $a > 2$ and $p > 2$. This concludes the argument. The statement about time localizations follows from the solution formula (11.41). □

11.3.3. A nonlinear application

As an application of Theorem 11.6 we now solve the nonlinear equation

$$i\partial_t \psi + \Delta\psi = \lambda|\psi|^{4/d}\psi \tag{11.44}$$

in d spatial dimensions, and with an arbitrary real-valued constant λ, for a small L^2 datum $\psi|_{t=0} = \psi_0$. First, it is not a priori clear what we mean by "solve", since the data are L^2, nor is it immediately clear why we should not consider smoother data. Second, we remark that the choice of power in (11.44) is not accidental. In fact, it is the unique power for which the rescaled functions

$$\lambda^{d/2}\psi(\lambda x, \lambda^2 t) \tag{11.45}$$

are still solutions. The relevance of this scaling law is that the L^2-invariant scaling of the data is $\lambda^{d/2}\psi_0(\lambda x)$, which leaves the smallness condition unchanged. Moreover, the induced rescaling of the solution is then given by (11.45). Hence the only setting in which we can hope for a meaningful global small L^2 data theory is (11.44).

As in the existence and uniqueness theory of ordinary differential equations, one reformulates (11.44) as an integral equation, as follows (this is the *Duhamel* formula):

$$\psi(t) = e^{it\Delta}\psi_0 + i\lambda \int_0^t e^{i(t-s)\Delta}\, |\psi(s)|^{4/d}\psi(s)\, ds. \tag{11.46}$$

One then seeks a solution of this integral equation, by contraction or Picard iteration, contained in the space $C_t([0, \infty); L_x^2(\mathbb{R}^d))$. This is very natural, as (11.44) conserves the L^2-norm for real λ (see the problems at the end of the chapter). However, this space is still too large to formulate a meaningful theory in, and so we restrict it further in a way that allows us to invoke Theorem 11.6.

Definition 11.7 By a global weak solution of (11.44) with data $\psi_0 \in L^2(\mathbb{R}^d)$ we mean a solution to the integral equation (11.46) that lies in the space

$$X := (C_t \cap L_t^\infty)([0, \infty); L_x^2(\mathbb{R}^d)) \cap L_t^{p_0}([0, \infty); L_x^{2p_0}(\mathbb{R}^d)) \tag{11.47}$$

with $p_0 := 1 + 4/d$.

We remark that the power p_0 in the definition occurs in a natural way. In fact, passing L^2-norms onto (11.46) yields

$$\|\psi(t)\|_2 \le C\left(\|\psi_0\|_2 + |\lambda| \int_0^t \|\psi(s)\|_{2p_0}^{p_0}\, ds\right),$$

which is finite at least for $\psi \in X$. Now we will show much more.

Corollary 11.8 *The nonlinear equation* (11.44) *in* $\mathbb{R}^{1+d}_{t,x}$ *admits a unique weak solution in the sense of Definition 11.7 for any data* $\psi_0 \in L^2(\mathbb{R}^d)$*, provided that* $\|\psi_0\|_2 \le \varepsilon_0(\lambda, d)$.

Proof For the sake of simplicity we set $d = 2$, which implies that the nonlinearity takes the form $\lambda|\psi|^2\psi$, and $p_0 = 3$. We leave the case of general dimensions to the reader. Let us first prove uniqueness. Thus, let ψ and φ be two weak solutions to (11.46). Taking the difference yields

$$(\psi - \varphi)(t) = i\lambda \int_0^t e^{i(t-s)\Delta}\, (|\psi(s)|^2\psi(s) - |\varphi(s)|^2\varphi(s))\, ds.$$

Therefore, applying the Strichartz estimates of the previous theorem locally in time on $I = [0, T)$ yields, with $X(I)$, as in (11.47), localized to I,

$$\begin{aligned}\|\psi - \varphi\|_{X(I)} &\le C|\lambda|\, \left\| |\psi(s)|^2\psi(s) - |\varphi(s)|^2\varphi(s)\right\|_{L^1([0,T);L^2(\mathbb{R}^2))} \\ &\le C|\lambda|(\|\psi\|_{X(I)} + \|\varphi\|_{X(I)})^2\, \|\psi - \varphi\|_{X(I)}.\end{aligned}$$

However, if I is small enough then $C|\lambda|(\|\psi\|_{X(I)} + \|\varphi\|_{X(I)})^2 < 1$, whence $\|\psi - \varphi\|_{X(I)} = 0$. Thus $\psi = \varphi$ on I. This argument shows that the set where $\psi = \varphi$ is both open and closed and, since it is nonempty (it contains $t = 0$), therefore equals the whole time line.

To prove the existence we set up a contraction argument in the space X, for small data. Thus, we define an operator A on X by the formula

$$(A\psi)(t) := e^{it\Delta}\psi_0 + i\lambda \int_0^t e^{i(t-s)\Delta}\, |\psi(s)|^2\psi(s)\, ds$$

and note that, by Theorem 11.6,

$$\|A\psi\|_X \le C(\|\psi_0\|_2 + |\lambda|\|\psi\|_X^3).$$

Therefore, if $\|\psi\|_X \le R$ and if

$$C(\|\psi_0\|_2 + |\lambda|R^3) < R \tag{11.48}$$

then we conclude that A takes an R-ball in X into itself. To satisfy (11.48) we simply choose ε small enough that $R = C\varepsilon$ will verify (11.48) for any $\|\psi_0\|_2 <$

ε. Taking differences as in the uniqueness proof then shows that A is in fact a contraction. Therefore there exists a unique fixed point ψ in the R-ball in X. But $A\psi = \psi$ is precisely the definition of a weak solution, and we are done. $\square$

Corollary 11.8 is one of the many results that can be obtained in this area. It is also clear that now many interesting questions arise, such as: the question of global existence for *large* data (the answer to which depends on the sign of λ), which is relatively easy for data in H^1 but hard for data in L^2; the persistence of regularity; the spatial decay of solutions; etc. We present some of the easier results in this direction in the problem section. Showing that an equation having a derivative nonlinearity, such as

$$i\partial_t \psi + \Delta\psi = \pm|\psi|^2\, \partial_1 \psi, \tag{11.49}$$

has global solutions for small data, in $\mathcal{S}$, say, *cannot be accomplished* by means of Strichartz estimates alone. Indeed, Strichartz estimates are not able to make up for the loss of a derivative on the right-hand side.

11.3.4. The wave equation

We conclude this section with a sketch of the Strichartz estimates for the wave equation in $\mathbb{R}^{1+d}_{t,x}$,

$$\begin{aligned} \Box u = \partial_{tt} u - \Delta u &= h, \\ u\big|_{t=0} = f, \quad \partial_t u\big|_{t=0} &= g. \end{aligned} \tag{11.50}$$

The solution to this Cauchy problem is

$$u(t) = \cos(t|\nabla|) f + \frac{\sin(t|\nabla|)}{|\nabla|} g + \int_0^t \frac{\sin((t-s)|\nabla|)}{|\nabla|} h(s)\, ds,$$

at least for Schwartz data, say. The meaning of the operators appearing here is based on the Fourier transform. In other words,

$$(\cos(t|\nabla|) f)(x) = \sum_{\pm} \frac{1}{2} \int_{\mathbb{R}^d} e^{2\pi i (x\cdot\xi \pm t|\xi|)}\, \hat{f}(\xi)\, d\xi.$$

Proceeding analogously to the Schrödinger equation case we are therefore led to interpret these integrals as the Fourier transforms of measures living on the double *light cone*

$$\Gamma := \big\{(\xi, \tau) \in \mathbb{R}^{d+1} \big|\, |\tau| = |\xi\,|\big\}.$$

This is different from the paraboloid in the Schrödinger equation in two important ways. First, Γ has one vanishing principal curvature, namely along the generators of the light cone. Second, Γ is singular at the vertex of the cone. To overcome the latter difficulty, we define

$$\Gamma_0 := \{(\xi, \tau) \in \mathbb{R}^{1+d} \,|\, 1 \le |\xi| = \tau \le 2\}$$

to be a section of the light cone in $\mathbb{R}^{1+d}$. Since the cone has one vanishing principal curvature, the Fourier transform of the surface measure on Γ_0 exhibits less decay than the paraboloid, which is reflected in a different Tomas–Stein exponent (in effect, the dimension d is reduced to $d-1$, reflecting the loss of one principal curvature).

Exercise 11.5

(a) By means of the stationary phase method from Chapter 4 show that

$$\left|\widehat{\varphi\sigma_{\Gamma_0}}(x,t)\right| \le C(1+|(x,t)|)^{-(d-1)/2} \tag{11.51}$$

for all $(x,t) \in \mathbb{R}^{d+1}$. Moreover, this is optimal for all directions (x,t) belonging to the dual cone Γ^* (which is equal to Γ if the opening angle is 90°). The dual cone is defined as the set of all normals to the cone.

(b) Check that the complex interpolation method from above therefore implies the following restriction estimate for Γ_0:

$$\|\hat{f} \restriction \Gamma_0\|_{L^2(\sigma_\Gamma)} \le C\|f\|_{L^p(\mathbb{R}^d)},$$

where $p = (2d+2)/(d+3)$ and $d \ge 2$ (for $d = 1$ this amounts to a trivial bound).

Via a rescaling and Littlewood–Paley theory we now arrive at the following analogue of the Strichartz bound (11.38) for the Schrödinger equation.

Theorem 11.9 *Let Γ be a double light cone in $\mathbb{R}^{d+1}_{\xi,\tau}$ with $d \ge 2$, equipped with the measure $\mu(d(\xi,\tau)) = d\xi/|\xi|$. Then*

$$\left(\int_\Gamma |\hat{f}(\xi,\tau)|^2\,\mu(d(\xi,\tau))\right)^{1/2} \le C\|f\|_{L^p(\mathbb{R}^{d+1})} \tag{11.52}$$

with $p = (2d+2)/(d+3)$.

Proof Let Γ_0 be the cone restricted to $1 \le |\tau| \le 2$ as above. Then

$$\left(\int_{\lambda\le|(\xi,\tau)|\le 2\lambda} |\hat{f}((\xi,\tau))|^2\,\mu(d(\xi,\tau))\right)^{1/2} \le \left(\int_{\Gamma_0} |\hat{f}(\lambda(\xi,\tau))|^2 \mu(d(\lambda(\xi,\tau)))\right)^{1/2}$$

$$\le C\lambda^{(d-1)/2}\left\|\frac{1}{\lambda^{d+1}} f\left(\frac{1}{\lambda}\cdot\right)\right\|_{L^p(\mathbb{R}^{d+1})}$$

$$= C\lambda^{(d-1)/2}\cdot\lambda^{-d-1}\cdot\lambda^{(d+1)p}\|f\|_{L^p(\mathbb{R}^{d+1})}$$

$$= C\|f\|_{L^p(\mathbb{R}^{d+1})}, \tag{11.53}$$

by a suitable choice of p. To sum up (11.53) for $\lambda = 2^j$ we use the Littlewood–Paley theorem together with Exercise 6.1. In fact, since $p < 2$ we may conclude that

$$\|\hat{f}\|_{L^2(\mu)} \le C \left(\sum_{j\in\mathbb{Z}} \|P_j f\|_{L^p}^2 \right)^{1/2} \le C \left\| \left(\sum_{j\in\mathbb{Z}} |P_j f|^2 \right)^{1/2} \right\|_{L^p} \le C \|f\|_{L^p(\mathbb{R}^{d+1})},$$

which is (11.52). □

For the wave equation this means the following.

Corollary 11.10 *Let $u(x,t)$ be a solution of* (11.50) *with $f, g \in \mathcal{S}(\mathbb{R}^d)$, $d \ge 2$. Then*

$$\|u\|_{L^{(2d+2)/(d-1)}(\mathbb{R}^{d+1})} \le C \left(\|f\|_{\dot{H}^{1/2}(\mathbb{R}^d)} + \|g\|_{\dot{H}^{-1/2}(\mathbb{R}^d)} \right)$$

with $C = C(d)$ a constant.

Proof For simplicity set $f = 0$. Then

$$\begin{aligned} u(x,t) &= \int_{\mathbb{R}^d} e^{2\pi i (t|\xi| + x\xi)} \hat{g}(\xi) \frac{d\xi}{4\pi i |\xi|} - \int_{\mathbb{R}^d} e^{2\pi i(-t|\xi| + x\cdot\xi)} \hat{g}(\xi) \frac{d\xi}{4\pi i |\xi|} \\ &= (F\mu)^{\vee}(x,t), \end{aligned}$$

where $F(\xi, \pm|\xi|) = \hat{g}(\xi)$ and $d\mu(\xi, \pm|\xi|) = d\xi/|\xi|$. By the dual to Theorem 11.9 we have

$$\|(F\mu)^{\vee}\|_{L^{p'}(\mathbb{R}^{d+1})} \le C \|F\|_{L^2(\mu)}, \tag{11.54}$$

where $p' = (2d+2)/(d-1)$. Clearly

$$\|F\|_{L^2(\mu)} = \left(\int_{\mathbb{R}^d} |\hat{g}(\xi)|^2 \frac{d\xi}{|\xi|} \right)^{1/2} = \|g\|_{\dot{H}^{-1/2}},$$

so that (11.54) implies

$$\|u\|_{L^{(2d+2)/(d-1)}(\mathbb{R}^{d+1})} \le C \|g\|_{\dot{H}^{-1/2}(\mathbb{R}^d)}$$

as claimed. □

Exercise 11.6 Derive estimates for the wave equation as in Corollary 11.10 but with $\|f\|_{\dot{H}^1} + \|g\|_{L^2}$ on the right-hand side.

11.4. Optimal two-dimensional restriction

We conclude this chapter by establishing Zygmund's restriction property in $\mathbb{R}^2$ in the optimal range given by (11.10).

Theorem 11.11 *Let $\gamma : I \to \mathbb{R}^2$ be a smooth curve with $\gamma' \neq 0$ and $\gamma'' \neq 0$ on some finite interval I. Let $4 < q \leq \infty$ and $3p' \leq q$. Then*

$$\left\| \int_I e^{i\gamma(t)\cdot\xi} \varphi(t)\, dt \right\|_{L^q(\mathbb{R}^2)} \leq C(q,p)\|\varphi\|_{L^p(I)} \tag{11.55}$$

for any $\varphi \in L^p(I)$.

Proof We may take I small enough that $(t,s) \mapsto \gamma(t) - \gamma(s) = x$ is invertible, with the Jacobian $J = \det \partial(t,s)/\partial x$ satisfying $|J| \simeq |t-s|^{-1}$. Then for any finite $q > 4$ one has, with $F(x) = \varphi(t)\bar{\varphi}(s)J$ and U the image of $I \times I$ under the aforementioned change of variables,

$$\begin{aligned}
\left\| \int_I e^{i\gamma(t)\cdot\xi} \varphi(t)\, dt \right\|_{L^q(\mathbb{R}^2)} &= \left\| \int_I \int_I e^{i(\gamma(t)-\gamma(s))\cdot\xi} \varphi(t)\bar{\varphi}(s)\, dt ds \right\|_{L^{q/2}(\mathbb{R}^2)} \\
&= \left\| \int_U e^{ix\cdot\xi} F(x)\, dx \right\|_{L^{q/2}(\mathbb{R}^2)} \\
&\leq \|F\|_{L^r(U)} \leq C \left(\int_I \int_I \frac{|\varphi(t)|^r |\varphi(s)|^r}{|t-s|^{r-1}}\, dt ds \right)^{1/r} \\
&\leq C\|\varphi\|_p^2.
\end{aligned}$$

The first inequality is the Hausdorff–Young inequality with $r' = q/2 > 2$; the final inequality follows by fractional integration since $1 < r < 2$, and

$$1 + \frac{1}{(p/r)'} \geq r - 1 + \frac{1}{p/r}.$$

The latter is the same as $3p' \leq q$, and we are done. □

Note that the Tomas–Stein theorem (for restrictions to regular curves of nonvanishing curvature) is a special case of Theorem 11.11. It is given by the choice $p = 2, q = 6$.

Notes

A standard introductory reference in this area is provided by Stein's Beijing lectures [110]. The first observation of the restriction phenomenon in the radial case is apparently due to Laurent Schwartz. Wolff's course notes [128] also cover much of this chapter and can serve as an introduction to the role played by Kakeya sets in the restriction conjecture. The original reference is Strichartz [113]. For an account summarizing more recent developments on the Kakeya and restriction problems, see Tao [115]. It is known, see [128], that the full restriction conjecture implies the *Kakeya conjecture*, which states that Kakeya sets in dimension $d \geq 3$ have (Hausdorff) dimension d. This latter conjecture appears to be very difficult.

The basic restriction theory as it appears here can be viewed as a subset of the theory of oscillatory integrals; see Sogge's book [105] on the topic, as well as Stein [111, Chapters VIII, IX]. This area encompasses some of the most challenging conjectures in harmonic analysis, such as the Bochner–Riesz and restriction conjectures in dimensions $d \geq 3$.

The area of Strichartz estimates is vast and continues to be developed, especially in the variable-coefficient setting; see for example Hassell, Tao, and Wunsch [53], Tataru [119], Marzuola, Metcalfe, Tataru, and Tohaneanu [81], as well as Bahouri, Chemin, and Danchin [5] for that aspect. A standard reference on dispersive evolution equations is Tao [116]. Keel and Tao [66] proved the endpoint for the Strichartz estimates, i.e., L_t^2-type estimates. For more on the Schrödinger equation, see Cazenave [17] as well as Sulem and Sulem [114]. For the wave equation, see Lindblad and Sogge [78] and Sogge's wave equation book [106]. For an introduction to nonlinear wave equations arising in geometry, see Shatah and Struwe [100]. Strichartz estimates for the Klein–Gordon equation, which behaves like the wave equation for large frequencies and the Schrödinger equation for small frequencies, are derived in Nakanishi and Schlag [86] (see the references therein for the original literature). For some of the original work by Ginibre and Velo on Strichartz estimates for various dispersive equations see [49]. For the solution theory of equations of the type (11.49), which require smoothing estimates for the Schrödinger equation, see Kenig, Ponce, and Vega [67].

As indicated in the final section of the chapter, which is based on Zygmund's work [132] and Fefferman's thesis [36], the two-dimensional case is more accessible – and has been settled as far as restriction and the Bochner–Riesz conjecture are concerned, owing to serendipity in the numbers. More specifically, in the restriction and Bochner–Riesz conjectures $p = 4$ appears as a natural endpoint. One then writes $4 = 2 \cdot 2$, which, for example, enables a reduction to Plancherel's theorem. A powerful geometric argument yielding the optimal two-dimensional Bochner–Riesz theorem was found by Córdoba [25]. It relies on Kakeya-type maximal function and almost orthogonality ideas going back to Fefferman [37]. The argument in [25] later inspired Mockenhaupt, Seeger, and Sogge [83] to obtain a partial result on the *local smoothing conjecture* for the wave equation in 2 + 1 dimensions. This conjecture was formulated by Sogge in [104] and is still open. See also Davies and Chang [28] for an exposition of Córdoba's work. For further comments on the early history of restriction theorems see also Fefferman [36].

Problems

Problem 11.1 Suppose that ϕ is a smooth function on $(-1, 1)$, with $\phi'(t) = 0$ if and only if $t = 0$. Assume further that $\phi''(0) = 0$ and $\phi'''(0) \neq 0$. Let $a \in C^\infty(-1, 1)$ with $\mathrm{supp}(a) \subset (-1, 1)$. Determine the sharp decay of

$$\int_{-1}^{1} e^{i\lambda\phi(t)}\, a(t)\, dt$$

as $\lambda \to \infty$.

Problem 11.2 Investigate the restriction of the Fourier transform of a function in $\mathbb{R}^3$ to a (section of a) smooth curve in $\mathbb{R}^3$. Assume that the curve has nonvanishing curvature and torsion. Try to generalize to other dimensions and codimensions.

Problem 11.3 This problem expands on Corollary 11.8.

(a) Show that if $\psi_0 \in H^k(\mathbb{R}^d)$ (one can again take $d = 1$ or $d = 2$ for simplicity), where k is a positive integer, then $\psi \in C_t([0, \infty); H^k(\mathbb{R}^d))$. Conclude via Sobolev embedding that if $\psi_0 \in \mathcal{S}(\mathbb{R}^d)$, say, then $\psi(x, t)$ is smooth in all variables and satisfies the nonlinear equation (11.44) in a pointwise sense.

(b) Show that if $\psi(x, t)$ is any solution (not necessarily small) of (11.44) with $\partial_t \psi \in C([0, T); L^2(\mathbb{R}^d))$ and $\psi \in C([0, T); H^2(\mathbb{R}^d))$ then one has the *conservation of mass* (recall that λ is real),

$$M(\psi) := \tfrac{1}{2}\|\psi(t)\|_2^2 = \tfrac{1}{2}\|\psi_0\|_2^2 \quad \forall\, 0 \leq t < T$$

and *energy*

$$E(\psi) := \int_{\mathbb{R}^d} \left(\tfrac{1}{2}|\nabla \psi(x, t)|^2 - \frac{\lambda}{2 + 4/d}|\psi(x, t)|^{2+4/d} \right) dx.$$

Hint: Differentiate M and E with respect to time, and integrate by parts.

(c) Prove the relation

$$e^{it\Delta} x = (x + 2it\nabla)e^{it\Delta}$$

for operators acting on Schwartz functions, and use this to show that if $\psi_0 \in \mathcal{S}(\mathbb{R}^d)$ then the solution constructed in Corollary 11.8 has the property that, for any time t, one has $\psi(t) \in \mathcal{S}(\mathbb{R}^d)$.

(d) Show that the solution of Corollary 11.8 with $\psi_0 \in L^2(\mathbb{R}^d)$ preserves the mass $M(\psi)$ and the energy provided that the data satisfy $\psi_0 \in H^1(\mathbb{R}^d)$.

Problem 11.4

(a) For any power $p \geq 2$ show that the equation

$$i\partial_t \psi + \partial_{xx}\psi = \lambda|\psi|^{p-1}\psi \tag{11.56}$$

on the line $\mathbb{R}$ has local in-time unique weak solutions for any datum $\psi_0 \in H^1(\mathbb{R})$. These solutions again need to be interpreted in the Duhamel integral equation sense, and the space in which to apply the contraction principle is $C([0, T); H^1(\mathbb{R}))$, where $T > 0$ can be chosen so that it is bounded below by a function of $\|\psi_0\|_{H^1(\mathbb{R})}$ alone. Show that these solutions conserve mass and energy.

(b) Show that if T_* is the maximal time for a weak solution to exist on $[0, T_*)$, and if $T_* < \infty$, then $\|\psi(t)\|_{H^1(\mathbb{R})} \to \infty$ as $t \to T_*$. Conclude that for $\lambda > 0$ one has global solutions, i.e., $T_* = \infty$. **Note:** For $\lambda < 0$ this is false. For example, solutions of negative energy blow up in finite time in the sense that $T_* < \infty$; see Cazenave [17] or Sulem and Sulem [114].

(c) For small data in $H^1(\mathbb{R})$ and powers $p \geq 5$, show that one has global solutions irrespective of the sign of λ.

Problem 11.5 For the one-dimensional wave equation $u_{tt} - u_{xx} = 0$ with smooth data $u|_{t=0} = f$ and $\partial_t u|_{t=0} = g$, show that the solution is given by

$$u(x,t) = \tfrac{1}{2}(f(x+t) + f(x-t)) + \tfrac{1}{2}\int_{x-t}^{x+t} g(y)\,dy$$

for all times. Conclude that such waves do not decay; this precludes any possibility of a Strichartz estimate other than one involving L_t^∞.

Problem 11.6 Carry out the Ginibre–Velo approach to Strichartz estimates for the wave equation, in analogy with Theorem 11.6 for the Schrödinger equation case. This needs to be done for a fixed piece P_0 of a Littlewood–Paley function and then rescaled and summed using the results of Chapter 10. This approach leads to Besov-space estimates that are stronger than Sobolev estimates (one can switch to the latter by means of Exercise 6.1, as we did in the proof of Corollary 11.10). *Hint:* See Sogge [106] and Shatah and Struwe [100] for details.

Problem 11.7 On compact manifolds such as tori it makes no sense to ask for dispersive estimates for Schrödinger or wave equations (why?). Nevertheless, the following "Strichartz estimates" for the periodic Schrödinger equation due to Bourgain show that there is some sort of analogue for the torus. Note that these estimates are periodic in space and time:

(a) Show that

$$\left\| \sum_{n\in\mathbb{Z}} a_n\, e^{2\pi i(nx+n^2t)} \right\|_4^2 \leq C \sum_n |a_n|^2, \tag{11.57}$$

where $\mathbb{T}^2 = \mathbb{R}^2/\mathbb{Z}^2$ and L^4 is $L^4_{x,t}(\mathbb{T}^2)$.

(b) For any $\varepsilon > 0$ and positive integer N one has

$$\left\| \sum_{(n,m)\in\mathbb{Z}^2} \chi_{[n^2+m^2\leq N^2]}\, a_{n,m}\, e^{2\pi i(nx+my+(n^2+m^2)t)} \right\|_4^2 \leq C_\varepsilon N^\varepsilon \sum_{n,m} |a_{n,m}|^2, \tag{11.58}$$

where L^4 stands for $L^4_{x,y,t}(\mathbb{T}^3)$. This requires a little number theory, namely the standard bound on the number of divisors of an integer n (it grows more slowly than any power of n; this may be taken for granted).

12

Introduction to the Weyl calculus

12.1. Motivation, definitions, basic properties

12.1.1. Quantization

The basic concept from which we would like to start is that of *quantization*. This name originates in the observation, at the turn of the twentieth century, that energy is exchanged in discrete units. Mathematically, quantization refers to a procedure by which one passes from functions on the phase space of classical mechanics in the Hamiltonian formulation (the cotangent bundle of a manifold) to operators on a Hilbert space, the phase space of quantum mechanics. As this is not a physics textbook, we will not motivate – let alone explore – this quantization problem in any generality or depth. Moreover, we assure the reader that this chapter is self-contained (up to knowing the Fourier transform and calculus) and that no knowledge of physics will be required to follow it. From a mathematical perspective the calculus of pseudodifferential operators (ΨDOs), which is a result of such a quantization procedure, is an essential tool in elliptic PDEs, through *microlocal techniques*, whereas Fourier integral operators (FIOs) arise naturally in hyperbolic PDEs.

The only information we start from is the following basic list of correspondences on the phase space $\mathbb{R}^{2d}$ for the variables (x, ξ) (we will not discuss here the motivation for these correspondences from physics):

$$\begin{aligned} x_j &\mapsto X_j, \\ \xi_j &\mapsto D_j, \\ 1 &\mapsto \mathrm{Id}. \end{aligned}$$

Here X_j is the (unbounded) operator on $L^2(\mathbb{R}^d)$ given by $X_j(f)(x) = x_j f(x)$ and

$$D_j := \frac{1}{2\pi i}\frac{\partial}{\partial x_j};$$

alternatively, $D_j f = (\xi_j \hat{f})^\vee$ at least for Schwartz functions. Throughout, we shall pay very little attention to operator-theoretic issues such as domains, the question of self-adjointness versus symmetry, and the like.

From the above list we would like to construct a way of passing from a smooth function $a(x, \xi)$ to an operator $a(x, D)$. Naturally, we would like the map $a(x, \xi) \mapsto a(x, D)$ to be linear:

$$(\alpha a + \beta b)(x, D) = \alpha a(x, D) + \beta b(x, D) \quad \forall \alpha, \beta \in \mathbb{C}.$$

The example $a(x, \xi) = x_j \xi_k$ shows the basic difficulty we face, namely the noncommutativity of X_j and D_k: one has

$$[D_j, X_k] = D_j X_k - X_k D_j = \frac{1}{2\pi i}\mathrm{Id} \quad \text{if } j = k;$$

the commutator is zero otherwise. Thus, should $x_j \xi_k$ correspond to $X_j D_k$ or $D_k X_j$ or something else? In the former case, one would necessarily have

$$\sum_{|\alpha| \le N} a_\alpha(x) \xi^\alpha \mapsto \sum_{|\alpha| \le N} a_\alpha(x) D^\alpha,$$

which gives the Kohn–Nirenberg calculus:

$$a(x, D) f(x) = \int_{\mathbb{R}^d} a(x, \xi) e^{2\pi i x \cdot \xi} \hat{f}(\xi)\, d\xi = \int_{\mathbb{R}^{2d}} a(x, \xi) e^{2\pi i (x-y)\cdot \xi} f(y)\, d\xi dy, \tag{12.1}$$

which we encountered in Chapter 9. Here we wish to proceed differently, namely via the Weyl calculus. To define it, we start from the requirement that exponentials of linear functionals transform "naturally", i.e., if

$$\ell_{q,p}(x, \xi) := q \cdot x + p \cdot \xi$$

then we demand that

$$e^{2\pi i \ell_{q,p}}(x, D) = e^{2\pi i \ell_{q,p}(x,D)} =: \rho(q, p), \tag{12.2}$$

where the right-hand side is to be interpreted as an infinite series. In the following lemma we shall elucidate precisely what this means when applied to a function. Let us merely note for now that, at least for $a \in \mathcal{S}(\mathbb{R}^{2d})$, we can then define the Weyl calculus as

$$a(x, D) f(x) = \int_{\mathbb{R}^{2d}} \hat{a}(q, p) \rho(q, p) f(x)\, dq dp. \tag{12.3}$$

What is the meaning of $\rho(q, p) f(x)$? The best way to approach this is via the differential equation

$$\begin{aligned} \partial_t u(t, x) &= 2\pi i \ell_{q,p} u(t, x) = 2\pi i q \cdot x u(t, x) + p \cdot \nabla u(t, x), \\ u(0, x) &= f(x). \end{aligned} \tag{12.4}$$

Indeed, one then has $\rho(q,p)f(x) = u(1,x)$. Owing to the $p\cdot\nabla$ term this equation has the form of a transport equation, which implies that setting $u(t,x) = v(t, x+tp)$ (or equivalently, $v(t,y) = u(t, y-tp)$) reduces the equation to the ODE

$$\partial_t v(t,y) = 2\pi i q\cdot(y-tp)v(t,y), \tag{12.5}$$

with solution

$$v(t,y) = e^{2\pi i(tq\cdot y - t^2 q\cdot p/2)}v(0,y) = e^{2\pi i(tq\cdot y - t^2 q\cdot p/2)}f(y).$$

Setting $y = x+tp$ yields

$$u(t,x) = e^{2\pi i(tq\cdot x + t^2 q\cdot p/2)}f(x+tp).$$

In other words, we arrive at the conclusion that

$$\left(e^{2\pi i \ell_{q,p}(x,D)}f\right)(x) = \rho(q,p)f(x) = e^{2\pi i(x\cdot q + q\cdot p/2)}f(x+p).$$

Inserting this into (12.3) defines the Weyl calculus. We now wish to remove the Fourier transform from a and obtain a formula in the spirit of (12.1).

Lemma 12.1 *If $a\in\mathcal{S}(\mathbb{R}^{2d})$ and $f\in\mathcal{S}(\mathbb{R}^d)$ then*

$$\begin{aligned} a(x,D)f(x) &= \int_{\mathbb{R}^{2d}} a\left(\frac{x+y}{2},\xi\right)e^{2\pi i\xi\cdot(x-y)}f(y)\,dyd\xi \\ &= \int_{\mathbb{R}^d} a\left(\frac{x+y}{2},\widehat{y-x}\right)f(y)\,dy. \end{aligned} \tag{12.6}$$

The integrals on the right-hand side are absolutely convergent.

Proof By (12.3) one has

$$a(x,D)f(x) = \int_{\mathbb{R}^{2d}} \hat{a}(q,p)\,e^{2\pi i(q\cdot x + q\cdot p/2)}f(x+p)\,dpdq.$$

Substituting $z := x+p$ into the right-hand side yields

$$\begin{aligned} a(x,D)f(x) &= \int_{\mathbb{R}^{2d}} \hat{a}(q,z-x)\,e^{2\pi i(q\cdot x + q\cdot(z-x)/2)}f(z)\,dzdq \\ &= \int_{\mathbb{R}^d} a\left(\frac{x+z}{2},\widehat{z-x}\right)f(z)\,dz \\ &= \int_{\mathbb{R}^{2d}} a\left(\frac{x+z}{2},\xi\right)e^{2\pi i\xi\cdot(x-z)}f(z)\,dzd\xi, \end{aligned}$$

as claimed. To pass to the second line we used the Fourier inversion theorem in the first variable; the notation $a(x,\hat{p})$ signifies a Fourier transform with respect to the second variable. □

Exercise 12.1 Let $a \in \mathcal{S}(\mathbb{R}^{2d})$ and let k be the kernel of $a(x, D)$. Show that

$$a(x,\xi) = 2^d \int_{\mathbb{R}^d} k(x - v, x + v) e^{4\pi i v\cdot\xi} \, dv. \tag{12.7}$$

The reader should note the similarity (as well as the difference) between (12.1) and (12.6). It suggests the possibility of quantization procedures of the form, for any fixed $0 \le t \le 1$,

$$a_t(x, D) f(x) := \int_{\mathbb{R}^{2d}} a(tx + (1-t)y, \xi)\, e^{2\pi i \xi\cdot(x-y)} f(y) \, dy d\xi; \tag{12.8}$$

the Weyl calculus results when $t = \frac{1}{2}$, and the Kohn–Nirenberg calculus results when $t = 1$. There is a way to switch between the different formulations, but we shall simply stick to the Weyl calculus. We wish to extend the representation derived in Lemma 12.1 to much larger classes of symbols $a(x, \xi)$.

12.1.2. Symbol classes

Definition 12.2 For any $m \in \mathbb{R}$, the symbol class S^m is the set of all $a \in C^\infty(\mathbb{R}^{2d})$ that satisfies the estimates

$$|\partial_x^\alpha \partial_\xi^\beta a(x,\xi)| \le C_{\alpha,\beta} \langle \xi \rangle^m \quad \forall (x,\xi) \in \mathbb{R}^{2d}, \tag{12.9}$$

for all α, β.

Note that S^m is closed under the taking of derivatives and satisfies the following multiplicative property: if $a \in S^m$ and $b \in S^n$ then $ab \in S^{m+n}$. Moreover, if $m < n$ then $S^m \subset S^n$. We set $S^\infty = \bigcup_m S^m$ and $S^{-\infty} := \bigcap_m S^m$. There are many alternative symbol classes, such as that where $\langle \xi \rangle^m$ in (12.9) is replaced with $\langle \xi \rangle^{m-|\beta|}$; see (12.14) below. However, for the present we shall work with S^m as defined above.

Lemma 12.3 *Let $m \in \mathbb{R}$ be arbitrary and let $a \in S^m$. Then for any $f \in \mathcal{S}(\mathbb{R}^d)$ one has*

$$a(x, D) f(x) = \int_{\mathbb{R}^d} \left(\int_{\mathbb{R}^d} a\left(\frac{x+y}{2}, \xi \right) e^{2\pi i \xi\cdot(x-y)} f(y) \, dy \right) d\xi \tag{12.10}$$

as an iterated integral. In fact, with a constant that depends on f, one has

$$\sup_x \left| \int_{\mathbb{R}^d} a\left(\frac{x+y}{2}, \xi \right) e^{2\pi i \xi\cdot(x-y)} f(y) \, dy \right| \le C_k \langle \xi \rangle^{-k}$$

for every $k \ge 0$. Finally, $a(x, D) f \in L^\infty$.

Proof Define

$$\mathcal{L}_\xi := 1 - \frac{\xi}{2\pi i} \cdot \nabla_y,$$

which is symmetric. Then

$$\mathcal{L}_\xi^\ell \, e^{2\pi i \xi\cdot(x-y)} = \langle\xi\rangle^{2\ell} e^{2\pi i\xi\cdot(x-y)},$$

whence

$$\begin{aligned}\int_{\mathbb{R}^d} a\left(\frac{x+y}{2},\xi\right) e^{2\pi i\xi\cdot(x-y)} f(y)\,dy \\ = \langle\xi\rangle^{-2\ell}\int_{\mathbb{R}^d}\mathcal{L}_\xi^\ell\left(a\left(\frac{x+y}{2},\xi\right)f(y)\right)e^{2\pi i\xi\cdot(x-y)}\,dy.\end{aligned}$$

Noting that the right-hand side is $O(\langle\xi\rangle^{-\ell+m})$ uniformly in x, we are done. □

We may also evaluate (12.10) distributionally, i.e., by taking Fourier transforms in ξ, which leads to the kernel formulation of (12.6). The integral there of course then needs to be interpreted in the sense of distributions. If the symbol does not depend on ξ, then

$$\begin{aligned}a(x,D)f(x) &= \int_{\mathbb{R}^{2d}} a\left(\frac{x+y}{2}\right) e^{2\pi i\xi\cdot(x-y)} f(y)\,d\xi dy \\ &= \int_{\mathbb{R}^d} a\left(\frac{x+y}{2}\right)\delta_0(x-y)f(y)\,dy = a(x)f(x),\end{aligned}$$

as expected. However, if a does not depend on x, say as for $a(x,\xi)=\xi_k$, then

$$\begin{aligned}a(x,D)f(x) &= \int_{\mathbb{R}^{2d}} \xi_k e^{2\pi i\xi\cdot(x-y)} f(y)\,d\xi dy \\ &= \int_{\mathbb{R}^{2d}} -\frac{\partial}{2\pi i\partial y_k}\left(e^{2\pi i\xi\cdot(x-y)}\right) f(y)\,d\xi dy \\ &= \int_{\mathbb{R}^{2d}} e^{2\pi i\xi\cdot(x-y)}\frac{\partial}{2\pi i\partial y_k} f(y)\,d\xi dy = D_k f(x).\end{aligned}$$

The previous calculation generalizes to ξ^β in an obvious way. Finally, for $a(x,\xi)=\chi(x)\cdot\xi$ where $\chi\in(C^\infty\cap L^\infty)(\mathbb{R}^d)$ one easily verifies that, for all $f\in\mathcal{S}(\mathbb{R}^d)$,

$$a(x,D)f(x) = \tfrac{1}{2}(\chi(x)\cdot D + D\cdot\chi(x))f(x).$$

We imposed the boundedness condition on χ for no other reason than to ensure that a is in the symbol class. It is evident, however, that exactly the same answer is obtained using the definition (12.6) for any $a\in C^1(\mathbb{R}^d)$.

12.1.3. Commutators

We now formulate a basic result on the commutators between a general S^m symbol and the operators X_j and D_k. The reader will easily be able to check the formulas in (12.11) for the special case of the commutator $[X_j, D_k]$.

Lemma 12.4 *If $a \in S^m$ and $f \in \mathcal{S}(\mathbb{R}^d)$ then*

$$\begin{aligned}[X_j, a(x,D)]f(x) &= -(D_{\xi_j} a)(x,D)f(x),\\ [D_j, a(x,D)]f(x) &= (D_{x_j} a)(x,D)f(x).\end{aligned} \tag{12.11}$$

Proof Using the symbol calculus of Lemma 12.3, one finds that

$$\begin{aligned}[X_j, a(x,D)]f(x) &= x_j \int_{\mathbb{R}^{2d}} a\left(\frac{x+y}{2}, \xi\right) e^{2\pi i \xi\cdot(x-y)} f(y)\,dy d\xi\\ &\quad - \int_{\mathbb{R}^{2d}} a\left(\frac{x+y}{2}, \xi\right) e^{2\pi i \xi\cdot(x-y)} y_j f(y)\,dy d\xi\\ &= \int_{\mathbb{R}^{2d}} a\left(\frac{x+y}{2}, \xi\right) \frac{\partial}{2\pi i \partial \xi_j}\left(e^{2\pi i \xi\cdot(x-y)}\right) f(y)\,dy d\xi\\ &= -\int_{\mathbb{R}^{2d}} (D_{\xi_j} a)\left(\frac{x+y}{2}, \xi\right) e^{2\pi i \xi\cdot(x-y)} f(y)\,dy d\xi.\end{aligned} \tag{12.12}$$

The reader will easily verify that these calculations are rigorous, for example distributionally,

For the second identity, we calculate

$$\begin{aligned}[D_j, a(x,D)]f(x) &= \frac{\partial}{2\pi i \partial x_j} \int_{\mathbb{R}^{2d}} a\left(\frac{x+y}{2}, \xi\right) e^{2\pi i \xi\cdot(x-y)} f(y)\,dy d\xi\\ &\quad - \int_{\mathbb{R}^{2d}} a\left(\frac{x+y}{2}, \xi\right) e^{2\pi i \xi\cdot(x-y)} \frac{\partial}{2\pi i \partial y_j} f(y)\,dy d\xi\\ &= \int_{\mathbb{R}^{2d}} (D_{x_j} a)\left(\frac{x+y}{2}, \xi\right) e^{2\pi i \xi\cdot(x-y)} f(y)\,dy d\xi;\end{aligned} \tag{12.13}$$

we integrated by parts in order to pass to the last line. □

As an immediate consequence we can show that $a(x,D)$ preserves the Schwartz space for any symbol class S^m.

Corollary 12.5 *For any $a \in S^m$ one has $a(x,D) : \mathcal{S}(\mathbb{R}^d) \to \mathcal{S}(\mathbb{R}^d)$.*

Proof It suffices to show that $x^\alpha D^\beta a(x,D) f \in L^\infty(\mathbb{R}^d)$ for any α, β. But this follows immediately by induction in $|\alpha| + |\beta|$ and repeated application of the commutation relations (12.11) as well as by the property that $a(x,D)f \in L^\infty$ for any $f \in \mathcal{S}(\mathbb{R}^d)$. □

12.1.4. Convergence of symbols

Next, we need to introduce the notion of convergence in S^m.

Definition 12.6 We say that $a_k \to a$ in S^m if and only if:

(i) $|\partial_x^\alpha \partial_\xi^\beta a_k(x,\xi)| \le C_{\alpha,\beta} \langle \xi \rangle^m$ for all α, β and all $\xi \in \mathbb{R}^d$, uniformly in k and x;

(ii) $a_k \to a$ uniformly on compact subsets of $\mathbb{R}^{2d}$ together with the same for all derivatives.

Obviously we would like to know whether $a_k(x, D) \to a(x, D)$ in that case. The following lemma shows that this holds in the sense of *strong convergence* on the Schwartz space.

Lemma 12.7 *If $a_k \to a$ in S^m then $a_k(x, D)f \to a(x, D)f$ in $\mathcal{S}(\mathbb{R}^d)$. Similarly, $a(x, D)f_k \to a(x, D)f$ for any sequence $f_k \to f$ in $\mathcal{S}(\mathbb{R}^d)$.*

Proof Recall that $f_k \to f$ in $\mathcal{S}(\mathbb{R}^d)$ if and only if, for all α, β, one has

$$x^\alpha D^\beta f_k \to x^\alpha D^\beta f$$

uniformly on $\mathbb{R}^d$ as $k \to \infty$. Let us begin with the first statement. By the commutator property of Lemma 12.4 and induction in $|\alpha| + |\beta|$ it suffices to show that

$$\|a_k(x, D)f - a(x, D)f\|_\infty \to 0 \text{ as } k \to \infty$$

for any $f \in \mathcal{S}(\mathbb{R}^d)$. However, this follows easily from the proof of Lemma 12.3. Indeed, one first integrates by parts sufficiently many times using $\mathcal{L}_\xi$; then, with $\ell > d + m$, we have

$$a_k(x, D)f(x) = \int_{\mathbb{R}^{2d}} \langle \xi \rangle^{-2\ell} \mathcal{L}_\xi^\ell \left(a_k\left(\frac{x+y}{2}, \xi\right) f(y)\right) e^{2\pi i \xi\cdot(x-y)}\, dy d\xi.$$

The integrand on the right-hand side is dominated uniformly in k by an $L^1_{y,\xi}(\mathbb{R}^{2d})$ function (see Definition 12.6(i)). This suffices to conclude the desired convergence uniformly on compact sets of x. To obtain the desired convergence uniformly on $\mathbb{R}^d$, we integrate by parts in ξ sufficiently many times using the symmetric operator

$$L_{x,y} = 1 + \frac{x-y}{2\pi i} \cdot \nabla_\xi,$$

which satisfies

$$L_{x,y}^n\, e^{2\pi i \xi\cdot(x-y)} = \langle x - y \rangle^{2n} e^{2\pi i \xi\cdot(x-y)}$$

for all integers $n \geq 0$. Therefore, for any such n,

$$a_k(x, D)f(x) = \int_{\mathbb{R}^{2d}} \langle x - y\rangle^{-2n} L_{x,y}^n \left(\langle \xi \rangle^{-2\ell} \mathcal{L}_\xi^\ell \left(a_k \left(\frac{x+y}{2}, \xi \right) f(y) \right) \right) e^{2\pi i \xi \cdot (x-y)}\, dy d\xi.$$

Taking $n \geq 1$ arbitrary shows that, for fixed $f \in \mathcal{S}(\mathbb{R}^d)$ and $\varepsilon > 0$, we may choose $R > 0$ sufficiently large that the integral on the right-hand side is less than ε in absolute value uniformly in $|x| > R$. The same holds for the contribution of $|y| > R$ and $|\xi| > R$ uniformly in $x \in \mathbb{R}^d$. Thus, we may pass to the uniform limit $k \to \infty$, which gives $a(x, D)f(x)$.

The second statement is proved in a completely analogous way. □

Lemma 12.7 is useful in conjunction with the following approximation fact.

Lemma 12.8 *If $a \in S^m$ then there exists $a_k \in \mathcal{S}(\mathbb{R}^{2d})$ with $a_k \to a$ in S^m.*

Proof Let $\varphi \in C^\infty(\mathbb{R}^{2d})$ be compactly supported with $\varphi(x, \xi) = 1$ on $|x| + |\xi| \leq 1$ and set $a_k(x, \xi) = \varphi(x/k, \xi/k)a(x, \xi)$ for any integer $k \geq 1$. Verifying that the conditions of Definition 12.6 are met is elementary, and we leave the details to the reader. □

The relevance of Lemmas 12.7 and 12.8 lies in the fact that it is sometimes preferable to work with (12.3) than with (12.6).

Exercise 12.2 Prove Lemma 12.8 for the symbol classes S_ρ^m, $0 \leq \rho \leq 1$, defined by

$$\left|\partial_x^\alpha \partial_\xi^\beta a(x, \xi)\right| \leq C_{\alpha,\beta} \langle \xi \rangle^{m-\rho|\beta|} \quad \forall (x, \xi) \in \mathbb{R}^{2d}, \tag{12.14}$$

for all α, β. Also formulate and prove the analogues for this symbol class of the results preceding that lemma.

12.2. Adjoints and compositions

We now investigate the adjoints and products of the Weyl operators $a(x, D)$.

12.2.1. Adjoints

Lemma 12.9 *If $a \in S^m$ and $\phi, \psi \in \mathcal{S}(\mathbb{R}^d)$ then*

$$\langle a(x, D)\phi, \psi\rangle = \langle \phi, \bar{a}(x, D)\psi\rangle.$$

In particular, if a is real-valued then $a(x, D)$ is a symmetric operator.

Proof We may assume that $a \in \mathcal{S}(\mathbb{R}^{2d})$. Then, with absolutely convergent integrals,

$$\begin{aligned}\langle a(x, D)\phi, \psi\rangle &= \int_{\mathbb{R}^d}\int_{\mathbb{R}^{2d}} a\left(\frac{x+y}{2}, \xi\right) e^{2\pi i\xi\cdot(x-y)}\phi(y)\,dyd\xi\,\overline{\psi(x)}\,dx \\ &= \int_{\mathbb{R}^d}\phi(x)\overline{\int_{\mathbb{R}^{2d}} \bar{a}\left(\frac{x+y}{2}, \xi\right) e^{2\pi i\xi\cdot(x-y)}\psi(y)\,dyd\xi}dx \\ &= \langle\phi, \bar{a}(x, D)\psi\rangle,\end{aligned}$$

as claimed. □

We remark that the previous lemma singles out the Weyl calculus amongst the quantization rules (12.8). Indeed, only for $t = \frac{1}{2}$ does one have the property that $a_t(x, D)$ is symmetric for real-valued $a(x, \xi)$.

12.2.2. Moyal product of Schwarz symbols

Next, we turn to the following question: what does the composition $a(x, D) \circ b(x, D)$ look like in the Weyl calculus? We may expect that this has something to do with $(ab)(x, D)$. Indeed, in the so-called semiclassical sense this is the "leading order" of $a(x, D) \circ b(x, D)$.

We begin by computing the composition $\rho(q, p) \circ \rho(q', p')$, which involves the symplectic form

$$\sigma((q, p), (q', p')) = p \cdot q' - q \cdot p' = \langle u, Jv\rangle, \quad u = (q, p),\ v = (q', p'),$$

where

$$J = \begin{bmatrix} 0 & -\mathrm{Id} \\ \mathrm{Id} & 0 \end{bmatrix}$$

is a $2d \times 2d$ matrix and Id is the identity on $\mathbb{R}^d$. Note that $J^* = -J$ and $J^2 = -\mathrm{Id}$.

Lemma 12.10 *For any $u, v \in \mathbb{R}^{2d}$ one has*

$$\rho(u) \circ \rho(v) = e^{\pi i\sigma(u,v)}\rho(u + v).$$

Proof One has

$$\begin{aligned}(\rho(u) \circ \rho(v))f(x) &= e^{2\pi i(q\cdot x + q\cdot p/2)}(\rho(v)f)(x + p) \\ &= e^{2\pi i(q\cdot x+q\cdot p/2)}e^{2\pi i(q'\cdot(x+p)+q'\cdot p'/2)}\, f(x + p + p') \\ &= e^{\pi i(p\cdot q' - q\cdot p')}e^{2\pi i(x\cdot(q+q')+(p+p')\cdot(q+q')/2)}\, f(x + p + p'),\end{aligned}$$

as claimed. □

Next, we compose Schwartz symbols.

Lemma 12.11 *If* $a, b \in \mathcal{S}(\mathbb{R}^{2d})$ *then*

$$a(x, D) \circ b(x, D) = (a \# b)(x, D),$$

where the Moyal product $a \# b \in \mathcal{S}(\mathbb{R}^{2d})$ *is defined as follows:*

$$(a \# b)(u) := 4^d \int_{\mathbb{R}^{4d}} a(u+v)b(u+w)e^{-4\pi i\sigma(v,w)}\, dv dw. \tag{12.15}$$

Furthermore,

$$\widehat{a \# b}(v) = \hat{a} \natural \hat{b}(v) := \int_{\mathbb{R}^{2d}} \hat{a}(u)\hat{b}(v-u)e^{\pi i\sigma(u,v)}\, du, \tag{12.16}$$

where the expression on the right-hand side is known as a twisted convolution.

Proof It is elementary to verify that $\hat{a}\natural\hat{b} \in \mathcal{S}(\mathbb{R}^{2d})$, and we leave this to the reader. Next, using Lemma 12.10 one obtains

$$\begin{aligned}
a(x, D) \circ b(x, D) f(x) &= \int_{\mathbb{R}^{4d}} \hat{a}(u)\hat{b}(v)\rho(u) \circ \rho(v) f(x)\, du dv \\
&= \int_{\mathbb{R}^{4d}} \hat{a}(u)\hat{b}(v)e^{\pi i\sigma(u,v)}\rho(u+v) f(x)\, du dv \\
&= \int_{\mathbb{R}^{4d}} \hat{a}(u)\hat{b}(v-u)e^{\pi i\sigma(u,v-u)}\rho(v) f(x)\, du dv \\
&= \int_{\mathbb{R}^{4d}} \hat{a}(u)\hat{b}(v-u)e^{\pi i\sigma(u,v)}\rho(v) f(x)\, du dv \\
&= k(x, D) f(x),
\end{aligned}$$

where $\hat{k}(v) := \hat{a} \natural \hat{b}(v)$. We now invert this Fourier transform:

$$\begin{aligned}
k(w) &= \int_{\mathbb{R}^{4d}} \hat{a}(u)\hat{b}(v-u)e^{\pi i\sigma(u,v)}\, e^{2\pi i v\cdot w}\, du dv \\
&= \int_{\mathbb{R}^{4d}} \hat{a}(u)\hat{b}(v-u)\, e^{2\pi i v\cdot(w-Ju/2)}\, du dv \\
&= \int_{\mathbb{R}^{2d}} \hat{a}(u)b(w - \tfrac{1}{2}Ju)e^{2\pi i u\cdot(w-Ju/2)}\, du
\end{aligned}$$

Changing variables according to $z = -\frac{1}{2}Ju$, $u = 2Jz$, $du = 4^d dz$ yields

$$\begin{aligned}
k(w) &= 4^d \int_{\mathbb{R}^{2d}} \hat{a}(2Jz)b(w+z)e^{4\pi i \langle Jz, w\rangle}\, dz \\
&= 4^d \int_{\mathbb{R}^{4d}} a(w+y)e^{-2\pi i \langle 2Jz, w+y\rangle} b(w+z)e^{4\pi i \langle Jz, w\rangle}\, dz dy \\
&= 4^d \int_{\mathbb{R}^{4d}} a(w+y)b(w+z)e^{4\pi i \sigma(z,y)}\, dz dy
\end{aligned}$$

as desired. □

In what sense is $(ab)(x, \xi)$ the "leading order" contribution to (12.15)? By inspection, $(v, w) \mapsto \sigma(v, w)$ on $\mathbb{R}^{4d}$ has a nondegenerate critical point at $(0, 0)$, and $(0, 0)$ is the only critical point. By the stationary phase method we therefore expect that the dominant contribution to the integral in (12.15) comes from the origin, which gives (at least up to constants) $(ab)(u)$. However, what does "dominant" or "leading order" mean here?

12.2.3. The semiclassical calculus

One common approach to this problem is to introduce a large parameter via the so-called "semiclassical" Weyl calculus (this corresponds to viewing the Planck constant $\hbar$ as a small parameter). Thus, with $0 < \hbar \leq 1$, we define

$$\begin{aligned}
a(x, \hbar D)f(x) &:= \hbar^{-d} \int_{\mathbb{R}^d}\int_{\mathbb{R}^d} a\left(\frac{x+y}{2}, \xi\right) e^{2\pi i \xi\cdot(x-y)/\hbar} f(y)\, dy\, d\xi \\
&= \int_{\mathbb{R}^d}\int_{\mathbb{R}^d} a\left(\frac{x+y}{2}, \hbar\xi\right) e^{2\pi i \xi\cdot(x-y)} f(y)\, dy\, d\xi.
\end{aligned}$$

If we set $\hbar = 0$ in the second integral, we obtain

$$\int_{\mathbb{R}^d}\int_{\mathbb{R}^d} a\left(\frac{x+y}{2}, 0\right) e^{2\pi i \xi\cdot(x-y)} f(y)\, dy\, d\xi = a(x, 0)f(x),$$

which corresponds to the "classical" picture in which symbols are smooth functions on the state space and the action is simply multiplication by the symbol. The semiclassical composition laws are of the form

$$\begin{aligned}
a(x, \hbar D) \circ b(x, \hbar D) &= (a \#_\hbar b)(x, \hbar D), \\
(a \#_\hbar b)(u) &:= \left(\frac{2}{\hbar}\right)^{2d} \int_{\mathbb{R}^{4d}} a(u+v)b(u+w)e^{-4\pi i \sigma(v,w)/\hbar}\, dv dw.
\end{aligned} \tag{12.17}$$

The basic correspondences underlying this calculus are

$$\begin{aligned} x_j &\mapsto X_j, \\ \xi_j &\mapsto \hbar D_j, \\ 1 &\mapsto \mathrm{Id}. \end{aligned}$$

The semiclassical analogues of the operators ρ take the form

$$(\rho_\hbar(q, p)f)(x) := e^{2\pi i(x\cdot q+\hbar q\cdot p/2)} f(x + \hbar p)$$

and satisfy

$$\begin{aligned} a(x, \hbar D)f &= \int_{\mathbb{R}^{2d}} \hat{a}(q, p)\rho_\hbar(q, p)f \, dq dp \\ \rho_\hbar(u) \circ \rho_\hbar(v) &= e^{\pi i\hbar\sigma(u,v)} \rho_\hbar(u + v), \end{aligned}$$

with Schwartz functions f and a.

Exercise 12.3 Verify the product relation in (12.17) as well as the statements concerning the operators $\rho_\hbar$. Also show that, for $a \in S^m$ fixed, one has the property that $a(x, \hbar D) : \mathcal{S}(\mathbb{R}^d) \to \mathcal{S}(\mathbb{R}^d)$ *uniformly* in $\hbar$, in other words, if $f \in \mathcal{S}(\mathbb{R}^d)$ is fixed then any Schwartz seminorm of $a(x, \hbar D)f(x)$ is bounded uniformly in $\hbar \in (0, 1]$. Generalize to symbols $a(x, \xi; \hbar)$ that belong to S^m uniformly in $\hbar$.

12.2.4. Stationary phase expansions

It is now natural to expand $a \#_\hbar b$ in powers of $\hbar$, and we expect that $(a \#_\hbar b) = ab + O(\hbar)$ in a suitable sense as $\hbar \to 0$. This is an application of the stationary phase method. However, the stationary phase estimates of Chapter 4 are insufficient, since we require an *expansion* in small quantities. Owing to the fact that the phase $\sigma(v, w)$ in $a \#_\hbar b$ is purely quadratic, this can be achieved in a fairly elementary fashion (via Fourier transforms of complex Gaussians). To demonstrate this technique, we start with an illustrative example and then apply it to $a \#_\hbar b$.

Lemma 12.12 *Let $a \in \mathcal{S}(\mathbb{R}^d)$ and $0 < \hbar \leq 1$. Then, for any integer $N \geq 0$, one has*

$$\hbar^{-d/2} e^{-\pi i d/4} \int_{\mathbb{R}^d} a(x) e^{\pi i|x|^2/\hbar} \, dx = \sum_{n=0}^{N} \frac{(-\pi i\hbar)^n}{n!} (D^{2n} a)(0) + O(\hbar^{N+1}) \tag{12.18}$$

as $\hbar \to 0$. In particular, for any $a \in \mathcal{S}(\mathbb{R})$,

$$\hbar^{-1/2}e^{-\pi i/4}\int_{-\infty}^{\infty} a(x)e^{\pi i|x|^2/\hbar}\,dx = a(0) + \frac{i\hbar}{4\pi}a''(0) + O(\hbar^2) \tag{12.19}$$

as $\hbar \to 0$.

Proof We shall first carry out the proof of (12.19). First, set

$$G_\hbar(x) := \hbar^{-1/2}e^{\pi i/4}e^{-\pi i x^2/\hbar}.$$

Then, by the Plancherel theorem (interpreted in the sense of an $\mathcal{S}$–$\mathcal{S}'$ pairing),

$$\int_{-\infty}^{\infty} a(x)\overline{G_\hbar(x)}\,dx = \int_{-\infty}^{\infty} \hat{a}(\xi)\overline{\widehat{G_\hbar}(\xi)}\,d\xi. \tag{12.20}$$

Note that the left-hand side coincides with that in (12.19). We now claim that $\widehat{G_\hbar}(\xi) = e^{\pi i\hbar\xi^2}$. To see this, let $\varepsilon > 0$ and compute the Fourier transform of the complex Gaussian with $\hbar = 1$, modified by a decaying Gaussian regularization:

$$\begin{aligned}\int_{\mathbb{R}} e^{-\pi i(1-i\varepsilon^2)x^2}e^{-2\pi i x\xi}\,dx &= \int_{\mathbb{R}} e^{-\pi i(1-i\varepsilon^2)(x+(1-i\varepsilon^2)^{-1}\xi)^2}\,dx\; e^{\pi i(1-i\varepsilon^2)^{-1}\xi^2} \\ &= (i(1-i\varepsilon^2))^{-1/2}\int_{\mathbb{R}} e^{-\pi y^2}\,dy\; e^{\pi i(1-i\varepsilon^2)^{-1}\xi^2}.\end{aligned}$$

The square roots are to be interpreted in the principal value sense. To pass to the second equality sign we used two deformations of the contour in the complex domain for holomorphic integrands: the first consists of a shift of the contour from $x + (1-i\varepsilon^2)^{-1}\xi$ to x, as shown in Figure 12.1. The second consists of a rotation of the line $t \mapsto (i(1-i\varepsilon^2))^{1/2}t$ into the real axis. We leave it to the reader to justify fully these applications of Cauchy's theorem. To complete the calculation we use $\int_{-\infty}^{\infty} e^{-\pi y^2}\,dy = 1$ and then pass to the limit $\varepsilon \to 0$ to obtain the $\mathcal{S}'$-valued Fourier transform:

$$\widehat{G_1}(\xi) = e^{\pi i\xi^2}$$

and, by rescaling, $\widehat{G_\hbar}(\xi) = e^{\pi i\hbar\xi^2}$ as claimed. Therefore, inserting this into (12.20) yields

$$e^{-\pi i/4}\int_{-\infty}^{\infty} a(x)\hbar^{-1/2}e^{\pi i x^2/\hbar}\,dx = \int_{\infty}^{\infty} \hat{a}(\xi)e^{-\pi i\hbar\xi^2}\,d\xi.$$

In order to derive an expansion in powers of $\hbar$, we Taylor-expand the exponential:

$$e^{-\pi i\hbar\xi^2} = \sum_{n=0}^{N}\frac{(-\pi i\hbar)^n}{n!}\xi^{2n} + R_N(\xi,\hbar),$$

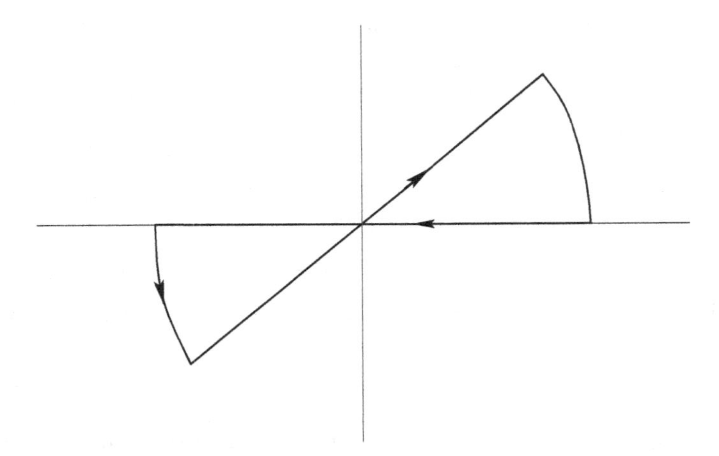

Figure 12.1. Integration contour.

where $|R_N(\xi, \hbar)| \leq C(N)\hbar^{N+1}\,\xi^{2N+2}$. Since $\xi^k \hat{a}(\xi) = \widehat{D^k a}(\xi)$, it follows that

$$\int_{\infty}^{\infty} \hat{a}(\xi)e^{-\pi i \hbar \xi^2}\, d\xi = \sum_{n=0}^{N} \frac{(-\pi i \hbar)^n}{n!}(D^{2n}a)(0) + \int_{\infty}^{\infty} \hat{a}(\xi)R_N(\xi, \hbar)\, d\xi$$

and (12.18) is proved for $d = 1$ and therefore also (12.19). For $d > 1$ one proceeds in an identical fashion using (with $\mathcal{F}$ the Fourier transform in $\mathcal{S}'(\mathbb{R}^d)$)

$$\mathcal{F}(\hbar^{-d/2}e^{d\pi i/4}e^{-\pi i x^2/\hbar})(\xi) = e^{\pi i \hbar |\xi|^2}$$

which reduces to the identity $d = 1$ that we have just derived by factorization. □

While we shall not make any direct use of Lemma 12.12, we now apply its proof scheme to $a \,\#_\hbar\, b$.

Proposition 12.13 *Let $a, b \in \mathcal{S}(\mathbb{R}^{2d})$. Then*

$$a \,\#_\hbar\, b = ab + \frac{\hbar}{4\pi i}\{a, b\} + \hbar^2 r_2(\cdot, \hbar), \tag{12.21}$$

where

$$\{a, b\}(x, \xi) = \frac{\partial a}{\partial \xi}\frac{\partial b}{\partial x}(x, \xi) - \frac{\partial a}{\partial x}\frac{\partial b}{\partial \xi}(x, \xi)$$

is the Poisson bracket *of a and b and $r_2(\cdot, \hbar) \in \mathcal{S}(\mathbb{R}^{2d})$ satisfies*

$$\sup_{u \in \mathbb{R}^{2d}} \left| u^\alpha \partial_u^\beta r_2(u, \hbar) \right| \leq C_{\alpha,\beta} \tag{12.22}$$

for all α, β, uniformly in $0 < \hbar \leq 1$. More generally,

$$a \#_\hbar b = \sum_{n=0}^{N-1} \frac{(\pi i \hbar)^n}{n!} L^{(n)}(ab) + \hbar^N r_N(\cdot, \hbar), \tag{12.23}$$

where $L^{(n)} := L^n \big|_{v=w}$ with $L := D_{v_2} D_{w_1} - D_{v_1} D_{w_2}$ and r_N satisfies (12.22).

Proof We start with the computation of the Fourier transform of $e^{4\pi i \sigma(v,w)/\hbar}$. First, in the distributional sense,

$$\begin{aligned}
&\widehat{e^{4\pi i \sigma(v,w)}}(\zeta, \eta) = \int_{\mathbb{R}^{4d}} e^{4\pi i \sigma(v,w)} e^{-2\pi i (v \cdot \zeta + w \cdot \eta)}\, dv dw \\
&= \int_{\mathbb{R}^{4d}} e^{4\pi i (p \cdot q' - q \cdot p')} e^{-2\pi i (q \cdot \zeta_1 + p \cdot \zeta_2 + q' \cdot \eta_1 + p' \cdot \eta_2)}\, dp dq dp' dq' \\
&= \int_{\mathbb{R}^{2d}} e^{4\pi i p \cdot q'} e^{-2\pi i (p \cdot \zeta_2 + q' \cdot \eta_1)}\, dp dq' \int_{\mathbb{R}^{2d}} e^{-4\pi i p' \cdot q} e^{-2\pi i (p' \cdot \eta_2 + q \cdot \zeta_1)}\, dp' dq \\
&=: I(\zeta_2, \eta_1) \overline{I(-\zeta_1, -\eta_2)}.
\end{aligned}$$

Setting $p_1 := (p+q)/2$ and $p_2 := (p-q)/2$, one has (where the integrals need to be interpreted in the appropriate sense, for example via the same Gaussian regularization procedure as in the previous proof)

$$\begin{aligned}
I(\zeta, \eta) &= \int_{\mathbb{R}^{2d}} e^{4\pi i p \cdot q} e^{-2\pi i (p \cdot \zeta + q \cdot \eta)}\, dp dq \\
&= 2^d \int_{\mathbb{R}^{2d}} e^{4\pi i (p_1^2 - p_2^2)} e^{-2\pi i ((p_1 + p_2) \cdot \zeta + (p_1 - p_2) \cdot \eta)}\, dp_1 dp_2 \\
&= 2^d \int_{\mathbb{R}^d} e^{4\pi i (p_1 - (\zeta + \eta)/4)^2}\, dp_1\, e^{-\pi i (\zeta + \eta)^2/4} \\
&\quad \times \int_{\mathbb{R}^d} e^{-4\pi i (p_2 + (\zeta - \eta)/4)^2}\, dp_2\, e^{\pi i (\zeta - \eta)^2/4} \\
&= 2^{-d} e^{\pi i ((\zeta - \eta)^2 - (\zeta + \eta)^2)/4} = 2^{-d} e^{-\pi i \zeta \cdot \eta}
\end{aligned}$$

and thus, with $u = (u_1, u_2) \in \mathbb{R}^{2d}$, $u_1, u_2 \in \mathbb{R}^d$,

$$
\begin{aligned}
(a \#_\hbar b)(u) &= 4^d \int_{\mathbb{R}^{4d}} \hat{a}(\zeta)\hat{b}(\eta) e^{2\pi i u\cdot(\zeta+\eta)} \overline{I\big(\sqrt{\hbar}\zeta_2, \sqrt{\hbar}\eta_1\big)} I\big(-\sqrt{\hbar}\zeta_1, -\sqrt{\hbar}\eta_2\big) d\zeta d\eta \\
&= \int_{\mathbb{R}^{4d}} \hat{a}(\zeta)\hat{b}(\eta) e^{2\pi i u\cdot(\zeta+\eta)}\, e^{\pi i \hbar(\zeta_2\eta_1 - \zeta_1\eta_2)}\, d\zeta d\eta \\
&= \int_{\mathbb{R}^{4d}} \hat{a}(\zeta)\hat{b}(\eta) e^{2\pi i u\cdot(\zeta+\eta)} \big(1 + \pi i \hbar(\zeta_2\eta_1 - \zeta_1\eta_2) \\
&\qquad + O\big(\hbar^2(\zeta_1\eta_2 - \zeta_2\eta_1)^2\big)\big)\, d\zeta d\eta \\
&= a(u)b(u) + \pi i \hbar \int_{\mathbb{R}^{4d}} \hat{a}(\zeta)\hat{b}(\eta) \bigg(\frac{\partial}{2\pi i \partial u_2} e^{2\pi i u\cdot\zeta} \frac{\partial}{2\pi i \partial u_1} e^{2\pi i u\cdot\eta} \\
&\qquad - \frac{\partial}{2\pi i \partial u_1} e^{2\pi i u\cdot\zeta} \frac{\partial}{2\pi i \partial u_2} e^{2\pi i u\cdot\eta} \bigg)\, d\eta d\zeta + O(\hbar^2) \\
&= (ab)(u) + \frac{\hbar}{4\pi i} \left(\frac{\partial a}{\partial u_2}\frac{\partial b}{\partial u_1} - \frac{\partial a}{\partial u_1}\frac{\partial b}{\partial u_2} \right)(u) + O(\hbar^2),
\end{aligned}
$$

where the $O(\hbar^2)$ term is easily seen to be a Schwartz function uniformly in $0 < \hbar \le 1$.

The general case follows in the same fashion, using higher-order Taylor expansions of the exponential. □

An immediate consequence of the preceding proposition is the following commutator result.

Corollary 12.14 *Let $a, b \in \mathcal{S}(\mathbb{R}^{2d})$. Then for any $f \in \mathcal{S}(\mathbb{R}^d)$ one has*

$$
\begin{aligned}
&[a(x, \hbar D), b(x, \hbar D)] f(x) \\
&\quad = \frac{\hbar}{2\pi i}\{a, b\}(x, \hbar D) f(x) + \hbar^2 g(x; \hbar) \quad \text{as } \hbar \to 0, \qquad (12.24)
\end{aligned}
$$

where $g(\cdot; \hbar) \in \mathcal{S}(\mathbb{R}^d)$ is bounded in the Schwartz space uniformly with respect to $\hbar \in (0, 1]$.

Proof This follows by taking differences in Proposition 12.13 and using Exercise 12.3. □

We see from Proposition 12.13 that the $O(\hbar^2)$ error in (12.24) does not appear if either of the symbols $a(x, \xi)$ or $b(x, \xi)$ is linear. Indeed, if $a(x, \xi) = x_j$ then we find that (with $\hbar = 1$, say)

$$
[a(x, D), b(x, D)] = -\frac{1}{2\pi i}\frac{\partial a}{\partial \xi_j}(x, D) = -(D_{\xi_j} a)(x, D)
$$

and, similarly, if $a(x,\xi) = \xi_j$ then one has

$$[a(x,D), b(x,D)] = \frac{1}{2\pi i}\frac{\partial a}{\partial x_j}(x,D) = (D_{x_j}a)(x,D),$$

which agrees with Lemma 12.4. Generally speaking, one can obtain an expansion to all orders of $\hbar$ in terms of the derivatives of a and b. The $\hbar^2$ term is determined by applying a second-order differential operator to a and b, which explains why it (as well as all other terms) vanish when either a or b is linear.

12.2.5. The composition law and expansions in general symbol classes

In order for the Weyl calculus to be of practical importance, we need to extend the composition law (12.15) to general symbol classes. We shall address this issue next, using approximation by Schwartz symbols.

Lemma 12.15 *Let $\ell, m \in \mathbb{R}$. Then with $\hbar \in (0,1]$ one has the following.*

(i) *For $a, b \in \mathcal{S}(\mathbb{R}^{2d})$, the $S^{\ell+m}$ seminorms of $a \#_\hbar b$ can be bounded in terms of finitely many S^ℓ seminorms of a and finitely many S^m-seminorms of b, uniformly with respect to $\hbar \in (0,1]$.*

(ii) *If $a_n, b_n \in \mathcal{S}(\mathbb{R}^{2d})$ are such that $a_n \to a$ in S^ℓ and $b_n \to b$ in S^m then $a_n \#_\hbar b_n \to a \#_\hbar b$ in $S^{\ell+m}$. The latter composition is given by (12.25) below provided that $k > 2d + \frac{1}{2}(|\ell| + |m|)$.*

Proof If $u, v, w \in \mathbb{R}^{2d}$, we write $u = (x,\xi)$, $v = (y,\eta)$, and $w = (z,\zeta)$. The entire lemma hinges on integrating by parts inside the integral

$$(a \#_\hbar b)(w) = \left(\frac{2}{\hbar}\right)^{2d} \int_{\mathbb{R}^{4d}} a(w+u)b(w+v)e^{-4\pi i\sigma(u,v)/\hbar}\, du dv.$$

By the stationary phase method, the main contribution to the integral comes from the region $|u|^2 + |v|^2 \le C\hbar$. Consequently, we define

$$L := 1 - \frac{\hbar}{16\pi^2}\Delta = 1 + \frac{1}{4}\left(\sum_{j=1}^{2d}\left(\sqrt{\hbar}D_{u_j}\right)^2 + \sum_{j=1}^{2d}\left(\sqrt{\hbar}D_{v_j}\right)^2\right),$$

where Δ is the Laplacian in $\mathbb{R}^{4d}$. By inspection,

$$L\big(e^{-4\pi i\sigma(u,v)/\hbar}\big) = \big(1 + \hbar^{-1}\big(|u|^2 + |v|^2\big)\big)\, e^{-4\pi i\sigma(u,v)/\hbar},$$

whence, with $g_\hbar(u,v) := 1 + \hbar^{-1}(|u|^2 + |v|^2)$ and $\mathcal{L}f := L(fg_\hbar^{-1})$,

$$\begin{aligned}
&(a \#_\hbar b)(w) \\
&= \left(\frac{2}{\hbar}\right)^{2d} \int_{\mathbb{R}^{4d}} (g_\hbar(u,v))^{-1} a(w+u)b(w+v) L\big(e^{-4\pi i\sigma(u,v)/\hbar}\big)\, du dv \\
&= \left(\frac{2}{\hbar}\right)^{2d} \int_{\mathbb{R}^{4d}} \mathcal{L}(a(w+u)b(w+v)) e^{-4\pi i\sigma(u,v)/\hbar}\, du dv
\end{aligned}$$

and thus, for any integer $k \geq 1$,

$$(a \#_\hbar b)(w) = \int_{\mathbb{R}^{4d}} \mathcal{L}^k(a(w+u)b(w+v)) e^{-4\pi i\sigma(u,v)/\hbar}\, du dv. \tag{12.25}$$

We now claim the following estimate:

$$\left|\mathcal{L}^k(a(w+u)b(w+v))\right| \leq C_k \langle\zeta\rangle^{\ell+m} g_\hbar(u,v)^{-k+(|\ell|+|m|)/2}. \tag{12.26}$$

To prove it note that, for any function $f(u,v)$, the expression $\mathcal{L}^k f$ is the sum of constants times terms of the form

$$\left(\sqrt{\hbar}D\right)^\alpha f \prod_{j=1}^k \left(\sqrt{\hbar}D\right)^{\beta^{(j)}} g_\hbar^{-1}, \tag{12.27}$$

where $\alpha, \beta^{(1)}, \ldots, \beta^{(k)}$ are multi-indices. Now a simple induction in k is required; we leave this to the reader. Next, by means of the product rule one may verify directly from the definition of the symbol classes that

$$\left|(\sqrt{\hbar}D)^\alpha \big(a(w+u)b(w+v)\big)\right| \leq C_\alpha (1 + |\zeta + \xi|)^\ell (1 + |\zeta + \eta|)^m \tag{12.28}$$

for all α, uniformly in $0 < \hbar \leq 1$. Clearly $\hbar = 1$ is the most difficult case. As far as the inverse powers of $g_\hbar$ are concerned, we claim the bound

$$|D^\alpha g_1^{-1}| \leq C_\alpha g_1^{-1}. \tag{12.29}$$

This follows from the fact that

$$D^\alpha g_1^{-1} = \frac{p_\alpha}{g_1^{|\alpha|+1}} \quad \forall \alpha, \tag{12.30}$$

where $p_\alpha(u,v)$ is a polynomial of degree $\leq |\alpha|$. This follows by induction in $|\alpha|$. Next, estimating (12.30) in a straightforward fashion yields

$$\left|D^\alpha g_1(u,v)^{-1}\right| \leq C_\alpha \frac{(1+|u|+|v|)^{|\alpha|}}{(1+|u|^2+|v|^2)^{|\alpha|}} g_1(u,v)^{-1} \leq C_\alpha\, g_1(u,v)^{-1},$$

which amounts to (12.29). Finally, we rescale that estimate to conclude that

$$\left|(\sqrt{\hbar}D)^{\alpha} g_{\hbar}^{-1}\right| \le C_{\alpha}\, g_{\hbar}^{-1} \tag{12.31}$$

uniformly in $0 < \hbar \le 1$. Indeed,

$$\begin{aligned}\left|(\sqrt{\hbar}D)^{\alpha} g_{\hbar}(u,v)^{-1}\right| &\le \left|\big(D^{\alpha} g_1^{-1}\big)\big(\hbar^{-1/2}u, \hbar^{-1/2}v\big)\right| \\ &\le C_{\alpha}\, g_1^{-1}\big(\hbar^{-1/2}u, \hbar^{-1/2}v\big) \le C_{\alpha}\, g_{\hbar}(u,v)^{-1},\end{aligned}$$

where the second inequality sign corresponds to (12.29). Finally, putting together (12.27), (12.28), and (12.31) yields (12.26):

$$\begin{aligned}\left|\mathcal{L}^k(a(w+u)b(w+v))\right| &\le C_k \langle \zeta+\xi\rangle^{\ell} \langle \zeta+\eta\rangle^{m}\, g_{\hbar}(u,v)^{-k} \\ &\le C_k \langle \zeta\rangle^{\ell+m} \langle \xi\rangle^{|\ell|} \langle \eta\rangle^{|m|}\, g_{\hbar}(u,v)^{-k} \\ &\le C_k \langle \zeta\rangle^{\ell+m}\, g_{\hbar}(u,v)^{-k+(|\ell|+|m|)/2},\end{aligned}$$

as claimed. To pass to the second inequality we used

$$1 + |\zeta+\xi| \le (1+|\zeta|)(1+|\xi|)$$

for $\ell \ge 0$,

$$1 + |\zeta| \le (1+|\zeta+\xi|)(1+|\xi|)$$

for $\ell < 0$, and similar relations for m.

We now turn to the bound on $a \#_{\hbar} b$. From (12.26) one obtains, for fixed $k > 2d + \frac{1}{2}(|\ell| + |m|)$,

$$|(a \#_{\hbar} b)(w)| \le C_k \hbar^{-2d} \langle \zeta\rangle^{\ell+m} \int_{\mathbb{R}^{4d}} g_{\hbar}(u,v)^{-k+(|\ell|+|m|)/2}\, du dv \le C_k\, \langle \zeta\rangle^{\ell+m},$$

where the final estimate follows on changing variables according to $u = \sqrt{\hbar}\, u'$, $v = \sqrt{\hbar}\, v'$.

For the convergence (again without derivatives), assume that $a_n \to a$ in S^{ℓ} and $b_n \to b$ in S^m. Then, with k sufficiently large,

$$\begin{aligned}&\left|(a_n \#_{\hbar} b_n)(w) - \left(\frac{2}{\hbar}\right)^{2d} \int_{\mathbb{R}^{4d}} \mathcal{L}^k(a(w+u)b(w+v)) e^{-4\pi i \sigma(u,v)/\hbar}\, du dv\right| \\ &\le C_k \left(\frac{2}{\hbar}\right)^{2d} \int_{\mathbb{R}^{4d}} \left|\mathcal{L}^k(a_n(w+u)b_n(w+v)) - \mathcal{L}^k(a(w+u)b(w+v))\right| du dv.\end{aligned}$$

The integrand converges to 0 uniformly on compact sets in (u, v, w)-space. Thus, with $M < \infty$ and $\hbar$ fixed,

$$\left(\frac{2}{\hbar}\right)^{2d} \int_{\mathbb{R}^{4d}} \chi_{[|u|+|v|>M]} \left|\mathcal{L}^k(a_n(w+u)b_n(w+v)) - \mathcal{L}^k(a(w+u)b(w+v))\right| \, dudv \to 0$$

as $n \to \infty$ uniformly on compact sets of w, say $|w| < R$. However, if $|w| < R$ then

$$\left(\frac{2}{\hbar}\right)^{2d} \int_{|u|+|v|>M} \left|\mathcal{L}^k(a_n(w+u)b_n(w+v)) - \mathcal{L}^k(a(w+u)b(w+v))\right| \, dudv$$
$$\leq C_R \hbar^{-2d} \int_{|u|+|v|>M-R} g_\hbar(u, v)^{-k+(|\ell|+|m|)/2} \, dudv < \varepsilon,$$

provided that M is sufficiently large. This proves that

$$(a_n \#_\hbar b_n)(w) \to \left(\frac{2}{\hbar}\right)^{2d} \int_{\mathbb{R}^{4d}} \mathcal{L}^k(a(w+u)b(w+v)) e^{-4\pi i \sigma(u,v)/\hbar}, \, dudv = (a \#_\hbar b)(w)$$

uniformly on $|w| < R$, as desired.

Finally, it remains to take derivatives into account. For part (i) this means bounding $D_z^\alpha D_\zeta^\beta (a \#_\hbar b)$. By differentiation under the integral sign, we find that this expression is the sum of terms of the form $D_x^\gamma D_\xi^\delta a \#_\hbar D_x^{\alpha-\gamma} D_\xi^{\beta-\delta} b$. Since the factors are in S^m and S^ℓ, respectively, the previous a priori bound applies. The convergence statement is treated analogously. □

Exercise 12.4 Adapt Lemma 12.15 to the case of symbols as in (12.14). Do the same with Lemma 12.16 below.

The following proposition is the desired composition law for general symbol classes. It of course holds for the $\hbar$-calculus, but for our purposes it suffices to set $\hbar = 1$.

Lemma 12.16 *The map $(a, b) \mapsto a \# b$ extends to a continuous map $S^\ell \times S^m \to S^{\ell+m}$. Furthermore, the extended $a \# b$ is given by* (12.25), *and one has $(a \# b)(x, D) = a(x, D) \circ b(x, D)$. The same applies to the more general symbol classes* (12.14).

Proof We have already shown that if $a_n \to a$ in S^ℓ and $b_n \to b$ in S^m as $n \to \infty$, where $a_n, b_n \in \mathcal{S}(\mathbb{R}^d)$, then $a_n \# b_n$ converges in $S^{\ell+m}$ and the limit is given by (12.25). We define $a \# b$ to be that limit. Note that this is well defined, in the sense that it does not depend on the choice of k as long as the latter is large enough. Now suppose that $a_n \to a \in S^\ell$ and $b_n \to b$ in S^m. Then choose $\tilde{a}_{n,j} \in \mathcal{S}$ with $\tilde{a}_{n,j} \to a_n$ in S^ℓ and $\tilde{b}_{n,j} \to b_n$ in S^m with $\tilde{b}_{n,j} \in \mathcal{S}$. Both limits are taken as $j \to \infty$. On the one hand, $\tilde{a}_{n,j} \# \tilde{b}_{n,j} \to a_n \# b_n$ as $j \to \infty$. On the other hand, there exists $j_n \to \infty$ with $\tilde{a}_{n,j_n} \to a$ in S^ℓ and $\tilde{b}_{n,j_n} \to b$ in S^m as $n \to \infty$. Therefore, $a_n \# b_n \to a \# b$ in $S^\ell \times S^m$ as $n \to \infty$.

We now prove the composition statement for the corresponding operators. First, suppose that $a \in \mathcal{S}$ and choose $b_n \in \mathcal{S}$ with $b_n \to b$ in S^m. Then, for any $f \in \mathcal{S}$, one has

$$\begin{aligned}
(a \# b)(x, D) f(x) &= \big(\lim_{n\to\infty} a \# b_n\big)(x, D) f(x) && (\text{convergence in } S^{\ell+m}) \\
&= \lim_{n\to\infty} (a \# b_n)(x, D) f(x) && (\text{convergence in } \mathcal{S}) \\
&= \lim_{n\to\infty} a(x, D)(b_n(x, D) f(x)) && (\text{convergence in } \mathcal{S}) \\
&= a(x, D) \circ b(x, D) f(x) && (\text{convergence in } \mathcal{S}).
\end{aligned} \tag{12.32}$$

Here we used Lemma 12.7 for the convergence in $\mathcal{S}$, and the composition law for Schwartz symbols to pass to the third line. Now let a, b be arbitrary and choose $a_n \in \mathcal{S}$ with $a_n \to a$ in S^ℓ. Then

$$\begin{aligned}
(a \# b)(x, D) f(x) &= \lim_{n\to\infty} (a_n \# b)(x, D) f(x) = \lim_{n\to\infty} a_n(x, D)(b(x, D) f(x)) \\
&= (a(x, D) \circ b(x, D)) f(x),
\end{aligned}$$

where we used (12.32) to pass to the second equality. □

Exercise 12.5 Let $a \in S^m$ and $b \in S^\ell$. Show that $\overline{a \# b} = \overline{b} \# \overline{a}$.

12.2.6. Expansions relative to the order

We would now like to revisit the question of extracting the leading terms from $a \# b$, but without involving $\hbar$ and for general symbols $a \in S^\ell_\rho$ and $b \in S^m_\rho$; see (12.14) for the definition of these symbol classes. Since we now have no small parameter at our disposal relative to which we can expand, we shall use the order of the symbol class to express the level of the expansion. For example, we shall prove that $a \# b - ab \in S^{\ell+m-\rho}$. Not surprisingly, the case $\rho = 0$ stands out here since in that case no improvement can be obtained (in other words, the "error" is indistinguishable from the main term, at least from the perspective of the symbol classes).

We start with a general discussion of asymptotic expansions of symbols. If $a, b \in S^\infty_\rho$ then we say that $a = b \bmod S^k_\rho$ if and only if $a - b \in S^k_\rho$. The point here is that symbols in S^k_ρ with k large and negative have very good smoothing properties and can thus be regarded as "error terms". We shall encounter this idea in the construction of parametrices for elliptic operators.

Definition 12.17 Suppose that $a \in S^m_\rho$, $a_j \in S^{m_j}_\rho$ for all $j \geq 0$, where m_j decreases to $-\infty$. If $a = \sum_{j=0}^{N-1} a_j \bmod S^{m_N}_\rho$ then we say that $\sum_{j=0}^\infty a_j$ is an asymptotic expansion of a and write $a \sim \sum_{j=0}^\infty a_j$.

Of course, $\sum_{j=0}^\infty a_j$ need not converge to a in any standard sense. In fact, if $a = b \mod S^{-\infty}_\rho$ and $a \sim \sum_j a_j$ then also $b \sim \sum_j a_j$. Conversely, if $a \sim \sum_{j=1}^\infty a_j$ and $b \sim \sum_{j=1}^\infty a_j$, then $a = b \bmod S^{-\infty}_\rho$. The following lemma states that every formal asymptotic expansion is the asymptotic expansion of some symbol (this fact is called the *Borel lemma*).

Lemma 12.18 *Suppose that* $\{m_j\}_{j=0}^\infty$ *decreases to* $-\infty$ *and* $a_j \in S^{m_j}_\rho$. *Then there exists* $a \in S^{m_0}_\rho$ *with* $a \sim \sum_j a_j$. *Throughout,* $0 \leq \rho \leq 1$ *is arbitrary but fixed.*

Proof Choose a function $\phi \in C^\infty(\mathbb{R}^d)$ of compact support with $\phi(\xi) = 1$ if $|\xi| \leq 1$. Define $\phi_t(\xi) := \phi(\xi/t)$ for $t > 0$. Note that $\{\phi_t\}_{t>1}$ is a bounded set of S^0_1 (in the sense that any seminorm of S^0_1 is bounded uniformly in $t > 0$): for all α, β one has

$$\left|D_x^\alpha D_\xi^\beta \phi_t(\xi)\right| \leq C_{\alpha,\beta} \langle \xi \rangle^{-|\beta|}.$$

The symbol a is defined as follows:

$$a = \sum_j (1 - \phi_{t_j}) a_j, \tag{12.33}$$

where $t_j \to \infty$ is some sequence that needs to be chosen. Note first that if t_j is any sequence that goes to ∞ then the sum in (12.33) converges in the C^∞-topology, i.e., together with all its derivatives it converges uniformly on compact subsets of (x, ξ). This is due to the fact that all but finitely many terms of that infinite series vanish on any given compact set. For the quantitative analysis we note the estimate

$$\left|D_x^\alpha D_\xi^\beta ((1 - \phi_{t_j}) a)(x, \xi)\right| \leq \begin{cases} 0, & |\xi| \leq t_j, \\ C^{(j)}_{\alpha,\beta} \langle \xi \rangle^{m_j - \rho|\beta|}, & |\xi| > t_j. \end{cases} \tag{12.34}$$

Here $C^{(j)}_{\alpha,\beta}$ is independent of the value of t_j but depends on j (through a_j). Accordingly, we may choose t_j so that (say) $C^{(j)}_{\alpha,\beta} \leq 2^{-j}(1 + t_j)$ for all α, β, with $|\alpha| + |\beta| \leq j$. In particular, from (12.34),

$$\left|D_x^\alpha D_\xi^\beta ((1 - \phi_{t_j}) a_j)(x, \xi)\right| \leq 2^{-j} \langle \xi \rangle^{m_j + 1 - \rho|\beta|} \quad \forall\, |\alpha| + |\beta| \leq j.$$

Now consider $D_x^\alpha D_\xi^\beta \sum_j (1 - \phi_{t_j}) a_j$. Since all but finitely many terms of this series vanish near any given point, we may interchange the summation and differentiation, whence by the preceding estimates we have

$$\left|D_x^\alpha D_\xi^\beta \sum_{j=N}^{\infty}(1-\phi_{t_j})a_j\right| \leq \sum_{\substack{N\leq j<|\alpha|+|\beta| \\ \text{or } m_j+1\geq m_N}} C_{\alpha,\beta}^{(j)}\langle\xi\rangle^{m_j-\rho|\beta|} + \sum_{\substack{j\geq|\alpha|+|\beta| \\ \text{and } m_j+1\leq m_N}} 2^{-j}\langle\xi\rangle^{m_j+1-\rho|\beta|}$$

$$\leq \sum_{\substack{j<|\alpha|+|\beta| \\ \text{or } m_j+1\geq m_N}} C_{\alpha,\beta}^{(j)}\langle\xi\rangle^{m_N-\rho|\beta|} + \sum_{j\geq|\alpha|+|\beta|} 2^{-j}\langle\xi\rangle^{m_N-\rho|\beta|}$$

$$= C(\alpha,\beta,N)\,\langle\xi\rangle^{m_N-\rho|\beta|}.$$

Setting $N=0$ we obtain $a \in S_\rho^{m_0}$, and furthermore

$$a-\sum_{j=0}^{N-1}a_j = a-\sum_{j=0}^{N-1}(1-\phi_{t_j})a_j - \sum_{j=0}^{N-1}\phi_{t_j}a_j$$

$$= \sum_{j=N}^{\infty}(1-\phi_{t_j})a_j - \sum_{j=0}^{N-1}\phi_{t_j}a_j.$$

The first term belongs to $S_\rho^{m_N}$, by the preceding estimates. The second term is in C^∞ with compact support, and thus lies in $S^{-\infty}$. □

We now turn to the problem of deriving an asymptotic expansion of $a\,\#\,b$ in the sense of Definition 12.17 and relative to the symbol classes S_ρ^m with $\rho>0$. We shall start with Schwartz symbols and then pass to the limit. This may appear strange since $\mathcal{S}(\mathbb{R}^{2d})\subset S^{-\infty}$, whence any Schwartz symbol has an identically vanishing asymptotic expansion. However, we shall rely on *uniform bounds in the symbol classes* and on the following lemma, which allows us to pass to limits in asymptotic expansions.

Lemma 12.19 *Let $0\leq\rho\leq 1$ be arbitrary but fixed. Suppose that $a_k, a\in S_\rho^m$, $m_j\to-\infty$, and $m_0=m$. Assume that for each k one has $a_k\sim\sum_j a_k^{(j)}$ with $a_k^{(j)}\in S_\rho^{m_j}$. Assume further that*

(i) *$a_k\to a$ in S_ρ^m,*
(ii) *for each j, $a_k^{(j)}\to a^{(j)}$ in $S_\rho^{m_j}$,*
(iii) *for each N, α, β we have*

$$\left|D_x^\alpha D_\xi^\beta\left(a_k-\sum_{j=0}^{N-1}a_k^{(j)}\right)\right| \leq C_{\alpha,\beta,N}\,\langle\xi\rangle^{m_N-\rho|\beta|}$$

with $C_{\alpha,\beta,N}$ independent of k.

Then $a\sim\sum_j a^{(j)}$.

Proof Properties (i) and (ii) imply that, for each fixed N,

$$a_k - \sum_{j=0}^{N-1} a_k^{(j)} \quad \to \quad a - \sum_{j=0}^{N-1} a^{(j)}$$

in the C^∞ topology. Therefore, by (iii),

$$\left| D_x^\alpha D_\xi^\beta \left(a - \sum_{j=0}^{N-1} a^{(j)} \right) \right| \leq C_{\alpha,\beta,N} \langle \xi \rangle^{m_N - \rho|\beta|},$$

whence $a \sim \sum_j a^{(j)}$. □

We can now formulate the asymptotic expansion of the Moyal product. In what follows,

$$D_x \cdot D_\eta := \sum_{i=1}^{d} D_{x_i} D_{\eta_i}$$

and similarly for $D_y \cdot D_\xi$.

Proposition 12.20 *Define*

$$L^{(j)}(a,b)(x,\xi) := (D_\xi \cdot D_y - D_x \cdot D_\eta)^j a(x,\xi) b(y,\eta)\Big|_{x=y,\ \xi=\eta}$$

for $j \geq 1$, *and set* $L^{(0)}(a,b) = ab$. *Suppose that* $a \in S_\rho^\ell$ *and* $b \in S_\rho^m$, *where* $0 < \rho \leq 1$ *is fixed. Then* $L^{(j)}(a,b) \in S_\rho^{\ell+m-\rho j}$ *and*

$$a \# b \sim \sum_{j=0}^{\infty} \frac{(\pi i)^j}{j!} L^{(j)}(a,b). \tag{12.35}$$

In particular, $ab \in S_\rho^{\ell+m}$, $\{a,b\} \in S_\rho^{\ell+m-\rho}$, *and*

$$a \# b = ab + \frac{1}{4\pi i}\{a,b\} \mod S_\rho^{\ell+m-2\rho}. \tag{12.36}$$

Proof We begin with the proof of (12.36). Assume first that $a, b \in \mathcal{S}$. Then

$$\widehat{a \# b}(v) = \int_{\mathbb{R}^{2d}} \hat{a}(u) \hat{b}(v-u) e^{\pi i \sigma(u,v)}\, du.$$

We expand the exponential as a Taylor series:

$$e^{\pi i \sigma(u,v)} = 1 + \pi i \sigma(u,v) + R(u,v),$$

$$R(u,v) = (\pi i \sigma(u,v))^2 \int_0^1 (1-\hbar) e^{\pi i \hbar \sigma(u,v)}\, d\hbar.$$

Since $\sigma(u, u) = 0$ we may write, expanding to second order,

$$\widehat{a\,\#\,b}(v) = \int_{\mathbb{R}^{2d}} \hat{a}(u)\hat{b}(v-u)\,du + \pi i \int_{\mathbb{R}^{2d}} \hat{a}(u)\hat{b}(v-u)\sigma(u, v-u)\,du$$
$$+ (\pi i)^2 \int_0^1 (1-\hbar)\,d\hbar \int_{\mathbb{R}^{2d}} \hat{a}(u)\hat{b}(v-u)(\sigma(u, v-u))^2 e^{\pi i \hbar \sigma(u,v)}\,du. \tag{12.37}$$

The first two integrals here are easily seen to give the desired expressions, so we will focus on the third term, which is

$$(\pi i)^2 \int_0^1 (1-\hbar) A_\hbar\,d\hbar,$$

where $A_\hbar$ is the inverse Fourier transform of the function

$$\widehat{A_\hbar}(v) = \int_{\mathbb{R}^{2d}} \hat{a}(u)\hat{b}(v-u)(\sigma(u, v-u))^2\, e^{\pi i \hbar \sigma(u,v)}\,du.$$

We claim that $A_\hbar \in S_\rho^{\ell+m-2\rho}$, with bounds that depend only on the S_ρ^ℓ norm of a and the S_ρ^m norm of b, respectively. First, note that $(\sigma(u, v-u))^2$ is a sum of terms of the form $\pm\xi_i(y_i - x_i)x_k(\eta_k - \xi_k)$ and $\pm\xi_i\xi_s x_i x_s$, and $(y_i - x_i)(y_j - x_j)(\eta_i - \xi_i)(\eta_j - \xi_j)$. It follows that $\widehat{A_\hbar}$ is a sum of terms of the form

$$\widehat{D_{\xi_i} D_{x_k} a} \natural_\hbar \widehat{D_{\xi_s} D_{x_j} b}$$

and similar terms with four derivatives operators, acting on either a or b. Lemma 12.15 and Exercise 12.4 now imply the desired bounds on $A_\hbar$, uniformly in $\hbar$. Integrating over $\hbar$ yields

$$a\,\#\,b - ab - \frac{1}{4\pi i}\{a, b\} \in S_\rho^{\ell+m-2\rho},$$

with the seminorms in that symbol class bounded in terms of the S_ρ^ℓ seminorms of a and the S_ρ^m seminorms of b, respectively. Recall that we have assumed that $a, b \in \mathcal{S}$, but no Schwartz seminorms entered our estimates.

In order to remove the Schwartz assumption, we invoke Lemma 12.19. Indeed, we approximate $a \in S_\rho^\ell$ and $b \in S_\rho^m$ by Schwartz symbols a_k and b_k, respectively, and note that $a_k b_k \to ab$ in $S_\rho^{\ell+m}$ and $\{a_k, b_k\} \to \{a, b\}$ in $S_\rho^{\ell+m-\rho}$. This concludes the proof of (12.36).

As for the general case, it proceeds in an identical fashion but using higher-order Taylor expansions. □

Exercise 12.6 Adapt the previous proof to the semiclassical context. That is, show that for a and b as in Proposition 12.20 one has

$$a \#_\hbar b \sim \sum_{j=0}^{\infty} \frac{(\pi i \hbar)^j}{j!} L^{(j)}(a, b)$$

in the sense that

$$a \#_\hbar b - \sum_{j=0}^{N-1} \frac{(\pi i \hbar)^j}{j!} L^{(j)}(a, b) \in \hbar^N S_\rho^{\ell+m-N\rho} \tag{12.38}$$

for each $N \geq 1$. The right-hand side in (12.38) signifies a function of the form $\hbar^N r_N(\cdot; \hbar)$ with the property that $r_N(\cdot; \hbar) \in S_\rho^{\ell+m-N\rho}$ uniformly in $\hbar$.

Symbols in the class $S^{-\infty}$ arise when one considers $a \# b$ where $ab = 0$, as the reader is asked to verify in the following exercise.

Exercise 12.7 Show that if $a \in S_\rho^m$ and $b \in S_\rho^\ell$ have disjoint supports, a being of compact support, then $a \# b \in S_\rho^{-\infty}$. As before, $0 < \rho \leq 1$.

12.2.7. Preservation of regularity

A most important property of pseudodifferential operators is that they *do not create singularities*. In order to formulate a rigorous statement to this effect, we begin with the simple observation that for any $a \in S^\infty$ one has

$$a(x, D) : \mathcal{S}' \to \mathcal{S}'.$$

This follows immediately by passing to adjoints in Corollary 12.5. In particular, if $f \in \mathcal{S}'$ then one defines $a(x, D) f$ via the identity

$$\langle a(x, D) f, \varphi \rangle = \langle f, \bar{a}(x, D) \varphi \rangle \quad \forall \varphi \in \mathcal{S}(\mathbb{R}^d). \tag{12.39}$$

We shall now demonstrate that operators with symbols in $S^{-\infty}$ are smoothing, in the sense that they arbitrarily regularize distributions. We begin with a simple statement to the effect that symbols in $S^{-\infty}$ yield *smoothing operators* under quantization.

Lemma 12.21 *If $a \in S^{-\infty}$ then $a(x, D)$ maps compactly supported distributions in $\mathcal{S}'$ to C^∞ functions.*

Proof The point is essentially that for $S^{-\infty}$ symbols one can evaluate the distribution $a(x, D) f(x)$ as in (12.6):

$$a(x, D) f(x) = \left\langle f, \; a\left(\frac{x + \cdot}{2}, \widehat{\cdot - x}\right)\right\rangle$$

where the pairing here is between a compactly supported tempered distribution and a smooth function. Smoothness is shown by an elementary integration by parts, which we leave to the reader. □

A compact support is needed here since we cannot guarantee any decay in x. Now we can formulate the aforementioned *preservation of regularity* property.

Proposition 12.22 *Let $a \in S_0^\infty$ and suppose that $f \in \mathcal{S}'$ with compact support equals a smooth function on some open set Ω. Then $a(x, D)f$ also equals a smooth function on Ω.*

Proof Let $\psi \in C^\infty(\Omega)$ have compact support in Ω. Then

$$a(x, D)f = a(x, D)(\psi f) + a(x, D)((1 - \psi)f).$$

The first term on the right-hand side is clearly smooth since $\psi f \in \mathcal{S}$. For the second term, let $\phi \in C^\infty(\Omega)$ have compact support so that $\phi\psi = \phi$. Using the composition of symbols we write

$$\phi a(x, D)((1 - \psi)f) = b(x, D)f,$$

where $b \in S^{-\infty}$ since $\phi(1 - \psi) = 0$. Since f has compact support we conclude from Lemma 12.21 that $b(x, D)f \in C^\infty$ as claimed. □

The assumption that f is compactly supported is a technical convenience (which, however, does not take anything essential away from Proposition 12.22) that can be avoided owing to the fact that ϕ has compact support. We leave this to the interested reader to pursue. Proposition 12.22 establishes the *pseudolocality* of ΨDO operators, which refers to the fact that these operators do not increase the singular support of a distribution. The latter is simply the complement of the largest open set on which a distribution equals a smooth function. Then the proposition states that for (compactly supported) $f \in \mathcal{S}'$ one has

$$\mathrm{singsupp}(a(x, D)f) \subset \mathrm{singsupp}(f)$$

for any $a \in S_0^\infty$.

12.2.8. Inverses of symbols

Pseudodifferential operators (ΨDOs) arise naturally in the study of partial differential operators of the elliptic type. This has to do with the fact that elliptic operators can be inverted in the class of ΨDOs, which is essential for many purposes. We begin by inverting the symbols of elliptic operators. For simplicity, we shall work with the symbol class S_1^m in this section although one can also consider S_ρ^m with $0 < \rho \leq 1$.

Definition 12.23 If $a \in S_1^m$ then we say that a is *mth-order elliptic* if there exists a constant C such that $a(x,\xi) \geq C^{-1}\langle\xi\rangle^m$ for all (x,ξ) with $|\xi| \geq C$. The operator $a(x,D)$ is also called mth-order elliptic.

Lemma 12.24 *Suppose that a is mth-order elliptic. Then there exists $b \in S_1^{-m}$ such that $a \# b = 1 \bmod S^{-\infty}$ and $b \# a = 1 \bmod S^{-\infty}$.*

Proof We first show that there is a sequence of symbols $\{b_j\}_{j=0}^{\infty} \subset S_1^{-m}$ with $b_j \in S_1^{-m-j}$ and such that for each N one has

$$a \# \sum_{j=0}^{N-1} b_j = 1 \mod S_1^{-N}. \tag{12.40}$$

Then we shall obtain b as an asymptotic sum of the b_j. The construction is iterative, starting with

$$b_0(x,\xi) := \frac{\phi(\xi)}{a(x,\xi)},$$

where $\phi(\xi) = 0$ for $|\xi| \leq C$ and $1 - \phi(\xi)$ is smooth with compact support. From these definitions one verifies that $b_0 \in S_1^{-m}$. Moreover, by the composition law, we have

$$a \# b_0 = ab_0 = \phi = 1 \mod S_1^{-1}, \tag{12.41}$$

which is (12.40) with $N = 1$. Now suppose that $\{b_j\}_{j=0}^{N-1}$ have been constructed with the desired properties. Define

$$b_N(x,\xi) := \phi(\xi)\frac{1 - a \# \sum_{j=0}^{N-1} b_j}{a(x,\xi)} = b_0(x,\xi)\left(1 - a \# \sum_{j=0}^{N-1} b_j\right).$$

We have $b_0 \in S_1^{-m}$ and $1 - a \# \sum_{j=0}^{N-1} b_j \in S_1^{-N}$, whence $b_N \in S_1^{-m-N}$. Furthermore,

$$\begin{aligned} a \# b_N &= \phi\left(1 - a \# \sum_{j=0}^{N-1} b_j\right) \mod S_1^{-N-1} \\ &= 1 - a \# \sum_{j=0}^{N-1} b_j \mod S_1^{-N-1}, \end{aligned}$$

where we have used (12.41) and the composition law. This concludes the induction step. Let $b \sim \sum_{j=0}^{\infty} b_j$. Then, for any N,

$$a \# b - 1 = \left(a \# \left(\sum_{j=0}^{N-1} b_j \right) - 1 \right) + a \# \left(b - \sum_{j=0}^{N-1} b_j \right)$$
$$\in S_1^{-N} + S_1^{m+(-m-N)} = S_1^{-N}$$

for any N, which means that $a \# b = 1 \bmod S_1^{-\infty}$. To show that $b \# a = 1 \bmod S_1^{-\infty}$, we argue as follows. First, a similar construction to that just carried out yields $\tilde{b} \in S_1^{-m}$ with $\tilde{b} \# a = 1 \bmod S_1^{-\infty}$. Thus,

$$\begin{aligned} \tilde{b} &= \tilde{b} \# \left(a \# b \right) \quad \bmod S_1^{-\infty} \\ &= (\tilde{b} \# a) \# b \quad \bmod S_1^{-\infty} \\ &= b \quad \bmod S_1^{-\infty}, \end{aligned} \tag{12.42}$$

and we are done. □

Exercise 12.8 In (12.42) we used the associativity of the Moyal product. Prove this property.

We are interested mainly in the elliptic *operators* $a(x, D)$ and less in their symbols. In the above proof we constructed a *parametrix* $b(x, D)$. This name refers to the fact that $b(x, D)$ is an approximate inverse, in the sense that $a(x, D) \circ b(x, D)$ is the identity modulo a *smoothing operator*, i.e.,

$$\begin{aligned} a(x, D) \circ b(x, D) &= \mathrm{Id} + E_1(x, D), \\ b(x, D) \circ a(x, D) &= \mathrm{Id} + E_2(x, D), \end{aligned} \tag{12.43}$$

where the symbols of $E_i(x, D)$ lie in $S^{-\infty}$, for $i = 1, 2$. We have already encountered the terminology *smoothing operator* in connection with the action of operators with $S^{-\infty}$-symbols on distributions. In the following section, in which we develop the mapping properties of ΨDO operators on L^2, we shall see that they improve regularity on the Sobolev scale by any desired amount.

12.3. The L^2 theory

The ΨDO calculus established thus far is very powerful in combination with some natural L^2-boundedness properties, to which we now turn. For example, we shall show that $a(x, D)$ is bounded on L^2 if $a \in S^0$ and that elliptic operators of order m take L^2 to the Sobolev space H^{-m}. In particular, if

$m < 0$ then regularity is gained. For example, in conjunction with the invariance properties of ΨDOs under coordinate transformations, these results lead to very general statements about the spectra of elliptic operators on compact manifolds.

We begin with a rather simple statement, which is far from optimal. Recall that $H^s(\mathbb{R}^d)$ are the standard Sobolev spaces with norms $\|f\|_{H^s(\mathbb{R}^d)} = \|\langle\Delta\rangle^{s/2} f\|_2$.

Lemma 12.25 *Suppose that $a \in S_0^m$ with $m < -d$. Then $a(x, D)$ is bounded on L^2. In particular, symbols in $S^{-\infty}$ yield L^2-bounded operators. In fact, if $a \in S^{-\infty}$ then $a(x, D) : L^2 \to H^s(\mathbb{R}^d)$ for any $s \geq 0$; in other words, $a(x, D)$ is smoothing.*

Proof We have

$$a(x, D)f(x) = \int_{\mathbb{R}^d} a\left(\frac{x+y}{2}, \widehat{y-x}\right) f(y)\, dy. \tag{12.44}$$

By our choice of symbol class,

$$a(z, \hat{u}) = \int_{\mathbb{R}^d} a(z, \zeta) e^{-2\pi i u \cdot \zeta}\, d\zeta \in L^\infty_{z,u}(\mathbb{R}^{2d}).$$

Integrating by parts improves this bound to

$$\sup_z |a(z, u)| \leq C_N \langle u \rangle^{-N} \quad \forall N \geq 0,$$

whence the kernel in (12.44) satisfies

$$\left| a\left(\frac{x+y}{2}, \widehat{y-x}\right) \right| \leq C_N \langle x - y \rangle^{-N}$$

for any N. By Schur's lemma it follows that $a(x, D)$ is bounded on L^2, as claimed.

It remains to show that operators with $S^{-\infty}$ symbols a are smoothing, in the sense that they take L^2 to H^s for any $s \geq 0$:

$$\|a(x, D)f\|_{H^s(\mathbb{R}^d)} \leq C(s, d)\, \|f\|_2.$$

However, the left-hand side is the same as $\|(\omega_s \# a)(x, D)f\|_2$, where $\omega_s(\xi) := \langle \xi \rangle^s$. By the composition law, $\omega_s \# a \in S^{-\infty}$ and we are done. □

We shall now drastically lower the requirement on the symbol, in fact optimally so (as far as the decay in ξ is concerned).

12.3.1. The Calderón–Vaillancourt theorem

In Chapter 9 we proved the L^2-boundedness of Kohn–Nirenberg operators of order 0. Here we do the same for Weyl operators, again basing the argument on Cotlar's lemma.

Let $a \in S^0 = S^0_0$ and pick a smooth compactly supported function χ on $\mathbb{R}^{2d}$ with

$$\sum_{k\in\mathbb{Z}^{2d}} \chi_k = 1, \qquad \chi_k := \chi(\cdot - k).$$

Define $a_k := \chi_k a$, whence $a = \sum_{k\in\mathbb{Z}^{2d}} a_k$. The main technical step is encoded in the following lemma.

Lemma 12.26 *For any $k, \ell \in \mathbb{Z}^{2d}$ one has the following estimates on $b_{k,\ell} := \overline{a_k} \# a_\ell$:*
(i) $|\partial_u^\gamma b_{k,\ell}(u)| \le C_{N,d,\gamma}(a)\, \langle k-\ell\rangle^{-N} \langle u - (k+\ell)/2\rangle^{-N}$ *for all N, γ;*
(ii) $\|b_{k,\ell}(x,D)\|_{2\to 2} \le C_{N,d}(a)\, \langle k-\ell\rangle^{-N}$. *The constant here satisfies*

$$C_{N,d}(a) \le C_{N,d} \sum_{|\gamma|\le 2N'+2d+1} \|\partial^\gamma a\|_{L^\infty(\mathbb{R}^{2d})},$$

where $N' := \max(N, 2d+1)$.

Proof We have, where χ_0 is a bump function around the origin in $\mathbb{R}^{4d}$ that equals 1 on $|v| + |w| \le 1$,

$$\begin{aligned} b_{k,\ell}(u) &= 4^d \int_{\mathbb{R}^{4d}} \overline{a_k}(u+v) a_\ell(u+w) e^{-4\pi i\sigma(v,w)}\, dv dw \\ &= 4^d \int_{\mathbb{R}^{4d}} \overline{a_k}(u+v) a_\ell(u+w) e^{-4\pi i\sigma(v,w)} \chi_0(v,w)\, dv dw \\ &\quad + 4^d \int_{\mathbb{R}^{4d}} \overline{a_k}(u+v) a_\ell(u+w) e^{-4\pi i\sigma(v,w)} (1-\chi_0(v,w))\, dv dw \\ &=: b^{(1)}_{k,\ell}(u) + b^{(2)}_{k,\ell}(u). \end{aligned} \tag{12.45}$$

One has $b^{(1)}_{k,\ell}(u) = 0$ unless $|u-k| \le C$ and $|u-\ell| \le C$. Hence $b^{(1)}_{k,\ell}(u)$ satisfies the desired bounds. To bound $b^{(2)}_{k,\ell}(u)$, we integrate by parts as in the proof of Lemma 12.15 using $L, \mathcal{L}$ as defined there. This yields, for N a large integer,

$$\begin{aligned} |b^{(2)}_{k,\ell}(u)| &= 4^d \Big| \int_{\mathbb{R}^{4d}} e^{-4\pi i\sigma(v,w)} \mathcal{L}^N \big((1-\chi_0(v,w)) \overline{a_k}(u+v) a_\ell(u+w) \big)\, dv dw \Big| \\ &\le C \int_{\mathbb{R}^{4d}} \langle |v| + |w| \rangle^{-2N} \chi_{[|u+v-k|\le C]} \chi_{[|u+w-\ell|\le C]}\, dv dw \\ &\le C\, \langle |u-k| + |u-\ell| \rangle^{-2N} \le C\, \langle k-\ell\rangle^{-N} \langle u - \tfrac12(k+\ell)\rangle^{-N}. \end{aligned}$$

To pass to the final estimate we used

$$|u-k|+|u-\ell| \geq \max(|k-\ell|, |u-\tfrac{1}{2}(k+\ell)/2|).$$

The estimate for the derivatives is analogous.

To prove the L^2 bound, we write

$$b_{k,\ell}(x,D)f(x) = \int_{\mathbb{R}^{2d}} \widehat{b_{k,\ell}}(v)\rho(v)f(x)\,dv,$$

which implies via the obvious fact $\|\rho(v)f\|_2 = \|f\|_2$ that

$$\begin{aligned}\|b_{k,\ell}(x,D)f\|_2 &\leq C\|\widehat{b_{k,\ell}}\|_1\|f\|_2 \leq C \sup_{|\gamma|\leq 2d+1} \|\partial^\gamma b_{k,\ell}\|_1\|f\|_2 \\ &\leq C\,\langle k-\ell\rangle^{-N}\|f\|_2,\end{aligned}$$

where the final estimate follows from part (i). □

The main result is now a simple consequence of part (ii) of the lemma.

Theorem 12.27 *For any $a \in S^0_0$ one has $\|a(x,D)f\|_2 \leq C(a,d)$. The constant depends on the $L^\infty(\mathbb{R}^{2d})$ norm of finitely many derivatives of a.*

Proof This follows immediately from Cotlar's lemma; see Section 9.1. □

Note that the proof of Theorem 12.27 gives the dependence of the constant on a as follows:

$$C(a,d) \leq C(d) \sum_{|\gamma|\leq 6d+3} \|\partial^\gamma a\|_{L^\infty(\mathbb{R}^{2d})},$$

where $C(d)$ depends only on the dimension.

We now present, by means of the following exercise, a somewhat different as well as less precise argument. It is inferior to the previous proof in so far as it only applies to symbols in S^0_ρ with $\rho > 0$, but for many applications this suffices. It is based on an what is customarily called *Hörmander's square root trick.*

Exercise 12.9 Let $a \in S^0_\rho$ for some $\rho > 0$. Show that there exists $b \in S^0_\rho$ with $b > 0$ and such that $(\lambda - \bar{a}\#a) - b\#b \in S^{-\infty}$, provided that $\lambda > 0$ is large enough. In other words, $b(x,D)$ is the square root modulo smoothing operators $\lambda\mathrm{Id} - a(x,D)^* \circ a(x,D)$. Conclude via Lemma 12.25 and positivity that $a(x,D)$ is bounded on L^2.

12.3.2. Mapping properties of elliptic operators

We now come to an important application of the above results, namely the mapping properties of elliptic operators on Sobolev spaces. We begin with a very elementary statement, which says that symbols of negative order $-m$ gain m derivatives.

Lemma 12.28 *Suppose that $m \geq 0$ and $a \in S_0^{-m}$. Then for any $s \geq 0$ one has an estimate of the form*

$$\|a(x, D)f\|_{H^{s+m}} \leq C\|f\|_s \tag{12.46}$$

for any $f \in \mathcal{S}$.

Proof As before, we let $\omega_s(\xi) := \langle \xi \rangle^s$. Clearly, $\|f\|_{H^s} = \|\omega_s(x, D)f\|_2$ and $\omega_s \in S_1^s$. By interpolation, we may assume that s is an integer and so we can use induction in s. If $s = 0$ then

$$\begin{aligned}\|a(x, D)f\|_{H^m} &= \|\omega_m(x, D) \circ a(x, D)f\|_2 \\ &= \|(\omega_m \# a)(x, D)f\|_2 \leq C\|f\|_2,\end{aligned}$$

by Theorem 12.27. Now let $s \geq 1$ be an integer and assume that the statement holds for integers up to $s - 1$. Then

$$\begin{aligned}\|a(x, D)f\|_{H^{m+s}} &= \|\omega_{m+s-1}(x, D) \circ \omega_1(x, D) \circ a(x, D)f(x)\|_2 \\ &\leq \|\omega_{m+s-1}(x, D) \circ a(x, D) \circ \omega_1(x, D)f(x)\|_2 \\ &\quad + \|\omega_{m+s-1}(x, D) \circ [\omega_1(x, D), a(x, D)]f(x)\|_2 \\ &= \|a(x, D) \circ \omega_1(x, D)f\|_{H^{m+s-1}} \\ &\quad + \|[\omega_1(x, D), a(x, D)]f(x)\|_{H^{m+s-1}}.\end{aligned}$$

The symbol in the commutator on the right-hand side belongs to S_0^{-m}, and the induction hypothesis implies that

$$\|a(x, D)f\|_{H^{m+s}} \leq C(\|\omega_1(x, D)f\|_{H^{s-1}} + \|f\|_{H^{s-1}}) \leq C\|f\|_{H^s}$$

as claimed. □

We now come to the main elliptic estimate of this chapter. It simply says that elliptic ΨDOs of order m gain m derivatives. After stating and proving this elliptic estimate we will apply it to differential operators in the ordinary sense, in Exercise 12.10.

Proposition 12.29 *Suppose that $m \geq 0$ and that $a \in S_\rho^m$, with $0 < \rho \leq 1$, is elliptic of order m. Then for any $s \geq 0$ one has the a priori estimate*

$$\|f\|_{H^{s+m}} \leq C(\|a(x, D)f\|_{H^s} + \|f\|_2) \tag{12.47}$$

for all $f \in \mathcal{S}(\mathbb{R}^d)$. The constant depends on m, d, s, and a.

Proof By Lemma 12.24 there exists $b \in S_\rho^{-m}$ with $a \# b = 1 \bmod S^{-\infty}$. Then we have

$$f = b(x, D)(a(x, D)f) - E(x, D)f,$$

where E has an $S^{-\infty}$-symbol. The estimate now follows from Lemma 12.28. □

The following exercise shows that elliptic differential operators (rather than ΨDOs) fall under the scope of Proposition 12.29.

Exercise 12.10 Let $L := \sum_{|\alpha| \le m} a_\alpha(x) D^\alpha$ be a differential operator with $a_\alpha \in C^\infty$. Assume moreover that the a_α together with all their derivatives are bounded. Prove that there is $a \in S_1^m$ such that

$$a(x, D)f(x) = Lf(x) \quad \forall f \in \mathcal{S}(\mathbb{R}^d).$$

Moreover, show that if L is elliptic, i.e., if there exists a constant C such that

$$\Big| \sum_{|\alpha|=m} a_\alpha(x)\xi^\alpha \Big| \ge C^{-1}|\xi|^m \quad \forall \xi \in \mathbb{R}^d$$

then a is an mth-order elliptic.

While Proposition 12.29 is stated as an a priori estimate, it suggests a corresponding regularity result of the following kind (under the same assumptions): if $f \in \mathcal{S}'$ satisfies $a(x, D)f \in H^s_{\mathrm{loc}}(\mathbb{R}^d)$ for some $s \ge 0$ then $f \in H^{s+m}_{\mathrm{loc}}(\mathbb{R}^d)$. This is indeed true, and can be obtained from (12.47) by means of a suitable regularization (see the end-of-chapter notes).

12.3.3. The sharp Gårding inequality

While it is not in general true that a nonnegative symbol $a \in S = S_0^0$ yields an operator $A = a(x, D)$ that is nonnegative, Gårding's inequality says that this is true, up to $O(\hbar)$ corrections, in a semiclassical formulation.

Theorem 12.30 *Let $a \in S$ satisfy $a \ge 0$. Then $a(x, \hbar D) \ge -O(\hbar)$, where the latter is in the sense of the L^2 operator norm. That is, for any $f \in L^2(\mathbb{R}^d)$ we have*

$$\langle a(x, \hbar D)f, f\rangle \ge -C\hbar \|f\|_2^2,$$

where C depends only on finitely many S seminorms of a.

Proof Define, with $u = (x, \xi) \in \mathbb{R}^{2d}$,

$$\tilde{a}(u; \hbar) := (\pi\hbar)^{-d} \int_{\mathbb{R}^{2d}} a(v) e^{-|u-v|^2/\hbar} \, dv. \tag{12.48}$$

Then it is elementary to check that $\|\tilde{a}(u; \hbar)\|_\infty \leq \|a\|_\infty$. Integrating by parts, one also verifies that $\widetilde{\partial^\alpha a} = \partial^\alpha \tilde{a}$, whence $\|\partial^\alpha \tilde{a}\|_\infty \leq \|\partial^\alpha a\|_\infty$. In summary, $\tilde{a} \in S$ uniformly in $0 < \hbar \leq 1$. We Taylor-expand, obtaining

$$a(v) = a(u) + (\partial a)(u)(v - u) + (u - v)^2 O(1),$$

where O is a bounded smooth function of $u, v \in \mathbb{R}^{2d}$. Inserting this into (12.48) yields

$$\|a - \tilde{a}(\cdot; \hbar)\|_\infty \leq C\hbar$$

as well as, for all α,

$$\|\partial^\alpha a - \partial^\alpha \tilde{a}(\cdot; \hbar)\|_\infty \leq C\hbar$$

by the aforementioned commutation of differentiation and regularization. By the L^2 theory we thus obtain that

$$\|a(x, \hbar D) - \tilde{a}(x, \hbar D; \hbar)\|_{2\to 2} \leq C\hbar; \tag{12.49}$$

the constant C depends on finitely many S seminorms of a. It will therefore suffice to show that $\langle \tilde{a}(x, \hbar D) f, f \rangle \geq 0$ for all $f \in \mathcal{S}(\mathbb{R}^d)$. To prove this, we compute (dropping the 2π factors in the Weyl calculus for simplicity)

$$\begin{aligned}\langle \tilde{a}(x, \hbar D) f, f \rangle &= \hbar^{-d} \int_{\mathbb{R}^d} \int_{\mathbb{R}^{2d}} \tilde{a}\left(\frac{x+u}{2}, \xi\right) e^{i(x-u)\cdot\xi/\hbar} f(u) \, du d\xi \, \bar{f}(x) \, dx \\ &= \pi^{-d} \hbar^{-2d} \int_{\mathbb{R}^d} \int_{\mathbb{R}^{2d}} \int_{\mathbb{R}^{2d}} a(y, \eta) e^{-(|(x+u)/2 - y|^2 + |\xi - \eta|^2)/\hbar} \, dy d\eta \\ &\qquad \times e^{i(x-u)\cdot\xi/\hbar} f(u) \, du d\xi \, \bar{f}(x) \, dx. \end{aligned} \tag{12.50}$$

The ξ-integral takes the form

$$\begin{aligned} &\int_{\mathbb{R}^d} e^{-|\xi-\eta|^2/\hbar} e^{i(x-u)\cdot\xi/\hbar} \, d\xi \\ &= \int_{\mathbb{R}^d} e^{-|\xi|^2/\hbar} e^{i(x-u)\cdot\xi/\hbar} \, d\xi \, e^{i(x-u)\cdot\eta/\hbar} \\ &= (\pi\hbar)^{d/2} e^{-|x-u|^2/4\hbar} \, e^{i(x-u)\cdot\eta/\hbar}. \end{aligned} \tag{12.51}$$

Inserting this into (12.50) yields

$$\langle \tilde{a}(x, \hbar D)f, f\rangle = \pi^{-d/2}\hbar^{-3d/2} \int_{\mathbb{R}^{2d}} \int_{\mathbb{R}^{2d}} a(y, \eta)e^{-|(x+u)/2-y|^2/\hbar}$$

$$e^{-|x-u|^2/4\hbar}\, e^{i(x-u)\cdot\eta/\hbar} e^{i(x-u)\cdot\xi/\hbar} f(u)\, du\, \bar{f}(x)\, dx dy d\eta$$

$$= \pi^{-d/2}\hbar^{-3d/2} \int_{\mathbb{R}^{2d}} a(y, \eta)\, e^{-|y|^2/\hbar}\, |U_\hbar(\eta)|^2\, dy\, d\eta, \tag{12.52}$$

where

$$U_\hbar(\eta) = \int_{\mathbb{R}^d} e^{-|x|^2/\hbar} e^{ix\cdot\eta/\hbar}\, f(x)\, dx.$$

Note that this calculation hinges on the fact that the nonlinear term $e^{-x\cdot u/2\hbar}$, which appears in the quadratic phase of (12.50), cancels with the term $e^{x\cdot u/2\hbar}$ that arises in the Gaussian Fourier transform (12.51). Since $a \geq 0$, the last line in (12.52) is nonnegative and we are done, in view of (12.49). □

The *Fefferman–Phong* inequality improves upon Theorem 12.30 by establishing the estimate

$$a(x, \hbar D) \geq -O(\hbar^2) \quad \Longleftrightarrow \quad \langle a(x, \hbar D)f, f\rangle \geq -C\hbar^2 \|f\|_2^2.$$

Without the semiclassical parameter, the sharp Gårding and Fefferman–Phong inequalities, respectively, take the following form:

$$\langle a(x, D)f, f\rangle \geq -C\|f\|_2^2, \tag{12.53}$$

where in the Gårding case we assume that $a \in S_1^1$ and, in the more difficult Fefferman–Phong case, that $a \in S_1^2$. The point is that no derivatives are required in the lower bound of (12.53).

12.4. A phase-space transform

12.4.1. The Bargman transform

This section provides a brief introduction to the so-called Bargman transform which is closely related to the Fourier–Bros–Iagolnitzer (FBI) transform and which is defined as follows.

Definition 12.31 For any $f \in \mathcal{S}'(\mathbb{R}^d)$, set

$$(Tf)(x, \xi) := 2^{-d/2}\pi^{-3d/4} \int_{\mathbb{R}^d} e^{-(x-y)^2/2} e^{i\xi\cdot(x-y)} f(y)\, dy, \tag{12.54}$$

where the integral is to be interpreted in the sense of the $\mathcal{S}'$–$\mathcal{S}$ pairing.

Note that we have departed here from the tradition of putting factors of π in the phases – we find it a little easier to omit them, at the expense of gaining certain (irrelevant) prefactors. This has the consequence that the Weyl calculus needs to be changed accordingly, but these changes are all trivial and can be ignored.

Exercise 12.11 Verify that $T : \mathcal{S}(\mathbb{R}^d) \to \mathcal{S}(\mathbb{R}^{2d})$.

The Bargman transform encodes properties of both f and $\hat{f}$ *simultaneously*. For example, we shall show next that if f is a Gaussian localized around x_0 with Fourier transform localized around ξ_0 then Tf is localized around (x_0, ξ_0). An important point here concerns the scale used to express "localization". If we use the scale $\sigma > 0$ in x-space then the uncertainty principle dictates that the localization scale in the Fourier transform is σ^{-1}. In particular, $\sigma = 1$ is the unique symmetric choice and it is also the only correct choice with respect to T as defined by (12.54); indeed, any other choice of σ would mean that (12.54) would need to be rescaled appropriately.

Lemma 12.32 *For any* $(x_0, \xi_0) \in \mathbb{R}^{2d}$ *define a* coherent state *as follows:*

$$f_{x_0,\xi_0}(y) := \pi^{-d/4} e^{-(y-x_0)^2/2} e^{i\xi_0\cdot(y-x_0)}.$$

Then

$$(Tf_{x_0,\xi_0})(x,\xi) = (2\pi)^{-d/2} e^{-((x-x_0)^2+(\xi-\xi_0)^2)/4}\, e^{i(x-x_0)\cdot(\xi+\xi_0)/2}.$$

Proof This is a typical "Gaussian" calculation:

$$\begin{aligned}
&(Tf_{x_0,\xi_0})(x,\xi)\\
&= 2^{-d/2}\pi^{-d} \int_{\mathbb{R}^d} e^{-((x-y)^2+(y-x_0)^2)/2} e^{-i(\xi-\xi_0)\cdot y}\, dy\, e^{i(\xi\cdot x-\xi_0\cdot x_0)}\\
&= 2^{-d/2}\pi^{-d} \int_{\mathbb{R}^d} e^{-(y-(x+x_0)/2)^2} e^{-i(\xi-\xi_0)\cdot y}\, dy\; e^{(x+x_0)^2/4} e^{-(x^2+x_0^2)/2} e^{i(\xi\cdot x-\xi_0\cdot x_0)}\\
&= 2^{-d/2}\pi^{-d} \int_{\mathbb{R}^d} e^{-y^2} e^{-i(y+(x+x_0)/2)\cdot(\xi-\xi_0)}\, dy\; e^{-(x-x_0)^2/4} e^{i(\xi\cdot x-\xi_0\cdot x_0)}\\
&= (2\pi)^{-d/2} e^{-((x-x_0)^2+(\xi-\xi_0)^2)/4} e^{i(x-x_0)\cdot(\xi+\xi_0)/2}
\end{aligned}$$

as claimed. □

By inspection, the (formal) adjoint $T^* : \mathcal{S}'(\mathbb{R}^{2d}) \to \mathcal{S}'(\mathbb{R}^d)$ is defined as

$$(T^*F)(y) = 2^{-d/2}\pi^{-3d/4} \int_{\mathbb{R}^{2d}} e^{-(x-y)^2/2} e^{-i\xi\cdot(x-y)}\, F(x,\xi)\, dx d\xi.$$

We shall now verify the important *inversion formula* $T^*T = \mathrm{Id}$.

Theorem 12.33 *One has* $T^*T = \mathrm{Id}$ *on* $\mathcal{S}'(\mathbb{R}^d)$*, and* T *is an isometry from* $L^2(\mathbb{R}^d) \to L^2(\mathbb{R}^{2d})$.

Proof It suffices to check that $(T^*T)f = f$ for $f \in \mathcal{S}(\mathbb{R}^d)$. For such f compute:

$$
\begin{aligned}
&(T^*Tf)(y) \\
&= 2^{-d}\pi^{-3d/2} \int_{\mathbb{R}^{2d}} \int_{\mathbb{R}^d} e^{-(x-y)^2/2} e^{-i\xi\cdot(x-y)} e^{-(x-z)^2/2} e^{i\xi\cdot(x-z)} f(z)\, dz\, dx d\xi \\
&= 2^{-d}\pi^{-3d/2} \int_{\mathbb{R}^d} e^{-(x-z)^2} \delta_0(y-z)(2\pi)^d f(z)\, dz dx \\
&= (2\pi)^{-d/2} \int_{\mathbb{R}^d} e^{-x^2/2}\, dx \int_{\mathbb{R}^d} \delta_0(y-z)\, f(z)\, dz = f(y),
\end{aligned}
\tag{12.55}
$$

as desired. Next,

$$
\|f\|_2^2 = \langle T^*Tf, f\rangle_{L^2(\mathbb{R}^{2d})} = \|Tf\|^2_{L^2(\mathbb{R}^d)} \tag{12.56}
$$

so that T is indeed an isometry.

Alternatively, and more generally, let $(T_\phi)f(x,\xi) := \int_{\mathbb{R}^d} \phi(x-y)e^{i\xi\cdot(x-y)} f(y)\, dy$, where $\phi \in L^2(\mathbb{R}^d)$. Then

$$
\begin{aligned}
&\int_{\mathbb{R}^{2d}} \left|(T_\phi)f(x,\xi)\right|^2 d\xi dx \\
&= \int_{\mathbb{R}^{2d}} \left| \int_{\mathbb{R}^d} \phi(u)e^{iu\cdot\xi} f(x-u)\, du \right|^2 d\xi dx \\
&= (2\pi)^d \int_{\mathbb{R}^{2d}} |\phi(u)|^2 |f(x-u)|^2\, dx du = (2\pi)^d \|\phi\|_2^2 \|f\|_2^2.
\end{aligned}
$$

To pass to the second equality we used Plancherel's theorem. In conclusion, as long as $(2\pi)^d\|\phi\|_2^2 = 1$ we have $\|T_\phi\|_{2\to 2} = 1$. In our case

$$
\phi(x) = 2^{-d/2}\pi^{-3d/4} e^{-|x|^2/2}
$$

does obey this normalization, and we recover the property that the Bargman transform is an isometry. Returning to (12.56), we conclude that $\langle T^*Tf, f\rangle = \|f\|_2^2$. Since T^*T is self-adjoint it must therefore be the identity. □

Exercise 12.12 Show that for any $f \in \mathcal{S}'(\mathbb{R}^d)$ the expression

$$
e^{\xi^2/2}(Tf)(x,\xi)
$$

defines a holomorphic function of $x - i\xi$. Conclude that Tf is smooth and satisfies

$$(\partial_x - i\partial_\xi)(Tf)(x,\xi) = i\xi(Tf)(x,\xi) \tag{12.57}$$

for all $(x,\xi) \in \mathbb{R}^{2d}$. This shows that the range of T on L^2, i.e., $T(L^2)$ as it appears in the previous theorem, is a subset of the space of functions satisfying these Cauchy–Riemann equations. Moreover, it also shows that the inversion formula given by the theorem is far from unique. In fact, it is known that $T(L^2(\mathbb{R}^d))$ coincides with those $C^\infty(\mathbb{C}^{2d})$ functions that solve (12.57) and are such that their restriction to $\mathbb{R}^{2d}$ belongs to $L^2(\mathbb{R}^{2d})$. This space is known as *Fock space*.

12.4.2. The S^0 calculus and the Bargman transform

It is not surprising that there should be a connection between the Bargman transform and the Weyl calculus of pseudodifferential operators. To be more precise, we might expect that the "leading order" of $T(a(x,D)f)$ equals $a(x,\xi)(Tf)(x,\xi)$. In other words, the action of $a(x,D)$ in the phase-space picture given by the Bargman transform equals, to leading order, multiplication by $a(x,\xi)$. This is indeed the case, at least in the semiclassical formulation where one has $Ta(x,\hbar D) - a(x,\hbar\xi)T = O(\hbar)$, in a suitable sense. In the absence of the semiclassical parameter one has corresponding statements, of which the "zero order" formulation is given by Proposition 12.34.

Denote by Ψ^0 the class of Weyl ΨDOs with S^0_0-symbols. That is, $A \in \Psi^0$ if and only if $A : \mathcal{S}(\mathbb{R}^d) \to \mathcal{S}'(\mathbb{R}^d)$ and there exists $a \in S^0_0$ with

$$(Af)(x) = \int_{\mathbb{R}^d} a\left(\frac{x+y}{2},\xi\right) e^{i(x-y)\cdot\xi} f(y)\,dy d\xi \quad \forall f \in \mathcal{S}(\mathbb{R}^d).$$

The Bargman transform then allows for the following characterization of Ψ^0.

Proposition 12.34 *The class Ψ^0 consists precisely of all operators $A : \mathcal{S}(\mathbb{R}^d) \to \mathcal{S}'(\mathbb{R}^d)$ with the property that the kernel of TAT^* decays rapidly away from the diagonal in the following sense:*

$$|K(x,\xi;y,\eta)| \le C_N \,\langle |x-y| + |\xi-\eta|\rangle^{-N} \quad \forall x,y,\xi,\eta \in \mathbb{R}^d \tag{12.58}$$

for all $N \ge 0$.

Proof Let $F \in \mathcal{S}(\mathbb{R}^{2d})$. One has

$$\begin{aligned}(TAT^*F)(x,\xi) &= 2^{-d/2}\pi^{-3d/4}\int_{\mathbb{R}^d} e^{-(x-u)^2/2}e^{i\xi\cdot(x-u)}(AT^*F)(z)\,dzd\zeta\,du\\ &= 2^{-d}\pi^{-3d/2}\int_{\mathbb{R}^d} e^{-(x-u)^2/2}e^{i\xi\cdot(x-u)}\int_{\mathbb{R}^{2d}} a\left(\frac{z+u}{2},\zeta\right)\\ &\quad\times\ e^{i\zeta\cdot(u-z)}\int_{\mathbb{R}^{2d}} e^{-(z-y)^2/2}e^{i\eta\cdot(z-y)}\,F(y,\eta)\,dyd\eta\,dzd\zeta du\\ &= \int_{\mathbb{R}^{2d}} k_A(x,\xi;y,\eta)F(y,\eta)\,dyd\eta,\end{aligned}$$

where the kernel is given by

$$\begin{aligned}k_A(x,\xi;y,\eta) := 2^{-d}\pi^{-3d/2}\int_{\mathbb{R}^{3d}} & e^{-((x-u)^2+(z-y)^2)/2}e^{i\xi\cdot(x-u)}e^{i\zeta\cdot(u-z)}\\ &\times\ a\left(\frac{z+u}{2},\zeta\right)e^{i\eta\cdot(z-y)}\,dzd\zeta du.\end{aligned} \tag{12.59}$$

While the z, u integration is absolutely convergent, the ζ integration does not exist in an absolute sense and requires integration by parts. In fact, integrating by parts in u, z, and ζ, respectively, leads to the bound

$$\begin{aligned}&|k_A(x,\xi;y,\eta)|\\ &\quad\le C_{N,d,a}\int_{\mathbb{R}^{3d}} e^{-((x-u)^2+(z-y)^2)/4}\langle u-z\rangle^{-N}\langle \xi-\zeta\rangle^{-N}\langle\zeta-\eta\rangle^{-N}\,dzd\zeta du\\ &\quad\le C_{N,d,a}\,\langle x-y\rangle^{-N}\langle\xi-\eta\rangle^{-N},\end{aligned} \tag{12.60}$$

provided that $N > d$. In passing we remark that the constant here is of the form

$$C_{N,a} = C_{N,d}\sum_{|\alpha|\le 3N}\|\partial_u^\alpha a(u)\|_{L^\infty(\mathbb{R}^{2d})}.$$

Conversely, suppose that TAT^* has Schwartz kernel k_A. Then the operator A has kernel

$$\begin{aligned}L_A(x,y) = 2^{-d}\pi^{-3d/2}\int_{\mathbb{R}^{4d}} & e^{-(x-x_1)^2/2}e^{i\xi\cdot(x-x_1)}\,k_A(x_1,\xi_1;x_2,\xi_2)\\ &\times\ e^{-(x_2-y)^2/2}e^{i\xi_2\cdot(x_2-y)}\,dx_1d\xi_1\,dx_2d\xi_2.\end{aligned}$$

Following Exercise 12.1, we now *define* the symbol $a(x,\xi)$ of A as

$$a(x,\xi) = \pi^{-d} \int_{\mathbb{R}^d} L_A(x-u, x+u) e^{2i\xi\cdot u}\, du$$

$$= c_d \int_{\mathbb{R}^{5d}} e^{-(x-u-x_1)^2/2} e^{i\xi_1\cdot(x-u-x_1)} k_A(x_1,\xi_1;x_2,\xi_2) e^{2i\xi\cdot u} \tag{12.61}$$

$$\times\, e^{-(x_2-x-u)^2/2} e^{i\xi_2\cdot(x_2-x-u)}\, du dx_1 d\xi_1 dx_2 d\xi_2;$$

see (12.7). Carring out the u-integral (and substituting $-u$ for u), we obtain

$$\int_{\mathbb{R}^d} e^{-((u+x-x_1)^2+(u+x_2-x)^2)/2}\, e^{iu\cdot(\xi_1+\xi_2)} e^{-2i\xi\cdot u}\, du$$

$$= \int_{\mathbb{R}^d} e^{-(u-(x_1-x_2)/2)^2} e^{iu\cdot(\xi_1+\xi_2-2\xi)}\, du\; e^{(x_1-x_2)^2/4} e^{-((x_1-x)^2+(x_2-x)^2)/2}$$

$$= c_d\, e^{-((\xi_1+\xi_2-2\xi)^2+(x_1+x_2-2x)^2)/4}\, e^{i(x_1-x_2)\cdot(\xi_1+\xi_2-2\xi)/2}.$$

Thus

$$a(x,\xi) = c_d \int_{\mathbb{R}^{4d}} e^{-((\xi_1+\xi_2-2\xi)^2+(x_1+x_2-2x)^2)/4}\, e^{i(x_1-x_2)\cdot(\xi_1+\xi_2-2\xi)/2}$$

$$\times\, e^{i(\xi_1-\xi_2)\cdot x} e^{i(\xi_2\cdot x_2-\xi_1\cdot x_1)}\, k_A(x_1,\xi_1;x_2,\xi_2)\, dx_1 d\xi_1 dx_2 d\xi_2.$$

From this and our assumptions on k_A it is immediate that

$$\|\partial_x^\alpha \partial_\xi^\beta a(x,\xi)\|_{L^\infty(\mathbb{R}^{2d})} \le C_{\alpha,\beta}$$

for all α, β. □

We refer to the kernels as in the proposition as *almost diagonal*. Note that the following nontrivial corollary does *not* use the composition calculus of symbols.

Corollary 12.35 *One has that Ψ^0 is an algebra of L^2-bounded operators.*

Proof We have just shown that $A \in \Psi^0$ if and only if TAT^* has an almost diagonal kernel. By Schur's lemma, any such kernel gives rise to an operator bounded on $L^2(\mathbb{R}^{2d})$. Hence

$$A = T^*(TAT^*)T \quad \text{is } L^2\text{-bounded},$$

where we used $T^*T = \mathrm{Id}$. If $A_1, A_2 \in \Psi^0$ then the kernel of

$$TA_1A_2T^* = TA_1T^*TA_2T^*$$

is the composition of the kernels of TA_1T^* and TA_2T^* and therefore also almost diagonal. Thus, $A_1A_2 \in \Psi^0$. □

Proposition 12.34 is only a beginning, in the sense that it does not expand TAT^* or explain to what extent multiplication by $a(x, \xi)$ gives the leading order of this operator's action. Without providing any details, we close this chapter by sketching how to proceed in this direction and offering an overview of other possible extensions of this theory. The reader will find references in the notes.

In the S_0^0 calculus, in the absence of an semiclassical parameter, an expansion of TAT^* is not very meaningful. One way to remedy this is through the symbol class $S_0^{(k)}$, which consists of all $a \in C^\infty(\mathbb{R}^{2d})$ that satisfy

$$\sup_{u \in \mathbb{R}^{2d}} |\partial^\alpha a(u)| \leq C_\alpha \quad \forall |\alpha| \geq k,$$

where $k \geq 1$ is some fixed integer. Note that these symbols are allowed to grow in x as well; we have previously excluded this possibility. It turns out that the Weyl calculus is flexible enough to accommodate such symbols, but we shall not dwell on this technical detail.

The main statement concerning leading orders in the $S_0^{(k)}$ calculus now reads as follows: for any $a \in S_0^{(k)}$ there exists a canonically defined operator A_k such that

$$Ta(x, D) = A_k T + R_k(a), \tag{12.62}$$

where the remainder $R_k(a)$ is bounded as a linear operator from $L^2(\mathbb{R}^d)$ to $L^2(\mathbb{R}^{2d})$. For $k = 1$ one has, as expected, $A_1 = a$, the operator corresponding to multiplication by $a(x, \xi)$, whereas for $k = 2$ the Hamiltonian vector field of a appears:

$$A_2 = a + i(a_x \partial_\xi - a_\xi(\partial_x - i\xi)). \tag{12.63}$$

The general formula is

$$A_k = \sum_{\substack{\alpha \leq \beta \\ |\alpha| + |\beta| < k}} 2^{|\beta| - |\alpha|} \frac{\partial^\alpha \bar{\partial}^\beta a(x, \xi)}{\alpha!(\beta - \alpha)!} (\partial - \tfrac{1}{2} i\xi)^{\beta - \alpha}, \tag{12.64}$$

where $\partial := \frac{1}{2}(\partial_x + i\partial_\xi)$ and $\bar{\partial} := \frac{1}{2}(\partial_x - i\partial_\xi)$.

To show the applicability of this type of phase-space analysis of ΨDOs, we now prove the following version of Gårding's inequality. Suppose that $a \in S_0^{(2)}$ is nonnegative. Then

$$\langle a(x, D)u, u\rangle = \langle Ta(x, D)u, Tu\rangle = \langle A_2 Tu, Tu\rangle + O(\|u\|_2^2).$$

We can rewrite A_2 as given by (12.63) in the following form:

$$A_2 = a + 2\bar{\partial} a(\partial - \tfrac{1}{2} i\xi).$$

Indeed, it follows from (12.57) that the actions of this form of (12.63) and that of (12.63) itself on the range of T are identical. Then, with $v = Tu$,

$$\begin{aligned}\langle A_2 v, v\rangle &= \langle av, v\rangle + \langle 2\bar{\partial}a(\partial - \tfrac{1}{2}i\xi)v, v\rangle \\ &= \langle (a - 2\partial\bar{\partial}a)v, v\rangle - \langle 2\bar{\partial}a\, v, (\bar{\partial} - \tfrac{1}{2}i\xi)v\rangle \\ &= \langle (a - 2\partial\bar{\partial}a)v, v\rangle\end{aligned}$$

where we used the Cauchy–Riemann equations (12.57) to pass to the last line. By assumption, $\langle av, v\rangle \geq 0$ and $\partial\bar{\partial}a \in L^\infty$, whence, by $\|v\|_2 = \|u\|_2$,

$$\langle a(x, D)u, u\rangle \geq -C\|u\|_2^2.$$

With considerably more work, one can relax the assumption here to $a \in S_0^{(4)}$. This corresponds to the Fefferman–Phong inequality. This is based on (12.64) with $k = 4$. As a final outlook, we note that Fourier integral operators (FIOs) can be defined via the Bargman transform as well. Indeed, in analogy with Proposition 12.34 we consider operators $A : \mathcal{S} \to \mathcal{S}'$ with the property that the kernel $K(u, v)$ of TAT^* satisfies

$$|K(u, v)| \leq C_N \langle v - \chi(u)\rangle^{-N},$$

where χ is a suitable *symplectomorphism* on $T^*\mathbb{R}^d$. This refers to a diffeomorphism that preserves the standard symplectic form $dx \wedge d\xi$. Without going into any details, we merely note a significant difference between ΨDOs and FIOs: the former does not move singularities, whereas the latter does move them, by a symplectomorphism. In many applications these maps are given by Hamiltonian flows, as defined for example by the vector field appearing in A_2; see above.

Notes

Pseudodifferential operators were introduced into analysis by Calderón; see his survey [13]. They arose naturally in the context of singular integral operators that are not translation invariant. Calderón applied them in his proof of the uniqueness of the Cauchy problem [12].

Weyl quantization was defined by Hermann Weyl in the early days of theoretical quantum mechanics as a way to pass from classical "observables" (i.e., functions $a(x, \xi)$ describing physical quantifies) to operators. The role of the Heisenberg group in this procedure was also recognized at that time as encoding the basic commutator relations. However, a calculus of such "Weyl operators" had not yet been developed.

In the context of the Kohn–Nirenberg quantization, such a calculus was established systematically during the 1960s by Calderón, Zgymund, and Kohn, Nirenberg; see Nirenberg [88]. A very general calculus for Weyl quantization was obtained by Hörmander [58]. As we observed in this chapter, Weyl calculus stands out owing to its symmetric treatment of x and ξ, which makes it preferable for certain applications.

There exist many excellent texts which introduce the reader to this vast area; see for example Taylor [120], Alinhac and Gérard [2], Grigis and Sjöstrand [50], and Folland [40]. The last-mentioned takes a more general phase-space-centered perspective and places the Weyl calculus in the context of the Heisenberg group. For example, it connects the operators $\rho(q, p)$ with unitary representations on that group (the Stone–von Neumann theorem). A very comprehensive treatment of both the ΨDO calculus and FIO calculus is the encyclopedic treatise by Hörmander [56, 59]. The survey article of Beals, Fefferman, and Grossman [6] can also serve as an introduction to a large amount of interesting analysis involving singular integrals and pseudodifferential operators.

For a concise yet broad introduction to the *semiclassical Weyl calculus* and many applications, especially to study of the eigenvalues of elliptic operators (the Weyl law) on compact manifolds or bounded domains, see Dimassi and Sjöstrand [29] as well as Evans and Zworski [35]; see also Zworski [130]. These publications develop the Weyl calculus exclusively in the semiclassical context and for symbols that may grow in both x and ξ. In fact, the entire symbol calculus is based on the very general notion of *order functions*.

For the Bargman transform we have followed Tataru's survey article [118], which can also serve as an introduction to Fourier integral operators. Folland's book [40] studies the Bargman, Gabor, and Wigner transforms and relates them to the Heisenberg group and the metaplectic representation. Tataru's article also provides details of the non-semiclassical version of both the sharp Gårding and Fefferman–Phong inequalities, which uses the Bargman transform, a sketch of which was provided towards the end of this chapter. The definition of an FIO is also taken from [118], and Egorov's theorem is proved there on the basis of this formalism.

Finally, Martinez [80] applies the Bargman (or, in that case, the FBI) transform in a semiclassical setting, with applications to differential operators having analytic coefficients.

Problems

Problem 12.1 If $u \in \mathbb{R}^{2d}$ and $G : \mathbb{R}^{2d} \to \mathbb{R}$ then let $G_u(v) := G(v - u)$ and $\widetilde{G}(v) := G(-Jv)$. Establish the following properties, where ρ is as in (12.2).

(a) If $u \in \mathbb{R}^{2d}$ then $\rho(u)$ extends uniquely to a unitary operator with $\rho(u)^{-1} = \rho(u)^* = \rho(-u)$.

(b) If $u \in \mathbb{R}^{2d}$, $v \in \mathbb{R}^{2d}$ then $\rho(u) \circ \rho(v) \circ \rho(u)^* = e^{\pi i \sigma(u,v)} \rho(v)$.

(c) If $u \in \mathbb{R}^{2d}$ then $\widehat{\rho(u) f} = \rho(-Ju)\hat{f}$ for all $f \in \mathcal{S}(\mathbb{R}^d)$.

(d) If $u \in \mathbb{R}^{2d}$, $G \in \mathcal{S}(\mathbb{R}^{2d})$ then $G_u(x, D) = \rho(Ju) \circ G(x, D) \circ \rho(-Ju)$; in particular, $G_u(x, D)$ and $G(x, D)$ are unitarily equivalent. Here J is the symplectic matrix from above; see the start of subsection 12.2.2 and Lemma 12.10.

(e) $\widehat{G(x, D) f} = \widetilde{G}(x, D)\hat{f}$ for all $f \in \mathcal{S}(\mathbb{R}^d)$ and $G \in \mathcal{S}(\mathbb{R}^{2d})$.

(f) For $a \in \mathcal{S}(\mathbb{R}^{2d})$ define $a^{(u)}(v) := e^{2\pi i \langle u, v\rangle} a(v)$. Then

$$a^{(u)}(x, D) = \rho\big(\tfrac{1}{2}u\big) a(x, D) \rho\big(\tfrac{1}{2}u\big).$$

Problem 12.2 Suppose that $f \in \mathcal{S}(\mathbb{R}^d)$. Show that

$$|\langle \rho(u)f, \rho(v)f \rangle| \le C_{N,f} \langle u - v \rangle^{-N} \quad \forall u, v \in \mathbb{R}^{2d}$$

for all $N \ge 0$.

Problem 12.3 Define $G : \mathbb{R}^{2d} \to \mathbb{R}$ by $G(u) := 2^d e^{-2\pi|u|^2}$, and $\gamma(u) : \mathbb{R}^d \to \mathbb{R}$ by $\gamma(u) := 2^{d/4} e^{-\pi|u|^2}$. Show that $\|\gamma\|_2 = 1$ and $G(x, D)$ is the orthogonal projection onto γ: $G(x, D)f = \langle f, \gamma \rangle \gamma$ for all $f \in L^2(\mathbb{R}^d)$.

Problem 12.4 Prove (12.62) with A_k as in (12.64).

Problem 12.5 Compute the free Schrödinger evolution of a coherent state.

References

[1] Adams, D. R. 1975. A note on Riesz potentials. *Duke Math. J.* **42**, 765–778.

[2] Alinhac, S. and Gérard, P. 2007. *Pseudo-Differential Operators and the Nash-Moser Theorem.* Graduate Studies in Mathematics, vol. 82, American Mathematical Society, Providence, RI.

[3] Antonov, N. Yu. 1996. Convergence of Fourier series. In Proc. XX Workshop on Function Theory (Moscow, 1995). *East J. Approx.* **2**, 187–196.

[4] Auscher, P., Hofmann, S., Muscalu, C., Tao, T., and Thiele, C. 2002. Carleson measures, trees, extrapolation, and $T(b)$ theorems. *Publ. Mat.* **46**, 257–325.

[5] Bahouri, H., Chemin, J.-Y., and Danchin, R. 2001. *Fourier Analysis and Nonlinear Partial Differential Equations.* Grundlehren der Mathematischen Wissenschaften, vol. 343, Springer, Heidelberg.

[6] Beals, M., Fefferman, C., and Grossman, R. 1983. Strictly pseudoconvex domains in $\mathbb{C}^n$. *Bull. Amer. Math. Soc.* (New Series) **8**, 125–322.

[7] Bergh, J. and Löfström, J. 1976. *Interpolation Spaces. An Introduction.* Grundlehren der Mathematischen Wissenschaften, vol. 223, Springer-Verlag, Berlin–New York.

[8] Bonami, A. and Demange, B. 2006. A survey on uncertainty principles related to quadratic forms. *Collect. Math.*, extra volume, 1–36.

[9] Bony, J.-M. 1981. Calcul symbolique et propagation des singularités pour les équations aux dérivées partielles non linéaires. *Ann. Sci. École Norm. Sup.* (4) **14**, 209–246.

[10] Bourgain, J. 1989. Bounded orthogonal systems and the $\Lambda(p)$-set problem. *Acta Math.* **162**, 227–245.

[11] Bourgain, J., Goldstein, M., and Schlag, W. 2001. Anderson localization for Schrödinger operators on $\mathbb{Z}$ with potentials given by the skew-shift. *Comm. Math. Phys.* **220**, 583–621.

[12] Calderón, A.-P. 1958. Uniqueness in the Cauchy problem for partial differential equations. *Amer. J. Math.* **80**, 16–36.

[13] Calderón, A.-P. 1966. Singular integrals. *Bull. Amer. Math. Soc.* **72**, 427–465.

[14] Calderón, A.-P. and Vaillancourt, R. 1972. A class of bounded pseudo-differential operators. *Proc. Nat. Acad. Sci. USA* **69**, 1185–1187.

[15] Calderón, A. P. and Zygmund, A. 1952. On the existence of certain singular integrals. *Acta Math.* **88**, 85–139.

[16] Carleson, L. 1966. On convergence and growth of partial sums of Fourier series. *Acta Math.* **116**, 135–157.

[17] Cazenave, T. 2003. *Semilinear Schrödinger Equations.* Courant Lecture Notes in Mathematics, vol. 10, New York University, Courant Institute of Mathematical Sciences, New York; American Mathematical Society, Providence, RI.

[18] Chandrasekharan, K. 1989. *Classical Fourier Transforms.* Universitext, Springer-Verlag, Berlin.

[19] Chandrasekharan, K. 1996. *A Course on Topological Groups.* Texts and Readings in Mathematics, vol. 9, Hindustan Book Agency, New Delhi.

[20] Chow, Y. S. and Teicher, H. 1997. *Probability Theory. Independence, Interchangeability, Martingales.* Third edition. Springer Texts in Statistics, Springer-Verlag, New York.

[21] Christ, M. 1990. *Lectures on Singular Integral Operators.* CBMS Regional Conference Series in Mathematics, vol. 77, published for the Conference Board of the Mathematical Sciences, Washington, DC, by the American Mathematical Society, Providence, RI.

[22] Christ, M. and Fefferman, R. 1983. A note on weighted norm inequalities for the Hardy–Littlewood maximal operator. *Proc. Amer. Math. Soc.* **87**, 447–448.

[23] Coifman, R. R., Jones, P. W., and Semmes, S. 1989. Two elementary proofs of the L^2 boundedness of Cauchy integrals on Lipschitz curves. *J. Amer. Math. Soc.* **2**, 553–564.

[24] Coifman, R. and Meyer, Y. 1997. *Wavelets. Calderón–Zygmund and Multilinear Operators.* Translated from the 1990 and 1991 French originals by David Salinger. Cambridge Studies in Advanced Mathematics, vol. 48, Cambridge University Press, Cambridge.

[25] Córdoba, A. 1979. A note on Bochner–Riesz operators. *Duke Math. J.* **46**, 505–511.

[26] Cotlar, M. 1955. A combinatorial inequality and its application to L^2 spaces. *Rev. Math. Guyana* **1**, 41–55.

[27] David, G. and Journé, J.-L. 1984. A boundedness criterion for generalized Calderón–Zygmund operators. *Ann. Math.* (2) **120**, 371–397.

[28] Davis, K. M. and Chang, Y.-C. 1987. *Lectures on Bochner Riesz Means.* London Mathematical Society Lecture Note Series, vol. 114, Cambridge University Press, Cambridge.

[29] Dimassi, M. and Sjöstrand, J. 1999. *Spectral Asymptotics in the Semi-Classical Limit.* London Mathematical Society Lecture Note Series, vol. 268, Cambridge University Press, Cambridge.

[30] Drury, S. W. 1970. Sur les ensembles de Sidon. *C. R. Acad. Sci. Paris Sér. A, B* **271**, A162–A163.

[31] Duoandikoetxea, J. 2001. *Fourier Analysis* (English summary). Translated and revised from the 1995 Spanish original by David Cruz-Uribe. Graduate Studies in Mathematics, vol. 29, American Mathematical Society, Providence, RI.

[32] Durrett, R. 2010. *Probability: Theory and Examples.* Fourth edition. Cambridge Series in Statistical and Probabilistic Mathematics, Cambridge University Press, Cambridge.

[33] Dym, H. and McKean, H. P. 1972. *Fourier Series and Integrals.* Probability and Mathematical Statistics, vol. 14, Academic Press, New York–London.

[34] Evans, L. C. 2010. *Partial Differential Equations.* Second edition. Graduate Studies in Mathematics, vol. 19, American Mathematical Society, Providence, RI.

[35] Evans, L. C. and Zworski, M. 2011. *Lectures on Semiclassical Analysis*, version 0.8, preprint.

[36] Fefferman, C. 1970. Inequalities for strongly singular convolution operators. *Acta Math.* **124**, 9–36.

[37] Fefferman, C. 1971. The multiplier problem for the ball. *Ann. Math.* (2) **94**, 330–336.

[38] Fefferman, C. 1983. The uncertainty principle. *Bull. Amer. Math. Soc.* **9**, 129–206.

[39] Fefferman, C. and Stein, E. M. 1972. H^p spaces of several variables. *Acta Math.* **129**, 137–193.

[40] Folland, G. B. 1989. *Harmonic Analysis in Phase Space.* Annals of Mathematics Studies, vol. 122, Princeton University Press, Princeton, NJ.

[41] Folland, G. B. 1995. *Introduction to Partial Differential Equations.* Second edition. Princeton University Press, Princeton, NJ.

[42] Frank, R. and Seiringer, R. 2008. Non-linear ground state representations and sharp Hardy inequalities. *J. Funct. Anal.* **255**, 3407–3430.

[43] Frazier, M., Jawerth, B., and Weiss, G. 1991. *Littlewood–Paley Theory and the Study of Function Spaces.* CBMS Regional Conference Series in Mathematics, vol. 79, published for the Conference Board of the Mathematical Sciences, Washington, DC by the American Mathematical Society, Providence, RI.

[44] Füredi, Z. and Loeb, P. 1994. On the best constant for the Besicovitch covering theorem. *Proc. Amer. Math. Soc.* **121**, 1063–1073.

[45] García-Cuerva, J. and Rubio de Francia, J. 1985. *Weighted Norm Inequalities and Related Topics.* North-Holland Mathematics Studies, vol. 116; Notas de Matemática, vol. 104, North-Holland, Amsterdam.

[46] Garnett, J. B. 2007. *Bounded Analytic Functions.* Revised first edition. Graduate Texts in Mathematics, vol. 236, Springer, New York.

[47] Garnett, J. B. and Marshall, D. E. 2008. *Harmonic Measure.* Reprint of the 2005 original. New Mathematical Monographs, vol. 2, Cambridge University Press, Cambridge.

[48] Gilbarg, D. and Trudinger, N. 1983. *Elliptic Partial Differential Equations of Second Order.* Second edition. Grundlehren der Mathematischen Wissenschaften [Fundamental Principles of Mathematical Sciences], vol. 224, Springer-Verlag, Berlin.

[49] Ginibre, J. and Velo, G. 1979. On a class of nonlinear Schrödinger equation. I. The Cauchy problems; II. Scattering theory, general case. *J. Func. Anal.* **32**, 33–71; 1985. Scattering theory in the energy space for a class of nonlinear Schrödinger equations. *J. Math. Pures Appl.* **64**, 363–401; 1985. The global Cauchy problem for the nonlinear Klein–Gordon equation. *Math. Z.* **189**, 487–505; 1985. Time decay of finite energy solutions of the nonlinear Klein–Gordon and Schrödinger equations. *Ann. Inst. H. Poincaré Phys. Théor.* **43**, 399–442.

[50] Grigis, A. and Sjöstrand, J. 1994. *Microlocal Analysis for Differential Operators. An Introduction.* London Mathematical Society Lecture Note Series, vol. 196, Cambridge University Press, Cambridge.

[51] Halmos, P. R. 1950. *Measure Theory.* Van Nostrand, New York.

[52] Han, Q. and Lin, F. 2011. *Elliptic Partial Differential Equations.* Second edition. Courant Lecture Notes in Mathematics, vol. 1, Courant Institute of Mathematical Sciences, New York; American Mathematical Society, Providence, RI.

[53] Hassell, A., Tao, T., and Wunsch, J. 2006. Sharp Strichartz estimates on nontrapping asymptotically conic manifolds. *Amer. J. Math.* **128**, 963–1024. 2005. A Strichartz inequality for the Schrödinger equation on nontrapping asymptotically conic manifolds. *Comm. Partial Diff. Eqs.* **30**, 157–205.

[54] Havin, V. and Jöricke, B. 1994. *The Uncertainty Principle in Harmonic Analysis.* Ergebnisse der Mathematik und ihrer Grenzgebiete, vol. 3, p. 28. Springer-Verlag, Berlin.

[55] Hoffman, K. 1988. *Banach Spaces of Analytic Functions.* Reprint of the 1962 original. Dover Publications, New York.

[56] Hörmander, L. 1990. *The Analysis of Linear Partial Differential Operators. I. Distribution Theory and Fourier Analysis.* Second edition. Springer Study Edition, Springer-Verlag, Berlin.

[57] Hörmander, L. 2005. *The Analysis of Linear Partial Differential Operators. II. Differential Operators with Constant Coefficients.* Reprint of the 1983 original. Classics in Mathematics, Springer-Verlag, Berlin.

[58] Hörmander, L. 2007. *The Analysis of Linear Partial Differential Operators. III. Pseudo-Differential Operators.* Reprint of the 1994 edition. Classics in Mathematics, Springer, Berlin.

[59] Hörmander, L. 2009. *The Analysis of Linear Partial Differential Operators. IV. Fourier Integral Operators.* Reprint of the 1994 edition. Classics in Mathematics, Springer-Verlag, Berlin.

[60] Jaming, P. 2007. Nazarov's uncertainty principles in higher dimensions. *J. Approx. Theory* **149**, 3041.

[61] Janson, S. 1978. Mean oscillation and commutators of singular integral operators. *Ark. Mat.* **16**, 263–270.

[62] Jerison, D. 1990. An elementary approach to local solvability for constant coefficient partial differential equations. *Forum Math.* **2**, 45–50.

[63] Kahane, J.-P. 1985. *Some Random Series of Functions.* Second edition. Cambridge Studies in Advanced Mathematics, vol. 5, Cambridge University Press, Cambridge.

[64] Katz, N. unpublished lecture notes.

[65] Katznelson, Y. 2004. *An Introduction to Harmonic Analysis.* Third edition. Cambridge University Press, Cambridge.

[66] Keel, M. and Tao, T. 1998. *Endpoint Strichartz estimates*, Amer. J. Math., 120, pp. 955–980.

[67] Kenig, C. E., Ponce, G., and Vega, L. 2007. The initial value problem for the general quasi-linear Schrödinger equation. In *Recent Developments in Nonlinear Partial Differential Equations*, Contemporary Mathematics, vol. 439, Amer. Math. Soc., Providence, RI.

[68] Knapp, A. W. and Stein, E. 1976. Intertwining operators for semi-simple groups. Ann. Math. **93**, 489–578.

[69] Kolmogorov, A. N. 1923. Une série de Fourier–Lebesgue divergente presque partout. *Fundamenta Mathematicae* **4**, 324–328.

[70] Konyagin, S. V. 2000. On the divergence everywhere of trigonometric Fourier series (in Russian). *Mat. Sb.* **191**, 103–126. Translation in *Sb. Math.* **191**, 97–120.

[71] Koosis, P. 1998. *Introduction to H_p Spaces.* Second edition. With two appendices by V. P. Havin. Cambridge Tracts in Mathematics, vol. 115, Cambridge University Press, Cambridge.

[72] Kovrijkine, O. 2001. Some results related to the Logvinenko–Sereda theorem. *Proc. Amer. Math. Soc.* **129**, 3037–3047.

[73] Lebedev, V. and Olevskiĭ, A. 1994. C^1 changes of variable: Beurling–Helson type theorem and Hörmander conjecture on Fourier multipliers. *Geom. Funct. Anal.* **4**, 213–235.

[74] Lefèvre, P. and Rodríguez-Piazza, L. 2009. Invariant means and thin sets in harmonic analysis with applications to prime numbers. *J. Lond. Math. Soc.* (2) **80**, 72–84.

[75] Levin, B. Ya. 1996. *Lectures on Entire Functions.* Translations of Mathematical Monographs, vol. 150, American Mathematical Society, Providence, RI.

[76] Li, D., Queffélec, H., and Rodríguez-Piazza, L. 2002. Some new thin sets of integers in harmonic analysis. *J. Anal. Math.* **86**, 105–138; 2008. On some random thin sets of integers. *Proc. Amer. Math. Soc.* **136**, 141–150.

[77] Lieb, E. H. and Loss, M. 2001. *Analysis.* Second edition. Graduate Studies in Mathematics, vol. 14, American Mathematical Society, Providence, RI.

[78] Lindblad, H. and Sogge, C. D. 1995. On existence and scattering with minimal regularity for semilinear wave equations. *J. Funct. Anal.* **130**, 357–426.

[79] Marcus, M. B. and Pisier, G. 1981. *Random Fourier Series with Applications to Harmonic Analysis.* Annals of Mathematics Studies, vol. 101, Princeton University Press, Princeton, NJ; University of Tokyo Press, Tokyo.

[80] Martinez, A. 2002. *An Introduction to Semiclassical and Microlocal Analysis.* Universitext, Springer-Verlag, New York.

[81] Marzuola, J., Metcalfe, J., Tataru, D., and Tohaneanu, M. 2010. Strichartz estimates on Schwarzschild black hole backgrounds. *Comm. Math. Phys.* **293**, 37–83.

[82] Meyer, Y. 1989. Wavelets and Operators. In *Analysis at Urbana I*. London Mathematical Society Lecture Note Series, vol. 137, Cambridge University Press, Cambridge, pp. 256–365.

[83] Mockenhaupt, G., Seeger, A., and Sogge, C. D. 1992. Wave front sets, local smoothing and Bourgain's circular maximal theorem. *Ann. Math.* (2) **136**, 207–218.

[84] Montgomery, H. L. 1994. *Ten Lectures on the Interface between Analytic Number Theory and Harmonic Analysis.* CMBS Lectures, vol. 84, American Mathematical Society, Providence, RI.

[85] Müller, P. F. X. 2005. *Isomorphisms between H^1 Spaces.* Instytut Matematyczny Polskiej Akademii Nauk. Monografie Matematyczne (New Series) [Mathematics Institute of the Polish Academy of Sciences Mathematical Monographs (New Series)], vol. 66, Birkhäuser, Basel.

[86] Nakanishi, K. and Schlag, W. 2011. *Invariant Manifolds and Dispersive Hamiltonian Evolution Equations.* European Mathematical Society.

[87] Nazarov, F. L. 1993. Local estimates for exponential polynomials and their applications to inequalities of the uncertainty principle type. *Algebra i Analiz* **5**, 3–66; 1994. Translation in *St Petersburg Math. J.* **5**, 663–717.

[88] Nirenberg, L. 1972. *Lectures on Linear Partial Differential Equations.* Expository lectures from the CBMS Regional Conference held at the Texas Technological University, Lubbock, 22–26 May.

[89] Opic, B. and Kufner, A. 1990. *Hardy-type Inequalities.* Pitman Research Notes in Mathematics Series, vol. 219, Longman Scientific & Technical.

[90] Paley, R. E. A. C., Wiener, N., and Zygmund, A. 1933. Notes on random functions. *Math. Z.* **37**, 647–668.

[91] Pommerenke, Ch. 1977. Schlichte Funktionen und analytische Funktionen von beschränkter mittlerer Oszillation. *Comment. Math. Helv.* **52**, 591–602.

[92] Rider, D. 1966. Gap series on groups and spheres. *Can. J. Math.* **18**, 389–398.

[93] Rider, D. 1975. Randomly continuous functions and Sidon sets. *Duke Math. J.* **42**, 759–764.

[94] Rudin, W. 1960. Trigonometric series with gaps. *J. Math. Mech.* **9**, 203–227.

[95] Rudin, W. 1962. *Fourier Analysis on Groups.* Interscience Tracts in Pure and Applied Mathematics, vol. 12, Interscience Publishers (a division of John Wiley and Sons), New York–London.

[96] Rudin, W. 1987. *Real and Complex Analysis.* Third edition. McGraw-Hill, New York.

[97] Rudin, W. 1991. *Functional Analysis.* Second edition. International Series in Pure and Applied Mathematics, McGraw-Hill, New York.

[98] Segovia, C. and Torrea, J. 1991. Weighted inequalities for commutators of fractional and singular integrals. In *Proc. Conf. on Mathematical Analysis (El Escorial, 1989), Publ. Mat.* **35**, 209–235.

[99] Semmes, S. 1994. A primer on Hardy spaces, and some remarks on a theorem of Evans and Müller. *Comm. Partial Diff. Eqs.* **19**, 277–319.

[100] Shatah, J. and Struwe, M. 1998. *Geometric Wave Equations.* Courant Lecture Notes in Mathematics, vol. 2, American Mathematical Society, Providence, RI.

[101] Sinai, Y. G. 1992. *Probability Theory. An Introductory Course.* Translated from the Russian and with a preface by D. Haughton. Springer Textbook, Springer-Verlag, Berlin.

[102] Sjölin, P. 1969. An inequality of Paley and convergence a.e. of Walsh–Fourier series. *Ark. Mat.* **7**, 551–570.

[103] Slavin, L. and Volberg, Al. 2007. Bellman function and the H^1-BMO duality. In *Harmonic Analysis, Partial Differential Equations, and Related Topics*, Contemporary Mathematics, vol. 428, American Mathematical Society, Providence, RI, pp. 113–126.

[104] Sogge, C. D. 1991. Propagation of singularities and maximal functions in the plane. *Invent. Math.* **104**, 349–376.

[105] Sogge, C. D. 1993. Fourier Integrals in Classical Analysis. Cambridge Tracts in Mathematics, vol. 105, Cambridge University Press, Cambridge.

[106] Sogge, C. D. 2008. *Lectures on Non-Linear Wave Equations.* Second edition. International Press, Boston, MA.

[107] Stein, E. M. 1961. On limits of sequences of operators. *Ann. Math.* (2) **74**, 140–170.

[108] Stein, E. M. 1970. *Singular Integrals and Differentiability Properties of Functions.* Princeton Mathematical Series, vol. 30, Princeton University Press, Princeton, NJ.

[109] Stein, E. M. 1970. *Topics in Harmonic Analysis Related to the Littlewood–Paley Theory.* Annals of Mathematics Studies, vol. 63, Princeton University Press, Princeton, NJ; University of Tokyo Press, Tokyo.

[110] Stein, E. M. 1986. Oscillatory integrals in Fourier analysis. In *Beijing Lectures in Harmonic Analysis*, Beijing, 1984, Annals of Mathematical Studies, vol. 112, Princeton University Press, Princeton, NJ, pp. 307–355.

[111] Stein, E. M. 1993. *Harmonic Analysis: Real-Variable Methods, Orthogonality, and Oscillatory Integrals.* With the assistance of T. S. Murphy. Princeton Mathematical Series, vol. 43, Monographs in Harmonic Analysis, III, Princeton University Press, Princeton, NJ.

[112] Stein, E. M. and Weiss, G. 1971. *Introduction to Fourier Analysis on Euclidean Spaces.* Princeton Mathematical Series, vol. 32, Princeton University Press, Princeton, NJ.

[113] Strichartz, R. S. 1977. Restrictions of Fourier transforms to quadratic surfaces and decay of solutions of wave equations. *Duke Math. J.* **44**, 705–714.

[114] Sulem, C. and Sulem, P-L. 1999. *The Nonlinear Schrödinger Equation. Self-Focusing and Wave Collapse.* Applied Mathematical Sciences, vol. 139, Springer-Verlag, New York.

[115] Tao, T. 2004. Some recent progress on the restriction conjecture. In *Applied Numerical Harmonic Analysis*, Birkhäuser, Boston, MA, pp. 217–243.

[116] Tao, T. 2006. *Nonlinear Dispersive Equations. Local and Global Analysis.* CBMS Regional Conference Series in Mathematics, vol. 106, American Mathematical Society, Providence, RI.

[117] Tataru, D. 2002. On the Fefferman–Phong inequality and related problems. *Comm. Partial Diff. Eqs.* **27**, 2101–2138.

[118] Tataru, D. 2004. *Phase Space Transforms and Microlocal Analysis. Phase Space Analysis of Partial Differential Equations, Vol. II*, pp. 505–524, Pubbl. Cent. Ric. Mat. Ennio Giorgi, Scuola Norm. Sup., Pisa.

[119] Tataru, D. 2008. Parametrices and dispersive estimates for Schrödinger operators with variable coefficients. *Amer. J. Math.* **130**, 571–634.

[120] Taylor, M. E. 1981. *Pseudodifferential Operators.* Princeton Mathematical Series, vol. 34, Princeton University Press, Princeton, NJ.

[121] Taylor, M. E. 1991. *Pseudodifferential Operators and Nonlinear PDEs.* Progress in Mathematics, vol. 100, Birkhäuser, Boston, MA.

[122] Taylor, M. E. 2000. *Tools for PDEs. Pseudodifferential Operators, Paradifferential Operators, and Layer Potentials.* Mathematical Surveys and Monographs, vol. 81, American Mathematical Society, Providence, RI.

[123] Torchinsky, A. 1986. *Real-Variable Methods in Harmonic Analysis.* Pure and Applied Mathematics, vol. 123, Academic Press, Orlando, FL.

[124] Uchiyama, A. 1982. A constructive proof of the Fefferman–Stein decomposition of BMO($\mathbf{R}^n$). *Acta Math.* **148**, 215–241.

[125] Wojtaszczyk, P. 1991. *Banach Spaces for Analysts.* Cambridge Studies in Advanced Mathematics, vol. 25, Cambridge University Press, Cambridge.

[126] Wolf, J. A. 2007. *Harmonic Analysis on Commutative Spaces.* Mathematical Surveys and Monographs, vol. 142, American Mathematical Society, Providence, RI.

[127] Wolff, T. H. 2001. A sharp bilinear cone restriction estimate. *Ann. Math.* (2) **153**, 661–698.

[128] Wolff, T. H. 2003. *Lectures on Harmonic Analysis.* With a foreword by C. Fefferman and preface by I. Łaba. (eds. I. Łaba and C. Shubin). University Lecture Series, vol. 29, American Mathematical Society, Providence, RI.

[129] Yafaev, D. 1999. Sharp constants in the Hardy–Rellich inequalities. *J. Funct. Anal.* **168**, 121–144.

[130] Zworski, M. 2012. *Semiclassical Analysis*, Graduate Studies in Mathematics, vol. 138, American Mathematical Society, Providence, RI.

[131] Zygmund, A. 1971. *Intégrales Singulières.* Lecture Notes in Mathematics, vol. 204, Springer-Verlag, Berlin–New York.

[132] Zygmund, A. 1974. On Fourier coefficients and transforms of functions of two variables. *Studia Math.* **50**, 189–201.

[133] Zygmund, A. 2002. *Trigonometric Series. Vols. I and II.* Third edition. With a foreword by Robert A. Fefferman. Cambridge Mathematical Library, Cambridge University Press, Cambridge.

Index

Printed in the United States
By Bookmasters